Conservation Agriculture

Global Prospects and Challenges

FSC
www.fsc.org
MIX
Paper from
responsible sources
FSC® C013604

Dedication

This book is dedicated to the global Conservation Agriculture movement but particularly to all the pioneer farmers, researchers and extension agents as well as all the champions in the public, private and civil sectors and in the donor community who are making Conservation Agriculture a global reality.

Acknowledgement

Editors are very grateful to Theodor Friedrich for his wholehearted support and guidance to edit this volume.

Conservation Agriculture

Global Prospects and Challenges

Ram A. Jat, Kanwar L. Sahrawat

International Crops Research Institute for the Semi-Arid Tropics, Patancheru, India and Directorate of Groundnut Research, Junagadh, India

and Amir H. Kassam

Food and Agriculture Organization of the United Nations, Rome, Italy and University of Reading, UK

www.cabi.org

CABI is a trading name of CAB International

CABI
Nosworthy Way
Wallingford
Oxfordshire OX10 8DE
UK

CABI
38 Chauncey Street
Suite 1002
Boston, MA 02111
USA

Tel: +44 (0)1491 832111
Fax: +44 (0)1491 833508
E-mail: info@cabi.org
Website: www.cabi.org

Tel: +1 800 552 3083 (toll free)
Tel: +1 617 395 4051
E-mail: cabi-nao@cabi.org

A catalogue record for this book is available from the British Library, London, UK.

Library of Congress Cataloging-in-Publication Data

Jat, Ram A.
Conservation agriculture : global prospects and challenges / Ram A. Jat, Kanwar L. Sahrawat, and Amir Kassam.
p. cm.
Includes bibliographical references and index.
ISBN 978-1-78064-259-8 (hbk)
1. Agricultural conservation. 2. Sustainable agriculture. I. Sahrawat, K. L. II. Kassam, A. H. III. Title.

S604.5.J38 2013
631.4′5--dc23

2013025174

ISBN-13: 978 1 78064 259 8

Commissioning editor: Sreepat Jain
Editorial assistant: Emma McCann
Production editor: Simon Hill

Typeset by SPi, Pondicherry, India.
Printed and bound in the UK by CPI Group (UK) Ltd, Croydon, CR0 4YY.

Contents

Contributors

Akmal Akramkhanov, Khorezm Rural Advisory Support Service, Khorezm, Uzbekistan. E-mail: api001@yahoo.com

Juliana Albertengo, Asociación Argentina de Productores de Siembra Directa (Aapresid), Dorrego 1639 - 2°A, 2000 – Rosario, Santa Fe, Argentina. E-mail: albertengo@aapresid.org.ar

Nadine Andrieu, CIRAD, UMR Innovation, Montpellier, France. E-mail: nadine.andrieu@cirad.fr

Frédéric Baudron, International Maize and Wheat Improvement Center (CIMMYT), PO Box 5689, Addis Ababa, Ethiopia. E-mail: F.Baudron@cgiar.org

Malik Bekenov, Ministry of Agriculture and Water Management, Bishkek, Kyrgyzstan. E-mail: mbekenov@yandex.ru

César Belloso, Asociación Argentina de Productores de Siembra Directa (Aapresid), Dorrego 1639 - 2°A, 2000 – Rosario, Santa Fe, Argentina. E-mail: belloso@aapresid.org.ar

Harouna Bougoum, Université Polytechnique de Bobo Dioulasso/Institut de Développement Rural (UPB/IDR), Burkina Faso. E-mail: bougountas@yahoo.fr

Stéphane Boulakia, Centre for International Cooperation in Agricultural Research and Development (France), Conservation Agriculture and Systems Engineering Research Unit, F-34398 Montpellier cedex 5, France; Conservation Agriculture Network in South-East Asia, c/o National Agriculture and Forestry Research Institute (Lao PDR), PO Box 7170, Vientiane, Lao PDR; and Support Project for the Development of Cambodian Agriculture – Ministry of Agriculture, Forestry and Fisheries/General Directorate of Agriculture, Phnom Penh, Cambodia. E-mail: stephane.boulakia@cirad.fr

Hakim Boulal, International Center for Agricultural Research in the Dry Areas (ICARDA), North Africa Program, Rabat, Morocco. E-mail: boulal@cgiar.org

Ademir Calegari, Agricultural Research Institute of Paraná State – IAPAR, Rodovia Celso Garcia Cid, Km 375, CEP-86047-902, Londrina, Paraná, Brazil. E-mail: calegari@iapar.br

Regis Chikowo, University of Zimbabwe, PO Box MP167, Mt Pleasant, Harare, Zimbabwe.

George W. Clayton, Agriculture and Agri-Food Canada, Lethbridge Research Center, 5303-1 Avenue South, Lethbridge, Alberta, Canada, T1J 4B1. E-mail: george.clayton@agr.gc.ca

Sandra Corsi, Plant Production and Protection Division, Food and Agriculture Organization of the United Nations, Viale delle Terme di Caracalla 00153 Rome, Italy and University of Teramo, Italy. E-mail: sandra.corsi@gmail.com

Antonio Costa, Agricultural Research Institute of Paraná State – IAPAR, Rodovia Celso Garcia Cid, Km 375, CEP-86047-902, Londrina, Paraná, Brazil. E-mail: antcosta@iapar.br

Tomás Coyos, Asociación Argentina de Productores de Siembra Directa (Aapresid). E-mail: coyos@aapresid.org.ar

Bill (W.L.) Crabtree, Crabtree Agricultural Consulting, 21 Brixton Street, Beckenham, Western Australia, WA 6107. E-mail: bill.crabtree@no-till.com.au

Jean-Marie Douzet, CIRAD, UR SCA, Ouagadougou, Burkina Faso. E-mail: jean-marie.douzet@cirad.fr

Patrick Dugué, CIRAD, UMR Innovation, Montpellier, France. E-mail: patrick.dugue@cirad.fr

Sjoerd W. Duiker, Penn State Cooperative Extension, Department of Plant Science, The Pennsylvania State University, 408 ASI Building, University Park, PA 16802, USA. E-mail: sduiker@psu.edu

Mohammed El Mourid, International Center for Agricultural Research in the Dry Areas (ICARDA), North Africa Program, Tunis, Tunisia. E-mail: elmourid@cgiar.org

Guillaume Essecofy, CIHEAM/IAM Montpellier, France. E-mail: escoffabio@yahoo.fr

D. Brian Fowler, Crop Development Center, University of Saskatchewan, College of Agriculture and Bioresources, 51 Campus Drive, Room 4D36 Agriculture Building, Saskatoon, Saskatchewan S7N5A8, Canada. E-mail: brian.fowler@usask.ca

Theodor Friedrich, Plant Production and Protection Division, Food and Agriculture Organization of the United Nations, Viale delle Terme di Caracalla, 00153 Rome, Italy. E-mail: theodor.friedrich@fao.org

Serge Ganou, Université de Ouagadougou, Burkina Faso. E-mail: ganouserge@yahoo.fr

Olivier Gilard, French Development Agency, Vientiane, BP 5923, Vientiane, Lao PDR. E-mail: gilardo@afd.fr

María Beatriz Giraudo, Asociación Argentina de Productores de Siembra Directa (Aapresid), Dorrego 1639 - 2°A, 2000 – Rosario, Santa Fe, Argentina. E-mail: pgiraudo@powervt.com.ar

Bram Govaerts, International Maize and Wheat Improvement Center (CIMMYT), Apdo. Postal 6-641, 06600 México DF, México. E-mail: B.Govaerts@cgiar.org

Augusto Guilherme de Araújo, Agricultural Research Institute of Paraná State – IAPAR, Rodovia Celso Garcia Cid, Km 375, CEP-86047-902, Londrina, Paraná, Brazil. E-mail: agaraujo@iapar.br

Raj Gupta, International Maize and Wheat Improvement Center, NAASC complex, New Delhi 110012, India. E-mail: rajgupta@cgiar.org

Atef Haddad, Diversification and Sustainable Intensification of Production System Research Program, International Center for Agricultural Research in the Dry Areas, Aleppo, Syria. E-mail: a.haddad@cgiar.org

Nasri Haddad, West Asia regional Program, International Center for Agricultural Research in the Dry Areas, Amman, Jordan. E-mail: N.Haddad@cgiar.org

Li Hongwen, Beijing Key Laboratory of Optimized Design for Modern Agricultural Equipment, College of Engineering, China Agricultural University, Beijing 100083, China. E-mail: lhwen@cau.edu.cn

Gao Huangwen, Beijing Key Laboratory of Optimized Design for Modern Agricultural Equipment, College of Engineering, China Agricultural University, Beijing 100083, China. E-mail: ghwbgs@cau.edu.cn

Ram A. Jat, RP1: Resilient Dryland Systems, International Crops Research Institute for the Semi-Arid Tropics, Patancheru 502 324, India; Directorate of Groundnut Research, Junagadh, 362001, Gujarat, India. E-mail: rajatagron@gmail.com

He Jin, Beijing Key Laboratory of Optimized Design for Modern Agricultural Equipment, College of Engineering, China Agricultural University, Beijing 100083, China. E-mail: hejin@cau.edu.cn

Ruy Casão Junior, Agricultural Research Institute of Paraná State – IAPAR, Rodovia Celso Garcia Cid, Km 375, CEP-86047-902, Londrina, Paraná, Brazil. E-mail: rcasao@iapar.br

Muratbek Karabayev, International Maize and Wheat Improvement Center, New Delhi, India. E-mail: M.Karabayev@CGIAR.ORG

Amir H. Kassam, Plant Production and Protection Division, Food and Agriculture Organization of the United Nations, Viale delle Terme di Caracalla 00153 Rome, Italy and School of Agriculture, Policy and Development, University of Reading, Reading RG6 6AR, UK. E-mail: kassamamir@aol.com; amirkassam786@googlemail.com

Habib Ketata, International Center for Agricultural Research in the Dry Areas (ICARDA), North Africa Program, Tunis, Tunisia. E-mail: ketata@cgiar.org

Yaseen Khalil, Diversification and Sustainable Intensification of Production System Research Program, International Center for Agricultural Research in the Dry Areas, Aleppo, Syria. E-mail: y.khalil@cgiar.org

Guy P. Lafond*, Agriculture and Agri-Food Canada, Indian Head Research Farm, RR#1 Gov Road, Box 760, Indian Head, Saskatchewan S0G2K0, Canada. *Deceased

John Lamers, Center for Development Research, Bonn, Germany. E-mail: jlamers@uni-bonn.de

Rafael Fuentes Lanillo, Agricultural Research Institute of Paraná State – IAPAR, Rodovia Celso Garcia Cid, Km 375, CEP-86047-902, Londrina, Paraná, Brazil. E-mail: rfuentes@iapar.br

Jean-Claude Legoupil, Centre for International Cooperation in Agricultural Research and Development (France), Conservation Agriculture and Systems Engineering Research Unit, F-34398 Montpellier, cedex 5, France; Conservation Agriculture Network in South East Asia, c/o National Agriculture and Forestry Research Institute (Lao PDR), PO Box 7170, Vientiane, Lao PDR; and National Agriculture and Forestry Research Institute (Lao PDR) – Conservation Agriculture and Land Development Centre, PO Box 7170, Vientiane, Lao PDR. E-mail: jean-claude.legoupil@cirad.fr

Pascal Lienhard, Centre for International Cooperation in Agricultural Research and Development (France), Conservation Agriculture and Systems Engineering Research Unit, F-34398 Montpellier, cedex 5, France; Conservation Agriculture Network in South East Asia, c/o National Agriculture and Forestry Research Institute (Lao PDR), PO Box 7170, Vientiane, Lao PDR; and National Agriculture and Forestry Research Institute (Lao PDR) – Conservation Agriculture and Land Development Centre, PO Box 7170, Vientiane, Lao PDR. E-mail: pascal.lienhard@cirad.fr

María Eugenia Magnelli, Asociación Argentina de Productores de Siembra Directa (Aapresid). E-mail: magnelli@aapresid.org.ar

Martín Marzetti, Asociación Argentina de Productores de Siembra Directa (Aapresid). E-mail: marzetti@aapresid.org.ar

Kizito Mazvimavi, International Crops Research Institute for the Semi-Arid Tropics, Patancheru 502 324, Andhra Pradesh, India. E-mail: k.mazvimavi@cgiar.org

Saidi Mkomwa, ACT, Nairobi, Kenya. E-mail: saidi.mkomwa@act-act.africa.org

Hafiz Muminjanov, Plant Production and Protection Division, Food and Agriculture Organization of the United Nations, Viale delle Terme di Caracalla, 00153 Rome, Italy. E-mail: Hafiz.Muminjanov@fao.org

Patrice Djamen Nana, ACT, Ouagadougou, Burkina Faso. E-mail: patrice.djamen@act-africa.org

Ali Nefzaoui, International Center for Agricultural Research in the Dry Areas (ICARDA), North Africa Program, Tunis, Tunisia.

Amos Ngwira, Department of Agricultural Research Services, Chitedze Research Station, PO Box 158, Lilongwe, Malawi. E-mail: amosingwira@yahoo.co.uk

Aziz Nurbekov, International Center for Agricultural Research in the Dry Areas (ICARDA) Central Asia and the Caucuses Regional Office, Tashkent, Uzbekistan. E-mail: A.Nurbekov@cgiar.org

Isaiah Nyagumbo, International Maize and Wheat Improvement Center (CIMMYT), PO Box MP163, Harare, Zimbabwe. E-mail: I.Nyagumbo@cgiar.org

Justice Nyamangara, International Crops Research Institute for the Semi-Arid Tropics, Matopos Research Station, PO Box 776, Bulawayo, Zimbabwe. E-mail: j.nyamangara@cgiar.org

Roberto Peiretti, Asociación Argentina de Productores de Siembra Directa (Aapresid), Dorrego 1639 - 2°A, 2000 – Rosario, Santa Fe, Argentina. E-mail: robertopeiretti@gmail.com

Hugo Permingeat, Asociación Argentina de Productores de Siembra Directa (Aapresid), Dorrego 1639 - 2°A, 2000 – Rosario, Santa Fe, Argentina; and Universidad Nacional de Rosario, Urquiza 1911, 2000 – Rosario, Santa Fe, Argentina. E-mail: permingeath@gmail.com

Colin Piggin, Australian Centre for International Agricultural Research (ACIAR), GPO Box 1571, Canberra, ACT 2601, Australia. E-mail: c.piggin@cgiar.org

Danilo Rheinheimer dos Santos, Soil Science Department, University of Santa Maria, Rio Grande do Sul (UFSM), Brazil. E-mail: danilo@ccr.ufsm.br

Jean-Francois (John) Rochecouste, Conservation Agriculture Alliance of Australia and New Zealand (CAA), PO Box 4866, Toowoomba, East Queensland, Australia 4350. E-mail: rochecouste@iinet.com.au

Leonard Rusinamhodzi, Centro Internacional de Agricultura Tropical, 12.5 km Peg Mazowe Road, PO Box MP228, Mt Pleasant, Harare, Zimbabwe. E-mail: leonard.rusinamhodzi@gmail.com

Kanwar L. Sahrawat, RP1: Resilient Dryland Systems, International Crops Research Institute for the Semi-Arid Tropics, Patancheru 502 324, India. E-mail: k.sahrawat@cgiar.org

Jules Benoît Da Sansan, ACT, Ouagadougou, Burkina Faso. E-mail: judalino4@yahoo.fr

Lucien Séguy, Agroecoriz, France. E-mail: seguyl@wanadoo.fr

Dossymbek Sydyk, South-Western Research Institute of Livestock and Crop Production, Chimkent, Kazakhstan. E-mail: nii-tassai@rambler.ru

Christian Thierfelder, International Maize and Wheat Improvement Center (CIMMYT), PO Box MP163, Harare, Zimbabwe. E-mail: C.Thierfelder@cgiar.org

Wade Thomason, Virginia Polytechnic Institute and State University, 185 Ag Quad Ln, 422, Smyth Hall (0404), Blacksburg, VA 24061, USA. E-mail: wthomaso@vt.edu

Jozef Turok, International Center for Agricultural Research in the Dry Areas (ICARDA) Central Asia and the Caucuses Regional Office, Tashkent, Uzbekistan. E-mail: j.turok@cgiar.org

Luis Wall, Universidad Nacional de Quilmes, Roque Saénz Peña 352, B1876BXD Quilmes, Buenos Aires, Argentina; and CONICET, Av. Rivadavia 1917, C1033AAJ Buenos Aires, Argentina. E-mail: lgwall@unq.edu.ar

Patrick Wall, Independent International Consultant, La Cañada 177, Sector O, Bahías de Huatulco, Oaxaca 70989, México. E-mail: pcwallmex@gmail.com

Ibrahima Zerbo, UPB/IDR, Burkina Faso. E-mail: pebayiba@yahoo.fr

Preface

The quality of the natural resource base, especially of soil and water, plays an extremely critical role in enhancing productivity and crop quality, and sustainability of various production systems. Moreover, even the agronomic potential of genetically improved crops or cultivars cannot be achieved in practical agriculture on a degraded soil resource base as a result of multiple soil-related physical, chemical and biological constraints. Hence, to meet the ever increasing demands for food, feed and fibre in a sustainable manner the maintenance of soil health is a prerequisite.

It is known that agricultural practices influence the quality and integrity of the natural resource base, especially soil and water quality and availability, which in turn impacts the sustainability of the production system and food quality. Over the last several decades, a general trend in the degradation of soil resource base has been observed. This degradation has been most severe in the developing nations, where the need for increased nutritious food is also the greatest. Lack of required investment in maintaining the quality of the soil resource base coupled with improper management of natural resources, has indeed led to large-scale soil degradation, which is further jeopardizing environmental quality and food security especially for smallholder, resource-poor farmers in the developing world.

However, it is not necessary that agricultural activities should lead to degradation of the natural resource base. In fact, agricultural practices that are focused on soil health and are in harmony with the ecosystem are sustainable in maintaining productivity at an enhanced level. Among the several practices used in diverse intensified production systems, especially in tropical agriculture, soil tillage and the lack of adequate organic matter input to the soil have a heavy toll in maintaining the integrity of the soil.

Nothing short of a new agricultural production paradigm is needed to sustainably enhance the soil resource base and productivity and simultaneously rehabilitate degraded soils. Conservation Agriculture has indeed provided an alternative way of agriculture that conserves and enhances soil and water resources, and thereby is helpful in maintaining soil health in the longer term while at the same time achieving the highest productivity. Of course the success or otherwise of conservation agriculture depends on numerous factors including those related to soil, climate and socio-economic condition of the farmers, to name a few. Nevertheless, Conservation Agriculture has been researched and applied in most regions of the globe.

The aim of this book is to provide an up-to-date state-of-the-art review on various aspects of Conservation Agriculture by reviewing the past and current research from various regions of the globe so that all others interested in Conservation Agriculture could benefit from experiences gained under different agroclimatic and socio-economic conditions across the globe. This review would aid in learning from the past experience regarding the success or otherwise of Conservation Agriculture. Knowledge gained from this volume should further help in the implementation of Conservation Agriculture and in the understanding of the role and importance of Conservation Agriculture to secure sustainable crop intensification for the benefit of future generations as well. The challenges in implementing Conservation Agriculture that need to be resolved through future research and development for a larger scale support and the spread of Conservation Agriculture are considered. We hope that this volume will further stimulate interest in advancing research and development as well as policy support on this new paradigm of agriculture.

Foreword

If you tell farmers to stop ploughing their land before they plant the next crop because it harms the soil, most of them will either laugh or give you that kind of look that implies you are crazy. More and more farmers, however, will nod their agreement.

Ploughing or digging the soil to turn it over has played a fundamental role in agriculture for thousands of years. It breaks up the soil, making it easy to create a fine tilth into which crops can easily be sown. It also reduces the extent to which weeds compete with crops by burying any vegetation and, in the process, may build up the level of organic matter in the soil. The invention of the plough made it possible for farmers to mechanize agriculture, first by harnessing oxen or horses and later by attaching ploughs to tractors, thereby enabling a family farmer to cultivate much larger areas of crops than was the case when he/she was restricted to manual labour.

The problem is that the rapidly growing demand for food has been pushing up the frequency with which land is cropped. Periods of fallow, which allow the organic matter content of soil to recover after several years of cropping, are getting shorter or have disappeared. When this happens, inversion tillage systems become a leading cause of soil degradation. With each movement of earth, soil particles become finer, allowing less moisture to enter the soil surface and less to be retained for uptake by crop roots. Rain tends to seal the soil surface, accumulate and run off, causing erosion and downstream flooding – and when the land dries out, the fine particles are picked up by the wind and carried away, as happened dramatically when the 'dust bowl' brought farming to a halt in the American prairies during the 1930s.

The structural damage to soils caused by their frequent inversion leads also to a progressive decline in their fertility and health. Organic matter content drops, and with it the extent of the biological activity that helps to make vital minerals and nutrients available to crops. The fertility decline is much faster in tropical than temperate areas because the higher temperatures lead to faster organic matter depletion.

This book shows how farmers all around the world – in both north and south – have woken up to the problems of excessive tillage and are abandoning their ploughs, spades or hoes. As a result of a movement that started in the 1960s, each year farmers now plant over 125 million ha of crops using no or minimum soil disturbance systems – and the area is growing rapidly. The various systems being applied are collectively known as Conservation Agriculture or 'CA'.

Conservation Agriculture offers an important set of technologies to help feed the world sustainably. This is a central element of the Food and Agriculture Organization of the United Nations (FAO) revised strategic framework that focuses the Organization's work on five crosscutting strategic objectives.

Our first strategic objective is to contribute to the eradication of hunger, food insecurity and malnutrition. The second strategic objective is to 'increase and improve the provision of goods and services from agriculture, forestry and fisheries *in a sustainable manner*' (the other strategic objectives are: reducing rural poverty, improving food systems and their fairness, and increasing resilience).

Our focus is, therefore, on sustainable agricultural intensification with the aim of raising agricultural productivity and output while enhancing and maintaining the health and resilience of agroecosystems. This shift has to take place at a time when farmers face the additional intersecting challenges of increasing competition for land and water, rising fuel and production input prices, and climate change.

In our *Save and Grow* approach to sustainable production intensification, we have made it clear that the present paradigm of intensive crop production based on tillage systems cannot meet the challenges of the new millennium. For agriculture to *grow* sustainably, we must learn to *save* by farming differently. Conservation Agriculture, by minimizing soil disturbance, protecting the soil surface with mulch and promoting cropping system diversification, is a central ingredient of *Save and Grow*, along with other good practices of crop, nutrient, pest and water management. Through their ability to harness nature, these can sustainably raise land productivity and efficiency of production while imparting ecological adaptability and resilience to rainfed and irrigated farming systems.

This is why, since 2001, FAO has been sponsoring and supporting the 'World Congress on Conservation Agriculture' process with national and international collaborators, and has played a strong and significant role in promoting CA globally as part of its general support for sustainable agriculture, food security, poverty alleviation, climate change adaptability and mitigation. Conservation Agriculture offers the prospect of a better future to both large-scale and smallholder farmers, and a means to raise productivity and secure economic and environmental benefits. The CA area is just about equally divided between developing countries and industrialized countries, and more recently, after a rapid spread in the Americas, adoption is increasing in Africa and Asia.

The aim of this book is to offer a state-of-the-art assessment of the status of CA in the various regions of the globe, including drawing lessons from past experience regarding its success or otherwise. This, along with new knowledge being generated through research and farmer practice, should help in promoting the further spread of CA in developing countries as well as globally.

I am sure that this volume will further stimulate the mobilization of local, national and international development support for this important approach to sustainable production intensification.

José Graziano da Silva
Director-General
Food and Agriculture Organization of the United Nations

Acronyms and Abbreviations

2,4-D:	2,4-dichlorophenoxyacetic acid
AAAID:	Arab Authority for Agricultural Investment and Development
AAPRESID:	Asociación Argentina de Productores en Siembra Directa – No-Till Argentinean Farmers Association
ABACO:	Agro-ecology based aggradation-conservation agriculture (Burkina Faso)
ABC Foundation:	Cooperative foundation, which integrates three cooperatives: Arapoti, Batavo and Castrolanda (Brazil)
ABS:	Australian Bureau of Statistics
ACIAR:	Australian Centre for International Agricultural Research
ACSAD:	Arab Center for the Study of Arid Zones and Dry Lands
ACT:	African Conservation Tillage Network
A-C-W:	Arid, cool winter, warm summer
ADAM:	Support Project to Conservation Agriculture Extension in Mountainous Areas of Vietnam
ADB:	Asian Development Bank
ADP:	Agricultural Diversification Project (Vietnam)
AEACSV:	Spanish CA Association for living soils – Asociación Española para Agricultura de Conservación – Suelos Vivos
AFD:	Agence Française de Développement (French Development Agency)
AIGACoS:	Associazione Italiana per la Gestione Agronomica e Conservativa del Suolo (Italy)
AFD:	French Development Agency
AIDS:	Acquired Immunity Deficiency Syndrome
A-K-W:	Arid, cold winter, warm summer
AN:	Ammonium nitrate
APAD:	Association to Promote Sustainable Agriculture – Association pour la Promotion d'une Agriculture Durable (France)
APOSOLO:	Portuguese Association for Conservation Tillage
APSIM:	Agricultural Production Systems simulator

APW1:	Australian Prime Hard Wheat 1
AREC:	Agricultural Research and Educational Center at AUB
ART:	Agricultural Research Trust
ASP:	Agroservicios Pampeanos (Argentina)
AUB:	American University of Beirut
AU-NEPAD:	African Union, The New Partnership for Africa's Development
AusAID:	Australian Agency for International Development
B. Wheat:	Bread wheat
Baldan:	Machinery manufacturer
Banco Do Brasil:	Brazilian Bank
BFS:	Bed planter furrow system
BIOSPAS:	Proyecto de Biología de suelo para una producción sustentable – Soil biology Project for Sustainable Production (Argentina)
Buffalo:	Machinery manufacturer
C:	carbon
CA:	Conservation Agriculture
CA2AFRICA:	Conservation Agriculture in Africa: Analysing and Foreseeing its Impact – Comprehending its Adoption
CAAANZ:	Conservation Agriculture Alliance of Australia and New Zealand
cm:	centimetre
CAAPAS:	Confederation of American Associations for Sustainable Agriculture (Brazil)
CACAARI:	Central Asian and Caucasus Association of Agricultural Research Institutes
CA-CS:	Conservation agriculture-based cropping system
CADP:	Community Agricultural Development Plans
CAIR:	CA Ireland
CANSEA:	Conservation Agriculture Network in South-east Asia
CA SARD:	Conservation Agriculture for Sustainable Agricultural Rural Development
CEIS:	Compagnie Européenne d'Intelligence Stratégique (France)
Cerrado:	Brazilian Savannah Region
CETAPAR:	Technological Centre for Agriculture, Paraguay
CF:	Conservation Farming
CFI:	Carbon Farming Initiative (Australian carbon market legislation)
CFU:	Conservation Farming Unit
CGIAR:	Consultative Group for International Agricultural Research
CIEC:	International Scientific Centre of Fertilizers (Italy)
CIMMYT:	Centro Internacional de Mejoramiento de Maíz y Trigo (International Maize and Wheat Improvement Center)
CIRAD:	Centre de Coopération Internationale en Recherche Agronomique pour le Développement (Centre for International Cooperation in Agricultural Research and Development – France)
CKARI:	Central Kazakh Agricultural Research Institute
Clube da Minhoca:	Earthworm Club (Brazil)

CLUSA:	The Cooperative League of the United States of America
CNPT/EMBRAPA:	Brazilian Wheat Research Centre, Rio Grande do Sul State (Brazil)
CO_2:	carbon dioxide
COMESA:	Common Market for Eastern and Southern Africa
CONAB:	Brazilian National Supplying Company
CONICET:	Consejo Nacional de Investigaciones Científicas y Técnicas – National Council of Scientific and Technical Research (Argentina)
ConvA:	Conventional agriculture
ConvT:	Conventional tillage
CORS:	Continuously Operating Reference Stations
CRS:	Catholic Relief Services
CSIRO:	Commonwealth Scientific and Industrial Research Organisation (Australia)
CT:	Conservation tillage
CTC:	Technical Cereal Center (Tunisia – now part of INGC: National Institute of Field Crops)
CTF:	Control Traffic Farming
CTIC:	Conservation Tillage Information Center (USA)
D. Wheat:	Durum wheat
Defra:	Department for Environment, Food and Rural Affairs (UK)
DNEA:	National Directorate for Agricultural Extension (Mozambique)
DPRK:	Democratic People Republic of Korea
DS:	Direct seeding
E&S Africa:	Eastern and Southern Africa
EC:	European Commission
EC:	electric conductivity
ECAF:	European Conservation Agriculture Federation
EMATER:	Rural State Extension Service, Brazil
EMBRAPA:	Brazilian Agricultural Research Corporation
EMBRAPA SOJA:	Soybean Research Brazilian Centre – Londrina, Paraná State
EPAGRI:	Research & Extension Institute of Santa Catarina State, Brazil
ESAK:	Academic Agricultural Education School at Kef (Tunisia)
ETH:	Eidgenössische Technische Hochschule (Zürich, Switzerland)
EU:	European Union
Fankhauser:	Machinery manufacturer
FAO:	Food and Agriculture Organization of the United Nations
FAT:	Eidgenössische Forschungsanstalt für Agrarwirtschaft und Landtechnik (Tänikon, Switzerland)
FEBRAPDP:	No-Till Brazilian Federation (Brazil)
FFS:	Farmer Field School
FINCA:	Finnish CA Association
Fitarelli:	Machinery manufacturer
FRDK:	Danish CA Association
FTC:	Farmer Training Centre
GAPs:	good agricultural practices

GART: Golden Valley Agricultural Research Trust (Zambia)
GHGs: greenhouse gases
GLS: Grey leaf spot
GM: gross margins
GMCC/gmcc: green-manure cover crops
GNSS: Global Navigation Satellite System
Gov: Government
GPS: Global Positioning System
Gralha Azul: First animal-drawn no-till planter prototype from IAPAR (Brazil)
GTZ: German Development Corporation
ha: hectare
HIV: Human Immunodeficiency Virus
IAARD: Indonesian Agency for Agriculture Research and Development
IACPA: Integrated Arable Crop Production Alliance (UK)
IAD: Institute for Sustainable Agriculture – Institut de l'Agriculture Durable (France)
IADEL: Machinery manufacturer (Brazil)
IAPAR: Agricultural Research Institute of Paraná State (Brazil)
ICAR: Indian Council for Agricultural Research
ICARDA: International Center for Agricultural Research in the Dry Areas
ICI: Imperial Chemical Industries
ICONA: Instituto Nacional para la Conservación de la Naturaleza (Spain)
ICRAF: International Council for Research in Agroforestry
ICRISAT: International Crop Research Institute for the Semi-Arid Tropics
IFAD: International Fund for Agricultural Development
IGME: Instituto Geológico y Minero de España (Spain)
IIAM: Mozambican Institute for Agricultural Research
IMASA: Machinery manufacturer
INE: Instituto Nacional de Estatística (Portugal)
INIA: National Institute for Agricultural Research (Mozambique)
INRA: Institut National de la Recherche Agronomique (National Institute of Agricultural Research – France)
INTA: Instituto Nacional de Tecnología Agropecuaria (National Institute of Agricultural Technology – Argentina)
IPCC: Intergovernmental Panel on Climate Change
IPM: Integrated Pest Management
IPNI: International Plant Nutrition Institute
IRD: French Research Institute for Development
ISFM: Integrated Soil Fertility Management
ISTRO: International Soil Tillage Research Organization
ITAIPU: Bi-national Hydroelectric Power Company (Brazil and Paraguay)
ITCF: Institut Technique des Cereales et Fourrages (France) (new name: Arvalis)
IWM: Integrated weed management
Jahnel: Machinery manufacturer

K: potassium
Knapik: Machinery manufacturer (Brazil)
KRIGF: Kazakh Research Institute of Grain Farming
KTBL: Kuratorium für Technik und Bauwesen in der Landwirtschaft (Germany)
KU: Kasetsart University (Thailand)
LEAF: Linking Environment and Farming (UK)
LFC: Soil light fraction carbon
LIFE: Less Intensive Farming Environment (UK project)
LKV: Verordnung über die Erhaltung der Lebensgrundlagen und der Kulturlandschaft (Switzerland)
LOP: Landwirtschaft ohne Pflug (Germany)
MAFF: Ministry of Agriculture, Fisheries and Food (UK); 2002 merged into Defra
MAF(F): Ministry of Agriculture and Forestry (and Fisheries)
Mafrense: Machinery manufacturer
MAGIC: Ministerio de Agricultura, Ganadería, Industria y Comercio – Argentinean Ministry of Agriculture, Cattle, Industry and Commerce
MAP: Monoammonium phosphate
Marchesan: Machinery manufacturer
MBC: Soil microbial biomass
MCPA: 4-chloro-2-methylphenoxyacetic acid
METAS: Group of institutions, companies and specialists that work with no-tillage system development in Brazil
Mg: magnesium
Mha: million hectare
mm: millimetre
MOA: Ministry of Agriculture
MOFA: Ministry of Food and Agriculture (Ghana)
Mt: megatonnes
N: nitrogen
NAFRI: National Agriculture and Forestry Research Institute (Lao PDR)
NGOs: non-government organizations
NIR: National Institute of Rubber (Vietnam)
NOMAFSI: Northern Mountainous Agricultural and Forestry Science Institute (Vietnam)
NPK: nitrogen, phosphorus and potassium
NSCP: National Soil Conservation Program (Canada)
NSW: New South Wales (Australian State)
NT: No-till/no-tillage/minimum tillage
NTA: No-till agriculture
NTCN: Controlled traffic with no tillage and full residue cover
NTG: No tillage, with grass mulch
NTL: No tillage, with legume mulch
NW: north-west
Offset ploughing: ploughing without driving in the furrow (for compaction control)
ORCATAD: Open Resource on Conservation Agriculture for Trade and Development (Lao PDR)

P: phosphorus
PADAC: Support Project for the Development of Cambodian Agriculture
PADER/BGN: Programme d'Appui au Développement Rural en Basse Guinée Nord
PAMPA: Multi-country Support Programme for Agroecology (AFD, France)
PASS: Development project for the South of Sayabouri Province (Lao PDR)
PB: permanent bed
PDRD: Programme de Développement Rural Durable
PES: Payment for Ecosystem Services
PHF: Rubber for Smallholder project (Cambodia)
PICOFA: Programme d'Investissement Communautaire en Fertilité Agricole
PIUCS: Integrated Programme of Soil Use and Conservation (Brazil)
PLUP: Participatory Land Use Planning
PMISA: Soil and Water Integrated Management Programme (Brazil)
PPILDA: Programme Promotion des Initiatives Locales de Développement à Aguie
PRB: permanent raised bed
PRECOP: Proyecto de eficiencia en cosecha y poscosecha de granos – Harvest and Postharvest Efficiency Project (Argentina)
PRODESSA: Project for the Development of the South of Sayabouri Province (Lao PDR)
PRODS/PAIA: Integrated Agricultural Production Systems as a Priority Area for Interdisciplinary Actions (PAIA) approach
Programa Paraná Rural: Paraná State Rural Development Programme (Brazil)
PRONAE: National Agroecology Programme (Lao PDR)
PROSA: Sector-based agroeology programme (Lao PDR)
PRP: Protracted Relief Programme
PSFI: permanent skip furrow irrigation
Qld: Queensland (Australian State)
QMS: Quality Management System
RELMA: Regional Land Management Unit of the Swedish International Development Agency
rpm: revolutions per minute
RT: Roto-tilling with straw cover
RTK: real-time kinematic
RTO: Refundable Tax Offset (tax terminology referring to depreciation of assets; Australia)
RUE: rainfall use efficiency
RWUE: rainwater use efficiency
RYC: Machinery manufacturer
S: sulfur
SA: South Australia (Australian State)
SA-C-W: Semi-arid, cool winter, warm summer
SA-K-W: Semi-arid, cold winter, warm summer
SAM: Mountainous Agrarian Systems Project (Vietnam)

SANREM CRSP:	Sustainable Agriculture and Natural Resource Management Collaborative Research Support Program (Cambodia, the Philippines)
SCAP:	Smallholder Conservation Agriculture Promotion project (Burkina Faso, Guinea and Niger)
S.D.:	standard deviation
SEA:	South-east Asia
SEAB-PR:	Secretary of Agriculture of Paraná State, Brazil
SEMEATO:	Machinery manufacturer (Brazil)
SFRI:	Soils and Fertilizers Research Institute (Vietnam)
SG2000:	Sasakawa Global 2000
Sgarbossa:	Machinery manufacturer
SIA:	Società Italiana d'Agronomia
SLM:	Sustainable Land Management
SMB:	soil microbial biomass
SMI:	Soil Management Initiative (UK)
SOC:	soil organic carbon
SOM:	soil organic matter
SON:	soil organic nitrogen
SOS:	Save Our Soils programme (Canada)
ST:	Subsoiling with straw cover
STCN:	Controlled Traffic with Shallow Tillage and Full Residue Cover
Teagasc:	'Learning' (Gaelic) – semi-state Agriculture and Food Development Authority (Ireland)
TLC:	Total Land Care
Triton:	Machinery manufacturer
UFRGS:	Universidade Federal do Rio Grande do Sul, Brazil
UFSM:	University of Santa Maria, Rio Grande do Sul State, Brazil
UQ:	University of Queensland (Australia)
USAID:	United States Aid
VAAS:	Vietnamese Academy of Agricultural Science
Vic:	Victoria (Australian State)
UK:	United Kingdom
UN:	United Nations
US$:	United States dollar
UZS:	Uzbek soum (national currency of Uzbekistan)
WA:	Western Australia (Australian State); West Asia
WANA:	West Asia and North Africa
WB:	World Bank
WCA:	West and Central Africa
Werner:	Machinery manufacturer
WESTCO:	Fertilizer company (Canada)
WFP:	World Food Programme
WHC:	water holding capacity
WUE:	water use efficiency
YAAS:	Yunnan Academy of Agricultural Science (China PRC)
ZCFU:	Zambia Conservation Farming Unit
ZNFU:	Zambia National Farmers Union
ZT:	zero-tillage

Keywords

Numbers indicate Chapter(s) in which keywords are used.

1 Conservation Agriculture for Sustainable and Resilient Agriculture: Global Status, Prospects and Challenges

Ram A. Jat,[1,2] Kanwar L. Sahrawat,[2] Amir H. Kassam[3,4] and Theodor Friedrich[3]
[1]Directorate of Groundnut Research, Junagadh, India; [2]International Crops Research Institute for the Semi-Arid Tropics, Patancheru, India; [3]Plant Production and Protection Division, Food and Agriculture Organization of the United Nations, Rome, Italy; [4]School of Agriculture, Policy and Development, University of Reading, Reading, UK

1.1 Introduction

Achieving food security for a burgeoning population, particularly in the less developed nations, and developing sustainable agricultural production systems are among the major challenges before the world in 21st century. The challenge is not only to ensure sufficient food for all the people but also to meet the ever increasing demand for meat, eggs, fruits and vegetables by the rapidly expanding middle class population in developing nations. The challenges are getting further confounded due to imminent climate change-related risks, the adverse effects of which have already started being experienced in one or other form in agricultural production systems in various parts of the globe. As more and more agricultural land is being diverted towards industrial and residential uses throughout the world, we have to produce more and more food from increasingly less-cultivated land. This will further strain the already fragile natural resource base, particularly land and water, making it more difficult to meet the food requirements of the world. Therefore, there is urgent need to conserve or even improve the natural resources from being degraded by water and wind erosion, which is accelerated manifold due to human activities.

> Although more than 99% of the world's food comes from the soil, experts estimate that each year more than 10 Mha of crop land are degraded or lost as rain and wind sweep away topsoil. An area large enough to feed Europe – 300 Mha, about ten times the size of the UK – has been so severely degraded it cannot produce food, according to UN figures (The Guardian, 2004).

Soil degradation is rampant both in developed and less developed nations. In fact the highest levels of land degradation are in Europe. 'Specifically degraded soils are found especially in semi-arid areas (Sub-Saharan Africa, Chile), areas with high population pressure (China, Mexico, India) and regions undergoing deforestation (Indonesia)' (Philippe Rekacewicz, UNEP/GRID-Arendal, 2007). The perception that land is an infinite natural resource has taken a heavy toll, leading to severe land degradation in many parts of the world. Every year millions of tonnes of sediments are discharged with runoff water throughout the world. This not only causes loss of agriculturally precious topsoil, but also affects aquatic ecosystems negatively by dumping nutrients and the silting of water bodies. Furthermore, widespread and severe decline of soil quality in almost all production regions also raises questions about the sustainability of current agricultural production practices (Verhulst *et al.*, 2010).

According to IPCC-based climate change predictions, most of the rainfall will occur in the form of high-intensity short-duration rain events due to global climate change effects (IPCC, 2007). If that becomes true, efficient use of rainwater through both *in situ* and *ex situ* moisture conservation practices will be imperative to achieve the objective of getting higher yields and conserving the natural resource base. This warrants that more proactive efforts should be made for developing and adopting resource-conserving technologies to increase global food production in a sustainable way amid the confounding challenges facing agriculture. Conservation Agriculture (CA), consisting of minimum mechanical soil disturbance, soil cover with plant biomass/cover crops and diversified crop rotations or associations, is viable and seems a more sustainable cultivation system than that presently practised. CA reduces soil erosion, improves soil quality, reduces soil compaction, improves rainwater use efficiency, moderates soil temperature, gives higher and stable yields, saves inputs, reduces cost of cultivation and helps in climate change mitigation and adaptation (Machado and Silva, 2001; Kassam *et al.*, 2009; Hobbs and Govaerts, 2010; Lal, 2010; Jat *et al.*, 2012b). CA principles are universally applicable to all agricultural landscapes and land uses with of course locally adapted practices (Kassam and Friedrich, 2012).

1.2 Conservation Agriculture: the Way Forward for Sustainable Agricultural Production

During the past few decades, rapid strides have been made all over the world to develop and disseminate CA practices. CA has emerged as a major way forward from the existing plough-based unsustainable conventional agriculture (ConvA), to protect the soil from water- and wind-led degradation processes and make agricultural production systems sustainable. Empirical evidences suggest that zero tillage-based agriculture along with crop residue retention and adoption of suitable crop rotations can be productive, economically viable and ecologically sustainable given that farmers are involved in all the stages of technology development and dissemination (Friedrich *et al.*, 2012). CA specifically aims to address the problems of soil degradation due to water and wind erosion, depletion of organic matter and nutrients from soil, runoff loss of water and labour shortage. Moreover, supporters of the CA movement claim that CA is able to address the negative consequences of climate change on agricultural production through improved rainwater use efficiency, moderating soil and plant canopy temperature and timely performance of agronomic operations (Gupta *et al.*, 2010; Jat *et al.*, 2012b). However, there is need to identify, evolve and disseminate region-specific CA practices through active involvement of farmers along with researchers, technicians, machinery manufacturers and policy makers (Fowler and Röckstrom, 2000).

1.3 Conservation Agriculture: Definition and Concept

According to the FAO, 'CA is an approach to managing agro-ecosystems for improved and sustained productivity, increased profits and food security while preserving and enhancing the resource base and the environment' (Friedrich *et al.*, 2012). CA has been designed on the principles of integrated management of soil, water and other agricultural resources in order to reach the objective of economically, ecologically and socially sustainable agricultural production.

CA is characterized by three major principles (FAO, 2012):

- Minimal mechanical soil disturbance by direct planting through the soil cover without seedbed preparation;
- Maintenance of a permanent soil cover by mulch or growing cover crops to protect the soil surface;

- Diversifying and fitting crop rotations and associations in the case of annual crops and plant associations in the case of perennial crops.

Usually, the retention of 30% surface cover by residues characterizes the lower limit of classification for CA. The concept of CA has evolved from the zero tillage (ZT) technique. In ZT, seed is put in the soil without any prior soil disturbance through any kind of tillage activity or only with minimum soil mechanical disturbance. In zero-tilled fields, with time, soil life takes over the functions of traditional soil tillage such as loosening the soil and mixing the organic matter. In CA, due to minimum soil disturbance, soil life and biological processes are not disturbed, which is crucial for a fertile soil supporting healthy plant growth and development. The soil surface is kept covered either by crop residues, cover crops or biomass sourced *ex situ* through agroforestry measures, which provide physical protection for the soil against agents of soil degradation; and equally importantly provides food for the soil life. The burning or incorporation of crop residues is strictly avoided in CA. At the same time varied crop rotations involving legumes in CA help to manage pest and disease problems and improve soil quality through biological nitrogen fixation and addition of organic matter (Baudron *et al.*, 2009).

1.4 Global History, Current Status and Prospects of Conservation Agriculture

The origin of the CA movement can be traced in the 1930s when the dustbowls devastated vast areas of the mid-west USA. The new concepts of reduced tillage were introduced, as against the conventional intensive tillage-based cultivation systems, so as to ensure minimum soil disturbance and to protect the soil from water and wind erosion. Seeding machinery was developed for seeding directly with minimum soil disturbance through the surface-lying residues to ensure optimum crop stand (Friedrich *et al.*, 2012). But it was not until the 1960s that CA could enter into the farming practices in the USA. At present, CA is practised over an area of 26.5 Mha in the USA, which constitutes only 16% of the cropland. Protecting soils from devastating soil erosion, moisture conservation and timely planting of crops have been the major incentives for development and spread of conservation tillage in the USA. The no-till system entered into Brazil in the early 1970s as a potential remedial measure to the severe problem of soil loss due to water erosion in the tropical and subtropical regions of Brazil. The no-till practice was further refined in Brazil to suit the local requirements with the active collaboration of researchers, extension workers, progressive farmers; and with government support. Subsequently, the principles of keeping the soil covered either with crop residues or cover crops, and the adoption of suitable crop rotations/associations were added with the principle of minimum soil disturbance, and the term CA was given to this new concept of farming (Denardin *et al.*, 2008). Brazil became the cradle for evolution of the CA movement.

> The expansion of NT area in Brazil occurred mainly due to the availability of no-till seeders, adapted and developed with the support of research institutions and with farmers' evaluations as well, the attractive agricultural investment financing, the farmers' interest in changing their farming system and the machinery industries' interest in expanding their market' (Calegari *et al.*, Chapter 3, this volume).

Currently, Brazil along with other Latin American countries of Argentina, Paraguay and Uruguay, is among the leading countries of the world having the largest area under CA of their total cropland. However, there are serious concerns about the quality of CA being practised in these countries; for example, due to market pressures farmers are practising monocropping of soybean without growing cover crops in between two successive crops of soybean, leading to heavy soil erosion and land degradation (Friedrich *et al.*, 2012). In Canada, even though no-till started in the 1970s, its rapid adoption started only in the early 1990s (see Lafond *et al.*, Chapter 4, this volume). The necessity to protect the soil against devastating wind erosion during the fallow dry season, the introduction of winter wheat in the Prairies of Canada, availability of cheaper and effective herbicides, determined efforts of progressive farmers, supportive government policies, knowledge transfer through farmers' associations, design and development of no-till seeders by the private manufacturers according to the needs of local farmers, were the major factors that contributed to the spread and successful adoption of CA in the Canadian Prairies. Today, with 13.5 Mha area under CA in Canada, with the highest being in Saskatchewan followed by Alberta, Canadian farmers are witnessing the benefits of CA in terms of reduced wind erosion, increased hectarage under winter wheat, improved soil quality and biodiversity, among others.

The CA movement in Australia started in the mid-1970s following the visit of Australian researchers and progressive farmers to the USA and the UK; this was ably supported subsequently with availability of herbicides, particularly glyphosate, at competitive rates by private manufacturers. The main incentives for shifting from conventional intensive tillage-based farming systems to CA-based systems in Australia were: soil protection against water erosion (in northern cropping zones) and wind erosion (in western and southern cropping zones), soil moisture conservation (particularly in the dry western parts of Australia) and timely sowing of the crops. CA adoption was led in northern, central, southern and western states of Australia by the farmers in the more marginal areas where benefits in terms of soil moisture conservation and timely crop sowing were initially more obvious. The Australian government has been proactively supporting the CA movement in their country by giving important incentives through programmes such as 'Care for our Country', 'The Carbon Farming Initiative' and 'Clean Energy Future Plan', which led to a steady increase in hectarage under CA in Australia since the early 1990s (see Rochecouste and Crabtree, Chapter 5, this volume). Currently, Australia and New Zealand together have 17.16 Mha area under CA, which constitutes 14% of global CA hectarage.

CA is not widespread in Europe; the no-till systems cover only 1% of arable cropland (Friedrich *et al.*, 2012). In Europe, ECAF (European Conservation Agriculture Federation) has been promoting CA since 1999. Spain (650,000 ha), France (200,000 ha), Finland (160,000 ha) and the UK (150,000 ha) are the leading countries in the adoption of CA in Europe. Other countries practising CA to some extent in Europe are Ireland, Portugal, Germany, Switzerland and Italy. The agricultural policies in the European Union such as direct payment to farmers and subsidies on certain commodities, moderate climate and interest groups opposing the introduction of CA are the main reasons for slower adoption of CA in Europe (see Friedrich *et al.*, Chapter 6, this volume).

In Russia, hectarage under CA as per FAO definition is 4.5 Mha, while conservation tillage is reported to be practised on 15 Mha. In Ukraine, area under CA has reached 600,000 ha.

In Central Asia, with the active support of development agencies such as FAO, CIMMYT and ICARDA, Kazakhstan and Uzbekistan have made good progress to successfully adopt CA in large areas of their croplands. In Kazakhstan, CA is mostly practised in northern dry steppes and has 10.5 Mha under reduced tillage and 1.6 Mha under real CA. The concentration of large land areas under agricultural joint-stock companies, which are the main adopters of CA practices, and government subsidies for adopting CA practices have helped in rapid spread of CA practices in northern Kazakhstan (Kazakhstan Farmers Union, 2011; Kienzler *et al.*, 2012).

In China, the CA movement started in the early 1990s and currently has an area of 3.1 Mha under CA. However, Wang *et al.* (2010) reported that the adoption of CA in China is still low; in particular, the full adoption of CA is almost zero. According to them, the main reasons for slow pickup of CA by Chinese farmers are the low labour cost and low share of machinery and fuel in the total cost of cultivation, which gives few incentives to farmers to adopt CA technology.

In the Indo-Gangetic plains in South Asia across India, Pakistan, Bangladesh and Nepal, no-till is practised in wheat in about 5 Mha (Friedrich *et al.*, 2012). However, the adoption of permanent no-till systems and full CA is only marginal. In South-east Asia, CA was introduced in the late 1990s with the help of developmental agencies and international research organizations such as AFD (French Development Agency), CIRAD, NAFRI and USAID, but still CA is limited mainly to the research sector with limited extension to farmers' fields.

In the WANA (West Asia and North Africa) region, work on CA has been started since the 1980s in countries including Morocco, Tunisia, Algeria, Syria, Lebanon, Jordan and Turkey. In this region, currently Syria has the largest hectarage under CA, followed by Tunisia and Morocco. In Tunisia, it is mainly the large estates that have adopted CA. The owners had access to information, enough money to import quality seeders from Brazil, France or Spain; and they could bear the risk of trying new practices (Kurt G. Steiner, Schönau/Germany, 2012, pers. comm.).

In Africa, despite nearly two decades of promotional efforts by the national extension programmes and numerous international developmental agencies, the adoption of CA has been very low. Currently, Africa has only 1.01 Mha under CA, which is the lowest among all the continents (Table 1.1). South Africa (368,000 ha), Zambia (200,000 ha), Mozambique (152,000 ha) and Zimbabwe (139,300 ha) are the leading countries in the adoption of CA in Africa. The main reasons for a slow adoption of CA in Africa are numerous, namely: a low degree of mechanization within the smallholder system; lack of appropriate implements; lack of appropriate soil fertility management options; problems of weed control under no-till systems; lack of access to credit; lack of appropriate technical information; blanket recommendations that ignore the resource status of rural households; competition for crop residues in the mixed crop–livestock systems; and limited availability of household labour (Twomlow *et al.*, 2006).

'In the last 11 years, the CA systems have expanded at an average rate of more than 7 Mha per year globally, showing the interest of farmers and national governments in this alternate production method' (Friedrich *et al.*, 2012). Table 1.2 presents area under CA in different countries of the world. Originally, the CA movement was started as a remedial measure against wind and water erosion (in the USA and Canada, and Brazil, respectively), drought (in Australia), to increase crop area (in Canada), but more recently, pressed again by the severity of soil erosion and land degradation in many agriculturally important regions, besides increase in the cost of energy and production inputs, CA is being promoted by national governments in many countries. With the entry of local manufacturers in making available CA machinery at affordable rates, the area under CA is spreading fast in several parts of the globe. Combining agroforestry with CA is an important viable option to augment biomass supply for CA, particularly in the rainfed tropics and subtropics where crop

Table 1.1. Area under Conservation Agriculture by continent (adapted from Friedrich *et al.*, 2012).

Continent	Area (ha)	Percentage of total CA area in world	CA as percentage of arable cropland
South America	55,464,100	45	57.3
North America	39,981,000	32	15.4
Australia and New Zealand	17,162,000	14	69.0
Asia	4,723,000	4	0.9
Russia and Ukraine	5,100,000	3	3.3
Europe	1,351,900	1	0.5
Africa	1,012,840	1	0.3
World	124,794,840	100	8.8

Table 1.2. Area (ha) under Conservation Agriculture in different countries of the world: the area with >30% ground cover qualified for CA (1000 ha) (from FAO: http://www.fao.org/ag/ca/6c.html).

Country	Area (year)
Argentina	25,553 (2009)
Australia	17,000 (2008)
Bolivia (Plurinational State of)	706 (2007)
Brazil	25,502 (2006)
Canada	13,481 (2006)
Chile	180 (2008)
China	3,100 (2011)
Colombia	127 (2011)
Democratic People's Republic of Korea	23 (2011)
Finland	160 (2011)
France	200 (2008)
Germany	5 (2011)
Ghana	30 (2008)
Hungary	8 (2005)
Ireland	0.1 (2005)
Italy	80 (2005)
Kazakhstan	1,600 (2011)
Kenya	33.1 (2011)
Lebanon	1.2 (2011)
Lesotho	2 (2011)
Madagascar	6 (2011)
Malawi	16 (2011)
Mexico	41 (2011)
Morocco	4 (2008)
Mozambique	152 (2011)
Namibia	0.34 (2011)
Netherlands	0.5 (2011)
New Zealand	162 (2008)
Paraguay	2,400 (2008)
Portugal	32 (2011)
Republic of Moldova	40 (2011)
Russian Federation	4,500 (2011)
Slovakia	10 (2006)
South Africa	368 (2008)
Spain	650 (2008)
Sudan and South Sudan	10 (2008)
Switzerland	16.3 (2011)
Syrian Arab Republic	18 (2011)
Tunisia	8 (2008)
Ukraine	600 (2011)
UK	150 (2011)
United Republic of Tanzania	25 (2011)
USA	26,500 (2007)
Uruguay	655.1 (2008)
Venezuela (Bolivarian Republic of)	300 (2005)
Zambia	200 (2011)
Zimbabwe	139.3 (2011)
Total	124,795

residues are used for cattle feeding and/or biomass production is low due to water stress and several other factors (Sims *et al.*, 2009). With the recent unfavourable changes in rainfall patterns in different parts of the globe and higher temperatures during critical crop growth stages, CA is becoming even more relevant to achieve food security and protect our environment (Kassam *et al.*, 2011a; Corsi *et al.*, 2012).

1.5 Research Results Reported

1.5.1 Soil and water conservation

Soil degradation by water and wind erosion, as well as a decline in soil physical, chemical and biological properties, can be linked to excessive levels of tillage, removal and/or burning of crop residues and fallow systems that are associated with conventional farming systems (Lumpkin and Sayre, 2009). Higher soil degradation in conventional farming systems is due to the fact that conventional tillage (ConvT) causes more physical disruption and less production of aggregate stabilizing materials (Bradford and Peterson, 2000). Moreover, incorporation of crop residues by tillage or their removal from field for cattle fodder or burning leaves soils exposed to the actions of rain, wind and heating by the sun, leading to enhanced rate of soil degradation. Higher aggregate stability in CA practices as compared to conventionally tilled fields results in lower soil erosion potential in CA (Derpsch *et al.*, 1991; Packer *et al.*, 1992; Uri *et al.*, 1999; Chan *et al.*, 2002; Hernanz *et al.*, 2002; Pinheiro *et al.*, 2004; López and Arrúe, 2005; Govaerts *et al.*, 2007c; Li *et al.*, 2007; Márquez *et al.*, 2008; Kassam *et al.*, 2011a). ZT with residue retention resulted in a high mean weight diameter and a high level of stable aggregates (considered as a parameter for predicting soil erodibility) in the rainfed systems of Mexico (Verhulst *et al.*, 2009). Presence of crop residues on the soil surface in CA leads to profound increase in microbial activity, leading to secretions of aggregate-binding chemicals in to the soil. As CA leaves more plant residues over the surface compared to ConvT, it protects soil from deleterious actions of rainfall, gusty winds and heating effects of the sun.

The soil erosion in CA fields is further reduced due to the reduced amount of runoff under CA conditions (Rao *et al.*, 1998; Rhoton *et al.*, 2002; Araya *et al.*, 2012). Maintenance of crop residues on the surface in CA prevents surface sealing, improving infiltration, which ultimately results in reduced soil erosion. Mulching, which is a part of CA, halts soil erosion by providing a protective layer to the soil surface, increasing resistance against overland flow and enhancing soil surface aggregate stability and permeability (Erenstein, 2003). Annual soil loss was 3.8 and 8.1 times greater without mulch when compared to mulching with 3 t ha^{-1} and 5 t ha^{-1} of crop residues in humid highlands of Kenya (Danga and Wakindiki, 2009). The corresponding decrease in runoff volume was 2.1 and 4.6 times compared to no mulching. The placement of straw over the surface also reduced runoff velocity along the slope, thereby decreasing the erosivity of runoff water, besides trapping the sediments carried by overland flow. Under CA, the 30% threshold for soil cover is expected to reduce soil erosion by 80%, but greater soil cover is expected to suppress soil erosion further (Erenstein, 2002). However, no-till fields, when residue cover is low, may be more vulnerable to runoff because no-till surfaces lack roughness and can experience soil compaction (Hansen *et al.*, 2012). Readers are referred to a review by Jat *et al.* (2012b) for a detailed discussion on the role of CA in controlling soil degradation.

1.5.2 Soil quality

Soil quality is 'the capacity of a specific kind of soil to function, within natural managed ecosystem boundaries, to sustain plant and animal productivity, maintain or enhance water and air quality, and support human health and habitation' (Karlen *et al.*, 1997). A simpler operational definition is given by Gregorich *et al.* (1994) as 'The degree of fitness of a soil for a specific use'. According to Verhulst *et al.* (2010), from an agricultural production point of view 'high soil quality equates to the ability of the soil to maintain a high productivity without significant soil or environmental degradation'. Evaluation of soil quality is based on physical, chemical and biological properties of the soil. 'With respect to biological soil quality, a high quality soil can be considered a "healthy" soil' (Verhulst *et al.*, 2010). A healthy soil is defined as a stable system with high levels of biological diversity and activity, internal nutrient cycling and resilience to disturbance (Rapport, 1995; Shaxson *et al.*, 2008).

Adoption of CA, following all the principles, for a sufficiently long period of time leads to significant improvement in soil quality, mainly in the surface layers (Hobbs, 2007; Mousques and Friedrich, 2007; Thomas *et al.*, 2007; Verhulst *et al.*, 2009; Lal, 2010). Soil structure is a key factor in soil functioning, and is an important factor in the evaluation of the sustainability of crop production systems (Verhulst *et al.*, 2010) and is often expressed as the degree of stability of aggregates (Bronick and Lal, 2005). ConvT results in reduced aggregation due to direct and indirect effects of tillage on aggregation (Beare *et al.*, 1997; Six *et al.*, 2000). Tillage breaks down the old aggregates and disrupts the process of new aggregate formation by fragmenting the plant roots and mycorrhizal hyphae, which are among the major binding agents for macro-aggregate formation, and also disrupts other biological activities in the soil. ZT with residue retention improves dry as well as wet aggregate size distribution compared to ConvT (Chan *et al.*, 2002; Filho *et al.*, 2002; Pinheiro *et al.*, 2004; Madari *et al.*, 2005; Govaerts *et al.*, 2007c; Li *et al.*, 2007; Lichter *et al.*, 2008; Verhulst *et al.*, 2009). In CA plots, increased microbial activity creates a stable soil structure through accumulation of organic matter due to retention of crop residues and addition of large amount of biomass by cover crops and legumes in rotation (De Gryze *et al.*, 2005; Lal, 2010; Verhulst *et al.*, 2010).

ConvT, for example, during long-term use of disc tillage equipment can cause compactness in soil subsurface layers leading to restricted root growth, waterlogging and poor aeration (Castro Filho *et al.*, 1991; Fageria *et al.*, 1997). CA has been reported to reduce soil compaction due to reduced traffic and growing of the deep-rooted cover crops or legumes in rotation, which break the compact layers in the subsurface (FAO, n.d. a; Kemper and Derpsch, 1981; Kayombo and Lal, 1993). CA has been found to reduce bulk density, particularly in surface layers, thereby facilitating better aeration and water retention (Machado and Silva, 2001; Nurbekov, 2008).

Residue retention and consequent greater microbial biomass and abundance of earthworms and macro-arthropods in soils under CA exert beneficial effects on soil fertility. CA leads to the stratification of nutrients, with higher amount of nutrients near the soil surface compared to deeper layers (Franzluebbers and Hons, 1996; Calegari and Alexander, 1998; Duiker and Beegle, 2006). As surface-placed residues decompose slowly, it may prevent rapid leaching of nutrients through the soil profile in CA fields (Kushwaha *et al.*, 2000; Balota *et al.*, 2004). CA may lead to lower nutrient availability because of greater immobilization by the residues left on the soil surface (Rice and Smith, 1984; Bradford and Peterson, 2000) in the initial years of adoption. But in the long run, as summarized by Verhulst *et al.* (2010), 'the net immobilization phase when CA is adopted is transitory, and the higher, but temporary immobilization of N in ZT systems reduces the opportunity for leaching and denitrification losses of mineral N'. The higher initial N-fertilizer requirement decreases over time because of reduced loss by erosion and the build-up of a larger pool of readily mineralizable organic N. Thomas *et al.* (2007) reported significantly higher total nitrogen in 0–30 cm soil depth and exchangeable K in 0–10 cm soil depth under no-till as compared to ConvT plots. Reduced tillage and addition of N by legumes in the cropping system increases total N in the soil under CA (Amado *et al.*, 1998).

The different cover crops have phosphorus (P)-recycling capacity; and this even further improves when the residues are retained on the surface (Calegari and Alexander, 1998). 'Numerous studies have reported higher extractable P levels in ZT than in tilled soil, largely due to reduced mixing of the fertilizer P with the soil, leading to lower P-fixation' (see Verhulst *et al.*, 2010). The organic acids resulting from the build-up of the soil organic matter may also increase P mobilization (Mousques and Friedrich, 2007). This helps enhance P-use efficiency when P is a limiting nutrient, but may cause environmental problems through loss of soluble P in runoff water when soil P levels are high (Duiker and Beegle, 2006). They also suggested that there may be less need for P starter fertilizer in long-term zero-tilled fields due to relatively high available P levels in the topsoil where the seed is placed. Micronutrients tend to be present in higher levels under CA compared to ConvT, especially extractable zinc and manganese near the soil surface due to the surface placement of crop residues (Franzluebbers and Hons, 1996).

The high organic matter contents in the surface soil layer, commonly observed under CA, can increase the cation exchange capacity of the surface layers (FAO, 2001; Duiker and Beegle, 2006). CA has been found to be effective in ameliorating sodicity and salinity in soils (Franzluebbers and Hons, 1996; Hulugalle and Entwistle, 1997; Sayre, 2005; Govaerts *et al.*, 2007c; Qadir *et al.*, 2007). For example, after 9 years of minimum tillage, the values of exchangeable Na, exchangeable sodium percentage and dispersion index were lower in an irrigated Vertisol compared to ConvT (Hulugalle and Entwistle, 1997). Thomas *et al.* (2007) also recorded lower exchangeable Na in surface layers due to no tillage (NT) compared to ConvT. The combination of ZT with sufficient crop residue retention reduces evaporation from the soil and salt accumulation on the soil surface (Nurbekov, 2008; Hobbs and Govaerts, 2010). Inclusion of legumes in crop rotations in CA may reduce the pH of alkaline soils due to intense nitrification followed by NO_3^- leaching, H_3O^+ excretion by legume roots (Burle *et al.*, 1997). Besides, in no-till all the N is placed on the soil surface and this leads to decrease in soil pH because of acidification following nitrification of the soil and applied N.

The soil microbial biomass (SMB) reflects the soil's ability to store and cycle plant nutrients (C, N, P and S) and organic matter (Dick, 1992; Carter *et al.*, 1999), and due to its dynamic character, SMB responds to changes in soil management often before effects can be measured in terms of organic C and N (Powlson and Jenkinson, 1981). SMB has a crucial role in plant nutrition. According to Weller *et al.* (2002), general soil-borne disease suppression is also related to total SMB, which competes with pathogens for resources or causes inhibition through more direct forms of antagonism. The rate of organic C addition from plant biomass is generally considered the most important factor determining the amount of SMB in the soil (Campbell *et al.*, 1997). In the subtropical highlands of Mexico, residue retention resulted in significantly higher amounts of SMB-C and N in the 0–15 cm layer compared to residue removal (Govaerts *et al.*, 2007b). Alvear *et al.* (2005) reported higher SMB-C and N in the 0–20 cm layer under ZT than under ConvT with discharrow in an Ultisol from southern Chile, and attributed this to the higher levels of C inputs available for microbial growth, better soil physical conditions and higher water retention under ZT. The favourable effects of ZT and residue retention on soil microbial population are mainly due to increased soil aeration, favourable temperature and moisture conditions, and higher C content in surface soil (Doran, 1980). Against this, each tillage operation increases organic matter decomposition with a subsequent decrease in SOM (Buchanan and King, 1992). Crop residue retention has been found to enhance enzymatic activities also mainly in soil surface layers (Alvear *et al.*, 2005; Roldán *et al.*, 2007; Nurbekov, 2008). Soil enzymes play an essential role in catalysing the reactions associated with organic matter decomposition and nutrient cycling.

Thus, it can be concluded that soils under CA are in general physically, chemically and biologically stratified with improved soil quality in surface layers.

1.5.3 Rainwater use efficiency

In rainfed agriculture, improving rainwater use efficiency (RWUE) is imperative to obtain higher yields. Other than rainfall pattern, the crops grown and management practices, RWUE is determined by the rate of water infiltration, water-holding capacity of soils and evaporative loss of water. CA has been found to improve RWUE by improving rainwater infiltration (Calegari and Alexander, 1998; Erenstein, 2002; Govaerts *et al.*, 2007a; Shaxon *et al.*, 2008; Verhulst *et al.*, 2009), water-holding capacity (Hudson, 1994; Acharya *et al.*, 1998; Govaerts *et al.*, 2007a, 2009; Mousques and Friedrich, 2007; Nurbekov, 2008) and reducing loss of water through evaporation (Erenstein, 2003; Scopel *et al.*, 2004; Nurbekov, 2008). According to Scopel and Findeling (2001), in the short run, residue heaps act as a succession of barriers giving the water more time to infiltrate; while in the long run (>5 years), retention of crop residues increases average infiltration rates up to 10 times compared to ConvT by preventing crust formation. Improved soil cohesion, pore continuity and aggregate stability, and the protection of the soil surface from direct impact of the raindrop, are the most important factors that contribute to improved water infiltration into the soil (Basch *et al.*, 2012). Large pores due to greater numbers of earthworms, termites, ants and millipedes combined with the channels created by decomposing plant roots and their higher density result in increased water infiltration in CA plots (Blevins *et al.*, 1983; Roth, 1985). Residues intercept the rainfall and release it more slowly afterwards, which helps to maintain higher moisture level in soil, leading to extended water supply for plants (Scopel and Findeling, 2001). Increase in SOM due to residue retention in CA fields increases water-holding capacity of soil. Hudson (1994) showed that for each 1% increase in SOM, the available water-holding capacity in the soil increased by 3.7%. Mulching in CA fields reduces loss of stored soil moisture by checking evaporation (Erenstein, 2003).

Changrong *et al.* (2009), while working in China, reported 1% to more than 20% increase in water availability in dryland fields due to zero or reduced tillage with residue retention compared to conventional farming. ZT with residue retention decreases the frequency and intensity of short mid-season droughts (Bradford and Peterson, 2000).

Thus, in CA plots most or all of the rainfall is harnessed as effective rainfall, with little runoff and no soil erosion, leading to longer and reliable moisture regime for crop growth, and improved drought proofing (Shaxson *et al.*, 2008).

1.5.4 Nutrient use efficiency

Reduced runoff and the use of appropriate deep-rooting cover crops contribute to reducing nutrient losses in CA fields (FAO, 2001) Crop residues release nutrients slowly, which help prevent nutrient losses by leaching and/or denitrification. Moreover, the immobilization of mineral N due to residue retention may also prevent potential losses due to NO_3-N leaching (Thomas *et al.*, 2007). In the short run, lower fertilizer use efficiency may be recorded as a result of immobilization of mineral nutrients by microorganisms. However, in the long-run, nutrient availability increases because of microbial activity and nutrient recycling (Carpenter-Boggs *et al.*, 2003).

Phosphorus use efficiency can be improved if crop residues are added to the soils (Iyamuremye and Dick, 1996; Sanchez *et al.*, 1997), which is further increased when combined with NT (Sidiras and Pavan, 1985; De Maria and Castro, 1993; Selles *et al.*, 1997). Thomas *et al.* (2007) also recorded higher levels of bicarbonate-extractable P in 0–10 cm layer under NT than ConvT. Greater available P levels in the upper layers of NT soils may be due to reduced mixing of fertilizer P, possibly increased quantities of organic P, and shielding of P adsorption sites (Weil *et al.*, 1988).

Inclusion of legumes in cropping systems increases the turnover and retention of soil N and other nutrients (Drinkwater *et al.*, 1998; Hansen *et al.*, 2012). Sisti *et al.* (2004) reported, from a 13-year study in southern Brazil, significant increase in soil N stocks when vetch, legume green manure crop, was included in rotation along with ZT compared to no legume green manure crop. Burle *et al.* (1997) found highest levels of exchangeable K, calcium (Ca) and magnesium (Mg) when pigeon pea and lablab (*Dolichos lablab*) were included in the systems. Increased aggregation and SOM at the soil surface also leads to increased nutrient use efficiency in CA fields (Franzluebbers, 2002). Hobbs and Gupta (2004) reported improved fertilizer use efficiency (10–15%) in the rice–wheat system, mainly as a result of better placement of fertilizer with the seed drill in CA fields as opposed to broadcasting in the conventional system.

1.5.5 Input use efficiency

In the long term, besides reducing the need for chemical fertilizers, CA may bring down demand for fuel, labour, machinery and pesticides as well as time (Zenter *et al.*, 2002; Fernandes *et al.*, 2008; SoCo, 2009; Freixial and Carvalho, 2010). As the knowledge and understanding of tenants about CA increases with time, the need for operations and off-farm inputs reduces (Derpsch, 1997). Direct sowing without or with minimum soil disturbance implies less labour, energy, time and machinery requirement. Fernandes *et al.* (2008), from a study conducted in Brazil, estimated a diesel saving of 6.4 l ha^{-1} by tractors when ConvT was replaced by NT; and the total energy budget was lower by 25.5 l diesel equivalent ha^{-1}. In DPRK (Democratic People's Republic of Korea) the adoption of CA resulted in input savings of 30–50% (Mousques and Friedrich, 2007). Omission of tillage operations in CA systems can help reducing labour requirements during a critical time in the agricultural calendar (Giller *et al.*, 2009), which makes it convenient for farmers to perform other operations such as the timely sowing of relatively large areas. Adoption of integrated weed management and mulching in CA could lead to lesser weed intensity, which reduces labour requirement for weeding in the long term. However, during initial years, the increased labour requirement due to higher weed intensity in CA plots compared to ConvT plots may outweigh the labour saving due to NT (Jat *et al.*, 2012a). Moreover, due to the higher weed problem in CA, the labour burden could be shifted on to the women, who traditionally are responsible for weeding, from the men, who are responsible for tillage (Giller *et al.*, 2009).

1.5.6 Insect-pest, disease and weed dynamics

Varying results of insect-pest dynamics in response to the adoption of CA have been reported in different studies from different parts of the globe. A review of 45 studies showed that 28% of the pest species increased with decreasing tillage, 29% showed no significant influence of tillage and 43% decreased with decreasing tillage (Stinner and House, 1990). Reduced tillage may lead to an increase in the number of insect-pests (Musick and Beasley, 1978), but it also tends to increase diversity of predators and parasites of crop-damaging insects (Stinner and House, 1990). Besides, crop rotations and plant associations, which are integral parts of CA, help break insect-pest cycles (FAO, n.d. b). Biological diversity processes and increased species and functional diversity due to reduced tillage, residue retention and crop rotations/plant associations in CA fields (Hobbs and Govaerts, 2010) also help keeping insect-pests and diseases under control. Therefore, better insect-pest management is possible in CA fields in the long term; none the less, higher incidence of insect-pests is quite possible during initial years of CA adoption when predators/parasites are not in sufficient number. Insect-pests may be harboured in the crop residues retained on soil surface (Hansen *et al.*, 2012) as well as in undisturbed soils in CA. The wheat stem sawfly (*Cephus cinctus* Norton) became a concern in the US Great Plains; and its spread is speculated to be associated with the spread of no-till area (Weaver *et al.*, 2009; Peairs *et al.*, 2010). However, these concerns were not confirmed and the pest occurrence was more related to wheat monocropping than to no-tillage (MANDAK, 2011).

As different pathogens have different survival strategies and life cycles, reduced tillage affects different plant pathogens in different ways (Bockus and Shroyer, 1998). Crop residue retention may directly affect the pathogens by changing composition of soil microbial community in favour of beneficial microorganisms; however, crop residues can carry over pathogens from one season to the next season. CA also affects pathogens indirectly through improved soil moisture, aeration and moderating soil temperatures (Krupinsky *et al.*, 2002). Crop rotations play a crucial role in CA to break disease cycles and neutralize the pathogen carry-over effects of residue retention and minimum mechanical disturbance of soils (Barker and Koenning, 1998). According to Forcella *et al.* (1994), due to one or more of the following mechanisms, the residues of some crops are able to reduce pathogen incidence: (i) leaching of inhibitory chemicals from decomposing residues; (ii) leaching of stimulatory chemicals from residues which promote populations of beneficial microbial control agents; (iii) enhanced populations of highly competitive non-pathogenic species in lieu of non-competitive pathogenic species due to high C:N ratios; and (iv) increased vigour of crops making them less susceptible to diseases due to higher soil water contents and improved soil quality. However, CA may increase or decrease disease incidence in different crops; for example, in maize, residue retention increased the incidence of root rot, while in wheat, residue decreased the incidence (Govaerts *et al.*, 2007a). Similarly, retention of wheat residues causes increased incidence of stem rot in groundnut.

Weed management is an important issue in promoting CA among smallholders. Muliokela *et al.* (2001) reported higher weed infestations with minimum tillage practices than ploughed fields in Zambia. Minimum tillage may lead to increased labour requirements for weeding, particularly during starting years of CA adoption if done gradually (Vogel, 1994; Haggblade and Tembo, 2003; Jat *et al.*, 2012a). Minimum tillage may lead to increased intensity of the perennial weed population in the long term (Vogel, 1994). For this reason, CA excludes minimum tillage by definition, since the level of soil disturbance in minimum tillage is still high enough to create weed problems (Friedrich and Kassam, 2012).

The net effect of crop residue retention in CA on weed control is somewhat contradictory. In some cases, crop residues suppress weed seed germination and/or seedling growth and thereby complement the effects of herbicides (Crutchfield *et al.*, 1986; Gill *et al.*, 1992; Vogel, 1994; Buhler *et al.*, 1996; Mashingaidze *et al.*, 2009). Gill *et al.* (1992) identified residue mulching as a practical method for early season weed control in minimum tillage systems for smallholder farmers in Zambia. They reported that the application of grass mulch at 5 t ha^{-1} significantly suppressed weed growth in the first 42 days of maize (*Zea mays*) grown under minimum tillage. In Zimbabwe, the retention of the previous season's maize residues significantly suppressed total dry weed biomass by more than 30% in the ripped plots compared to no mulching (Vogel, 1994).

However in some other cases, crop residue retention lessened the herbicide's efficacy (Erbach and Lovely, 1975; Forcella *et al.*, 1994; Jat *et al.*, 2012a). However, rainfall may wash the intercepted herbicides by crop residues into the soil and efficacy may remain high (Johnson *et al.*, 1989). Sometimes, weed suppression occurs only when relatively high rates of crop residues are applied, which makes it impractical for smallholders in the developing countries where biomass production is low or it has competing alternate uses (e.g. for cattle fodder).

In the long run, when appropriate weed control practices are adopted and the weed seed bank becomes exhausted the weed problem may reduce in CA fields (Blackshaw *et al.*, 2001; Nurbekov, 2008). Some cereal crop residues have been reported to inhibit the germination of some weed seeds due to their allelopathic properties (Steinsiek *et al.*, 1982; Lodhi and Malik, 1987; Jung *et al.*, 2004) and depriving weed seeds of sunlight (Ross and Lembi, 1985).

1.5.7 Crop productivity

Short-term effects of CA on crop yield vis-à-vis ConvT remain variable depending on the initial soil fertility status, climate, rainfall received in the season, tenants' management practices and the type and amount of crop residues retained, among others. Therefore, the short-term effects of CA on crop yield may be positive, neutral or negative (Gill and Aulakh, 1990; Mousques and Friedrich, 2007; Nurbekov, 2008; Lumpkin and Sayre, 2009; Jat *et al.*, 2012a). However, in the long term CA has been reported to increase crop yields due to associated benefits such as prevention of soil degradation, improved soil quality, better moisture regimes, timely field operations (mainly sowing) and crop rotational benefits. Over time, the benefits from reduced soil degradation and improved soil physical, chemical and biological properties due to mulching and legumes in rotations accumulate, resulting into higher and stable yields in CA fields (Erenstein, 2003; Sisti *et al.*, 2004). Under rainfed situations in dry climates where soil moisture is the most limiting factor, CA helps improve crop yields due to improved through increased infiltration, reduced evaporation loss and higher water-holding capacity of the soil. Moreover, CA gives more stable yields compared to ConvT due mainly to timely planting, maintenance of favourable soil moisture regime, improved soil quality, less soil erosion, and less incidence of diseases and insect-pests (FAO, 2001; Hobbs and Govaerts, 2010). Crop rotation, which is one of the underlying principles of CA, helps in better performance of crops compared to when the same crop is grown in the same field year after year (FAO, n.d. b; Kasasa *et al.*, 1999; Giller, 2001).

In dry climates, timely sowing is important to obtain higher yields as the window of sowing after first occurrence of rains remains short. Moreover, many smallholders may not have sufficient sources of traction and machinery for timely sowing of the crops during the critical period of sowing after the first rains (Twomlow *et al.*, 2006). This may lead to delays in crop sowing leading to yield penalties. CA may help to sow larger areas in the given sowing window span by removing the need for tilling the land before sowing. In light-textured soils where surface crusting is an important constraint, crop residue retention on the soil surface in CA can assist in better germination and emergence of seedlings (LeBissonnais, 1996; Lal and Shukla, 2004). Mulching in CA fields maintains more favourable temperatures for crop plants and soil life, favouring better plant growth and development (Bot and Benites, 2005; Fabrizzi *et al.*, 2005).

However, some studies have reported that yield benefits due to CA are conspicuous only during dry years and yields are low during normal or above rainfall years (Giller *et al.*, 2009; Wang *et al.*, 2011). This is because rain water conservation effects of CA are more pronounced during dry years.

1.5.8 Climate change mitigation and adaptation

Conventional agriculture generally contributes more to climate change by greater emissions of carbon dioxide (CO_2) and nitrous oxide (N_2O) at various stages of input production, transportation and during and after their application in the field. Emission of CO_2 in ConvA occurs due to tilling of land, mixing of crop residues and burning of biomass (FAO, 2001; Hobbs and Govaerts, 2010).

CA can help to mitigate climate change through carbon sequestration and reduced emission of CO_2 and N_2O and probably of methane (CH_4). CA leads to carbon sequestration due to reduced decomposition of soil organic matter and addition of biomass as mulch (Corbeels *et al.*, 2006; Giller *et al.*, 2009) and through crop rotations followed in CA (Sidiras and Pavan, 1985; Calegari *et al.*, 2008). Reduced soil disturbance may also lead to higher carbon sequestration in CA fields due to slower decomposition and oxidation of SOM (Jat *et al.*, 2012b). Besides, greater micro-aggregation and aggregate stability due to CA (Lal, 1997; Six *et al.*, 2000; Verhulst *et al.*, 2009) may lead to higher carbon sequestration in the CA fields. Because crop residues are retained on the soil surface in CA, it avoids emission of CO_2 due to burning of crop residues. Due to direct sowing and avoidance of tillage operations, CA saves a considerable amount of fuel and thus leads to reduced CO_2 emissions (West and Marland, 2002; Hobbs and Gupta, 2004; Wang and Dalal, 2006; Erenstein *et al.*, 2008). N_2O emission may be lower in CA fields in the long term due to reduced need of nitrogenous fertilizers as a result of improved soil fertility status. Moreover, higher SOM and the presence of crop residues in CA fields leads to the immobilization of externally applied nitrogen, leading to decreased availability of NO_3^--N for denitrification. Depending on whether CA improves or worsens soil aeration under a particular set of agro-climatic and management conditions, it may increase or decrease CH_4 emission from the soil (Hütsch, 1998; Omonode *et al.*, 2007). Direct sowing or transplanting of young rice seedlings under aerobic soil conditions could reduce both CH_4 (Hobbs and Govaerts, 2010) and N_2O emissions (Kassam *et al.*, 2011b).

At the same time, CA can help adapt to climate change mainly through better soil moisture status, moderating extreme soil temperatures, timely farm operations and better health of crops in CA fields. ZT with residue retention generally increases surface soil water contents compared to tilled soils (Govaerts *et al.*, 2007b), and consequently decreases the frequency and intensity of short mid-season droughts (Blevins *et al.*, 1971; Bradford and Peterson, 2000). Due to improved soil quality and better plant nutrition, CA imparts greater resilience to crop plants against climatic variability (Hobbs and Govaerts, 2010). Moreover, CA has been reported to moderate extreme temperatures in the soil (Acharya *et al.*, 1998; Oliveira *et al.*, 2001) and reduces air temperature around the crop canopy (Jacks *et al.*, 1955; Gupta *et al.*, 2010). Hansen *et al.* (2012) reported that the inclusion of annual forage crops can improve precipitation use efficiency and resilience under climate change in the Great Plains of the USA.

1.5.9 Benefits at ecosystem level

Under CA, the minimal mechanical soil disturbance, maintenance of biomass on the soil surface, use of cover crops and adoption of crop rotations naturally favours abundance and diversity of both below- and above-ground flora and fauna (Nuutinen 1992; Chan and Heenan, 1993; Hartley *et al.*, 1994; Karlen *et al.*, 1994; Buckerfield and Webster, 1996; FAO, 2001; Clapperton, 2003; Govaerts *et al.*, 2007b; Verhulst *et al.*, 2010). Zero or reduced tillage, unlike ConvT, does not disturb activity and the habitats of soil-inhabiting organisms (Doran, 1980; Linn and Doran, 1984; Buchanan and King, 1992; Angers *et al.*, 1993; Chan and Heenan, 1993; Ferreira *et al.*, 2000). Retention of biomass provides sufficient food and creates a supporting microclimate to enable communities of organisms such as bacteria, fungi, actinomycetes, earthworms, arthropods, etc. to flourish in CA fields. Cover crops and residues moderate soil temperature. Several studies have reported greater abundance and diversity of earthworms and arthropods in the CA fields due to no or lesser soil mechanical disturbance and supply of abundant food (Chan and Heenan, 1993; Acharya *et al.*, 1998; Kladivko, 2001; Rodriguez *et al.*, 2006; Verhulst *et al.*, 2010). Thus, CA fields have near natural conditions for the biological communities to flourish therein. Cover crops and crop rotations favour several species of symbiotic microorganisms with crop plants (Hungria *et al.*, 1997; Ferreira *et al.*, 2000). CA has been found to improve above-ground biodiversity also by providing habitats and food for birds, mammals, reptiles and insects among others (FAO, 2001). Mousques and Friedrich (2007) reported a significant increase in the numbers and diversity of beneficial fauna in CA fields in DPRK.

CA has been reported to provide many ecological benefits in its surroundings, for example, recharge of groundwater bodies, reduced flooding in downstream areas, reduced siltation and chemical pollution of watercourses (Kassam *et al.*, 2011c). Improved macro-porosity in CA fields due to higher earthworm numbers and their activities and continuity of channels created by decay of deep roots of legumes such as pigeon pea lead to greater percolation of rainwater, which helps recharge aquifers (Barley, 1954; Disparte, 1987; Green *et al.*, 2003). This also helps reduce soil erosion, flooding in the catchment areas and the siltation of rivers and water reservoirs or other water bodies. As crops under CA are healthier due to improved moisture availability and improved soil quality, they require less fertilizers and pesticides to feed and protect them, which leads to reduced emission of chemicals into the environment at both input production and field level (FAO, 2008; Kassam *et al.*, 2011c).

However the environmental cost, if no-till is applied without the additional elements of CA, due to total reliance on herbicides for weed control, can be high, which is another argument for integrated weed control approaches under CA, differentiating CA from other no-till and from minimum tillage practices.

1.5.10 Farm profitability

Depending on the length of adoption of CA and management skills of individual farmers, profit gains due to CA may be neutral, positive or negative. During initial years of CA adoption, the net profits may remain unchanged or may even decrease. In CA, the cost saving due to reduced/zero tillage may be outweighed by increased cost of weeding and possible slight yield reductions in initial years compared to ConvT (Jat *et al.*, 2012a). Moreover, farmers need to invest in the form of new machinery for CA, which may put some financial burden on smallholders when they start to adopt CA. However, in the long term, when the positive impacts of CA on soil and water conservation, soil quality, input use efficiency, etc., start to accumulate and farmers become more acquainted with CA technologies, net profits due to CA are higher compared to CovT. Many studies have reported a significant decrease in the cost of cultivation in CA fields due mainly to less input (fuel, labour, time, etc.) use (FAO, 1998; Hobbs and Gupta, 2004; Sangar *et al.*, 2004;

Hobbs, 2007; Mousques and Friedrich, 2007; Changrong *et al.*, 2009).

1.6 Challenges in Up-Scaling and Out-Scaling CA Worldwide

Even though CA is known to provide numerous benefits at the field, ecosystem and society level, its adoption has not been widespread globally except in a few countries, despite about eight decades since the start of the reduced tillage movement in the USA in the 1930s. However, Mercosur countries of Argentina, Brazil, Paraguay and Uruguay, and Australia, the USA, Canada, Ukraine etc. have made good progress in adopting CA due to consistent efforts and coordination among farmers, scientific community and policy makers. The more common factors that hinder the widespread adoption of CA in different parts of globe include tillage mindset and lack of awareness of how ConvT leads to soil degradation, lack of sufficient biomass for mulching, need for new implements and operating skills for CA, weed menace in CA fields, probable initial yield reductions, and the lack of sufficient research and government policies in many countries. Although soil degradation due to soil erosion is widespread in both developed and less-developed nations, it seems there is a lack of a sense of urgency on the part of both farmers and policy makers to check soil degradation probably due to its slow, creeping and often unnoticeable nature. Farmers and policy makers in general do not recognize how CA can contribute to reverse the rampant process of soil degradation and thereby lead to sustainable agricultural intensification. Moreover, there is a prevailing feeling among farmers that to obtain good crop yields, tilling the land is essential. As Hobbs and Govaerts (2010) pointed out, overcoming this mindset about tillage is probably the most important factor in the large scale promotion of CA. It is difficult to convince famers, particularly in less developed countries, about the potential benefits of CA, except about cost reductions due to zero/reduced tillage. Further, probable yield reductions during the initial years of the adoption of CA may dampen the spirits of smallholders. In CA fields, higher weed intensity due to no/reduced tillage (Mousques and Friedrich, 2007; Jat *et al.*, 2012a), nutrient immobilization (Abiven and Recous, 2007; Giller *et al.*, 2009), and higher number of insect pests (Mousques and Friedrich, 2007; Giller *et al.*, 2009) and disease (Cook *et al.*, 1978; Hinkle, 1983) during the conversion phase may cause slight yield reductions compared to ConvT. Weed management is a major challenge in the successful adoption of CA. Zero tillage and no mechanical inter-cultivation can lead to heavy weed infestation (Jat *et al.*, 2012a). Herbicides alone do not provide proper weed control in the presence of crop residues on the soil surface. Moreover, intermittent rains that reduce the efficacy of applied herbicides and the lack of availability of herbicides, particularly for local popular intercropping systems, further make it difficult to achieve successful weed control in CA fields. Retention of fresh biomass, mainly cereal residues with high C:N ratio as mulch in CA, results in net immobilization of plant nutrients, especially N (Abiven and Recous, 2007). This is more evident during the early years of CA adoption and may lead to nutrient deficiency in crop plants unless extra amount of nutrients are applied externally (Nurbekov, 2008). Many farmers, mainly in tropical and subtropical countries, due to their cash-crunch situation are not able to make new investments for CA machinery (rippers, zero seed drill etc.). As CA is a paradigm change in production technology, farmers need to learn and equip themselves with new skills and even do experiments and innovate at their individual level in their specific set of operating conditions. This is where many farmers hesitate to take risks to venture into a new field for them.

Maintaining soil cover with crop residues or growing cover crops is essential to obtain the benefits of CA, but supply of crop residues is a limiting factor in successfully promoting CA in the tropics and subtropics. Not only are current biomass production levels are low, but also priority is given to the use of crop residues as cattle fodder due to high economic and cultural importance of livestock for smallholders. Prevalence of communal grazing and termite menace are other major hurdles in maintaining residue mulch in many African and Asian countries (Giller *et al.*, 2009; Umar *et al.*, 2011). Moreover, resource-poor farmers in the less developed countries are not in a position to grow cover crops during the fallow season because it requires extra inputs, but no direct economic returns are received (Ali and Narciso, 1996). It has been found that farmers do not follow all the principles of CA due to reasons such as the shortage of crop residues, lack of sufficient resources and input supply (herbicides), market pressures, labour constraints, etc. (Baudron *et al.*, 2007; Shetto and Owenya, 2007). However, problems of high residue supply and its management, particularly in temperate climates, are also not uncommon (see Duiker and Thomason, Chapter 2, this volume). Further, there is lack of sufficient research on weed control, suitable machinery, cropping systems and cover crops for CA, and on the long-term effects of CA on yield and soil quality (soil acidity, alkalinity, compaction, nutrient behaviour, etc.), particularly in the context of less-developed nations. For a detailed discussion on various factors limiting widespread adoption of CA, readers are referred to a recent review by Jat *et al.* (2012b).

To ensure sufficient biomass for use in CA, particularly in tropics and subtropics, there is a need to improve total biomass yield of the production systems. Additional sources of biomass could also be explored, for example, by integrating agroforestry systems with CA. Plants such as *Cassia tora, Gliricidia maculata, Leucaena leucocephala*, which grow and produce relatively large biomass in the low rainfall areas, could be appropriate plants for this purpose. These and other plants used for providing additional biomass could be grown on field bunds, wastelands and around water bodies.

1.7 Conclusions

To promote CA, a two-pronged strategy is needed. First, efforts should be made to share information, and discuss and make farmers aware about the benefits of the CA, especially in the longer term, and convince them on 'why they should follow CA'. Second, from the point of initiation, an active participation of all the concerned stakeholders needs to be ensured. In an effort to promote CA and its relevance among farmers, it is necessary to educate them on the link of excessive tillage and residue removal with soil quality sustainability problems, and as to how these problems can be reduced or alleviated through the adoption of CA (Lumpkin and Sayre, 2009). Once farmers become convinced and are ready to adopt CA, there should be active involvement of researchers, farmers, policy makers, input suppliers, NGOs and others in promoting CA. Governments can facilitate in CA adoption by providing subsidy for purchasing zero-till machinery and by making credit available on easy terms to tenants; besides, of course, protecting the tenants' rights. Active participation of equipment manufacturers is essential so as to help design and supply machinery, which is best suitable to the local conditions and meets the requirements of different categories of farmers. The NGOs can facilitate linking farmers with other stakeholders including researchers, input suppliers and government agencies. NGOs can also target specific potential areas for CA to begin with, and facilitate to the formation of farmers' self-help groups, organize farmers' visits, workshops, provide information on input supply, credit lines and take new technological advancements to the farmers' doorsteps. To make CA attractive to farmers, research should be undertaken to make CA profitable in the shorter term also. Developing an economic weed control strategy remains a major challenge for the successful adoption of CA. This also needs to be seen in the light of the fact that a total reliance on the use of herbicides for weed control in CA could lead to heavy environmental costs. Therefore, there is need to develop an economic and effective weed control strategy that is based on integrated weed management for the site-specific implementation as a component of CA (Friedrich and Kassam, 2009).

References

Abiven, S. and Recous, S. (2007) Mineralization of crop residues on the soil surface or incorporated in the soil under controlled conditions. *Biology and Fertility of Soils* 43, 849–852.

Acharya, C.L., Kapur, O.C. and Dixit, S.P. (1998) Moisture conservation for rainfed wheat production with alternative mulches and conservation tillage in the hills of north-west India. *Soil and Tillage Research* 46, 153–163.

Alvear, M., Rosas, A., Rouanet, J.L. and Borie, F. (2005) Effects of three soil tillage systems on some biological activities in an Ultisol from southern Chile. *Soil and Tillage Research* 82,195–202.

Ali, M. and Narciso, J.H. (1996) Farmers' perception and economic evaluation of green manure use in rice-based farming systems. *Tropical Agriculture* (Trinidad) 73, 148–154.

Amado, T.J.C., Fernandez, S.B. and Mielniczuk, J. (1998) Nitrogen availability as affected by ten years of cover crop and tillage systems in southern Brazil. *Journal of Soil and Water Conservation* 53(3), 268–271.

Angers, D.A., Bissonette, N. and Legere, A. (1993) Microbial and biochemical changes induced by rotation and tillage in a soil under barley production. *Canadian Journal of Soil Science* 73, 39–50.

Araya, T., Cornelis, W.M., Nyssen, J., Govaerts, B., Getnet, F., Bauer, H., Amare, K., Raes, D., Haile, M. and Deckers, J. (2012) Medium-term effects of conservation agriculture based cropping systems for sustainable soil and water management and crop productivity in the Ethiopian highlands. *Field Crops Research* 132, 53–62.

Balota, E.L., Colozzi, A., Andrade, D.S. and Dick, R.P. (2004) Long-term tillage and crop rotation effects on microbial biomass and C and N mineralization in a Brazilian Oxisol. *Soil and Tillage Research 77*, 137–145.

Barker, K.R. and Koenning, S.R. (1998) Developing sustainable systems for nematode management. *Annual Reviews in Phytopathology* 36, 165–205.

Barley, K.P. (1954) Effects of root growth and decay on the permeability of a synthetic sandy soil. *Soil Science* 78, 205–210.

Basch, G., Kassam, A., Friedrich, T., Santos, F.L., Gubiani, P.I., Calegari, A., Reichert, J.M. and Dos Santos, D.R. (2012) Sustainable soil water management systems. In: Lal, R. and Stewart, B.A. (eds) *Soil Water and Agronomic Productivity Advances in Soil Science*. CRC Press, Boca Raton, Florida, pp. 229–288.

Baudron, F., Mwanza, H.M., Triomphe, B. and Bwalya, M. (2007) *Conservation agriculture in Zambia: A case study of southern province*. African Conservation Tillage Network, Centre de Cooperation Internationale de Recherche Agronomique pour le Dévelopment, Food and Agricultural Organization of the United Nations, Nairobi, Kenya.

Baudron, F., Corbeels, M., Monicat, F. and Giller, K.E. (2009) Cotton expansion and biodiversity loss in African savannahs, opportunities and challenges for conservation agriculture: a review paper based on two case studies. *Biodiversity Conservation* 18, 2625–2644.

Beare, M.H., Hu, S., Coleman, D.C. and Hendrix, P.F. (1997) Influences of mycelial fungi on soil aggregation and organic matter storage in conventional and no-tillage soils. *Applied Soil Ecology* 5(3), 211–219.

Blackshaw, R.E., Larney, F.J., Lindwall, C.W., Watson, P.R. and Derksen, D.A. (2001) Tillage intensity and crop rotation affect weed community dynamics in a winter wheat cropping system. *Canadian Journal of Plant Science* 81, 805–813.

Blevins, R.L., Cook, D., Phillips, S.H. and Philips, R.E. (1971) Influence of no-tillage on soil moisture. *Agronomy Journal* 63, 593–596.

Blevins, R.L., Smith, M.S., Thomas, G.W. and Frye, W.W. (1983) Influence of conservation tillage on soil properties. *Journal of Soil Science and Water Conservation* 38(3), 301–305.

Bockus, W.W. and Shroyer, J.P. (1998) The impact of reduced tillage on soil-borne plant pathogens. *Annual Reviews of Phytopathology* 36, 485–500.

Bot, A. and Benites, J. (2005) Creating drought-resistant soil. In: *The Importance of Soil Organic Matter: key to drought-resistant soil and sustained food production*. FAO soils bulletin 80, FAO land and plant nutrition management service, Food and Agriculture Organization of the United Nations, Rome, pp. 35–40.

Bradford, J.M. and Peterson, G.A. (2000) Conservation tillage. In: Sumner, M.E. (ed.) *Handbook of Soil Science*. CRC Press, Boca Raton, Florida, pp. G247–G269.

Bronick, C.J. and Lal, R. (2005) Soil structure and management: a review. *Geoderma* 124, 3–22.

Buchanan, M. and King, L.D. (1992) Seasonal fluctuations in soil microbial biomass carbon, phosphorus, and activity in no-till and reduced-chemical-input maize agroecosystems. *Biology and Fertility of Soils* 13, 211–217.

Buckerfield, J.C. and Webster, K.A. (1996) Earthworms, mulching, soil moisture and grape yields: Earthworm responses to soil management practices in vineyards, Barossa Valley, South Australia. *Australia and New Zealand Wine Industry Journal* 11, 47–53.

Buhler, D.D., Mester, T.C. and Kohler, K.A. (1996) The effect of maize residues and tillage on emergence of *Setaria faberi, Abutilon theophrasti, Amaranthus retroflexus* and *Chenopodium album Weed Research* 36,153–165.

Burle, B., Mielniczuk, J. and Focchi, S. (1997) Effect of cropping systems on soil chemical characteristics, with emphasis on soil acidification. *Plant Soil* 190, 309–316.

Calegari, A. and Alexander, I. (1998) The effects of tillage and cover crops on some chemical properties of an oxisol and summer crop yields in southwestern Paraná, Brazil. *Advances in Geo Ecology* 31, 1239–1246.

Calegari, A., Hargrove, W.L., Rheinheimer, D.D.S., Ralisch, R., Tessier, D., Tourdonnet, S. and Guimarães, M. de F. (2008) Impact of long-term no-tillage and cropping system management on soil organic carbon in an Oxisol: a model for sustainability. *Agronomy Journal* 100(4), 1013–1019.

Campbell, C.A., Janzen, H.H. and Juma, N.G. (1997) Case studies of soil quality in the Canadian prairies: long-term field experiments. In: Gregorich, E.G. and Carter, M.R. (eds) *Soil Quality for Crop Production and Ecosystems Health*. Elsevier, Amsterdam, the Netherlands, pp. 351–397.

Carpenter-Boggs, L., Stahl, P.D., Lindstrom, M.J. and Schumacher, T.E. (2003) Soil microbial properties under permanent grass, conventional tillage, and no-till management in south Dakota. *Soil and Tillage Research* 71, 15–23.

Carter, M.R., Gregorich, E.G., Angers, D.A., Beare, M.H., Sparling, G.P., Wardle, D.A. and Voroney, R.P. (1999) Interpretation of microbial biomass measurements for soil quality assessment in humid temperate regions. *Canadian Journal of Soil Science* 79, 507–520.

Castro Filho, C., Henklain, J.C., Vieira, M.J. and Casão Jr, R. (1991) Tillage methods and soil and water conservation in southern Brazil. *Soil and Tillage Research* 20, 271–283.

Chan, K.Y. and Heenan, D.P. (1993) Surface hydraulic properties of a red earth under continuous cropping with different management practices. *Australian Journal of Soil Research* 31, 13–24.

Chan, K.Y., Heenan, D.P. and Oates, A. (2002) Soil carbon fractions and relationship to soil quality under different tillage and stubble management. *Soil and Tillage Research* 63, 133–139.

Changrong, Y., Wenqing, H., Xurong, M., Dixon, J., Qin, L., Shuang, L. and Enke L. (2009) Critical research for dryland conservation agriculture in the Yellow river basin, China: recent results. In: *Proceedings of 4th World Congress on Conservation Agriculture on Innovations for Improving Efficiency, Equity and Environment*. New Delhi, India, pp. 51–59.

Clapperton, M.J. (2003) Increasing soil biodiversity through conservation agriculture: Managing the soil as a habitat. In: *Proceedings of 2nd World Congress on Conservation Agriculture on Producing in Harmony with Nature*. Iguassu Falls, Parana-Brazil. FAO, Rome (CD).

Cook, R.J., Boosalis, M.G. and Doupnik, B. (1978) Influence of crop residues on plant diseases. In: Oshwald, W.R. (ed.) *Crop Residue Management Systems*. ASA Special Publication 31, Madison, Wisconsin, pp. 147–163.

Corbeels, M., Scopel, E., Cardoso, A., Bernoux, M., Douzet, J.M. and Neto, M.S. (2006) Soil carbon storage potential of direct seeding mulch-based cropping systems in the Cerrados of Brazil. *Global Change Biology* 12, 1773–1787.

Corsi, S., Friedrich, T., Kassam, A., Pisante, M. and de Moraes Sà, J.C. (2012) Soil Organic Carbon Accumulation and Greenhouse Gas Emission Reductions from Conservation Agriculture: A literature review. *Integrated Crop Management* Vol. 16-2012, AGP/FAO, Rome.

Crutchfield, D.A., Wicks, G.A. and Burnside, O.C. (1986) Effect of winter wheat (*Triticum aestivum*) straw mulch level on weed control. *Weed Sciences* 34, 110–114.

Danga, B.O. and Wakindiki, I.I.C. (2009) Effect of placement of straw mulch on soil conservation, nutrient accumulation, and wheat yield in a humid Kenyan highland. *Journal of Tropical Agriculture* 47(1–2), 30–36.

De Gryze, S., Six, J., Brits, C. and Merckx, R. (2005) A quantification of short-term macroaggregate dynamics: influences of wheat residue input and texture. *Soil Biology and Biochemistry* 37, 55–66.

De Maria, I.C. and Castro, O.M. (1993) Fósforo, potássio e material orgânica em um Latossolo Roxo, sob sistemas de manejo com milho e soja. *Rev bras Ci Solo* 17, 471–477.

Denardin, J.E., Kochhann, R.A., Bacaltchuk, B., Sattler, A., Denardin, N.D'a., Faganello, A. and Wiethölter, S. (2008) Sistema plantio direto: fator de potencialidade da agricultura tropical brasileira. In: Albuquerque, A.C.S. and Silva, A.G. da (eds) *Agricultura tropical: quatro décadas de inovações tecnológicas, institucionais e políticas*. Embrapa Informação Tecnológica, vol. 1, cap. 1, Brasília, DF, pp. 1251–1273.

Derpsch, R. (1997) Importance of the direct seeding for the sustainability of agricultural production. In: *Proceedings of 5th National Congress*. AAPRESID, Silver Sea, Argentina.

Derpsch, R., Roth, C.H., Sidiras, N. and Kopke, U. (1991) *Controle da erosãono Paraná, Brasil: Sistemas de cobertura de solo, plantio direto e prepare conservacionista do solo*. GTZ, Eschborn, Alemanha e IAPAR, Londrina, Brasil.

Dick, R.P. (1992) A review – long-term effects of agricultural systems on soil biochemical and microbial parameters. *Agriculture Ecosystems and Environment* 40, 25–36.

Disparte, A.A. (1987) Effect of root mass density on infiltration among four Mediterranean dryland forages and two irrigated forage legumes. MSc thesis, University of California, Riverside, California.

Doran, J.W. (1980) Soil microbial and biochemical-changes associated with reduced tillage. *Soil Science Society of America Journal* 44, 765–771.

Drinkwater, L.E., Wagoner, P. and Sarrantonio, M. (1998) Legume-based cropping systems have reduced carbon and nitrogen losses. *Nature* 396, 262–264.

Duiker, S.W. and Beegle, D.B. (2006) Soil fertility distributions in long-term no-till, chisel/disk and mouldboard plow/disk systems. *Soil and Tillage Research* 88, 30–41.

Erbach, D.C. and Lovely, W.G. (1975) Effect of plant residue on herbicide performance in no-tillage corn. *Weed Science* 23, 512–515.

Erenstein, O. (2002) Crop residue mulching in tropical and semi-tropical countries: an evaluation of residue availability and other technological implications. *Soil and Tillage Research* 67, 115–133.

Erenstein, O. (2003) Smallholder conservation farming in the tropics and sub-tropics: a guide to the development and dissemination of mulching with crop residues and cover crops. *Agriculture, Ecosystems and Environment* 100, 17–37.

Erenstein, O., Farooq, U., Malik, R.K. and Sharif, M. (2008) On-farm impacts of zero tillage wheat in south Asia's rice-wheat systems. *Field Crops Research* 105, 240–252.

Fabrizzi, K.P., Garcia, F.O., Costa, J.L. and Picone, L.I. (2005) Soil water dynamics, physical properties and corn and wheat responses to minimum and no-tillage systems in the southern Pampas of Argentina. *Soil and Tillage Research* 81, 57–69.

Fageria, N.K., Baligar, V.C. and Wright, R.J. (1997) Soil environment and root growth dynamics of field crops. *Recent Research Developments in Agronomy* 1, 15–58.

FAO (1998) Press release 98/42. FAO: Conventional tillage severely erodes the soil; new concepts for soil conservation required. Available at: http://www.fao.org/WAICENT/OIS/PRESS_NE/PRESSENG/1998/pren9842.htm (accessed 17 January 2012).

FAO (2001) *Conservation agriculture case studies in Latin America and Africa*. Introduction. FAO Soils Bulletin No. 78. FAO, Rome.

FAO (2008) Integrated Crop management. In: *Proceedings of An International Technical Workshop on Investing in Sustainable Crop Intensification: The Case for Improving Soil Health*. Vol.6-2008, FAO, Rome. Available at: http://www.fao.org/ag/ca/doc/WORKSHOP-LR.pdf (accessed 8 December 2010).

FAO (2012) What is Conservation Agriculture. Available at: http://www.fao.org/ag/ca/1a.html (accessed 15 December 2012).

FAO (n.d. a) The importance of cover crops in conservation agriculture. Available at: http://www.fao.org/ag/ca/2a.html (accessed 15 December 2012).

FAO (n.d. b) The main principles of conservation agriculture. Available at: http://www.fao.org/ag/ca/1b.html (accessed 15 December 2012).

Fernandes, H.C., Silveira, J.C.M. and Rinaldi, P.C.N. (2008) Avaliação do custo energético de diferentes operações agrícolas mecanizadas. Ciência e Agrotecnologia 32 (5),1582-1587. Available at: http://www.scielo.br/scielo.php? (accessed 15 January 2013).

Ferreira, M.C., Andrade, D.S., Chueire, L.M.O., Takemura, M. and Hungria, M. (2000) Tillage method and crop rotation effects on the population sizes and diversity of bradyrhizobia nodulating soybean. *Soil Biology and Biochemistry* 32, 627–637.

Filho, C.C., Lourenco, A., Guimaraes, M.D.F. and Fonseca, I.C.B. (2002) Aggregate stability under different soil management systems in a red latosol in the state of Parana, Brazil. *Soil and Tillage Research* 65, 45–51.

Forcella, F., Buhler, D.D. and McGiffen, M.E. (1994) Pest management and crop residues. In: Hatfield, J.L. and Stewart, B.A. (eds) *Crop Residue Management. Advances in Soil Science*. CRC Press, Boca Raton, Florida, pp. 173–189.

Fowler, R. and Rockstrom, J. (2000) Conservation tillage for sustainable agriculture: an agrarian revolution gathers momentum in Africa. Keynote address, ISTRO 2000, Fort Worth, USA.

Franzluebbers, A.J. (2002) Water infiltration and soil structure related to organic matter and its stratification with depth. *Soil and Tillage Research* 66, 197–205.

Franzluebbers, A.J. and Hons, F.M. (1996) Soil-profile distribution of primary and secondary plant available nutrients under conventional and no tillage. *Soil and Tillage Research* 39, 229–239.

Freixial, R. and Carvalho, M. (2010) Aspetos práticos fundamentales en la implantacion de la Agricultura de Conservacion/Siembra Direta en el sur de Portugal. In: *Proceedings of the European Congress on Conservation Agriculture: Towards Agro-Environmental Climate and Energetic Sustainability*. Madrid, Spain, pp. 361–369.

Friedrich, T. and Kassam, A. (2009) Adoption of Conservation Agriculture Technologies: Constraints and Opportunities. In: ICAR (Indian Council for Agricultural Research) (eds) *Proceedings of the 4th World Congress on Conservation Agriculture*, Lead Papers, New Delhi, 4–7 February 2009, pp. 257–264.

Friedrich, T. and Kassam, A. (2012) No-till farming and the environment: Do no-till systems require more chemicals? *Outlooks on Pest Management* August 2012, pp. 153–157.

Friedrich, T., Derpsch, R. and Kassam, A. (2012) Overview of the global spread of conservation agriculture. Field Actions Science Reports, Special issue 6 (2012) on Reconciling Poverty Eradication and Protection of the Environment. Available at: http://factsreports.revues.org/1941 (accessed 5 December, 2012).

Gill, K.S. and Aulakh, B.S. (1990) Wheat yield and soil bulk-density response to some tillage systems on an Oxisol. *Soil and Tillage Research* 18, 37–45.

Gill, K.S., Arshad, M.A., Chivundu, B.K., Phiri, B. and Gumbo, M. (1992) Influence of residue mulch, tillage and cultural practices on weed mass and corn yield from three field experiments. *Soil and Tillage Research* 24, 211– 223.

Giller, K.E. (2001) *Nitrogen Fixation in Tropical Cropping Systems*. CAB International, Wallingford, UK.

Giller, K.E., Witter, E., Corbeels, M. and Tittonell, P. (2009) Conservation agriculture and smallholder farming in Africa: The heretics' view. *Soil and Tillage Research* 114, 23–34.

Govaerts, B., Fuentes, M., Mezzalama, M., Nicol, J.M., Deckers, J., Etchevers, J.D., Figueroa-Sandoval, B. and Sayre, K.D. (2007a) Infiltration, soil moisture, root rot and nematode populations after 12 years of different tillage, residue and crop rotation managements. *Soil and Tillage Research* 94, 209–219.

Govaerts, B., Mezzalama, M., Unno, Y., Sayre, K.D., Luna-Guido, M., Vanherck, K., Dendooven, L. and Deckers, J. (2007b) Influence of tillage, residue management, and crop rotation on soil microbial biomass and catabolic diversity. *Applied Soil Ecology* 37, 18–30.

Govaerts, B., Sayre, K.D., Lichter, K., Dendooven, L. and Deckers, J. (2007c) Influence of permanent raised bed planting and residue management on physical and chemical soil quality in rain fed maize/wheat systems. *Plant and Soil* 291, 39–54.

Govaerts, B., Sayre, K.D., Goudeseune, B., De Corte, P., Lichter, K., Dendooven, L. and Deckers, J. (2009) Conservation agriculture as a sustainable option for the central Mexican highlands. *Soil and Tillage Research* 103, 222–230.

Green, T.R., Ahuja, L.R. and Benjamin, J.G. (2003) Advances and challenges in predicting agricultural management effects on soil hydraulic properties. *Geoderma* 116, 3–27.

Gregorich, E.G., Carter, M.R., Angers, D.A., Monreal, C.M. and Ellert, B.H. (1994) Towards a minimum data set to assess soil organic-matter quality in agricultural soils. *Canadian Journal of Soil Science* 74, 367–385.

Gupta, R., Gopal, R., Jat, M.L., Jat, R.K., Sidhu, H.S., Minhas, P.S. and Malik, R.K. (2010) Wheat productivity in Indo-Gangetic plains of India during 2010: Terminal heat stress and mitigating strategies. *Conservation agriculture newsletter*, Getting agriculture to work for people and the environment, PACA, New Delhi, India.

Haggblade, S. and Tembo, G. (2003) *Conservation farming in Zambia*. International Food Policy Research Institute, Washington, DC.

Hansen, N.C., Allen, B.L., Baumhardt, R.L. and Lyon, D.J. (2012) Research achievements and adoption of no-till, dryland cropping in the semi-arid US Great Plains. *Field Crops Research* 132, 196–203.

Hartley, M.J., Ragman, A. and Popay, A.J. (1994) Use of mulches and herbicide in an apple orchard. In: *Proceedings of New Zealand Plant Protection Conference* 47, 320–324.

Hernanz, J.L., Lopez, R., Navarrete, L. and Sanchez-Giron, V. (2002) Long-term effects of tillage systems and rotations on soil structural stability and organic carbon stratification in semiarid central Spain. *Soil and Tillage Research* 66,129–141.

Hinkle, M.K. (1983) Problems with conservation tillage. *Journal of Soil and Water Conservation* 38(3), 201–206.

Hobbs, P.R. (2007) Conservation agriculture: what is it and why is it important for future sustainable food production? *Journal of Agricultural Sciences* 145, 127–137.

Hobbs, P.R. and Govaerts, B. (2010) How conservation agriculture can contribute to buffering climate change. In: Reynolds, M.P. (ed.) *Climate Change and Crop Production*. CAB International, Wallingford, UK, pp. 177–199.

Hobbs, P.R. and Gupta, R.K. (2004) Problems and challenges of no-till farming for the rice-wheat systems of the Indo-Gangetic Plains in South Asia. In: Lal, R., Hobbs, P., Uphoff, N. and Hansen, D.O. (eds) *Sustainable Agriculture and the Rice-Wheat System*. Ohio State University, Columbus, Ohio, and Marcel Dekker, New York, pp. 101–119.

Hudson, B.D. (1994) Soil organic matter and available water capacity. *Journal of Soil and Water Conservation* 49(2), 189–194.

Hulugalle, N.R. and Entwistle, P. (1997) Soil properties, nutrient uptake and crop growth in an irrigated vertisol after nine years of minimum tillage. *Soil and Tillage Research* 42, 15–32.

Hungria, M., Andrade, D.S., Balota, E.L. and Collozzi-Filho, A. (1997) Importância do sistema de semeadura directa na populaçâo microbiana do solo. *Comunicado Técnico* 56, p.9. EMBRAPA-CNPSo, Londrina, Brazil.

Hütsch, B.W. (1998) Tillage and land use effects on methane oxidation rates and their vertical profiles in soil. *Biology and Fertility of Soils* 27, 284–292.

IPCC, Climate Change (2007) The Physical Science Basis, Contribution of Working Group I to the Fourth Assessment Report of the Intergovernmental Panel on Climate Change, Cambridge University Press, Cambridge, UK, p. 1009.

Iyamuremye, F. and Dick, R.P. (1996) Organic amendments and phosphorus sorption by soils. *Advances in Agronomy* 56, 139–185.

Jacks, G.V., Brind, W.D. and Smith, W. (1955) *Mulching*. Technical Communication No. 49. Commonwealth Bureau of Soils, UK.

Jat, R.A., Wani, S.P., Singh, P., Pathak, P., Srinivas, K., Kumar, U., Pavani, E. and Velmurgan, R. (2012a) Effect of conservation agriculture on productivity and economics of different cropping systems under rainfed condition in the semi-arid tropics. In: *Proceedings of 3rd International Agronomy Congress*, New Delhi, India, pp. 888–890.

Jat, R.A., Wani, S.P. and Sahrawat, K.L. (2012b) Conservation agriculture in the semi-arid tropics: prospects and problems. In: Sparks D.L. (ed.) *Advances in Agronomy* 117, 191–273.

Johnson, M.D., Wyse, D.L. and Lueschen, W.E. (1989) The influence of herbicide formulation on weed control in four tillage systems. *Weed Science* 37, 239–249.

Jung, W.S., Kim, K.H., Ahn, J.K., Hahn, S.J. and Chung, I.M. (2004) Allelopathic potential of rice (*Oryza sativa* L.) residues against *Echinochloa crusgalli*. *Crop Protection* 23, 211–218.

Karlen, D.L., Wollenhaupt, N.C., Erbach, D.C., Berry, E.C., Swan, J.B., Eash, N.S. and Jordahl, J.L. (1994) Crop residue effects on soil quality following 10-years of no-till corn. *Soil and Tillage Research* 31, 149–167.

Karlen, D.L., Mausbach, M.J., Doran, J.W., Cline, R.G., Harris, R.F. and Schuman, G.E. (1997) Soil quality: a concept, definition, and framework for evaluation. *Soil Science Society of America Journal* 61, 4–10.

Kasasa, P., Mpepereki, S., Musiyiwa, K., Makonese, F. and Giller, K. (1999) Residual nitrogen benefits of promiscuous soybeans to maize under field conditions. *African Crop Science Journal* 7, 375–382.

Kassam, A. and Friedrich, T. (2012) An ecologically sustainable approach to agricultural production intensification: Global perspectives and developments, Field Actions Science Reports [Online], Special Issue 6, 17 April 2012. Available at: http://factsreports.revues.org/1382 (accessed on 21 June 2012).

Kassam, A., Freidrich, T., Shaxson, F. and Pretty, J. (2009) The spread of Conservation Agriculture: Justification, sustainability and uptake. *International Journal for Agricultural Sustainability* 7(4), 292–320.

Kassam, A., Friedrich, T., Shaxson, F., Reeves, T., Pretty, J. and Moraes Sá, J.C. de (2011a) Production Systems for Sustainable Intensification Integrating Productivity with Ecosystem Services. *Technikfolgenabschätzung – Theorie und Praxis* 20. Jg., Heft 2, July 2011, pp. 38–45.

Kassam, A.H., Stoop, W. and Uphoff, N. (2011b) Review of SRI modifications in rice crop and water management and research issues for making further improvements in agricultural and water productivity. *Paddy Water Environments* 9, 163–180.

Kassam, A., Mello, I., Goddard, T., Friedrich, T., Laurent, F., Reeves, T. and Hansmann, B. (2011c) *5th World Congress of Conservation Agriculture incorporating 3rd Farming Systems Design Conference*, September 2011, Brisbane, Australia.

Kayombo, B. and Lal, R. (1993) Tillage systems and soil compaction in Africa. *Soil and Tillage Research* 27, 35–72.

Kazakhstan Farmers Union (2011) Stop the tractor! I till no more! Available at http://sfk.kz/index.php?id=9&kid=10 (accessed 7 September 2011) (in Russian).

Kemper, B. and Derpsch, R. (1981) Soil compaction and root growth in Paraná. In: Russel, R.S., Igue, K. and Mehta, Y.R. (eds) *The Soil Root System in Relation to Brazilian Agriculture*. Instituto Agronômico do Paraná, Londrina, PR, Brazil, pp. 81–101.

Kienzler, K.M., Lamers, J.P.A., McDonald, A., Mirzabaev, A., Ibragimov, N., Egamberdiev, O., Ruzibaev, E. and Akramkhanov, A. (2012) Conservation agriculture in central Asia – what do we know and where do we go from here? *Field Crops Research* 132, 95–105.

Kladivko, E.J. (2001) Tillage systems and soil ecology. *Soil and Tillage Research* 61, 61–76.

Krupinsky, J.M., Bailey, K.L., McMullen, M.P., Gossen, B.D. and Turkington, T.K. (2002) Managing plant disease risk in diversified cropping systems. *Agronomy Journal* 94, 198–209.

Kushwaha, C.P., Tripathi, S.K. and Singh, K.P. (2000) Variations in soil microbial biomass and N availability due to residue and tillage management in a dryland rice agro-ecosystem. *Soil and Tillage Research* 56,153–166.

Lal, R. (1997) Residue management, conservation tillage and soil restoration for mitigating greenhouse effect by CO_2-enrichment. *Soil and Tillage Research* 43, 81–107.

Lal, R. (2010) A dual response of conservation agriculture to climate change: reducing CO_2 emissions and improving the soil carbon sink. Opening address, European congress on conservation agriculture. Madrid, Spain. Available at: http://www.marm.gob.es/es/ministerio/servicios-generales/publicaciones/Opening_address_tcm7-158491.pdf (accessed 7 October, 2011).

Lal, R. and Shukla, M.J. (2004) *Principles of Soil Physics*. Marcel Dekker, New York.

LeBissonnais, Y. (1996) Aggregate stability and assessment of soil crustability and erodibility. 1. Theory and methodology. *European Journal of Soil Science* 47, 425–437.

Li, H.W., Gao, H.W., Wu, H.D., Li, W.Y., Wang, X.Y. and He, J. (2007) Effects of 15 years of conservation tillage on soil structure and productivity of wheat cultivation in northern China 1. *Australian Journal of Soil Research* 45, 344–350.

Lichter, K., Govaerts, B., Six, J., Sayre, K.D., Deckers, J. and Dendooven, L. (2008) Aggregation and C and N contents of soil organic matter fractions in a permanent raised-bed planting system in the highlands of central Mexico. *Plant Soil* 305, 237–252.

Linn, D.M. and Doran, J.W. (1984) Effect of water filled pore space on carbon dioxide and nitrous oxide production in tilled and non-tilled soils. *Soil Science Society of America Journal* 48, 1267–1272.

Lodhi, M.A.K. and Malik, K.A. (1987) Allelopathy in agroecosystems: wheat phytotoxicity and its possible role in crop rotation. *Journal of Chemical Ecology* 13, 1881–1889.

López, M.V. and Arrúe, J.L. (2005) Soil tillage and wind erosion in fallow lands of Central Aragon (Spain): an overview. In: Faz-Cano, A., Ortiz, R. and Mermut, A.R. (eds) *Sustainable Use and Management of Soils – Arid and Semiarid Regions. Advances in GeoEcology* 36, Catena-Verlag, Reiskirchen, pp. 93–102.

Lumpkin, T.A. and Sayre, K. (2009) Enhancing resource productivity and efficiency through conservation agriculture. In: *Proceedings of 4th World Congress on Conservation Agriculture on Innovations for Improving Efficiency, Equity and Environment*, New Delhi, India, pp. 3–8.

Machado, P.L.O.A. and Silva, C.A. (2001) Soil management under no tillage systems in the tropics with special reference to Brazil. *Nutrient Cycling in Agroecosystems* 61, 119–130.

Madari, B., Machado, P.L.O.A., Torres, E., de Andrade, A.G. and Valencia, L.I.O. (2005) No tillage and crop rotation effects on soil aggregation and organic carbon in a Rhodic Ferralsol from southern Brazil. *Soil and Tillage Research* 80, 185–200.

MANDAK (Manitoba–North Dakota Zero Tillage Farmers Association) (2011) Beyond the Beginning – The Zero Till Evolution. Manitoba-North Dakota Zero Tillage Farmers Association, Farming for Tomorrow, Regina, Canada, 58 pp.

Márquez, F., Giráldez, J.V., Repullo, M., Ordóñez, R., Espejo, A.J. and Rodríguez, A. (2008) Eficiencia de las cubiertas vegetales como método de conservación de suelo y agua en olivar. *Simposio del Agua en Andalucía*, pp. 631–641.

Mashingaidze, N., Twomlow, S.J. and Hove, L. (2009) Crop and weed responses to residue retention and method of weeding in first two years of a hoe-based minimum tillage system in semi-arid Zimbabwe. *Journal of SAT Agricultural Research* 7, 1–11.

Mousques, C. and Friedrich, T. (2007) Conservation agriculture in China and the Democratic People's Republic of Korea. FAO crop and grassland service working paper, Food and Agriculture Organization of the United Nations, Rome.

Muliokela, S.W., Hoogmed, W.B., Steven, P. and Dibbits, H. (2001) Constraints and possibilities of conservation farming in Zambia. In: Garcia-Torres, L., Berutes, J. and Martinez-Vilela, A. (eds) *Conservation Agriculture: A World Challenge*. Vol. II: *Offered Contributions, Environment, Farmers' Experiences, Innovations, Socio-economic Policy*. XUL Avda, Medina, Spain, pp. 61–65.

Musick, G.J. and Beasley, L.E. (1978) Effect of crop residue management system on pest problems in field corn (*Zea mays* L.) production. In: Oshwald, W.R. (ed.) *Crop Residue Management Systems*. ASA special publication 31, Madison, Wisconsin, pp. 173–186.

Nurbekov, A. (2008) *Manual on conservation agriculture practices in Uzbekistan*. Tashkent, Uzbekistan, 40 pp.

Nuutinen, V. (1992) Earthworm community responses to tillage and residue management on different soil types in southern Finland. *Soil and Tillage Research* 23, 221–239.

Oliveira, J.C.M., Timm, L.C., Tominaga, T.T., Cassaro, F.A.M., Reichardt, K., Bacchi, O.O.S., Dourado-Neto, D. and Camara, G.M.D. (2001) Soil temperature in a sugarcane crop as a function of the management system. *Plant Soil* 230, 61–66.

Omonode, R.A., Vyn, T.J., Smith, D.R., Hegymegi, P. and Gal, A. (2007) Soil carbon dioxide and methane fluxes from long-term tillage systems in continuous corn and corn-soybean rotations. *Soil and Tillage Research* 95, 182–195.

Packer, I.J., Hamilton, G.J. and Koren, T.B. (1992) Runoff, soil loss and soil physical properties changes of light textured surface soils from long-term tillage treatments. *Australian Journal of Soil Research* 30, 789–806.

Peairs, F.B., Hein, G.L. and Brewer, M.J. (2010) High Plains integrated pest management: wheat stem sawfly. Available at: http://wiki.bugwood.org/HPIPM:Wheat Stem Sawfly (accessed 5 June 2011).

Philippe Rekacewicz, UNEP/GRID-Arendal (2007) Severity of land degradation. Available at: http://www.grida.no/graphicslib/detail/severity-of-land-degradation_d197 (accessed 12 March 2013).

Pinheiro, E.F.M., Pereira, M.G. and Anjos, L.H.C. (2004) Aggregate distribution and soil organic matter under different tillage systems for vegetable crops in a Red Latosol from Brazil. *Soil and Tillage Research 77*, 79–84.

Powlson, D.S. and Jenkinson, D.S. (1981) A comparison of the organic-matter, biomass, adenosine-triphosphate and mineralizable nitrogen contents of ploughed and direct-drilled soils. *Journal of Agricultural Sciences* 97, 713–721.

Qadir, M., Oster, J.D., Schubert, S., Noble, A.D. and Sahrawat, K.L. (2007) Phytoremediation of sodic and saline-sodic soils. *Advances in Agronomy* 96, 197–247.

Rao, K.P.C., Steenhuis, T.S., Cogle, A.L., Srinivasan, S.T., Yule, D.F. and Smith, G.D. (1998) Rainfall infiltration and runoff from an Alfisol in semi-arid tropical India. I. No-till systems. *Soil and Tillage Research* 48, 51–59.

Rapport, D.J. (1995) Ecosystem health – more than a metaphor. *Environmental Values* 4, 287–309.

Rhoton, F.E., Shipitalo, M.J. and Lindbo, D.L. (2002) Runoff and soil loss from mid-western and southeastern US silt loam soils as affected by tillage practice and soil organic matter content. *Soil and Tillage Research* 66, 1–11.

Rice, C.W. and Smith, M.S. (1984) Short-term immobilization of fertilizer nitrogen at the surface of no-till and plowed soils. *Soil Science Society of America Journal* 48, 295–297.

Rodriguez, E., Fernandez-Anero, F.J., Ruiz, P. and Campos, M. (2006) Soil arthropod abundance under conventional and no- tillage in a Mediterranean climate. *Soil and Tillage Research* 85, 229–233.

Roldân, A., Salinas-Garcia, J.R., Alguacil, M.M. and Caravaca, F. (2007) Soil sustainability indicators following conservation tillage practices under subtropical maize and bean crops. *Soil and Tillage Research* 93, 273–282.

Ross, M.A. and Lembi, C.A. (1985) *Applied Weed Science*. Macmillan Publishing Company, New York.

Roth, C.H. (1985) Infiltrabilität von Latosolo-Roxo-Böden in Nordparaná, Brasilien, in Feldversuchen zur Erosionskontrolle mit verschiedenen Bodenbearbeitungs-systemen und Rotationen. *Göttinger Bodenkundliche Berichte* 83, 1–104.

Sanchez, P.A., Shepherd, K.D., Soule, M.J., Place, F.M., Buresh, R.J., Izac, A.M., Mokwunye, A.U., Kwesiga, F.R., Ndiritu, C.G. and Woomer, P.L. (1997) Soil fertility replenishment in Africa: an investment in natural resource capital. In: Buresh, R.J., Sanchez, P.A. and Calhoun, F. (eds) *Replenishing Soil Fertility in Africa*. Soil Science Society of America, Madison, Wisconsin, pp. 1–46.

Sangar, S., Abrol, I.P. and Gupta, R.K. (2004) Conference report – conservation agriculture: conserving resources – enhancing productivity. Centre for Advancement of Sustainable Agriculture, National Agriculture Science Centre (NASC) Complex DPS Marg, Pusa Campus, New Delhi, India.

Sayre, K.D. (2005) Conservation agriculture for irrigated production systems permanent bed planting technologies. In*:* Morgounov, A., McNab, A., Campbell, K.G. and Paroda, R. (eds) *Proceedings of 1st Central Asian Wheat Conference on Wheat Production in Central Asia Through Science and Cooperation*, International Maize and Wheat Improvement Center (CIMMYT), Almaty, Kazakhstan, pp. 158–163.

Scopel, E. and Findeling, A. (2001) Conservation tillage effects on runoff reduction in rainfed maize of semi-arid zones of western Mexico. In: Garcia-Torres, L., Benites, J. and Martinez-Vilela, A. (eds) *Proceedings of 1st World Congress on Conservation Agriculture: Conservation Agriculture – A Worldwide Challenge*. Madrid, XUL, Cordoba, Spain, pp. 179–184.

Scopel, E., Da Silva, F.A.M., Corbeels, M., Affholder, F.O. and Maraux, F. (2004) Modelling crop residue mulching effects on water use and production of maize under semi-arid and humid tropical conditions. *Agronomie* 24, 383–395.

Selles, F., Kochann, R.A., Denardin, J.E., Zentner, R.P. and Faganello, A. (1997) Attribution of phosphorus fractions in a Brazilian oxisol under different tillage systems. *Soil and Tillage Research* 11, 23–31.

Shaxson, F., Kassam, A., Friedrich, T., Boddey, B. and Adekunle, A. (2008) Underpinning conservation agriculture's benefits: the roots of soil health and function. In: *Proceedings of an International Technical Workshop on Investing in Sustainable Crop Intensification: The case for improving soil health*. FAO, Rome, *Integrated Crop Management* 6, 69–116.

Shetto, R. and Owenya, M. (2007) Conservation agriculture as practiced in Tanzania: three case studies. African Conservation Tillage Network, Centre de Coopération Internationale de Recherche Agronomique pour le Développement, Food and Agriculture Organization of the United Nations, Nairobi, Kenya.

Sidiras, N. and Pavan, M.A. (1985) Influencia do sistema de manejo do solo no seu nıvel de fertilidade. *Revista Brasileira de Ciência do Solo* 9, 249–254.

Sims, B., Friedrich, T., Kassam, A. and Kienzle, J. (2009) Agroforestry and Conservation Agriculture: Complementary practices for sustainable development. *Agriculture for Development* 8, 13–20.

Sisti, C.P.J., Santos, H.P. dos, Kohhann, R., Alves, B.J.R., Urquiaga, S. and Boddey, R.M. (2004) Change in carbon and nitrogen stocks in soil under 13 years of conventional or zero tillage in southern Brazil. *Soil and Tillage Research* 76, 39–58.

Six, J., Paustian, K., Elliott, E.T. and Combrink, C. (2000) Soil structure and soil organic matter: I. Distribution of aggregate size classes and aggregate associated carbon. *Soil Science Society of America Journal* 64, 681–689.

SoCo (2009) Final Report on the Project 'Sustainable Agriculture and Soil Conservation (SoCo)'; European Commission Directorate-General for Agriculture and Rural Development, Luxemburg 2009; EU23820EN.

Steinsiek, J.W., Oliver, L.R. and Collins, F. (1982) Allelopathic potential of wheat (*Triticum aestivum*) straw on selected weed species. *Weed Science* 30, 495–497.

Stinner, B.R. and House, G.J. (1990) Arthropods and other invertebrates in conservation-tillage agriculture. *Annual Reviews in Entomology* 35, 299–318.

The Guardian (2004) Tim Radford in Seattle. Soil erosion as big a problem as global warming, say scientists. The Guardian, Saturday 14 February 2004. Available at: http://www.guardian.co.uk/world/2004/feb/14/science.environment (accessed 25 November 2012).

Thomas, G.A., Dalal, R.C. and Standley, J. (2007) No-till effects on organic matter, pH, cation exchange capacity and nutrient distribution in a Luvisol in the semi-arid subtropics. *Soil and Tillage Research* 94, 295–304.

Twomlow, S.J., Steyn, J.T. and du Preez, C.C. (2006) Dryland farming in southern Africa. In: *Dryland Agriculture*, 2nd edn. Agronomy Monograph No. 23. American Society of Agronomy, Madison, Wisconsin, pp. 769–836.

Umar, B.B., Aune, J.B., Johnsen, F.H. and Lungu, O.I. (2011) Options for improving smallholder conservation agriculture in Zambia. *Journal of Agriculture Sciences* 3(3), 50–62.

Uri, N.D., Atwood, J.D. and Sanabria, J. (1999) The environment benefit and cost of conservation tillage. *Environmental Geology* 38, 111–125.

Verhulst, N., Govaerts, B., Verachtert, E., Kienle, F., Limon-Ortega, A., Deckers, J., Raes, D. and Sayre, K.D. (2009) The importance of crop residue management in maintaining soil quality in zero tillage systems: a comparison between long-term trials in rainfed and irrigated wheat systems. In: *Proceedings of the 4th World Congress on Conservation Agriculture on Innovations for Improving Efficiency, Equity and Environment*. Indian Council of Agricultural Research (ICAR), New Delhi, India, pp. 71–79.

Verhulst, N., Govaerts, B., Verachtert, E., Castellanos-Navarrete, A., Mezzalama, M., Wall, P., Deckers, J. and Sayre, K.D. (2010) Conservation agriculture, improving soil quality for sustainable production systems? In: Lal, R. and Stewart, B.A. (eds) *Advances in Soil Science: Food Security and Soil Quality*. CRC Press, Boca Raton, Florida, pp. 137–208.

Vogel, H. (1994) Weeds in single crop conservation farming in Zimbabwe. *Soil and Tillage Research* 31, 169–185.

Wang, J., Huang, J., Zhang, L., Rozelle, S. and Farnsworth, H.F. (2010) Why is China's Blue Revolution so 'Blue'? The determinants of conservation tillage in China. *Journal of Soil and Water Conservation* 65(2), 113–129.

Wang, W.J. and Dalal, R.C. (2006) Carbon inventory for a cereal cropping system under contrasting tillage, nitrogen fertilization, and stubble management practices. *Soil and Tillage Research* 91, 68–74.

Wang, X., Dai, K., Zhang, D., Zhang, X., Wang, Y., Zhao, Q., Cai, D., Hoogmoed, W.B. and Oenema, O. (2011) Dryland maize yields and water use efficiency in response to tillage/crop stubble and nutrient management practices in China. *Field Crops Research* 120, 47–57.

Weaver, D.K., Buteler, M., Hofland, M.L., Runyon, J.B., Nansen, C., Talbert, L.E., Lamb, P. and Carlson, G.R. (2009) Cultivar preferences of ovipositing wheat stem sawflies as influenced by the amount of volatile attractant. *Journal of Economic Entomology* 102, 1009–1017.

Weil, R.R., Benedetto, P.W., Bandel, V.A. and Sikora, L.J. (1988) Influence of tillage practices on phosphorus distribution and forms in three ultisols. *Agronomy Journal* 80, 503–509.

Weller, D.M., Raaijmakers, J.M., Gardener, B.B.M. and Thomashow, L.S. (2002) Microbial populations responsible for specific soil suppressiveness to plant pathogens. *Annual Review of Phytopathology* 40, 309–348.

West, T.O. and Marland, G. (2002) A synthesis of carbon sequestration, carbon emissions, and net carbon flux in agriculture, comparing tillage practices in the United States. *Agriculture, Ecosystems and Environment* 91, 217–232.

Zenter, R.P., Wall, D.D., Nagy, C.N., Smith, E.G., Young, D.L., Miller, P.R., Campbell, C.A., Mcconkey, B.G., Brandt, S.A., Lafond, G.P., Johnston, A.M. and Derksen, D.A. (2002) Economics of crop diversification and soil tillage opportunities in the Canadian Prairies. *Agronomy Journal* 94, 216–230.

2 Conservation Agriculture in the USA

Sjoerd W. Duiker[1] and Wade Thomason[2]
[1]*Penn State Cooperative Extension, Department of Plant Science, The Pennsylvania State University, Pennsylvania;* [2]*Virginia Polytechnic Institute and State University, Blacksburg, Virginia, USA*

2.1 Introduction

2.1.1 A short history of agriculture in the USA

Pre-colonial crop production in the USA had a low impact on the environment (Fig. 2.1). It was predominantly practised on soft soils along streams and rivers, and was characterized by shallow or no-tillage (NT), intercropping and long fallow periods (Haystead and Fite, 1955; Hurt, 1994). This was to change dramatically with European colonization, characterized by 'one-crop agriculture', cash-crop orientation and intensive tillage, which led to phenomenal soil degradation (Haystead and Fite, 1955).

The first major cash crop, widely grown in the Chesapeake Bay area from 1600 onward, was tobacco (Hurt, 1994). Tobacco quickly depleted the soil in 3 to 4 years followed by maize, wheat, or barley until the unproductive land was left fallow, often for 20 years (Hurt, 1994). After the invention of the cotton gin by Eli Whitney in 1793 cotton became the dominant crop of the south (Haystead and Fite, 1955; Hurt, 1994). Similar to tobacco, after a few years of cotton the land was left in degraded state. As a result, topsoil was lost and thousands of miles of gullies dissected the South (Trimble, 1973). Slowly cotton production moved to the western end of the Deep South leaving behind unproductive lands overtaken by brush, grassland or pine plantations.

In search of new lands, farmers started to till the deep, fertile prairie soils of the Midwest. The steel mouldboard plough, developed by John Deere in 1837, enabled termination of the tough prairie sod and stimulated decomposition of organic matter, which released massive amounts of plant nutrients. Research from the Morrow plots in Illinois shows release of at least 27–52 Mg ha^{-1} carbon, 1900–3700 kg ha^{-1} N, 250–480 kg ha^{-1} P and 250–480 kg ha^{-1} S in a 50-year period (assuming C:N:P:S ratios of organic matter of 140:10:1.3:1.3) (Darmody and Peck, 1997; Stevenson and Cole, 1999). In 100 years of unfertilized maize monoculture on original tall grass prairie soil at the Sanborn Field in Missouri 3428 kg ha^{-1} N, 445 kg ha^{-1} P and 3665 kg ha^{-1} K was removed (Buyanovsky *et al.*, 1997). Mixed farming with hogs and dairy dominated the Midwest until the Second World War, but with the availability of cheap nitrogen fertilizer and the development of soybeans, maize–soybean became the signature rotation.

Serious settlement of the Great Plains did not begin until the late 1870s (Hurt, 1994). A wheat–bare fallow rotation to conserve moisture became common practice. Periods of plentiful rainfall and high wheat

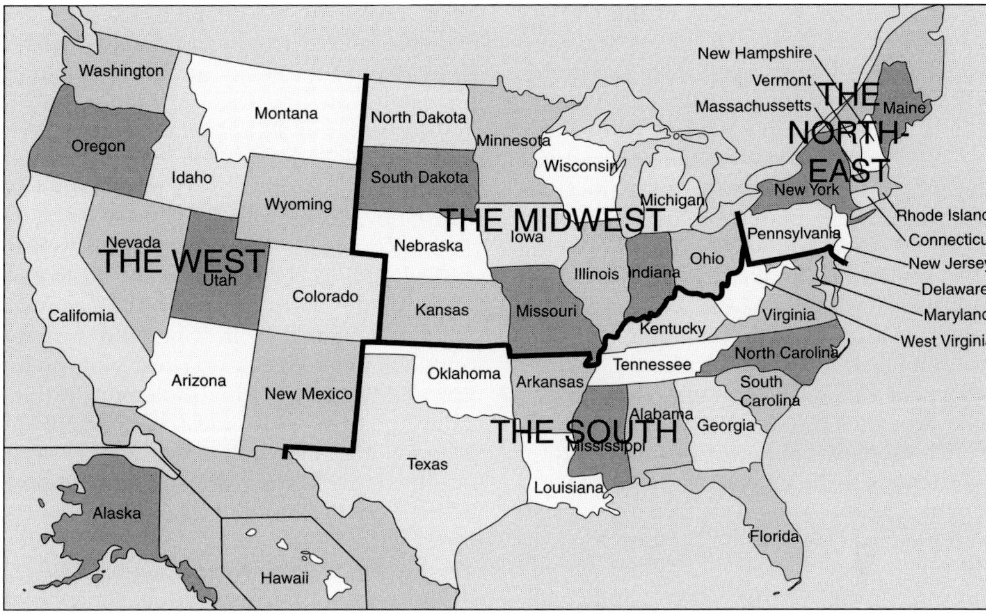

Fig. 2.1. The regions of the USA.

prices encouraged widespread cultivation of the prairies, but recurrent drought caused devastation. A drought period from 1917 to 1921, for example, caused 60% of northern Plains farmers to go bankrupt, whereas the southern Plains were hit by drought in the 1930s, causing the infamous Dust Bowl and the greatest population displacement in US history (Hurt, 1994). Similarly, wheat was and still is grown on highly productive, deep but steeply undulating loess soils of the Palouse. Conventional tillage (ConvT) in combination with winter precipitation and snowmelt on frozen soil led to high rates of water erosion. Since the Palouse was first cultivated all of the original topsoil has been lost from about 10% of the cropland, one-fourth to three-fourths of the original topsoil has been lost from another 60%, and organic matter content has been reduced by 50% (Veseth, 1985; Rasmussen and Smiley, 1997).

2.1.2 The advent of Conservation Agriculture

Realization of the enormous damage caused by prevailing soil management practices, of intensive tillage and one-crop agriculture led to today's concept of Conservation Agriculture (CA) based on the principles of minimum soil disturbance, continuous organic matter cover by crop residue or cover crops, and diverse crop rotations (Kassam *et al.*, 2010). Hugh Bennett, the 'Father of Soil Conservation', alerted the nation to the destruction caused by soil erosion, which eventually led to establishment of the Soil Conservation Service (SCS – today's Natural Resources Conservation Service) and Soil Conservation Districts (Helms, 2010). The SCS and Conservation Districts collaborated with the Cooperative Extension Service of the Land-Grant Universities to educate and assist farmers in the implementation of conservation practices (Helms, 2010). Intensive tillage, however, was still the predominant way of soil preparation for crop production. In the 1940s, Edward Faulkner, an Ohio farmer, proposed 'thrash farming' – growing heavy cover crops and discing those into the surface soil as an alternative to mouldboard ploughing (Faulkner, 1943). Isolated research trials were conducted investigating aspects of CA. In 1937 a 'stubble mulch tillage' trial was started with specially designed subsurface tillage equipment at the University of Nebraska in which weeds and crop residues were left at the surface to combat erosion (Van Es and Notier, 1988). As a result, stubble mulch farming spread throughout the Great Plains and western Canada, and was used on 18 million acres in 1961 (McCalla *et al.*, 1962). The practice was not adopted in the Midwest, however, where concern with erosion on the deep prairie soils was minimal. Additionally, a lack of weed control alternatives to tillage and unavailability of equipment to plant into heavy residue and tough soil were problems that needed to be overcome (McCalla *et al.*, 1962).

Development of herbicides made no-till (NT) farming in the modern sense of the word possible. Sinox (sodium dinitro-o-cresylate) was introduced in the 1930s in North America and 2,4D was registered in 1945 (Holm and Johnson, 2009). Shortly after, Land-Grant-University researchers in New Jersey and Connecticut started to experiment with NT to renovate pasture (Van Es and Notier, 1988). In the early 1950s K.C. Barron of Dow Chemical Company drilled small grains and planted maize and soybeans into killed ladino clover and obtained good yields (Van Es and Notier, 1988). However, perennial grass control became a problem in the absence of effective herbicides to control them. The advent of Dalapon enabled better, but not complete, perennial grass control. A group of farmers in Christian County, Kentucky started using this system in the late 1950s to produce maize but reverted back to tillage when certain weeds that were not controlled effectively became problematic (Hyup, 1979). The development of Paraquat by Chevron, registered in the early 1960s in the USA, finally offered an effective burn-down programme (Hyup, 1979).

As more herbicides became available in the early 1960s, NT trials were started at Land-Grant Universities in New York, Ohio and Virginia (Van Es and Notier, 1988). In Virginia in 1960–1965, NT maize was hand planted with good results. In New York, Free used a mechanical planter to plant maize into killed lucerne/grass sod. However, yields were depressed 10% compared with ConvT and herbicide residues did not allow cover crops to be planted. Researchers Triplett and Van Doren in Ohio started NT maize work at the same time with encouraging results on low organic matter, highly erodible soils but depressed yields on poorly drained lake-bed soils (Van Doren *et al.*, 1976). Farmer interest in NT now started for real. One reason was the introduction in 1966 of the Allis-Chalmer planter, which worked well in NT soil (Van Es and Notier, 1988). In the early 1960s, the NT pioneer Harry M. Young Jr in Christian County, Kentucky planted small NT test plots on his farm after visiting NT maize demonstrations at the University of Illinois (Hyup, 1979; Van Es and Notier, 1988). By adopting NT, North Carolina farmers were able to maintain profitability in the face of declining tobacco prices with double-cropped soybeans after wheat or barley (Van Es and Notier, 1988).

Recognizing the multiple challenges such as equipment, soil management, crop selection, weed, pest, disease and residue management, researchers at some Land-Grant Universities started to form interdisciplinary teams to develop a whole-systems approach to NT. Two important books published as a result were *No-Tillage Agriculture* written by researchers and extension specialists from the University of Kentucky (Phillips and Phillips, 1984) and *No-Tillage and Surface Tillage Agriculture: The Tillage Revolution*, written by a group of researchers from several Land-Grant Universities and other organizations (Sprague and Triplett, 1986). Another important development was the creation of NT organizations, mainly farmer-led, starting in the early 1970s (Table 2.1). Many of these organizations have annual conferences to exchange new ideas and research information, to network and organize field days to demonstrate and discuss CA practices. These venues are important vehicles for interaction between farmers, industry, researchers and policy makers.

2.1.3 Current status of Conservation Agriculture in the USA

After a transition from ConvT to reduced tillage, there has been a trend since 1990

Table 2.1. Conservation agriculture organizations/venues in the USA.

Organization	Regional focus	Leadership
Conservation Technology Information Center (http://www.ctic.purdue.edu)	North America	Government, industry, research and extension
No-Till Farmer (http://www.no-till farmers.com)	Midwest	Farmers
No-Till on the Plains (http://www.notill.org)	Plains	Farmers
Pacific Northwest Direct Seed Association (http://www.directseed.org/)	Northwest	Farmers
Southern Conservation Tillage Conference	Southern USA	Researchers, extension, government service providers
Conservation Tillage Workgroup (http://casi.ucanr.edu/)	California	University, farmers, government, private industry
Delta Conservation Demonstration Center (http://www.dcdcfarm.org)	Mississippi	Farmers
Ohio No-Till Council	Ohio	Farmers/extension
Pennsylvania No-Till Alliance (http://www.panotill.org)	Pennsylvania	Farmers
Southern Plains Agricultural Resources Coalition (http://www.ctic.org/resourcedisplay/84)	Oklahoma	Farmers to consumers
Georgia Conservation Tillage Alliance (http://www.gcta-ga.org)	Georgia	Farmers
Dakota Lakes Research Farm (http://www.dakotalakes.com)	South Dakota	Farmers/Land-Grant University
South Dakota No-Till Association (http://www.sdnotill.com)	South Dakota	Farmers
Manitoba North Dakota Zero Tillage Farmers Association (http://www.mandakzerotill.org)	North Dakota	Farmers
Colorado Conservation Tillage Association (http://www.highplains.notill.com)	Colorado	Farmers/industry/Land-Grant University
Virginia No-Till Alliance (http://www.virginianotill.com)	Virginia	Farmers/industry
South Central Kansas Residue Alliance (http://www.sckra.org)	Kansas	Farmers/industry/Land-Grant University

toward increasing NT seeding of the major crops grown in the USA (Fig. 2.2).

The majority of double-crop soybeans, those planted immediately follow a small grain crop, are planted NT because of moisture conservation and because seeding of the next crop can occur immediately after harvest, avoiding any delays in planting. Since 1994, NT full season soybeans have also increased dramatically to nearly 40% of the total crop. Moisture conservation has driven this trend in many areas, but this change has also been greatly aided by the increased availability and ownership of seeders that can successfully plant in high residue conditions. Over 85% of the soybeans that are reported to be grown under conservation tillage (a general term indicating any tillage system leaving more than 30% residue cover after planting; CTIC, 2013) are NT (Fig. 2.3).

This trend also holds true for some other crops such as cotton but others, such as spring-seeded small grains, report a much lower proportion of NT production as a percentage of conservation tillage. In general, conservation tillage, especially NT, has increased in popularity in the recent decade.

However, cropping system, climate and soil type all influence the benefits and challenges of CA. For example, DeFelice *et al.* (2006) found that the maize yield increase in response to NT production was much greater in the southern and western regions than in the northern USA (Fig. 2.4). The authors speculate that this lack of positive response in more northern areas is due to cooler soils and slower early season crop growth with high residue systems.

Crop diversity, continuous soil cover and the use of cover crops to maintain a

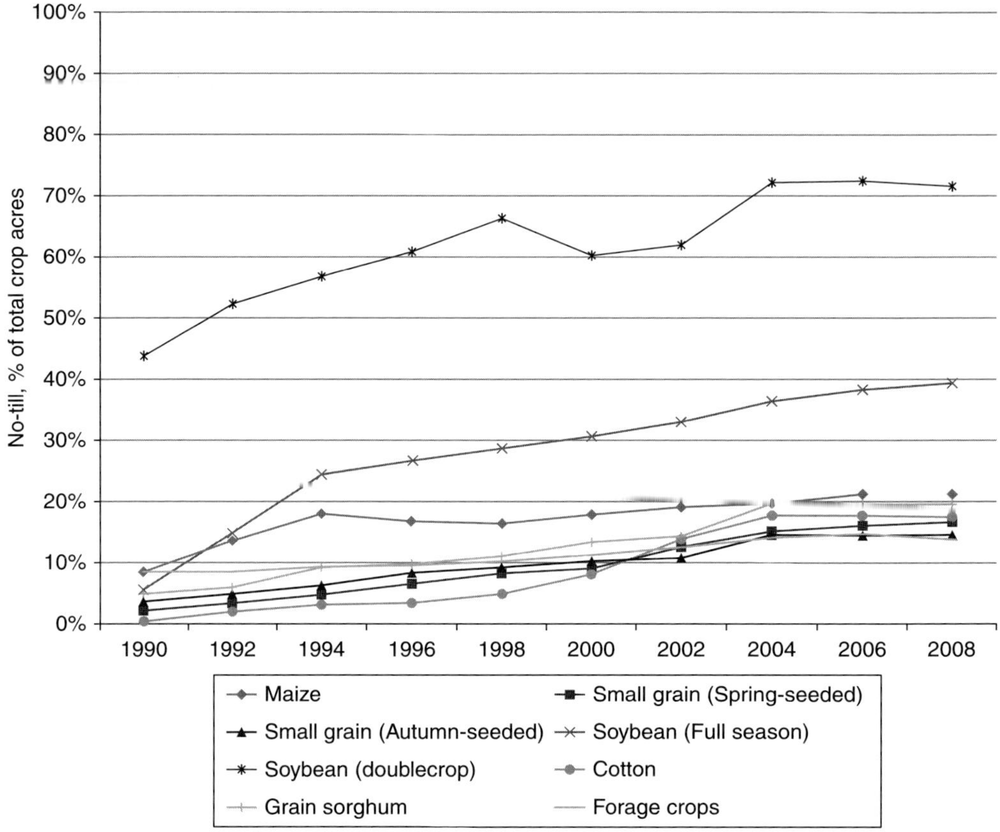

Fig. 2.2. Major US crops, percentage in no-till farming systems, 1990–2008 (Source: Conservation Technology Information Center).

green and growing plant cover in crop fields for as much of the year as possible, has received considerably less attention than reduced tillage in the USA (Clark, 2007). Since 2000, maize plantings in the USA have risen at a rate of more than 500,000 ha $year^{-1}$ (Table 2.2). At the same time, other rotation crops such as barley, oats and sorghum have declined in their extent. All this results into decreased crop and agroecosystem diversity in many cropping situations. However, many producers and scientists have recently recognized advantages that can be gained by reintroduction of cover crops and increasing the biodiversity within the crop rotation as exemplified by the CA concept (Clark, 2007; Magdoff and Van Es, 2010).

Higher cash crop yields from many crops including maize (Wagger, 1989), cotton (Daniel *et al.*, 1999) and vegetables (Wyland *et al.*, 1996) have been reported in response to the introduction of various cover crops. Other indirect benefits such as weed suppression (Akemo *et al.*, 2000), nutrient recycling (Reicosky and Forcella, 1998) and reduced nematode pressure (McSorley and Gallaher, 1994; Abawi and Widmer, 2000) are also frequently cited in the literature. In addition, Smith *et al.* (2008) reported that the benefits of including diverse cover crops extend not only to higher yield of cash crops, but also to improved broader ecosystem functions and services.

Many cover crop advocates are supporting multi-species mixtures of cover crops in an effort to increase productivity and resiliency (Tilman *et al.*, 1998). Wortman *et al.* (2012b) investigated eight cover crops and

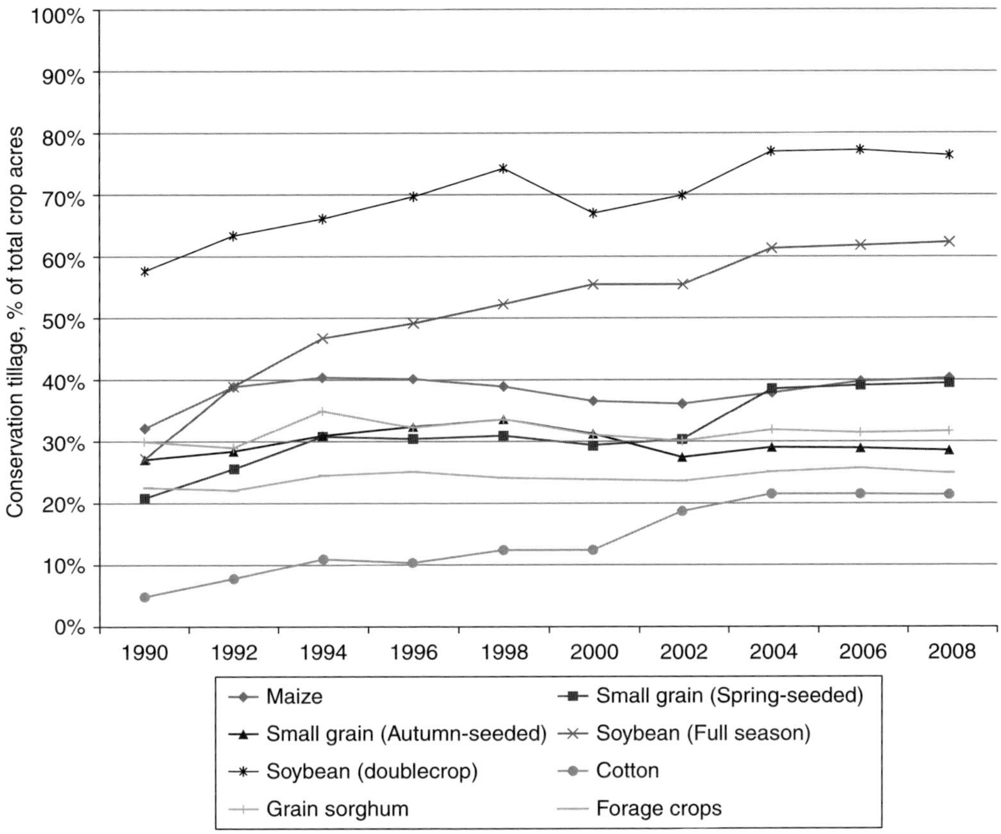

Fig. 2.3. Major US crops, percentage in conservation tillage farming systems, 1990–2008 (CTIC, 2013).

mixtures of species belonging to the Fabaceae or Brassicaceae plant families in the western US corn belt. They found that mixtures of species were more productive than the individual component species grown alone. They attributed this to the resiliency of mixtures in the face of extreme and variable weather. One of the most important functions of cover crops is as a source of readily available C as a food source for soil flora and fauna (Reicosky *et al.*, 1995). High diversity cover crop mixtures contribute to increased soil biological diversity (Snapp *et al.*, 2005), which in turn strengthens nutrient cycling. The future expansion of cover crops within CA systems in many portions of the USA will likely depend on broader appeal and an understanding of the relationship between soil biology and productivity. Cover crops will need to do more than just scavenge nutrients or reduce sediment losses. Recognition of the many benefits of diverse mixtures, alternative species, adaptable cover crop systems and seeding techniques will be needed to broaden the further appeal and adoption of cover crops.

2.2 Research Findings on Conservation Agriculture in the USA

2.2.1 Soil quality

Building soil organic matter content will be critical to restore US soils that have lost excessive amounts of organic matter and surface soil, even including prairie soils (Olson *et al.*, 2005). The first concern is to stop erosion causing loss of surface organic

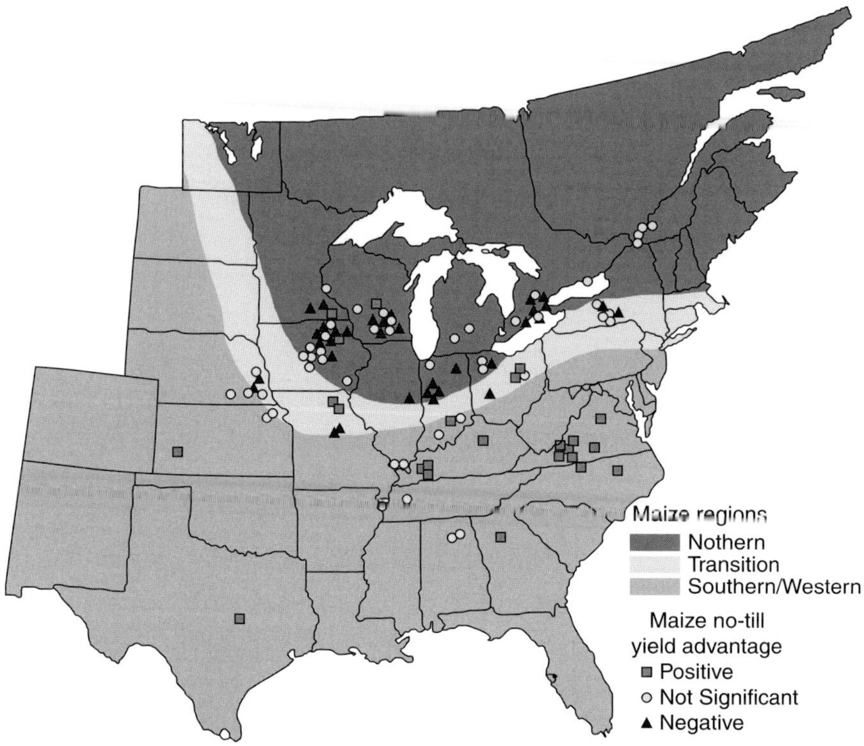

Fig. 2.4. No-till maize yield advantage in various regions of the USA (DeFelice *et al.*, 2006).

matter, reduction of soil depth to bedrock, exposure of high-clay subsoil (Mendoza *et al.*, 2008), carbonates (Papiernik *et al.*, 2009), acid subsoil, fragipans and duripans. Elimination of inversion tillage will help restore surface organic matter content and water-stable aggregation (Duiker and Beegle, 2006). Additionally, perennials in the rotation and maximum live-root activity increase aggregate stability (Grover, 2008). The long-term NT soil has a firm matrix interspersed by a network of macro- and micro-pores stimulating drainage, aeration and deep root penetration (Hill and Cruse, 1985; Franzluebbers *et al.*, 1995; Kemper *et al.*, 2011). CA systems with straight- or bent-leg subsoilers in combination with cover crops have been developed for naturally compacted Coastal Plain soils in the south-eastern USA (Camp *et al.*, 1984; Busscher *et al.*, 1986; Siri-Prieto *et al.*, 2007). If subsoil structure is well developed, no deep tillage is necessary (Sene *et al.*, 1985), so it may be possible in rotations with deep-rooting, soil-ameliorating crops to eliminate tillage altogether even on these soils. Application of CA leads to a soil profile resembling that under permanent sod, characterized by surface protection by living or dead vegetation, high surface organic matter content and stable aggregates, high biological activity, permanent macro-pores formed by earthworms and decomposing roots penetrating into the subsoil, and absence of tillage pans (Fig. 2.5).

Concerns about the lack of incorporation of applied nutrients in continuous NT have proven unfounded. High surface organic matter content helps to buffer pH and supply nutrients and surface acidity is neutralized by regular surface lime application (Duiker and Beegle, 2006). Nutrient stratification in CA has not proven to be an obstacle to high yields. Immobile and somewhat immobile nutrients such as phosphorus and potassium applied to the soil surface are taken up by fine crop roots and mycorrhizae,

Table 2.2. Planted hectares of major crops in the USA, 2000–2012 (USDA-NASS, 2013).

Year	Maize	Barley	Oats	Sorghum	Sugarbeet	Sunflower	Cotton	Soybean	Wheat
2000	32,206,883	2,348,583	1,810,931	3,722,672	633,279	1,149,798	6,282,267	30,067,206	25,323,482
2001	30,648,583	2,004,453	1,781,781	4,148,988	552,753	1,065,992	6,384,008	29,989,879	24,061,538
2002	31,940,891	2,027,530	2,022,267	3,882,186	577,854	1,044,939	5,650,972	29,944,534	24,420,243
2003	31,823,077	2,165,182	1,861,134	3,813,765	552,794	948,988	5,457,328	29,718,219	25,158,300
2004	32,764,777	1,832,794	1,653,846	3,030,769	544,777	758,300	5,529,798	30,448,583	24,147,368
2005	33,108,907	1,568,826	1,719,028	2,612,955	526,235	1,096,761	5,767,368	29,162,753	23,163,563
2006	31,711,336	1,397,571	1,686,640	2,640,486	553,117	789,474	6,183,806	30,575,709	23,212,146
2007	37,865,182	1,626,721	1,523,482	3,122,267	513,684	838,057	4,383,482	26,210,931	24,477,733
2008	34,810,526	1,719,028	1,314,575	3,353,846	441,579	1,018,826	3,834,413	30,655,061	25,584,211
2009	34,972,470	1,444,130	1,378,138	2,685,425	480,081	821,862	3,704,251	31,356,680	23,954,656
2010	35,705,263	1,162,753	1,270,445	2,187,854	474,453	790,081	4,442,996	31,337,652	21,697,571
2011	37,214,980	1,036,032	1,010,526	2,219,028	499,069	624,696	5,965,749	30,354,656	22,027,935
2012	39,030,364	1,489,069	1,111,741	2,514,170	503,441	730,567	5,115,385	30,801,619	22,678,947

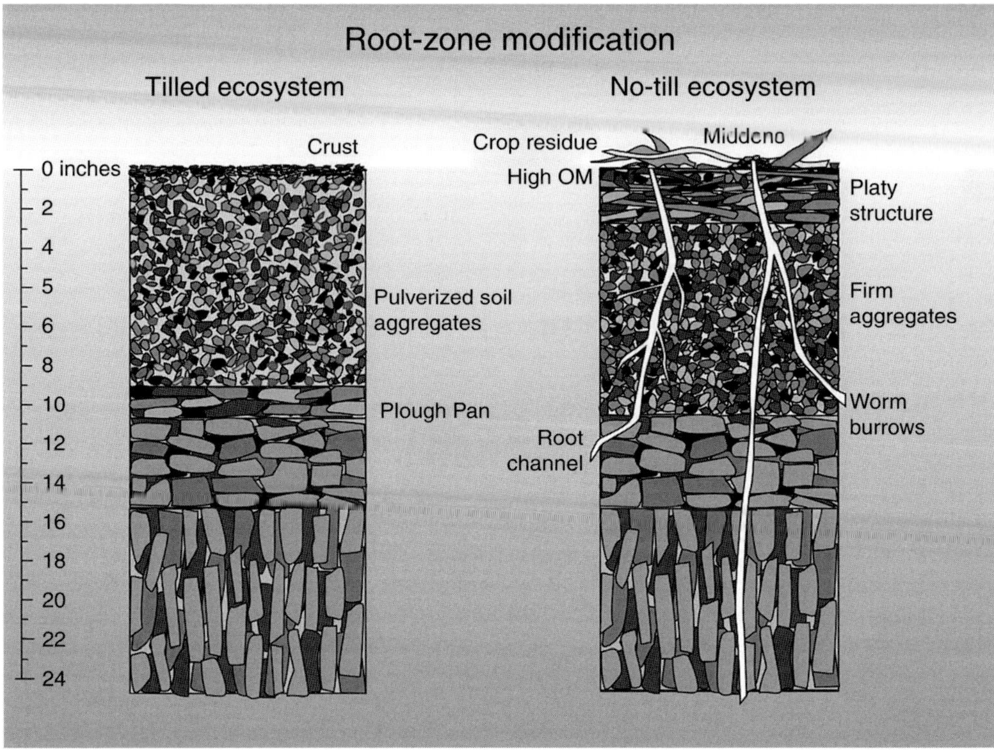

Fig. 2.5. Soil profile of a soil subject to annual inversion tillage contrasted with that of a soil under Conservation Agriculture (no-till) (Duiker and Myers, 2006).

which proliferate in the surface NT soil (Vyn *et al.*, 2002). High moisture content under the surface mulch extends the availability of surface nutrients. The lack of mixing of phosphorus with soil improves its availability, even leading to concerns with soluble phosphorus runoff (Staver and Brinsfield, 1994; Verbree *et al.*, 2010; Sharpley *et al.*, 2012). The greatest concern with fertilization of NT fields is potential gaseous and leaching losses of nitrogen. Surface application of urea or manure may lead to losses up to 30% of nitrogen due to ammonia volatilization (Fox *et al.*, 1996). To reduce volatilization, nitrogen fertilizer or manure may be injected into the soil with low-disturbance methods at planting or at side-dress time (Dell *et al.*, 2012), banded on the surface, or formulated with urease inhibitors that stabilize urea (Fox *et al.*, 1996; Slaton *et al.*, 2011). Injection can increase nitrate leaching, but cover crops help reduce this loss pathway (Cambardella *et al.*, 2010). Nitrous oxide loss is not different in NT compared to tillage systems on light-textured, well-drained soils, but may be greater on poorly drained, heavy clay soils (Rochette *et al.*, 2009). Ploughing in leguminous cover crops is not recommended in CA due to the importance of surface mulch, and does not normally affect nitrogen recovery by the following crops (Craig, 1987; Levin *et al.*, 1987; Varco *et al.*, 1989).

CA also affects biological soil quality. The effect of CA on macro-invertebrates has been known for a few decades now (Edwards and Lofty, 1982). Anecic, deep-burrowing earthworms are especially favoured by CA, while endogeic, topsoil-dwelling earthworms are less affected by tillage, and epigeic, surface-dwelling earthworms are scarce in crop fields independent of tillage (Kladivko, 2001). Greater numbers of deep burrowing earthworms in long-term NT have been found to increase deep root penetration compared with clean tilled fields (Kemper *et al.*, 2011). Crop diversity (legumes, sod) and organic amendments (manure) also benefit earthworm numbers (Table 2.3).

Microbial biomass is typically higher near the soil surface in NT than in ConvT, but at depth the reverse may be the case (Franzluebbers *et al.*, 1994; Jangid *et al.*, 2011). A study in several US states suggested that a larger proportion of microbial biomass is composed of fungi than bacteria in long-term NT soil (Frey *et al.*, 1999).

2.2.2 Carbon sequestration

The effect of tillage on carbon sequestration was believed to be straightforward: early research showed NT led to sequestration while ConvT led to loss of organic carbon (Reicosky *et al.*, 1997; Duiker and Lal, 1999; Halvorson *et al.*, 2002; West and Post, 2002; Lal *et al.*, 2003). This view has recently been challenged, based on the suggestion that many studies did not sample carbon to sufficient depth (Baker *et al.*, 2007; Chatterjee and Lal, 2009). On the other hand, high subsoil organic carbon variability may obscure real near-soil tillage differences (Syswerda *et al.*, 2011). Crop rotation and cover crops also have an impact on carbon by means of the quantity and quality of crop residue and root mass returned to the soil. The analysis by West and Post (2002) showed the positive effects of diverse crop rotations on carbon sequestration (with the exception of continuous maize compared to maize–soybean) and the negative effects of fallow periods. Drinkwater *et al.* (1998) showed the value of leguminous cover crops and compost in increasing soil organic carbon content. Manure application can also increase organic carbon content but has not been widely studied in the USA (Min *et al.*, 2003). The mechanisms and required sampling procedures of carbon sequestration are still poorly understood as is evidenced by the recent controversy over tillage effects.

2.2.3 Crop yield

Many trials have compared the effect of tillage on crop yield. DeFelice *et al.* (2006) summarized results of 61 maize trials in North America representing 687 site-years of data and 43 full-season soybean trials representing 455 site-years of data. Yields in continuous NT with minimal soil disturbance at planting and/or fertilizer application were compared with ConvT systems varying from mouldboard ploughing + secondary tillage to chisel ploughing + secondary tillage. Results for maize are shown in Table 2.4. Although average maize and soybean yields varied little between tillage systems, there were regional differences. In the southern/western regions where water shortage and high summer temperatures are common, maize and soybean yields were higher with NT than with ConvT, while in the northern regions maize and soybean yields were lower with NT than with ConvT. No-tillage maize and soybean performed better on well-drained soils than on poorly drained soils. Crop rotation was an important practice to improve crop yields with NT, especially in northern regions and/or on poorly drained soils, hence

Table 2.3. Effect of crop and tillage practice on endogeic and anecic earthworm numbers (Adapted from Mackay and Kladivko, 1985).

Crop	Tillage	Adult	Juveniles (m^{-2})	Total
Maize	Plough	5	3	8
Maize	No-till	8	8	6
Soybean	Plough	35	26	62
Soybean	No-till	58	83	141
Clover/ryegrass	Pasture	258	213	470
Clover/ryegrass + manure	Pasture	811	486	1298

Table 2.4. Effect of soil drainage and crop rotation on change in maize yield (%) due to no-tillage compared to conventional tillage (DeFelice *et al.*, 2006).

	Soil drainage		Crop rotation	
	Moderate/well	Poor	Monoculture	Rotation
South/western USA	12.9	7.0	12.3	13.1
Transition zone	−0.7	−2.6	−4.0	1.9
Northern USA	−4.8	−8.1	−6.2	−4.1

the need to practise all three components of CA for success.

The increased yields with NT in areas experiencing frequent water deficit and high temperatures confirm the effect of mulch cover to reduce soil temperature, increase water infiltration and reduce evaporation, resulting in reduced heat and drought stress (Drury *et al.*, 1999). Reduced yields in the northern USA can be explained by slower early season growth in NT soils due to colder soil temperatures which can translate in reduced crop yield in areas with short growing seasons lacking significant drought stress. When soils are poorly drained, water excess and anaerobic conditions may be prolonged under a mulch cover resulting in suboptimal conditions for crop growth. Besides anaerobic conditions, poor aeration may also increase root and seedling pest and disease pressure.

Although it is easy to see the preference for NT in the southern USA, NT has also great potential in the northern USA for the following reasons.

1. No-tillage saves costs and increases farming efficiency. Farmers are able to prepare fields quickly and plant when soil conditions are fit. They can also own less and smaller equipment that consumes less fuel because of the removal of high-power-demanding tillage operations.
2. No-tillage enables more intensive crop production while maintaining environmental function.
3. Expertise and technology improvements result in better results today than when these studies were performed with equipment, technology and knowledge developed under a tillage philosophy.

2.2.4 Runoff, infiltration, soil water content and soil conservation

Many trials conducted in the USA have shown that CA with high mulch cover dramatically lowers wind, water and tillage erosion (Shipitalo and Edwards, 1998; Li *et al.*, 2008; Raczkowski *et al.*, 2009). Although elimination of tillage alone reduces tillage erosion, mulch cover is essential to limit wind and water erosion. Research has shown that at least 30% soil cover is needed to significantly reduce wind and water erosion (Lyon *et al.*, 2000). Improvement of infiltration in NT depends on a number of factors (Table 2.5). Mulch cover is essential to protect the soil surface from sealing and crusting, and to improve surface aggregation and provide habitat to deep-burrowing earthworms that create water-conducting macro-pores (Edwards *et al.*, 1990; Shipitalo and Edwards, 1998). Low mulch cover explains the lack of success with NT in continuous cotton and wheat–fallow crop production in the semi-arid Great Plains (Unger and Baumhardt, 2001; Baumhardt and Jones, 2002b). Greater infiltration with NT can be expected on well- or moderately well-drained soils sensitive to sealing and crusting (Fig. 2.6). On poorly drained soils, the infiltration benefit of NT may not be seen (Kleinman *et al.*, 2008; Verbree *et al.*, 2010). Similarly, infiltration benefits of NT are small on non-crusting, coarse-textured Coastal Plain soils with shallow water table (Staver and Brinsfield, 1994).

Infiltration benefits of NT are greatest during high intensity rainfall events (Kleinman *et al.*, 2008). Finally, greater infiltration can be expected over time in NT

Table 2.5. The effects of conservation agriculture on runoff and infiltration compared with conventional tillage systems as affected by several factors.

Factor	Effect
Soil drainage class	Infiltration on well-drained soils is increased more than on poorly drained soils
Soil texture	Infiltration on fine- or medium textured soils increases more than on non-sealing coarse textured soils
Slope	Infiltration on steeply sloping soils is increased more than on flat soils
Rainfall characteristics	In areas with highly erosive rains benefits are greater than in areas with low erosivity
Mulch cover	The higher the mulch cover, the greater the benefit of CA on infiltration
Time after tillage	Infiltration immediately after tillage may be greater than in no-tillage, but this is reversed as soil consolidates
Continuity of no-tillage	Long-term no-tillage has greater numbers of continuous macropores

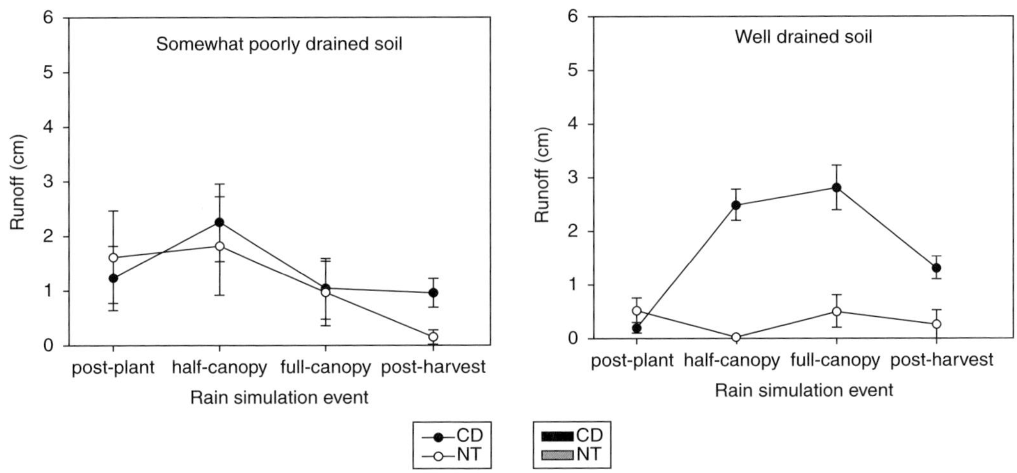

Fig. 2.6. Runoff from no-tillage (NT) compared with chisel-disc tillage (CD) is significantly reduced on well-drained soils but not on somewhat poorly drained soil as shown in this rainfall simulation study in maize on a Hagerstown silt loam and a Buchanan gravelly loam in Pennsylvania. Despite variation in infiltration benefits, soil erosion was reduced significantly with NT in both cases (Verbree *et al.*, 2010).

because of the time it takes for soil-improvement to take effect (Dick *et al.*, 1989).

2.2.5 Climate change mitigation and adaptation

The three greenhouse gases causing radiative forcing are carbon dioxide, methane (global warming potential 21 times greater than that of CO_2) and nitrous oxides (global warming potential 310 times greater than that of CO_2) (Lal *et al.*, 1998). Crop production affects emissions of these gases, and is impacted by a changing climate due to their increasing concentration in the atmosphere. The US Global Change Program expects US temperatures will be higher, growing seasons longer, evaporation greater, heavy downpours more frequent and snow cover to be less (Karl *et al.*, 2009). A USDA study suggested hotter and drier conditions could lead to a loss of US\$1.5 billion to an increase of US\$3.6 billion in farm income, depending on adaptation strategies (Malcolm *et al.*, 2012). CA can play an important role to adapt to a changing climate, and can reduce the net emissions of greenhouse gases from US agriculture (Lal *et al.*, 2011). Surface mulch cover in CA is probably the most important single attribute to build resilience against future climate change into the cropping systems of the future. High mulch cover will help moderate sub-optimally high soil temperatures (Johnson and Lowery, 1985), increase infiltration and reduce erosion, especially during intense precipitation events (Shipitalo and Edwards, 1998; Dabney *et al.*, 2004; Verbree *et al.*, 2010), reduce evaporation and improve water use efficiency (Wagger and Denton, 1992; Baumhardt and Jones, 2002a; Bauer *et al.*, 2010), and improve aggregation (Duiker and Beegle, 2006) and continuous macropores created by anecic earthworms (Edwards *et al.*, 1990). Crop diversity practised in CA helps deal with greater climate variability and exploits periods in the year when temperature conditions are favourable for crop production as well as make better use of water (Bordovsky *et al.*, 1994; Farahani *et al.*, 1998; Schlegel *et al.*, 2002). Integration of cover crops or perennial forages and livestock with grain crop production can help increase crop yields and improve soil quality (Franzluebbers, 2007; Maughan *et al.*, 2009).

Crop production impacts atmospheric greenhouse gas concentrations when fossil fuel is combusted to produce machinery and inputs, for field operations, postharvest processing and storage, and transportation. Additionally, atmospheric greenhouse gas concentrations are affected when soil organic carbon content changes, when nitrous oxide is released in the process of denitrification, or when methane is either released or absorbed. Reduced use of machinery and elimination of tillage in CA reduces fossil fuel needs and hence CO_2 emissions compared with ConvT systems (Uri, 1998) and improves energy efficiency of crop production (Gelfand *et al.*, 2010). Integration of leguminous (cover) crops in CA crop rotations reduces the need for nitrogen fertilizers, which represents up to 30% of fossil fuel consumption in crop production (Pimentel, 2009). Carbon sequestration obtainable with CA, at least on certain soil types as discussed above, is a way to mitigate atmospheric CO_2 increases. Nitrous oxide is released from agricultural lands and is impacted by the application of organic and inorganic nitrogen sources and the degree of denitrification. By using leguminous (cover) crops in diverse CA crop rotations, nitrogen fertilizer use can be reduced, reducing nitrous oxide emissions (Ebelhar *et al.*, 1984; Drinkwater *et al.*, 1998). Deep-rooted non-leguminous cover crops can reduce nitrate leaching, further reducing the potential for denitrification (Meisinger and Delgado, 2002). Nitrous oxide emissions are also reduced with long-term use of NT and placement of fertilizer below the soil surface (Kessel *et al.*, 2013). Major agricultural activities releasing methane are paddy rice and animal manure (Lal *et al.*, 1998). Production of crops on aerated soils has been found to be a sink for methane (Kern *et al.*, 2012). Diversifying from continuous barley to a barley–pea–wheat rotation helped increase methane absorption (Sainju *et al.*, 2012), but no effect of long-term NT on methane absorption was observed in another study (Bayer *et al.*, 2012). In conclusion, CA can play an important role in adapting to and mitigating climate change.

2.2.6 Insect-pest and disease dynamics

Research into reduced tillage cotton production in the mid-south USA reported an indirect effect of tillage system on insect dynamics due to differences in crop phenology (Pettigrew and Jones, 2001). High residue cover alters soil and lowers crop canopy temperatures, resulting in different crop growth patterns, which may place the crop in slightly different development stage than the conventional comparison. No difference in insect pressure between tillage systems was shown for soybean in the US corn belt (Lam and Pedigo, 1998).

CA necessarily results in increased surface residue in the field at the time of crop seeding. This residue provides habitat for both pests and beneficial insects. Especially in the early years of NT cropping systems, increased pest pressure was reported compared with ConvT (Gregory and Musick, 1976). This necessitated adaptation and innovation of integrated pest management (IPM) approaches. In particular, seedling and early season pests must be carefully monitored. High residue IPM systems have been employed in most areas of the USA and include killing the cover crop well ahead of planting, seed treatment or at-planting insecticides, and early season scouting and treatment options.

Previous crop residue of a similar crop may harbour plant disease pathogens that can then readily infest the cash crop. For example, foliar diseases of wheat such as tan spot (*Pyrenophora tritici-repentis*) typically increase in severity when wheat stubble is present (Bockus and Claassen, 1992). In the Pacific Northwest increased disease pressure from *Rhizoctonia solani* and *Pythium* species have been documented in reduced tillage (Cook *et al.*, 2002). Similarly, grey leaf spot (*Cercospora zeae-maydis*) in maize is typically more troublesome when significant maize residue remains in the field from a previous crop (Payne *et al.*, 1987). Also, some very damaging diseases result from pathogens harboured by a different previous crop. Probably, the most important disease in row crops affected by tillage is fusarium head blight (*Fusarium graminearum*), which is hosted by previous maize crop residue and can infect the spikes of wheat and barley.

Understanding crop rotation effects and the value of diverse crops in rotation is important in dealing with diseases carried over by crop residue. For example, small grain cover crops are often used in pumpkin production to reduce the level of powdery mildew (*Podosphaera xanthii*) infestation (Everts, 2002), and in the suppression of phytophthora blight (*Phytophthora capsici*) in bell peppers (Ristaino *et al.*, 1997). A greater understanding of the multiple ways in which synergy in disease control and prevention can be achieved via cover crops and rotations is still needed to fully capitalize on these benefits.

2.2.7 Input use efficiency

Labour, fuel and machinery costs are typically reduced with a reduction in the intensity and frequency of tillage (Uri, 1998; Zentner *et al.*, 2002). These authors also report increases in profitability of farm operations from more diversified crop rotations and higher cropping intensity due to higher value crop products and improved land productivity. In cases where yields are increased with CA systems, such as those where water-holding capacity is increased, increases in nutrient and water use efficiency are also generally noted (Roygard *et al.*, 2002). Similarly, the impacts of diverse cover crops in rotation can increase cash crop yields in maize and soybean (Davis *et al.*, 2012). Including legume cover crops can increase available nitrogen for the following cash crop, thereby reducing the need for supplemental N fertilizer (Bruulsema and Christie, 1987). In the mid-Atlantic USA, Clark *et al.* (1994) have demonstrated that hairy vetch (*Vicia vilosa* L.) can typically supply the equivalent of 150 kg N ha^{-1} to maize the following summer. Similar reports exist for the mid-south (Blevins *et al.*, 1990) and corn belt regions (Power *et al.*, 1991) of the USA (Fig. 2.7).

These increases in input use efficiency must generally be accompanied by an increase in the level of management, however (Davis *et al.*, 2012). Seeding cover crops in a timely manner requires an adequate labour and machinery component on the farm, but is essential for achieving optimum benefits from cover crops. Similarly, timing cover-crop termination to achieve all desired benefits but early enough to avoid soil moisture depletion or problems with cash crop seeding takes experience. Many farmers have struggled in the early years of CA crop production. This has been attributed to a number of reasons but adequate equipment (Epplin *et al.*, 1982) and the changing dynamics of the soil–plant system (Blevins *et al.*, 1977) are generally reported to be the major hurdles. These factors can be overcome, as evidenced by the widespread adoption of CA farming techniques in many areas. The opportunity to increase intensive use of the land with CA is evidenced by the increase in double-crop soybean systems in the southeast USA, where the need for timely planting after small grain harvest has stimulated growers to adopt NT seeding (Camper *et al.*, 1972).

2.2.8 Economic returns

Many authors suggest reasons for increased profitability for NT over ConvT cropping

Fig. 2.7. Maize planted into a mulch of hair vetch cover crop.

systems including reduced equipment, fuel and labour costs. In the mid-Atlantic USA NT had a gross margin of over US$200 ha^{-1} and carried less risk than the reduced tillage system (Lu *et al.*, 1999). No-till production is often reported to be more profitable than the comparable, full-tillage system in the USA. In the mid-Atlantic region Cavigelli *et al.* (2009) reported that net returns for individual crops and rotations in 2- and 3-year crop rotations were greater and risks were lower for NT than for ConvT. A recent study conducted in the US corn belt comparing maize–soybean with rotations including small grains and forage crops found similar profitability among the systems with fewer negative effects on the broader ecosystem with the more diverse cropping systems (Davis *et al.*, 2012). Other authors have argued that these ecosystem services are undervalued and that CA-based cropping systems are much more valuable than conventional systems to society overall (Lyson and Welsh, 1993).

2.3 Problems Encountered in Scaling-up Conservation Agriculture in the USA

2.3.1 Residue management and supply

The benefits of CA for soil and water conservation and soil improvement are primarily due to high crop residue cover as shown above. In much of the USA, specialization has caused an uncoupling of grain and livestock production; and crop residue is usually left in the field instead of it being used for livestock feed or bedding. Although from a soil management point of view this is a positive development, high amounts of residue may pose a challenge at harvest and planting time. This is a special problem of maize because it produces a lot of crop residue. Residue needs to be evenly distributed over the entire width of the combine head (Allmans *et al.*, 1985; Smith *et al.*, 2000; Smith, 2008). Eleven-metre-wide small grain/soybean combine headers or 13.5-m-wide maize combine headers are not uncommon today, and very powerful residue spreaders are needed to spread the residue over this width. A better solution would be to just strip the grain from the plant and leave the residue in place. Industry is developing solutions to these problems. Strip headers are available for small grains that only strip the grain from the stalk, or maize headers that snap the cob from the plant (Neale *et al.*, 1987). Combine headers can also process maize stalks to speed up their decomposition. This can range from crimping the maize stalks to completely shredding them with knives right at the combine header (Wehrspann, 2010). Residue shredders are available on the back of combines to create fine particles from residue that has passed through the combine so that it decomposes more quickly (Parsons, 1995).

After the farmer has made sure residue has been uniformly distributed, planting equipment needs to place the seeds at the proper depth and spacing and close the seed slot (Jasa, 2000). Planting equipment (Fig. 2.8) is continuously being improved better to deal with high residue amounts (Morrison *et al.*, 1988). 'Hairpinning' is a problem when planting through heavy residue that may be somewhat moist: instead of cutting the residue the residue is stuffed into the seed-slot and poor seed-to-soil contact ensues. This causes deficiencies in plant populations and poor seedling development because decomposing residue may release damaging compounds to the developing seedling. Therefore, many different models of residue cleaners are available to move residue from the seed row to allow proper placement of seed. Coulters, which may be smooth, fluted, or rippled (bubble coulters are not recommended for NT), have been a standard in the industry to cut through crop residue and loosen soil so that following double-disc openers can open a slot for optimal seed placement (Jasa, 2000).

Fig. 2.8. Set-up of planter to plant maize or soybean with no coulter, residue cleaners, off-set double disc openers, and seed slot closing mechanism.

However, some believe coulters are an inheritance from the past when planters and drills were developed for tilled soil. Increasingly, planters and drills that have no coulters in front of the seed opener discs are becoming popular. With heavier designs and better quality steel, the single or (sometimes slightly offset) double disc openers can cut through residue and provide optimal seed placement. As surface soil tilth improves over time in CA, problems with smearing and packing the soil with the openers become less. Seed firmers have recently become popular in CA; they gently push the seed into the bottom of the seed slot to guarantee uniformity in seed depth placement. After the seed has been placed in the seed slot, closing wheels facilitate good seed-to-soil contact and a closed seed slot. Closing wheels can be solid steel or rubber, fingered closing wheels that fracture the wall of the seed slot, or 'posy close' wheels, which provide less dense soil on top of the seed, or can be a combination of small concave discs that push soil into the seed slot followed by a wide packing wheel. Sometimes chains follow the closing wheel to break up any clods created by coulters or closing wheels (Parsons, 1995; Jasa, 2000).

Despite concerns with too much residue, there looms a larger threat of excessive residue removal. In 2005, the USDA and USDOE released the 'Billion ton' study, which was updated in 2011 (Perlack and Stokes, 2011). This study suggests that 400 Mt of crop residues could be harvested to produce cellulosic biofuel. The authors assume that with the use of NT and cover crops soil quality can be maintained, but there remains much concern that residue removal would compromise the functionality of CA (Johnson *et al.*, 2006; Wilhelm *et al.*, 2008).

2.3.2 Tillage mindset and skills of farmers

Tillage has traditionally been used to prepare the soil for planting, to eliminate weeds and previous vegetation and to control insect-pests and diseases. To eliminate tillage requires a different mindset for the farmer and support personnel (researchers, extension agents, industry staff, crop advisors, etc.). Often, knowledge and practice of NT is more common among younger farmers, less steeped in tradition (Vitale *et al.*, 2011). Chemicals were often looked upon to replace tillage, creating concern about a pesticide treadmill in NT, although a survey of Midwest production practices did not show increased pesticide use in NT, and quality-adjusted herbicide use in soybean did not increase with the adoption of conservation tillage (Fuglie, 1999; Fernandez-Cornejo *et al.*, 2012). None the less, leading NT farmers and advisors realize that an ecological approach to CA is necessary. Diverse crop rotations and cover crops need to be considered an integral part of CA or the disasters of 'one crop agriculture' may be repeated. By not abiding by these principles, farmers are faced with serious problems such as herbicide-resistant weeds and problem insects and diseases that may threaten the future of CA (Shaw *et al.*, 2012). An example in case is the development of glyphosate-resistant horseweed (*Conyza canadensis*), which started in monocultures of roundup-ready soybeans in which only glyphosate was used for weed control (VanGessel, 2001). Monotonous crop rotations such as maize–soybean are another example – in parts of the upper Mid-West western corn rootworm (*Diabrotica virgifera virgifera*) has evolved that can survive 1 year of soybean, necessitating control of this pest in maize (Cullen *et al.*, 2008). No-till adoption has stalled in parts of the Great Plains in continuous wheat systems due to weed, pest and disease problems (Vitale *et al.*, 2011). A new emphasis on crop diversity is therefore imperative. In combination with cultural practices such as high residue cover, fertilizer placement and cover crops, herbicide use can be reduced as much as 50% and resistance development avoided (Anderson, 2008; Beckie, 2011; Vencill *et al.*, 2012). Crop rotation is also one of the most effective means of controlling crop insect-pests and diseases (Curl, 1963). Further, ecological soil management (for example cover cropping) can help develop disease-suppressive soils (Mazzola, 2002). It is clear that dealing with ecological practices such as diverse crop rotation and cover crops increases the level of skill required to manage an operation. Farmers will have to acquire this knowledge either themselves or they need to rely on crop consultants to assist them. The CCA (Certified Crop Adviser) programme in the USA has been instrumental in guaranteeing quality, independent advisors for farmers (Petersen, 1999). Whether farmers develop their own crop management plan or use crop advisors, the Land-Grant University cooperative extension system will be crucial to keep CA on the cutting edge in the USA.

2.3.3 Weed infestation

No-tillage has been reported to decrease the prevalence of annual weeds over time (Davis *et al.*, 2012) especially those, like crabgrass (*Digitaria sanginalis*), that favour disturbed environments (Doub *et al.*, 1988). However, perennial weeds, especially deep-rooted perennials, often increase in prevalence in NT fields (Koskinen and McWhorter, 1986).

Many NT cropping systems rely on selective herbicides or on genetically modified crops that are tolerant to non-selective herbicides in order to achieve acceptable weed control. In the latter case, it is often the same non-selective herbicide that would be used, alone or in combination with other chemicals, to kill a cover crop. Extensive use of some common herbicides such as atrazine and glyphosate has resulted in significant selection pressure on weed populations (Vencill *et al.*, 2012). This, along with the ability of some weeds to metabolize chemicals or to mutate rapidly to resistant biotypes, has resulted in a significant increase in herbicide-tolerant and -resistant weeds in recent years. In a recent summary of this situation, Moss (2002) reports that the number of reported resistant species has increased in recent years and that using herbicides with multiple modes of action, using diverse crop rotations and avoiding continuous cropping with the same species are all useful in reducing the impact of resistant weeds.

Cover crops have been reported to prevent or slow the emergence of weed seedlings (Hartwig and Ammon, 2002). These authors also state that cover crops and living mulch can help fight the constant battle of an ever-changing weed spectrum. Vigorous cover crops control weeds while growing by competing with weeds, and after termination by providing heavy mulch cover (Teasdale *et al.*, 1991; Anderson, 2005). In addition, cover crops have been shown to provide beneficial habitat for insects that practise herbivory on weeds (Hartwig and Ammon, 2002).

2.3.4 Yield reduction

Researchers and practitioners have often reported reduced yield in the early years of adoption of CA systems (Carter and Barnett, 1987; Edwards *et al.*, 1988). Some have attributed this to the learning curve associated with managing cover crops (Wagger, 1989) or to lack of experience for planting in NT conditions. The presence of large quantities of residue on the soil surface has been shown to immobilize greater amounts of N fertilizer when compared to systems with little crop or cover-crop residue (Wagger, 1989). The solution often proposed to solve this dilemma has been to increase total N application rate to make up for that which is immobilized or to use cover-crop species or mixtures that will provide N as well as C (Wagger, 1989). Residue with high C to N ratios will tend to immobilize soil N, while those with lower C to N ratios tend to result in net N mineralization (Havlin *et al.*, 1999). Yield reductions are more frequently associated with NT on some soil types, notably those with high clay content (Cosper, 1983). Also, several reports indicate that maize yield response to NT is often negative in northern areas of the USA (DeFelice *et al.*, 2006). This is generally attributed to cooler and wetter soils associated with high surface residues that reflect heat and hold moisture (DeFelice *et al.*, 2006). In general, however, crop yields have been similar under ConvT and NT systems; and higher yields are recorded in NT in areas subject to frequent in-season drought stress (Roygard *et al.*, 2002).

2.3.5 Insect-pest and disease problems

Increased surface residue, either from cash or cover crops, often results in a favourable environment for certain plant insect-pests and diseases. Dick and Van Doren (1985) report on increased severity of phytophthora root rot in continuous NT soybean, but that when cultivars with greater resistance were used, yields were similar to ConvT. When changes are made to the crop production systems, often a different set of problems emerges but in general these challenges are not impossible to overcome as exemplified by the use of disease-resistant cultivars in the first example. Some other pests common in high residue systems are more difficult to manage. Slugs, for example, often use residue as shelter in crop fields from which they emerge under favourable conditions and feed on young crops (Hammond *et al.*, 1999). While there are effective molluscicides available, they are less than ideal in terms of ease of application and in concern over toxicity to other organisms. An integrated management approach combining crop diversity, cover crops and other cultural practices with continuous NT, will be necessary to successfully manage pests and diseases in CA (Anderson, 2008).

2.4 Efforts/Policies Required for Scaling-up Conservation Agriculture in the USA

The demands for food, feed, fibre and fuel are expected to continue to increase rapidly in the coming decades, but it will be increasingly difficult to meet these challenges without undesirable impacts on the natural environment (Government Office of Science, 2010). CA systems are a way to meet the need for increased crop yields while sustaining the resource base (Kassam *et al.*, 2010). In addition to high-yield agriculture, CA allows sustainable cropping intensification and sustainable use of marginal soils such as highly erodible and droughty soils (Meyer *et al.*, 1999; Rhoton *et al.*, 2002). CA is based on ecological principles, many of which still need elucidation, and yet it is already leading to innovations in crop production. For example, water savings with NT and discovery of the inefficiency of the bare fallow to save water have led to cropping diversification and intensification in the Great Plains region (Farahani *et al.*, 1998). Double or even triple cropping in forage-based rotations in Pennsylvania has been made possible, something thought impossible before (Fouli *et al.*, 2012). Cover-crop mixtures are being investigated (Brennan *et al.*, 2011; Wortman *et al.*, 2012a). The integration of crops with livestock also needs to be reconsidered; the move to increased specialization has led to a divorce between livestock and crops, but this has led to many undesirable externalities such as monoculture, concentration of manure in certain areas, weed resistance, disease issues etc. The use of cover crops for forage in CA systems may open up new opportunities for crop–livestock integration (Franzluebbers, 2007; Maughan *et al.*, 2009). The use of crop residue to generate biofuel while maintaining functionality of the CA system urgently needs to be investigated; cover crop options need to be developed (Johnson *et al.*, 2006).

Machinery for CA is a special need, which has been neglected for too long (Erbach *et al.*, 1983; Morrison, 2002). In many cases, engineers have divulged responsibility for agricultural equipment to equipment companies or innovative farmers. However, completely new ideas have to be explored. Currently available seeding machinery needs to be re-evaluated (Chen *et al.*, 2004). It is mostly a result of gradual adaptation of equipment suited for tilled soil, and it would be beneficial to consider dramatically different designs with improved performance in high-residue systems (Baker *et al.*, 1996). Equipment to establish crop mixtures and relay-crops need to be developed (ASA, 2012). Fertilizer, manure and pesticide handling and application machinery needs to be developed to accommodate the needs of CA, such as either fertilizer and manure treatment for surface application or injection with minimal residue and soil disturbance (Singer *et al.*, 2008). Machinery needs to be considered that can harvest multiple crops to accommodate diverse crop rotations and crop mixtures. Harvesting machinery to collect biomass while leaving enough residue also needs to be developed (Siemens and Hulick, 2008). There is also a need for new weed, insect-pest and disease control options, and a better understanding of their ecology (such as life cycle, the effect of crop rotations, how to manage natural enemies).

These new opportunities for cropping system innovations need solid support from government at the national, state and local level. Support is needed for research as well as education to teach students about the fundamentals, opportunities and newest findings in CA. The Cooperative Extension Service of the Land-Grant Universities needs to receive more vigorous support to facilitate adoption and improvement of CA. On the other hand, policies that favour tillage, monoculture and separation of crops and livestock need to be discontinued. CA is knowledge-intensive and agronomy educators, working in teams with state specialists, have been at the foundation of CA revolution in the USA. We have observed that, where these educators have good support from specialists and administrators, and where they have good relationships with other important players in the field such as NT alliances, Conservation Districts, USDA-NRCS, agricultural businesses and leading CA farmers, adoption of CA and its continued adaptation to changing circumstances has been very rapid. If, on the other hand, these actors all work on diverging agendas CA adoption has been low. Information dissemination on CA is still highly reliant on field demonstrations (Fig. 2.9), newsletters, fact sheets, winter meetings, CA conferences and newspaper and professional journal articles. Although use of new technology such as the Internet and mobile phone networks also needs to be explored, we feel that the proven ways that have shown their impact should not be ignored.

2.5 Concluding Remarks

US crop production was characterized for almost 300 years after colonization by 'one crop agriculture' and intensive tillage, resulting in enormous soil degradation and human suffering. When visionary scientists, practitioners and policy makers realized the very future of the society was in danger, the conservation movement was born. Modern NT farming, originating in the USA, was a result of this movement. It was made possible by a combination of human ingenuity and persistence and the joining of hands of Land-Grant University agronomists, agro-industry colleagues, federal, state and county government employees and innovative farmers. No-tillage is increasing in popularity, but it is now recognized that it needs to be complemented by diverse crop rotations and use of cover crops during fallow periods in the crop rotation, as encompassed in the concept of CA. If this is not taken seriously, insurmountable problems may arise that may endanger the future of NT in the USA. CA enables meeting future production needs by increasing cropping

Fig. 2.9. Field days, a meeting point of cooperative extension specialists and educators, farmers, agribusiness and government personnel, continue to be fundamental for continued Conservation Agriculture adoption and improvement.

intensification and sustainable use of marginal soils such as highly erodible lands. But its future can only to be safeguarded by vigorous government support for the research, extension and education system, which underlays the development and expansion of CA systems in the USA. This support is only guaranteed if the general public is also well-informed about CA and its underlying principles.

References

Abawi, G.S. and Widmer, T.L. (2000) Impact of soil health management practices on soilborne pathogens, nematodes and root diseases of vegetable crops. *Applied Soil Ecology* 15, 37–47.

Akemo, M.C., Regnier, E.E. and Bennett, M.A. (2000) Weed suppression in spring-sown rye (*Secale cereale*)-pea (*Pisum sativum*) cover crop mixes. *Weed Technology* 14, 545–549.

Allmans, R.R., Douglas, C.L., Jr., Rasmussen, P.E. and Baarstad, L.L. (1985) Distribution of small grain residue produced by combines. *Agronomy Journal* 77, 730–734.

Anderson, R.L. (2005) A multi-tactic approach to manage weed population dynamics in crop rotations. *Agronomy Journal* 97, 1579–1583.

Anderson, R.L. (2008) Diversity and no-till: keys for pest management in the US great plains. *Weed Science* 56, 141–145.

ASA (2012) Machine gives winter cover crops a summer jumpstart [Online]. Available at: https://www.agronomy.org/story/2012/oct/fri/machine-gives-winter-cover-crops-a-summer-jumpstart (accessed 6 February 2013).

Baker, C.J., Saxton, K.E. and Ritchie, W.R. (1996) *No-Tillage Seeding: science and practice*. CAB International, Wallingford, UK.

Baker, J.M., Ochsner, T.E., Venterea, R.T. and Griffis, T.J. (2007) Tillage and soil carbon sequestration – what do we really know? *Agriculture Ecosystems and Environment* 118, 1–5.

Bauer, P.J., Fortnum, B.A. and Frederick, J.R. (2010) Cotton responses to tillage and rotation during the turn of the century drought. *Agronomy Journal* 102, 1145–1148.

Baumhardt, R.L. and Jones, O.R. (2002a) Residue management and paratillage effects on some soil properties and rain infiltration. *Soil and Tillage Research* 65, 19–27.

Baumhardt, R.L. and Jones, O.R. (2002b) Residue management and tillage effects on soil-water storage and grain yield of dryland wheat and sorghum for a clay loam in Texas. *Soil and Tillage Research* 68, 71–82.

Bayer, C., Gomes, J., Vieira, F.C.B., Zanatta, J.A., Piccolo, M.D.C. and Dieckow, J. (2012) Methane emission from soil under long-term no-till cropping systems. *Soil and Tillage Research* 124, 1–7.

Beckie, H.J. (2011) Herbicide-resistant weed management: focus on glyphosate. *Pest Management Science* 67, 1037–1048.

Blevins, R.L., Thomas, G.W. and Cornelius, P.L. (1977) Influence of no-tillage and nitrogen fertilization on certain soil properties after 5 years of continuous corn. *Agronomy Journal* 69, 383–386.

Blevins, R.L., Herbek, J.H. and Frye, W.W. (1990) Legume cover crops as a nitrogen source for no-till corn and grain sorghum. *Agronomy Journal* 82, 769–772.

Bockus, W.W. and Claassen, M.M. (1992) Effects of crop rotation and residue management practices on severity of tan spot of winter wheat. *Plant Disease* 79, 633–636.

Bordovsky, J. P., Lyle, W.M. and Keeling, J.W. (1994) Crop rotation and tillage effects on soil water and cotton yield. *Agronomy Journal* 86, 1–6.

Brennan, E.B., Boyd, N.S., Smith, R.F. and Foster, P. (2011) Comparison of rye and legume-rye cover crop mixtures for vegetable production in California. *Agronomy Journal* 103, 449–463.

Bruulsema, T.W. and Christie, B.R. (1987) Nitrogen contribution to succeeding corn from alfalfa and red clover. *Agronomy Journal* 79, 96–100.

Busscher, W.J., Sojka, R.E. and Doty, C.W. (1986) Residual effects of tillage on coastal plain soil strength. *Soil Science* 141, 144–148.

Buyanovsky, G.A., Brown, J.R. and Wagner, G.H. (1997) Sanborn field: Effect of 100 years of cropping on soil parameters influencing productivity. In: Paul, E.A., Elliott, E.T., Paustian, K. and Cole, C.V. (eds) *Soil Organic Matter in Temperate Agroecosystems: Long-term experiments in North America.* CRC Press, Boca Raton, Florida, pp. 73–83.

Cambardella, C.A., Moorman, T.B. and Singer, J.W. (2010) Soil nitrogen response to coupling cover crops with manure injection. *Nutrient Cycling in Agroecosystems* 87, 383–393.

Camp, C.R., Christenbury, G.D. and Doty, C.W. (1984) Tillage effects on crop yield in coastal plain soils. *Transactions of the American Society of Agricultural Engineers* 27, 1729–1733.
Camper, H.M., Genter, C.F. and Loope, K.E. (1972) Double cropping following winter barley harvest in eastern Virginia. *Agronomy Journal* 64, 1–3.
Carter, P.R. and Barnett, K.H. (1987) Corn-hybrid performance under conventional and no-tillage systems after thinning. *Agronomy Journal* 79, 919–926.
Cavigelli, M.A., Hima, B.L., Hanson, J.C., Teasdale, J.R., Conklin, A.E. and Lu, Y.C. (2009) Long-term economic performance of organic and conventional field crops in the mid-Atlantic region. *Renewable Agriculture and Food Systems* 24, 102–119.
Chatterjee, A. and Lal, R. (2009) On farm assessment of tillage impact on soil carbon and associated soil quality parameters. *Soil and Tillage Research* 104, 270–277.
Chen, Y., Tessier, S. and Irvine, B. (2004) Drill and crop performances as affected by different drill configurations for no-till seeding. *Soil and Tillage Research* 77, 147–155.
Clark, A. (2007) *Managing Cover Crops Profitably*, 3rd edn. Sustainable Agriculture Network, Beltsville, Maryland.
Clark, A.J., Decker, A.M. and Meisinger, J.J. (1994) Seeding rate and kill date effects on hairy vetch-cereal rye cover crop mixtures for corn production. *Agronomy Journal* 86, 1065–1070.
Cook, R.J., Weller, D.M., El-Banna, A.Y., Vakoch, D. and Zhang, H. (2002) Yield responses of direct-seeded wheat to rhizobacteria and fungicide seed treatments. *Plant Disease* 86, 780–784.
Cosper, H.R. (1983) Soil suitability for conservation tillage. *Journal of Soil and Water Conservation* 38, 152–155.
Craig, P.H. (1987) Effect of soil tillage on residual nitrogen contribution by a red clover green manure crop to subsequent corn crops. MSc thesis, The Pennsylvania State University, University Park, Pennsylvania.
CTIC (Conservation Technology Information Center) (2013) National crop residue management survey [online]. Available at: http://www.ctic.purdue.edu/crm_results (accessed 6 February 2013).
Cullen, E.M., Stute, J.K., Raymond, K.L. and Boyd, H.H. (2008) Farmers' perspectives on IPM field scouting during a period of insect pest range expansion: a case study of variant western corn rootworm (Coleoptera: Chrysomelidae) in Wisconsin. *American Entomologist* 54, 170–178.
Curl, E.A. (1963) Control of plant diseases by crop rotation. *Botanical Review* 29, 413–479.
Dabney, S.M., Wilson, G.V., Mcgregor, K.C. and Foster, G.R. (2004) History, residue, and tillage effects on erosion of loessial soil. *American Society of Agricultural Engineers* 47, 767–775.
Daniel, J.B., Abaye, A.O., Alley, M.M., Adcock, C.W. and Maitland, J.C. (1999) Winter annual cover crops in a Virginia no-till cotton production system: II. Cover crop and tillage effects on soil moisture, cotton yield, and cotton quality. *Journal of Cotton Science* 3, 84–94.
Darmody, R.G. and Peck, T.R. (1997) Soil organic carbon changes through time at the University of Illinois Morrow Plots. In: Paul, E.A., Elliott, E.T., Paustian, K. and Cole, C.V. (eds) *Soil Organic Matter in Temperate Agroecosystems*. CRC Press, Boca Raton, Florida, pp. 161–169.
Davis, A.S., Hill, J.D., Chase, C.A., Johanns, A.M. and Liebman, M. (2012) Increasing cropping system diversity balances productivity, profitability and environmental health. *PLoS ONE* 7(10): e47149. doi:10.1371/journal.pone.0047149
DeFelice, M.S., Carter, P.R. and Mitchell, S.B. (2006) Influence of tillage on corn and soybean yield in the United States and Canada. *Crop Management*. doi:10.1094/CM-2006-0626-01-R
Dell, C.J., Kleinman, P.J.A., Schmidt, J.P. and Beegle, D.B. (2012) Low-disturbance manure incorporation effects on ammonia and nitrate loss. *Journal of Environmental Quality* 41, 928–937.
Dick, W.A. and Van Doren, D.M. (1985) Continuous tillage and rotation combinations effects on corn, soybean, and oat yields. *Agronomy Journal* 77, 459–465.
Dick, W.A., Roseberg, R.J., Mccoy, E.L., Edwards, W.M. and Haghiri, F. (1989) Surface hydrologic response of soils to no-tillage. *Soil Science Society of America Journal* 53, 1520–1526.
Doub, J.P., Wilson, H.P., Hines, T.E. and Kriton, K.H. (1988) Consecutive annual applications of alachlor and metolachlor to continuous no-till corn (*Zea mays*). *Weed Science* 36, 340–344.
Drinkwater, L.E., Wagoner, P. and Sarrantonio, M. (1998) Legume-based cropping systems have reduced carbon and nitrogen losses. *Nature* 396, 262–265.
Drury, C.F., Tan, C., Welacky, T.W., Oloya, T.O., Hamill, A.S. and Weaver, S.E. (1999) Red clover and tillage influence on soil temperature, water content, and corn emergence. *Agronomy Journal* 91, 101–108.
Duiker, S.W. and Beegle, D.B. (2006) Soil fertility distributions in long-term no-till, chisel/disk and moldboard plow/disk systems. *Soil and Tillage Research* 88, 30–41.

Duiker, S.W. and Lal, R. (1999) Crop residue and tillage effects on carbon sequestration in a Luvisol in central Ohio. *Soil and Tillage Research* 52, 73–81.

Duiker, S.W. and Myers, J.C. (2006) *Steps Toward a Successful Transition to No-till.* The Pennsylvania State University, University Park, Pennsylvania.

Ebelhar, S.A., Frye, W.W. and Belvins, R.L. (1984) Nitrogen from legume cover crops for no-tillage corn. *Agronomy Journal* 76, 51–55.

Edwards, C.A. and Lofty, J.R. (1982) The effect of direct drilling and minimal cultivation on earthworm populations. *Journal of Applied Ecology* 19, 723–734.

Edwards, J.H., Thurlow, D.L. and Eason, J.T. (1988) Influence of tillage and crop rotation on yields of corn, soybean, and wheat. *Agronomy Journal* 80, 76–80.

Edwards, W.M., Shipitalo, M.J., Owens, L.B. and Norton, L.D. (1990) Effect of *Lumbricus terrestris* L. burrows on hydrology of continuous no-till corn fields. *Geoderma* 46, 73–84.

Epplin, F.M., Tice, T.F., Baquet, A.E. and Handke, S.J. (1982) Impacts of reduced tillage on operating inputs and machinery requirements. *American Journal of Agricultural Economics* 64, 1039–1046.

Erbach, D.C., Morrison, J.E. and Wilkins, D.E. (1983) Equipment modification and innovation for conservation tillage. *Journal of Soil and Water Conservation* 38, 182–185.

Everts, K.L. (2002) Reduced fungicide applications and host resistance for managing three diseases in pumpkin grown on a no-till cover crop. *Plant Disease* 86, 1134–1141.

Farahani, H.J., Peterson, G.A. and Westfall, D.G. (1998) Dryland cropping intensification: A fundamental solution to efficient use of precipitation. *Advances in Agronomy* 64, 197–223.

Faulkner, E.H. (1943) *Plowman's Folly*. Grosset and Dunlap, New York.

Fernandez-Cornejo, J., Hallahan, C., Nehring, R., Wechsler, S. and Grube, A. (2012) Conservation tillage, herbicide use, and genetically engineered crops in the United States: the case of soybeans. *AgBioForum* 15, 231–241.

Fouli, Y., Duiker, S.W., Fritton, D.D., Hall, M.H., Watson, J.E. and Johnson, D.H. (2012) Double cropping effects on forage yield and the field water balance. *Agricultural Water Management* 115, 104–117.

Fox, R.H., Piekielek, W.P. and Macneal, K.E. (1996) Estimating ammonia volatilization losses from urea fertilizers using a simplified micrometeorological sampler. *Soil Science Society of America Journal* 60, 596–601.

Franzluebbers, A.J. (2007) Integrated crop-livestock systems in the southeastern USA. *Agronomy Journal* 99, 361–372.

Franzluebbers, A.J., Hons, F.M. and Zuberer, D.A. (1994) Seasonal changes in microbial biomass and mineralizable C and N in wheat management systems. *Soil Biology and Biochemistry* 26, 1469–1475.

Franzluebbers, A.J., Hons, F.M. and Zuberer, D.A. (1995) Tillage and crop effects on seasonal dynamics of soil CO2 evolution, water content, temperature, and bulk density. *Applied Soil Ecology* 2, 95–109.

Frey, S.D., Elliott, E.T. and Paustian, K. (1999) Bacterial and fungal abundance and biomass in conventional and no-tillage agroecosystems along two climatic gradients. *Soil Biology and Biochemistry* 31, 573–585.

Fuglie, K.O. (1999) Conservation tillage and pesticide use in the Cornbelt. *Journal of Agricultural and Applied Economics* 31, 133–147.

Gelfand, I., Snapp, S.S. and Robertson, G.P. (2010) Energy efficiency of conventional, organic, and alternative cropping systems for food and fuel at a site in the US Midwest. *Environmental Science and Technology* 44, 4006–4011.

Government Office of Science (2010) *The Future of Food and Farming: challenges and choices for global sustainability.* Government Office for Science, London.

Gregory, W.W. and Musick, G.J. (1976) Insect management in reduced tillage systems. *Bulletin of the ESA* 22, 302–304.

Grover, K.K. (2008) Long-term cropping systems effects on soil aggregate stability, corn grain yields, and yield stability. PhD thesis, The Pennsylvania State University, University Park, Pennsylvania.

Halvorson, A., Wienhold, B.J. and Black, A.L. (2002) Tillage, nitrogen, and cropping system effects on soil carbon sequestration. *Soil Science Society of America Journal* 66, 906–912.

Hammond, R.B., Beck, T., Smith, J.A., Amos, R., Barker, J., Moore, R., Siegrist, H., Slates, D. and Ward, B. (1999) Slugs in conservation tillage corn and soybeans in the eastern corn belt. *Journal of Entomological Science* 34, 467–478.

Hartwig, N.L. and Ammon, H.U. (2002) Cover crops and living mulches. *Weed Science* 50, 688–699.

Havlin, J.L., Beaton, J.D., Tisdale, S.L. and Nelson, W.L. (1999) *Soil Fertility and Fertilizers: an introduction to nutrient management*, 6th edn. Prentice-Hall, Upper Saddle River, New Jersey.

Haystead, L. and Fite, G.C. (1955) *The Agricultural Regions of the United States*. University of Oklahoma, Norman, Oklahoma.

Helms, D. (2010) Hugh Hammond Bennett and the creation of the Soil Conservation Service. *Journal of Soil and Water Conservation* 65, 37A–47A.

Hill, R.L. and Cruse, R.M. (1985) Tillage effects on bulk density and soil strength of two mollisols. *Soil Science Society of America Journal* 49, 1270–1273.

Holm, F.A. and Johnson, E.N. (2009) The history of herbicide use for weed management on the prairies. *Prairie Soils and Crops Journal* 2, 1–11.

Hurt, R.D. (1994) *American Agriculture: A brief history*. Iowa State University Press, Ames, Iowa.

Hyup, C. (1979) Socioeconomic aspects of no-tillage agriculture: a case study of farmers in Christian County, Kentucky. Agricultural Experiment Station. RS, No. 63. Lexington, Kentucky.

Jangid, K., Williams, M.A., Franzluebbers, A.J., Schmidt, T.M., Coleman, D.C. and Whitman, W.B. (2011) Land-use history has a stronger impact on soil microbial community composition than aboveground vegetation and soil properties. *Soil Biology and Biochemistry* 43, 2184–2193.

Jasa, P. (2000) No-till planting equipment. In: Reeder, R. (ed.) *Conservation Tillage Systems and Management*, 2nd edn. Midwest Plan Service, Ames, Iowa, pp. 201–216.

Johnson, J.M.F., Reicosky, D., Allmaras, R., Archer, D. and Wilhelm, W. (2006) A matter of balance: conservation and renewable energy. *Journal of Soil and Water Conservation (Ankeny)* 61, 120A–125A.

Johnson, M.D. and Lowery, B. (1985) Effect of three conservation tillage practices on soil temperature and thermal properties. *Soil Science Society of America Journal* 49, 1547–1552.

Karl, T.R., Melillo, J.M. and Peterson, T.C. (2009) *Global Climate Change Impacts in the United States*. Cambridge University Press, New York.

Kassam, A., Friedrich, T. and Derpsch, R. (2010) Conservation Agriculture in the 21st Century: a paradigm of sustainable agriculture. In: *Proceedings of the European Congress on Conservation Agriculture, Towards Agro-Environmental Climate and Energetic Sustainability*, Madrid, Spain, 4–7/10/2010. Asociación Española Agricultura de Conservación Suelos Vivos, Ministerio de Medio Ambiente y Medio Rural y Marino, Secretaria General Técnica, Centro de Publicaciones, pp. 19–68.

Kemper, W.D., Schneider, N.N. and Sinclair, T.R. (2011) No-till can increase earthworm populations and rooting depths. *Journal of Soil and Water Conservation* 66, 13A–17A.

Kern, J., Hellebrand, H.J., Gömmel, M., Ammon, C. and Berg, W. (2012) Effects of climatic factors and soil management on the methane flux in soils from annual and perennial energy crops. *Biology and Fertility of Soils* 48, 1–8.

Kessel, C.V., Venterea, R., Six, J., Adviento-Borbe, M.A., Linquist, B. and Groenigen, K.J.V. (2013) Climate, duration, and N placement determine N_2O emissions in reduced tillage systems: a meta-analysis. *Global Change Biology* 19, 33–44.

Kladivko, E.J. (2001) Tillage systems and soil ecology. *Soil and Tillage Research* 61, 61–76.

Kleinman, P., Srinivasan, M.S., Dell, C.J., Schmidt, J.P., Sharpley, A.N. and Bryant, R.B. (2008) Role of rainfall intensity and hydrology in nutrient transport via surface runoff. *Journal of Environmental Quality* 35, 1248–1259.

Koskinen, W.C. and McWhorter, C.G. (1986) Weed control in conservation tillage. *Journal of Soil and Water Conservation* 41, 365–370.

Lal, R., Kimble, J.M., Follett, R.F. and Cole, C.V. (1998) *The Potential of US Cropland to Sequester Carbon and Mitigate the Greenhouse Effect*. Sleeping Bear Press, Chelsea, Michigan.

Lal, R., Follett, R.F. and Kimble, J.M. (2003) Achieving soil carbon sequestration in the United States: a challenge to the policy makers. *Soil Science* 168, 827–845.

Lal, R., Delgado, J.A., Groffman, P.M., Millar, N., Dell, C. and Rotz, A. (2011) Management to mitigate and adapt to climate change. *Journal of Soil and Water Conservation* 66, 276–285.

Lam, W.-K.F. and Pedigo, L.P. (1998) Response of soybean insect communities to row width under crop-residue management systems. *Environmental Entomology* 27, 1069–1079.

Levin, A., Beegle, D.B. and Fox, R.H. (1987) Effect of tillage on residual nitrogen availability from alfalfa to succeeding corn crops. *Agronomy Journal* 79, 34–38.

Li, S., Lobb, D.A., Lindstrom, M.J. and Farenhorst, A. (2008) Patterns of water and tillage erosion on topographically complex landscapes in the North American Great Plains. *Journal of Soil and Water Conservation* 63, 37–46.

Lu, Y.-C., Watkins, B. and Teasdale, J. (1999) Economic analysis of sustainable agricultural cropping systems for mid-Atlantic states. *Journal of Sustainable Agriculture* 15, 77–93.

Lyon, D., Smith, J. and Fryrear, D. (2000) Wind erosion. In: Reeder, R. (ed.) *Conservation Tillage Systems and Management*, 2nd edn. MidWest Plan Services, Ames, Iowa, pp. 11–16.

Lyson, T.A. and Welsh, R. (1993) The production function, crop diversity, and the debate between conventional and sustainable agriculture. *Rural Sociology* 58, 424–439.

Mackay, A.D. and Kladivko, E.J. (1985) Earthworms and rate of breakdown of soybean and maize residues in soil. *Soil Biology and Biochemistry* 17, 851–857.

Magdoff, F. and Van Es, H. (2010) *Building Soils for Better Crops*, 3rd edn. Sustainable Agriculture Network, Burlington, Vermont.

Malcolm, S., Marshall, E., Aillery, M., Heisey, P., Lingston, M. and Day-Rubenstein, K. (2012) Agricultural adaptation to a changing climate: Economic and environmental implications vary by US region. *Economic Research Report* No 136, Economic Research Service, USDA, Washington, DC.

Maughan, M.W., Flores, J.P.C., Anghinoni, I., Bollero, G., Fernández, F.G. and Tracy, B.F. (2009) Soil quality and corn yield under crop-livestock integration in Illinois. *Agronomy Journal* 101, 1503–1510.

Mazzola, M. (2002) Mechanisms of natural soil suppressiveness to soilborne diseases. *Antonie van Leeuwenhoek* 81, 557–564.

McCalla, T.M., Army, T.J. and Whitfield, C.J. (1962) Stubble-mulch farming. *Journal of Soil and Water Conservation* 17, 204.

McSorley, R. and Gallaher, R.N. (1994) Effect of tillage and crop residue management on nematode densities on corn. *Journal of Nematology* 26, 669–674.

Meisinger, J.J. and Delgado, J.A. (2002) Principles for managing nitrogen leaching. *Journal of Soil and Water Conservation* 57, 485–498.

Mendoza, R.B., Franti, T.G., Doran, J.W., Powers, T.O. and Zanner, C.W. (2008) Tillage effects on soil quality indicators and nematode abundance in loessial soil under long-term no-till production. *Communications in Soil Science and Plant Analysis* 39, 2169–2190.

Meyer, L.D., Dabney, S.M., Murphree, C.E., Harmon, W.C. and Grissinger, E.H. (1999) Crop production systems to control erosion and reduce runoff from upland silty soils. *American Society of Agricultural Engineers* 42, 1645–1652.

Min, D.H., Islam, K.R., Vough, L.R. and Weil, R.R. (2003) Dairy manure effects on soil quality properties and carbon sequestration in alfalfa-orchardgrass systems. *Communications in Soil Science and Plant Analysis* 34, 781–799.

Morrison, J.E., Jr (2002) Development and future of conservation tillage in America. *Journal of Research and Applications in Agricultural Engineering* 47, 5–13.

Morrison, J.E., Jr, Allen, R.R., Wilkins, D.E., Powell, G.M., Grisso, R.D., Erbach, D.C., Herndon, L.P., Murray, D.L., Formanek, G.E., Pfost, D.L., Herron, M.M. and Baumert, D.J. (1988) Conservation planter, drill and air-type seeder selection guideline. *Applied Engineering in Agriculture* 4, 300–309.

Moss, S.R. (2002) Herbicide-resistant weeds. In: Naylor, R.E. (ed.) *Weed Management Handbook*. Wiley-Blackwell, New York, pp. 225–252.

Neale, M.A., Klinner, W.E. and Arnold, R.E. (1987) A new stripping header for combine harvesters. *Agricultural Engineer* 42, 9–14.

Olson, K.R., Fenton, T.E., Smeck, N.E., Hammer, R.D., Ransom, M.D., Zanner, C.W., Mcleese, R. and Sucik, M.T. (2005) Identification, mapping, classification, and interpretation of eroded Mollisols in the US Midwest. *Soil Survey Horizons* 46, 23–35.

Papiernik, S.K., Schumacher, T.E., Lobb, D.A., Lindstrom, M.J., Lieser, M.L., Eynard, A. and Schumacher, J.A. (2009) Soil properties and productivity as affected by topsoil movement within an eroded landform. *Soil and Tillage Research* 102, 67–77.

Parsons, S.D. (1995) Crop residue distributors for grain combines. Purdue University Cooperative extension, West Lafayette, Indiana.

Payne, G.A., Duncan, H.E. and Adkins, C.R. (1987) Influence of tillage on development of gray leaf spot and number of airborne conidia of *Cercospora zeae-maydis*. *Plant Disease* 71, 329–332.

Perlack, R.D. and Stokes, B.J. (2011) US billion-ton update: Biomass supply for a bioenergy and bioproducts industry. *ORNL/TM-2011/224*. Oak Ridge National Laboratory, Oak Ridge, Tennessee.

Petersen, P.M. (1999) ICCA meets the challenge for the future in agriculture. *Comunica* 4, 62–66.

Pettigrew, W.T. and Jones, M.A. (2001) Cotton growth under no-till production in the lower Mississippi river valley alluvial flood plain. *Agronomy Journal* 93, 1398–1404.

Phillips, R.E. and Phillips, S.H. (1984) *No-tillage Agriculture*. Van Nostrand Reinhold, New York.

Pimentel, D. (2009) Energy inputs in food crop production in developing and developed nations. *Energies* 2, 1–24.

Power, J.F., Doran, J.W. and Koerner, P.T. (1991) Hairy vetch as a winter cover crop for dryland corn production. *Journal of Production Agriculture* 4, 62–67.

Raczkowski, C.W., Reyes, M.R., Reddy, G.B., Busscher, W.J. and Bauer, P.J. (2009) Comparison of conventional and no-tillage corn and soybean production on runoff and erosion in the southeastern US Piedmont. *Journal of Soil and Water Conservation* 64, 53–60.

Rasmussen, P.E. and Smiley, R.W. (1997) Soil carbon and nitrogen change in long-term agricultural experiments at Pendleton, Oregon. In: Paul, E.A., Elliott, E.T., Paustian, K. and Cole, C.V. (eds) *Soil Organic Matter in Temperate Agroecosystems: Long-term experiments in North America*. CRC Press, Boca Raton, Florida, pp. 353–360.

Reicosky, D.C. and Forcella, F. (1998) Cover crop and soil quality interactions in agroecosystems. *Journal of Soil and Water Conservation* 53, 224–229.

Reicosky, D.C., Kemper, W.D., Langdale, G.W., Douglas, J.C.L. and Rasmussen, P.E. (1995) Soil organic matter changes resulting from tillage and biomass production. *Journal of Soil and Water Conservation* 50(3), 253–261.

Reicosky, D.C., Dugas, W.A. and Torbert, H.A. (1997) Tillage-induced soil carbon dioxide loss from different cropping systems. *Soil and Tillage Research* 41, 105–118.

Rhoton, F.E., Shipitalo, M.J. and Lindbo, D.L. (2002) Runoff and soil loss from midwestern and southeastern US silt loam soils as affected by tillage practice and soil organic matter content. *Soil and Tillage Research* 66, 1–11.

Ristaino, J.B., Parra, G. and Campbell, C.L. (1997) Suppression of Phytophthora blight in bell pepper by a no-till wheat cover crop. *Phytopathology* 87, 242–249.

Rochette, P., Angers, D.A., Chantigny, M.H. and Bertrand, N. (2009) Nitrous oxide emissions respond differently to no-till in a loam and heavy clay soil. *Soil Science Society of America Journal* 72, 1363–1359.

Roygard, J.K.F., Alley, M.M. and Khosla, R. (2002) No-till corn yields and water balance in the mid-Atlantic Coastal Plain. *Agronomy Journal* 94, 612–623.

Sainju, U.M., Stevens, W.B., Caesar-Tonthat, T. and Liebig, M.A. (2012) Soil greenhouse gas emissions affected by irrigation, tillage, crop rotation, and nitrogen fertilization. *Journal of Environmental Quality* 41, 1774–1786.

Schlegel, A.J., Dumler, T.J. and Thompson, C.R. (2002) Feasibility of four-year crop rotations in the central High Plains. *Agronomy Journal* 94, 509–517.

Sene, M., Vepraskas, M.J., Naderman, G.C. and Denton, H.P. (1985) Relationships of soil texture and structure to corn yield response to subsoiling. *Soil Science Society of America Journal* 49, 422–427.

Sharpley, A., Richards, P., Herron, S. and Baker, D. (2012) Case study comparison between litigated and voluntary nutrient management strategies. *Journal of Soil and Water Conservation* 67, 442–450.

Shaw, D.R., Culpepper, S., Owen, M., Price, A. and Wilson, R. (2012) Herbicide-resistant weeds threaten soil conservation gains: finding a balance for soil and farm sustainability. *CAST Issue Paper*. American Society of Agronomy, Ames, Iowa.

Shipitalo, M.J. and Edwards, W.M. (1998) Runoff and erosion control with conservation tillage and reduced-input practices on cropped watersheds. *Soil and Tillage Research* 46, 1–12.

Siemens, M.C. and Hulick, D.E. (2008) A new grain harvesting system for single-pass grain harvest, biomass collection, crop residue sizing, and grain segregation. *Transactions of the ASABE* 51, 1519–1527.

Singer, J.W., Cambardella, C.A. and Moorman, T.B. (2008) Enhancing nutrient cycling by coupling cover crops with manure injection. *Agronomy Journal* 100, 1735–1739.

Siri-Prieto, G., Reeves, D.W. and Raper, R.L. (2007) Tillage requirements for integrating winter-annual grazing in cotton production: plant water status and productivity. *Soil Science Society of America Journal* 71, 197–205.

Slaton, N.A., Norman, R.J. and Kelley, J. (2011) Winter wheat yield response to a urea amended with a urease inhibitor and fertilization time. *Crop Management* doi:10.1094/CM-2011-0126-01-RS.

Smith, J. (2008) *Distribution of crop residue a requirement for conservation tillage*. University of Nebraska Cooperative Extension, Lincoln, Nebraska.

Smith, J., Hofman, V. and Taylor, R. (2000) Residue management at harvest. In: Reeder, R. (ed.) *Conservation Tillage Systems and Management*, 2nd edn. Midwest Plan Service, Ames, Iowa.

Smith, R., Gross, K. and Robertson, G. (2008) Effects of crop diversity on agroecosystem function: Crop yield response. *Ecosystems* 11, 355–366.

Snapp, S.S., Swinton, S.M., Labarta, R., Mutch, D., Black, J.R., Leep, R., Nyiraneza, J. and O'Neil, K. (2005) Evaluating cover crops for benefits, costs and performance within cropping system niches. *Agronomy Journal* 97, 322–332.

Sprague, M.A. and Triplett, G.B. (1986) *No-Tillage and Surface-Tillage Agriculture: The tillage revolution*. John Wiley and Sons, New York.

Staver, K.W. and Brinsfield, R.B. (1994) The effect of erosion control practices on phosphorus transport from coastal plain agricultural watersheds. In: *Proceedings, Conference Towards a Sustainable Coastal Watershed: The Chesapeake Experiment. 1–3 June 1994*. Chesapeake Research Consortium, Norfolk, Virginia, pp. 215–225.

Stevenson, F.J. and Cole, M.A. (1999) *Cycles in Soils: Carbon, nitrogen, phosphorus, sulfur, micronutrients*. John Wiley and Sons, New York.

Syswerda, S.P., Corbin, A.T., Mokma, D.L., Kravchenko, A.N. and Robertson, G.P. (2011) Agricultural management and soil carbon storage in surface vs. deep layers. *Soil Science Society of America Journal* 75, 92–101.

Teasdale, J.R., Beste, C.E. and Potts, W.E. (1991) Response of weeds to tillage and cover crop residue. *Weed Science* 39, 195–199.

Tilman, D., Lehman, C.L. and Bristow, C.E. (1998) Diversity-stability relationships: Statistical inevitability or ecological consequence? *American Naturalist* 151, 277–282.

Trimble, S.W. (1973) *A Geographic Analysis of Erosive Land Use on the Southern Piedmont*. University of Georgia, Athens, Georgia.

Unger, P.W. and Baumhardt, R.L. (2001) Historical development of conservation tillage in the southern Great Plains. In: Stiegler, J.H. (ed.) *Proceedings of the 24th Annual Southern Conservation Tillage Conference for Sustainable Agriculture*, Oklahoma City, 9–11 July 2001. Oklahoma State University, Oklahoma City, Oklahoma.

Uri, N.D. (1998) Conservation tillage and the use of energy and other inputs in US agriculture. *Energy Economics* 20, 389–410.

USDA-NASS (2013) Quick Stats [Online]. Available: http://www.nass.usda.gov/Quick_Stats/Lite/ (Accessed 6 February 2013).

Van Doren, D.M., Triplett, G.B. and Henry, J.E. (1976) Influence of long term tillage, crop rotation, and soil type combinations on corn yield. *Soil Science Society of America Journal* 40, 100–105.

Van Es, J.C. and Notier, P. (1988) No-till farming in the United States: research and policy environment in the development and utilization of an innovation. *Society and Natural Resources* 1, 93–107.

VanGessel, M.J. (2001) Glyphosate-resistant horseweed from Delaware. *Weed Science* 49, 703–705.

Varco, J.J., Frye, W.W., Smith, M.S. and Mackown, C.T. (1989) Tillage effects on nitrogen recovery by corn from a nitrogen-15 labeled legume cover crop. *Soil Science Society of America Journal* 53, 822–827.

Vencill, W.K., Nichols, R.L., Webster, T.M., Soteres, J.K., Mallory-Smith, C., Burgos, N.R., Johnson, W.G. and Mcclelland, M.R. (2012) Herbicide resistance: toward an understanding of resistance development and the impact of herbicide-resistant crops. *Weed Science* 60, 2–30.

Verbree, D.A., Duiker, S.W. and Kleinman, P.J.A. (2010) Runoff losses of sediment and phosphorus from no-till and cultivated soils receiving dairy manure. *Journal of Environmental Quality* 39, 1762–1770.

Veseth, R. (1985) Erosion impacts on the Palouse misunderstood. In: Kok, H. (ed.) *PNW Conservation Tillage Handbook*. University of Idaho, Moscow, Idaho. [Online]. Available at: http://pnwsteep.wsu.edu/tillagehandbook/index.htm (Accessed 15 February 2013).

Vitale, J.D., Godsey, C., Edwards, J. and Taylor, R. (2011) The adoption of conservation tillage practices in Oklahoma: findings from a producer survey. *Journal of Soil and Water Conservation (Ankeny)* 66, 250–264.

Vyn, T.J., Galic, D.M. and Janovicek, K.J. (2002) Corn response to potassium placement in conservation tillage. *Soil and Tillage Research* 67, 159–169.

Wagger, M.G. (1989) Cover crop management and nitrogen rate in relation to growth and yield of no-till corn. *Agronomy Journal* 81, 533–538.

Wagger, M.G. and Denton, H.P. (1992) Crop and tillage rotations: grain yield, residue cover, and soil water. *Soil Science Society of America Journal* 56, 1233–1237.

Wehrspann, J. (2010) 9 new stalk-chopping corn heads [Online]. Available at: http://farmindustrynews.com/combines/9-new-stalk-chopping-corn-heads (Accessed 15 February 2013).

West, T.O. and Post, W.M. (2002) Soil organic carbon sequestration rates by tillage and crop rotation: a global data analysis. *Soil Science Society of America Journal* 66, 1930–1946.

Wilhelm, W.W., Johnson, J.M.F., Karlen, D.L. and Lightle, D.T. (2008) Corn stover to sustain soil organic carbon further constrains biomass supply. *Agronomy Journal* 99, 1665–1667.

Wortman, S.E., Francis, C.A. and Lindquist, J.L. (2012a) Cover crop mixtures for the western corn belt: opportunities for increased productivity and stability. *Agronomy Journal* 104, 699–705.

Wortman, S.E., Francis, C.A. and Lindquist, J.L. (2012b) Cover crop mixtures for the western corn belt: Opportunities for increased productivity and stability. *Agronomy Journal* 104, 699–705.

Wyland, L.J., Jackson, L.E., Chaney, W.E., Klonsky, K., Koike, S.T. and Kimple, B. (1996) Winter cover crops in a vegetable cropping system: Impacts on nitrate leaching, soil water, crop yield, pests and management costs. *Agriculture, Ecosystems and Environment* 59, 1–17.

Zentner, R.P., Wall, D.D., Nagy, C.N., Smith, E.G., Young, D.L., Miller, P.R., Campbell, C.A., Mcconkey, B.G., Brandt, S.A., Lafond, G.P., Johnston, A.M. and Derksen, D.A. (2002) Economics of crop diversification and soil tillage opportunities in the Canadian prairies. *Agronomy Journal* 94, 216–230.

3 Conservation Agriculture in Brazil

Ademir Calegari,[1] Augusto Guilherme de Araújo,[1] Antonio Costa,[1] Rafael Fuentes Lanillo,[1] Ruy Casão Junior[1] and Danilo Rheinheimer dos Santos[2]

[1]Agricultural Research Institute of Paraná State – IAPAR, Londrina, Paraná; [2]Soil Science Department, University of Santa Maria, Rio Grande do Sul, Brazil

3.1 Introduction

In agricultural areas all over the world, land in tropical and subtropical regions in general is intensively cultivated and, in most cases, the soil and water management systems are not practised using an integrated and sustainable approach. In many farming systems where no orderly crop diversification, including cash crops and cover crops in a crop rotation system, are followed and a continuous soil disturbance cannot provide an adequate addition of organic carbon to the system, the organic matter decomposition processes are accelerated, which causes a severe decrease in the productive potential of the agricultural soil of these regions.

Current evidence shows that greenhouse effects and climatic changes result in alterations in the precipitation distributions and levels, tending in many regions to reduce the number of rain events and to increase the intensity, which certainly will lead to bigger risks of erosion and, in consequence, loss of soil particles, water and nutrients. These climatic change observations have been made mainly in the last 3–5 years in Brazil. To mitigate this, agricultural soils should be cultivated under soil-conserving agricultural systems. As a part of conservation agriculture (CA), the soil surface must be covered with crop residues, there should be minimum soil disturbance (no-tillage or direct drilling of the crops) and the soil profile should be receptive to water infiltration, including a harmonic integration of soil and water conservation methods including the use of cover crops/green manure, crop rotation and the implementation of soil and water conservation practices.

In Brazil, mainly in tropical and subtropical regions, water erosion has been considered the greatest environmental problem of the agricultural sector, and the execution of government programmes, having mechanical conservation practices as the main feature of these actions, were insufficient to control soil erosion. Sorrenson and Montoya (1989) reported that in Paraná State (south Brazil) average soil loss of 10–40 t ha^{-1} $year^{-1}$ of fertile soil has been observed under a traditional soil tillage system. These results created increased awareness of the problem and led growers to search for alternative ways to conserve soil and water resources.

To face this strong challenge, in the 1970s the no-till (NT) system (NTS) started in Brazil, and this important soil management system was followed by many challenges at the farmer, extension and researcher level. The expansion of NT area in Brazil occurred mainly due to the availability of no-till seeders adapted and developed with the support of research institutions and through farmers' evaluation.

Following the availability of a few NTS seeders to medium- and large-scale farmers, the results obtained by research and validation by farmers mainly with soybean, maize, wheat and cotton contributed to the shifting from conventional to the NTS in different parts of Brazil, mainly in south and later in the savannah region (cerrado).

Conversely, smallholders adopted the NTS and observed great reduction in soil erosion, workload, saved time, increased crop yields and they also diversified their activities, mainly through higher value adding through this system. This has had a positive impact on the improvement of the quality of life of smallholder farm families and also contributed positively to the diffusion and adoption of the NTS.

Following this, other successful results were achieved at the research and farm level, through the addition of organic matter to the soil, and keeping plant residues on the soil surface was an important measure to preserve and foster organic matter balance in the soil. Thus, plants used as cover crops, given their high capacity to produce biomass (shoot and roots), and direct and indirect positive effects on the soil–water–plant systems, play a fundamental role when they are part of the orderly rotation systems with profitable crops, and these results obtained also contributed to increased adoption of NTS in Brazil.

The positive results consistently obtained in the south (subtropical region) mainly in savannah areas of Brazil proved that cover crops and cropping rotations, comprising a NTS, are economically feasible as well as ecologically sustainable. NTS not only increased crop productivity but also conserved and maintained soil fertility, biologic balance in the soil, and decreased the incidence of insect-pests and/or diseases. In other words, it represents a promising strategy for sustainable fertility management. At the present time, NT occupies almost 6 Mha in Paraná and estimates show that NT covers more than 26 Mha in Brazil.

3.2 History and Development of No-till in Brazil

The European immigrants in the 1820s (German) and in 1870s (Italian) started to open areas for agricultural production in Rio Grande do Sul, Santa Catarina and Paraná states by employing the technological model that came from their original lands. This was based on the use of human power (hand jab planter), conventional animal traction machinery (plough and disc-harrowing), and disc ploughs and heavy harrows powered by tractors for the incorporation of crop biomass and for weed control. Such techniques were frequently preceded by residue burning for the purpose of reducing the volume of vegetative biomass and facilitating the use of machinery.

The rapid expansion of the agriculture area, which grew from 800,000 ha cultivated in 1969 to 4 Mha in 1977, based on the conventional system (ploughing and harrowing), caused soil erosion losses of up to 10 Mg ha^{-1} for each tonne of grain produced.

As a consequence, soil degradation in this region during the 1970s and mid-1980s compromised the gains in crop productivity resulting from the technological advances in plant genetics, effective and efficient use of chemical inputs, and improved machines (Amado and Eltz, 2003).

At the end of the 1960s following soil degradation and the need to open new lands for cultivation, the agricultural border in the extreme south expanded to new regions such as the western and eastern parts of the state of Paraná, and the same environmental problems occurred again. The exposure of soil to rain and its compaction by conventional management, which reduced water infiltration capacity, resulted in huge losses of the soil by water erosion.

From that time, farmers, technicians and researchers started to look for new crop establishment systems with reduced soil mechanical disturbance. The search suggested two possible routes. The first was the use of subsoil ploughs, mainly in the west of Paraná, and the other focused on the implementation of NT as a system with minimal soil disturbance.

However, at the beginning of the 1980s soil erosion in the western and northern regions of Paraná still constituted the main problem for many farmers. In order to alter this, some organizations (professional associations, farmers' cooperatives, research institutions, public universities, rural extension services, banks and others) came together and decided to constitute the first Municipal Soil Commission with the objective of convincing farmers to implement contour terraces, because up to this point soybean, the main crop, was cultivated predominantly at 90° to the contour.

During the last 50 years, a lack of planning in the colonization process of Paraná reduced the state's natural vegetative cover from 87% to 10%, bringing about serious consequences for the management of soil and water quality (Vieira, 1991). Soil erosion by water had the main and most visible destructive effect as a result of this inadequate management of natural resources. The destruction was a consequence of the degradation of the soil structure, which was exacerbated by the impact of raindrops, followed by runoff and along with it the transport and deposition of soil sediments.

In Paraná, the challenge really started when the conservation organizations began to articulate actions to integrate rural development programmes with soil management activities, including soil conservation. These programmes were promoted during the 1980s and 1990s by successive state governments with the support of international organizations. The actions were, to a great extent, defined by local entities (including farmers) and developed and executed with the financial support of the national programmes. Beyond the construction of civil facilities, actions also included the training and capacity development of technicians and farmers through lectures, field days, courses, regional and state meetings on soil management and the publication of technical manuals.

The severe environmental problems experienced in southern Brazil acted as the catalyst for scientific studies on improved soil management practices with a conservation focus. In 1969 at the experimental station of the Ministry of Agriculture in Não-Me-Toque, Rio Grande do Sul State, a pioneer plot of NT seeded sorghum was established on crop residues with the use of 'Buffalo' North American machinery. By 1973, there were experiments being conducted in Ponta Grossa, Paraná, with different soil preparation systems, including cultivation without soil inversion. The results of this work were published in 1974 and constitute the first research record on conservation soil management in Brazil. Still within the ambit of the Ministry of Agriculture in the early 1970s, experiments were conducted in the north of Paraná. These NT studies were only possible after the 1961 launch of the contact herbicide Paraquat by Imperial Chemical Industries (ICI). In 1971, ICI made the first demonstrations of NT in the north of Paraná with the soybean–wheat rotation and, in 1974, the work to implement the practice in the plateau of Rio Grande do Sul was started (Muzilli, 1981). In São Paulo State, the first experiments by ICI were established in 1976 in the Ribeirão Preto region. In 1977, a Technical Cooperation Project among IAPAR, Brazilian Government, and GTZ, German Government, coordinated by Rolf Derpsch and Soil Research Area of IAPAR, was conceived that initiated studies on testing different genotypes of cover-crop species from different parts of the world, crop rotation, NT and chisel plough, mainly in the north of Paraná State at the IAPAR Experimental Station, cooperative farms and on farmers' fields. These results and definitions of main cover-crop species validated in farm conditions formed important basic information to leverage the NT system in Brazil (Derpsch and Calegari, 1985; Derpsch *et al.*, 1991).

The increase of the area under NT in Brazil occurred in three distinct periods with respect to adoption rates. The first period was up to 1979, and was discussed in the previous section. Between 1980 and

1991, agricultural research had proven the effectiveness of NT for controlling soil erosion by water; soil losses are reduced by a factor of five when compared to conventional tillage practices. Furthermore, the main principles of NT were consolidated during this period, that is, minimal soil disturbance, permanent organic soil cover with crop residues or live plants, and crop rotations (Denardin *et al.*, 2008).

In this period, IAPAR increased its research on NT, including trials and validation of systems with farmers. Also during this time, the cover-crop technology studied by IAPAR began to spread all over Brazil, into other Latin-American countries and throughout the world. Adoption rates increased following the positive results achieved; and there were numerous meetings, talks, field days and demonstrations in agricultural shows, highlighting the advantages of the NT system from an economic perspective as it reduced the use of fertilizers and herbicides both in annual and perennial crops. However, many researchers still claimed that there was a need to have a transition phase of minimal cultivation before the implementation of the full NTS. This discussion lasted for almost a decade. Some would also say that a rotation was necessary in soil tillage, which was completely disproved by the results of studies made, so long as the NTS was properly implemented.

There was a huge effort by IAPAR and Agricultural Secretary of Paraná (SEAB-PR) to make the information on NTS widely available, which led to the publication of the following documents among others:

- *No-Till in Paraná State* (IAPAR, 1981);
- *Guide to herbicides and their appropriate use in no-till and conventional system* (Almeida and Rodrigues, 1985);
- *Winter cover crops guide* (Derspsch and Calegari, 1985);
- *Allelopathy and plants* (Almeida, 1988);
- *Winter green manure plants in southwestern Paraná* (Calegari, 1990);
- *Manual of the soil management and conservation sub-programme* (SEAB, 1994);
- *Summer cover crops in Paraná* (Calegari, 1995b).

However, the biggest obstacle for the expansion of NT, chemical weed control, still required appropriate technical solutions.

During the 1970s there were a few products such as Paraquat and Diquat (contact desiccant herbicides), Glyphosate (still little used due to its high cost) and a few soil-applied herbicides such as Atrazine, 2,4-D and Trifluralin. The biggest problem was the low efficiency of the herbicides when applied on straw-covered soil, as the post-emergence chemicals had been developed for application on soils without cover. Apart from that, in NT it was common to have weeds in different development stages and with deep roots at the time of control. In 1984, Glyphosate started to be produced in Brazil, resulting in reduction in the cost of the chemical. By that time, there was already a wide variety of crop rotations and many options of cover-crop plants available.

Permanent soil cover with straw started to be considered an important component for weed control, given that the use of herbicides still presented limitations. The increase in the amount of straw covering the soil began to be an important objective in NT.

However, this was sometimes difficult, mainly in warm regions, because while it was possible to achieve 7 t ha^{-1} $year^{-1}$ or more of straw in southern Paraná due to its mild climate, in the northern area no more than 2 t ha^{-1} $year^{-1}$ was achieved. At the time, a mixture of Glyphosate and 2,4-D was effective in desiccating the cover crops; however, with the legal prohibition of 2,4-D use in many municipalities of the country, the reduction in the price of Glyphosate and increasingly easy access to it, Glyphosate started to be used alone or in a mixture with post-emergence residual herbicides.

In the early 1980s, soil erosion was still a big challenge in Paraná, and that made agricultural organizations, such as agronomists' associations, extension service (EMATER), research service (IAPAR), agricultural cooperatives and the Brazilian Bank (Banco do Brasil), to work together and constitute the so called 'soil commissions' aimed at promoting the use of contour terraces, as it was still common to sow soybean 'downhill'.

Facing the challenge commenced with efforts by a regional organization through state government rural development programmes, which were strongly supported by the World Bank. Such programmes, which focused on soil management and conservation, were implemented all over Paraná in the 1980s and 1990s; their main strategies were to increase water infiltration into the soil profile and to reduce surface runoff.

In 1982, the PMISA (Soil and Water Integrated Management Programme) was implemented, with its focus on soil and water management in micro-catchments (micro-watershed) with emphasis on the integration of terraces between adjacent properties, chisel ploughing, soil acidity amendment, set-up of road route and gully erosion control. At that time, the municipalities of Maringá and Toledo, north-western and western Paraná, were the pioneers in these activities. From 1988 to 1993, soil management and conservation interventions were developed under the Programa Paraná Rural (Rural Development Programme of Paraná). The focus was on hydrographic catchments, though it also involved integrated actions of NT promotion, research, extension and farmers' organizations, among others. The strategic objectives, however, were the same, that is to avoid surface runoff and increase water infiltration. Programme assessments indicated high levels of adoption of NT by farmers. Similar programmes aimed at controlling severe natural resource degradation were started in Rio Grande do Sul. Outstanding examples included: the integrated project of soil use and conservation named PIUCS (1979); the Saraquá project on the basaltic hill slopes from 1980 onwards; and the hydrographic micro-catchment programme started in 1984. A series of conservation practices were disseminated, such as the elimination of wheat residue burning, reduction in soil disturbance frequency and intensity, terracing, contour planting, gully elimination, soil-cover evaluation, rural roads relocation, diffusion of minimal soil disturbance, minimum tillage and NT. However, in 1993, there were only 300,000 ha of NTS in Rio Grande do Sul, and several public and private entities decided to stimulate its expansion with emphasis on the development capacity of farmers and technicians. At that time, basic research was conducted by EMBRAPA, UFRGS, UFSM and other universities of Rio Grande do Sul State; these institutions were also involved in training and capacity building on different topics including soil liming, phosphate fertilization, small farms' NT machines, micronutrient application, and inoculation of crops with efficient microbial species including rhizobia. The training and capacity building involved the extension service (EMATER). A project named METAS was very active in promoting these activities to promote and develop NTS in Rio Grande do Sul; private seed, chemical, and fertilizer and machinery companies were also involved. Some agriculture cooperatives as well at that time were involved in farmers' capacity development.

In the early 1990s, only 13% of the farmers in Paraná had adopted NTS (EMATER, 1996), and several initiatives were aimed at increasing the area of adoption. One of these initiatives was by ITAIPU, Hydro-electrical Company, in the extreme west of the state, that has a reservoir with a flooded area of 1350 km^2. Concerned with minimizing the runoff and sedimentation from the conventional agricultural systems areas into their reservoir, ITAIPU diagnosed the situation with 280 farmers and decided to implement technological validation actions to improve NTS quality with IAPAR's support. Such activities were based on the identification of farmers as partners who were willing to test and validate new technologies together with a group of researchers following the discussions of the problems they faced in their farms. The technologies introduced in these areas were related to crop rotation, cover crops, input reduction, adequate NT planters and cover-crop seed production. Over a period of 5 years, a multidisciplinary team from IAPAR interacted directly with farmers, ITAIPU technicians, cooperatives, municipalities and EMATER, along with other regional agents on the board of ITAIPU. The team organized several courses, field days, meetings, working machine demonstrations, production and distribution of cover-crop seeds, among other activities to large numbers of farmers and farmers' associations as well. The main results of this work are reported in the book *No-till System with Quality*, published by IAPAR and ITAIPU in 2006 (Casão Junior *et al.*, 2006).

In another report, Casão Junior *et al.* (2012) highlighted the main factors involved in the evolution of both NTS and conservation farming mechanization in southern Brazil. The highlights of this study are as follows.

3.2.1 Soil erosion

The conservation concern in southern Brazil stemmed from the severe soil erosion problems, which motivated several initiatives by communities associated with farming to solve the problem. Farmers, governmental and international support programmes, research institutions, rural extension initiatives, universities, cooperatives, farmers' associations and agricultural industries all participated individually or collectively to overcome the big challenge.

3.2.2 Governmental integrated soil management programmes

In a tenuous way during the 1970s and more intensively in the 1980s, several integrated soil management and conservation programmes were implemented in southern Brazil. Some of them were financed by international organizations and, within a period of only a decade, played a fundamental role in shifting from conventional soil tillage, with ploughs and harrows, to minimum tillage with the use of chisel ploughs, which provided reduced soil disturbance and retained crop residues on the soil surface.

3.2.3 Pioneers' leadership in the 1980s

The pioneer farmers' leadership in the search for solutions and knowledge dissemination during the 1980s provoked ample discussions on NTS and motivated several initiatives, not only related to the adaptation of machines in regional commercial workshops, but also in the use of cover-crop species to provide permanent soil protection (IAPAR had contributed strongly on this issue), beyond initiatives in other aspects connected to NTS.

Simultaneously, the determination of the NTS pioneer machinery industries, based on research results and interactions with pioneer farmers, enabled the development of the first national NT planters, which were predominantly continuous flow seed drills, rather than precision planters.

The results of pioneer scientific institutions' efforts showed the viability of NTS. These were very often supported by international organizations and multinational companies interested in market expansion, mainly with new herbicides. This work was fundamental to the consolidation of NTS principles, especially concerning cover-crop species, crop rotations, and allelopathic, chemical weed control and soil fertility management. Such efforts provided the technical conditions for the evolution and the adoption of NTS in the region.

3.2.4 Beginning of no-till system expansion

The irreversible expansion in the adoption of NTS occurred after the mid-1980s due to a combination of factors, such as:

- The economic and energy crises at that time demanded that farmers looked for alternatives to reduce production costs and NTS met such a demand as it required fewer machine hours with significant fuel economy;
- The reduction in the price of the herbicide Glyphosate, which started to be produced in Brazil in 1985, along with more availability of other pre- and post-emergent herbicides in the market, making weed control easier;
- The availability of NTS technology for all the main annual crops;

- The market availability of several precision planter and seed drill models manufactured by agricultural machinery companies; several machine adaptation workshops were conducted that provided direct good experience in the use of farm machinery.

3.2.5 No-till system consolidation after 1993

The NTS cultivated area in Brazil expanded impressively from 1 Mha in 1992 to 25 Mha in 2007. This expansion was possible due to the availability of NT planters in the national market that were appropriate for the range of soil types in Brazil.

Several agricultural machinery manufacturers believed in the market expansion and invested heavily in the improvement of NT seeders. They also counted on the support of research institutions through their efforts in comparative testing of the commercial models.

After 1995, agricultural financing, mainly for investment, started to have lower and fixed interest rates, which resulted in an increase in machinery acquisition throughout the country. Agricultural machinery fairs proliferated, mainly the ones in which there were dynamic exhibitions of NT machines. In this way, such fairs turned into reference events for the launching of new machine models by the industry.

3.2.6 No-till in small farms

The government's family agriculture support policy in Paraná, implemented in the early 1980s, resulted in the development of animal-powered NT machines. In the 1990s, the main NTS technologies were validated in small farms and, in later years, were responsible for the wide adoption of the system, as well as for the appearance of small manual and animal-powered equipment manufacturers mainly in Santa Catarina and Rio Grande do Sul states, which enabled the change to NT production systems of small farms in southern Brazil.

The wide dissemination of the good results obtained by means of the technical events, the attractive agricultural investment financing facilities, the farmers' interest in changing the production system and the machinery, and the input industries' interest in expanding their market provided fertile ground for the adoption of NTS on small farms.

3.3 History and Advances in No-Till Mechanization in Brazil

At the beginning of the 1970s, attempts were made to work with minimum soil disturbance in Rio Grande do Sul (RS) and Paraná (PR), but the pioneer farmer Herbert Bartz from Rolândia (PR) was the first farmer to use NT with an imported Allis Chalmers seeder in 1972. A larger NT acceptance was achieved in Ponta Grossa (PR) region from 1976 with the leadership of Frank Dijkstra and Manoel Henrique Pereira (Mr Nonô Pereira), two other NT pioneers. As a result Earthworm Club, ABC Foundation, FEBRAPDP and CAAPAS were created and at that moment IAPAR and CNPT/EMBRAPA started a programme for systematic research on NT. In 1981, IAPAR published its first book (IAPAR, 1981), supported by ICI.

CNPT/EMBRAPA in Rio Grande do Sul focused on NT seeder furrower development and these studies became a stimulus for industries that start producing the first machines for seeding. They were inspired by the British drill seeding machinery model Bettinson-3D and in the Canadian staggered double-discs system produced by SEMEATO (Casão Junior *et al.*, 2012).

The only available commercial machinery in Brazil during the 1970s was ROTACASTER, which was characterized by its high soil disturbance and low operational efficiency. At the end of the decade, the market started to offer machinery for NT, mainly models from SEMEATO, IMASA, Fankhauser, Marchesan and Baldan.

The 1980s was a decade of research and new developments, but at that time there was no clear definition of all the desirable requirements that an NT seeder was expected to meet. Adaptation of machines was very common; and cutting discs and devices for furrow opening were introduced to the conventional seeders.

The main setbacks to NT expansion in the 1980s were the lack of efficient herbicides and adapted machines to work on clayey soils. With the developments in research and farmers' experiences, the principle of crop rotation was set and the use of cover crops started to be incorporated in the NT conception (Muzilli, 2006).

The development of the precision seeder model PAR of SEMEATO, at the beginning of the 1990s, was a very important issue as the old model TD and other seeder machinery presented many drawbacks (Casão Junior *et al.*, 2012). At that time, the myth of needing heavy seeder machinery for NT was still very strong mainly because staggered double discs were commonly used and the penetration into the soils which were not previously mobilized was very difficult. In 1992, NT machinery market was strengthened with several industries releasing new seeder models especially the precision-type ones.

Also many industries, 'local artisans', blacksmiths, garages and farmers from different regions started to use tines instead of double discs, which permitted the expansion of NT to heavy clayey soils.

CNPT/EMBRAPA started field evaluations of NT seeders in Rio Grande do Sul from 1993 to 1997 and by IAPAR, in Paraná State, from 1996 to 2003. This work promoted interactions among researchers and industries, and IAPAR spread the results during expositions held in the north and west of Paraná. Eighteen industries took part in these evaluations and expositions with about 150 different models of NT seeders (Casão Junior and Siqueira, 2003, 2004).

By the mid-1990s, large expositions in the main agricultural regions of the country started to include dynamic field presentations of NT machinery, spreading the adoption of system. It was an important period, when industries multiplied the number of machinery models to supply regional and international demands. As a result, currently there are more than 300 different commercial models of NT seeders available in Brazil.

Since this period, NT area for annual crop production experienced an explosive increase from 1 Mha in 1992 to 25 Mha at the end of first decade of the 21st century.

3.3.1 History and advances in no-till mechanization for small farmers

During the 1980s, NT was commonly associated with large-scale farms (big farmers) with general requirement of intensive use of inputs (herbicides, fertilizers, etc.) and because only heavy machinery was available at that time. Despite this fact, IAPAR started in 1985 a research project to develop a NT animal traction seeder and to evaluate the feasibility of NT with light machinery specially for seeding.

The IAPAR project resulted in the development of a prototype named 'Gralha Azul', a NT animal-traction seeder and, with the support of Paraná State and FEBRAPDP, started the diffusion of this machine among farmers together with the state extension service (EMATER). This process led to an increase in the number of industries interested to manufacture the equipment.

At the initial stage the diffusion process covered 31 farmers and extension technicians in different zones of Paraná. The group received training, cover-crop seeds, fertilizers, herbicides and NT animal-traction seeder machinery. They were evaluated during 3 years by a group of researchers from a technical and economical point of view (Darolt, 1998).

The results were very clear to the farmers, machinery dealers, manufactures and technicians. In most regions, farmers reduced labour time and costs, increased yields and cropped land, and also devoted more time to higher income activities. There was a significant improvement in the quality of life of small farmers.

According to Bolliger *et al.* (2006), one very important implement innovation that has been refined through adaptive experimentation and trials by Brazilian farmers is the 'knife-roller' ('roller-crimper') designed to crush or break and roll cover crops and also selected weeds. This important tool was invented by a small farmer in Paraná State, who at that time adapted a tree trunk with some longitudinal knives (made from old truck springs). Although a knife roller commonly comprises a cylinder with blades to be drawn by an animal or a tractor, versions in Brazil range from simple pieces of wood that crush plant stands when towed through them (mainly useful when plant biomass is not very high), to complex cylinder-and-disc systems attached to the front or rear of tractors (Araújo *et al.*, 1993, 1998; Freitas, 2000; Ashford and Reeves, 2003). Apart from reducing the reliance on herbicides to desiccate cover crops, rolling also has the advantage that residues fall down in the direction of rolling, thereby facilitating planting, and also that the whole plant remains almost intact or in large pieces covering the soil surface. 'This important and useful tool "knife-roller" is being used worldwide now' (Bolliger *et al.*, 2006). This protected the residues from attack by soil organisms, and also prevented dispersal of loose residue by wind and during planting operations and decreased residue decomposition rate, consequently extending the effectiveness of the residue cover, reduced evaporation from the soil and also suppressed weed growth. However, the timing of the rolling operation is crucial to its success, as most plant species can regenerate if they are rolled or slashed prematurely, while mature seeds of the cover crop or weeds may germinate if elimination is carried out too late (Skora Neto and Darolt, 1996; Skora Neto, 1998; Teasdale *et al.*, 2007). Trials to this aspect indicated that the best time to roll grasses is at the milky grain phase, while in legumes this is best done at the beginning of pod formation or full flowering, depending on the species (Calegari, 1998b; Ashford and Reeves, 2003).

Several manufacturers of NT animal-traction equipment emerged in Paraná, Santa Catarina (SC) and Rio Grande do Sul including industries such as Mafrense, RYC, Buffalo, Triton, Werner, Fitarelli, IADEL, Knapik, Jahnel and Sgarbossa among others. Some manufacturers specialize in sprayers like Guarani and Scotton and the traditional manufacture of hand jab planters such as Krupp.

Nowadays, small farmers in the south of Brazil are experiencing a transition process, changing from animal traction to medium tractor mechanization using small NT machinery as well renting services for seeding and spraying.

3.3.2 NT expansion determinants in Brazil: mechanization inputs

One important reason for the strong expansion of NT cultivated area in Brazil was the availability of NT seeders in the national market that were appropriate for use in the range of soil types in Brazil. Since the 1980s the Brazilian agricultural machinery industry has played a key role in adapting and developing machinery with the support of research institutions through their studies and comparative testing of commercial models with farmers, which promoted the development of initial models of drill seeders and improvements in many other models, especially precision seeders (Casão Junior *et al.*, 2012).

As a result, by the mid-1990s, Brazil already had a mature industry for NT machinery with different alternative types and a consolidated market, which also presented an impressive growth in more recent decades.

After 1995 agricultural financing, mainly for investment, started to present lower and fixed interest rates, which had a large influence in increasing machinery acquisition throughout the country.

For small farmers, the adoption of NT expanded due to the wide dissemination of good results obtained by research studies and validation activities, the attractive agricultural investment financing, the farmers' interest in changing their farming system (especially to reduce drudgery and increase work productivity) and the machinery industries' interest in expanding their market.

As a consequence, small farmers have managed to control soil erosion, reduce their workload, save time, increase crop yields and diversify their activities, mainly with higher aggregated value activities. This has had a positive impact on the improvement of the quality of life of smallholder farm families.

3.4 Considerations and Development of the No-Till 'Revolution' in Brazil

The history of the Brazilian NT 'revolution' is well documented in the literature. According to Bolliger *et al.* (2006), NT development precipitated out of the widespread soil degradation in the 1960s–1970s in subtropical southern Brazil (especially Paraná) and also in the Savannah area (Cerrado Region); it spread from here to Paraguay and tropical Brazil in the early 1980s, as well as serving as an example to other countries all over the world.

The results of the land area distribution under various land uses and the potential land area available for agricultural purposes in Brazil are presented in Figs 3.1 and 3.2.

Brazil has a huge potential area to be developed and expanded in agriculture when compared to all other countries in the world. It is considered that around 50 Mha or more of pasture area show a certain degree of degradation, and this area could be shifted to annual grain crops in the future, without use of the Amazon Biome to grow pasture or other land uses and just using the Savannah (Cerrado) area. Thus for the agricultural area, the NT system has a high potential for development to produce grain, fibre, vegetable and animal protein, pasture, oil, wood, etc.

In Brazil (Figs 3.1, 3.2) during the 1960s, a significant expansion of the area under soybean (*Glycine max* L. Merryl) and winter wheat (*Triticum aestivum* L.) occurred in southern Brazil, and later on in the Savannah area. The intensive ploughing and discing, residue burning and downhill seeding regimes widely adopted for growing these crops exposed the bare soils to intensive rainfall, which in turn led to extensive soil erosion and concomitant economic loss through soil and nutrient loss, and pollution of the natural resource base, especially surface and groundwaters.

The process of soil occupation started more or less during the 1820s (German) and in the 1870s (Italian) without adequate occupation planning and use of the territory, leading to serious consequences in the degradation of soil and water resources.

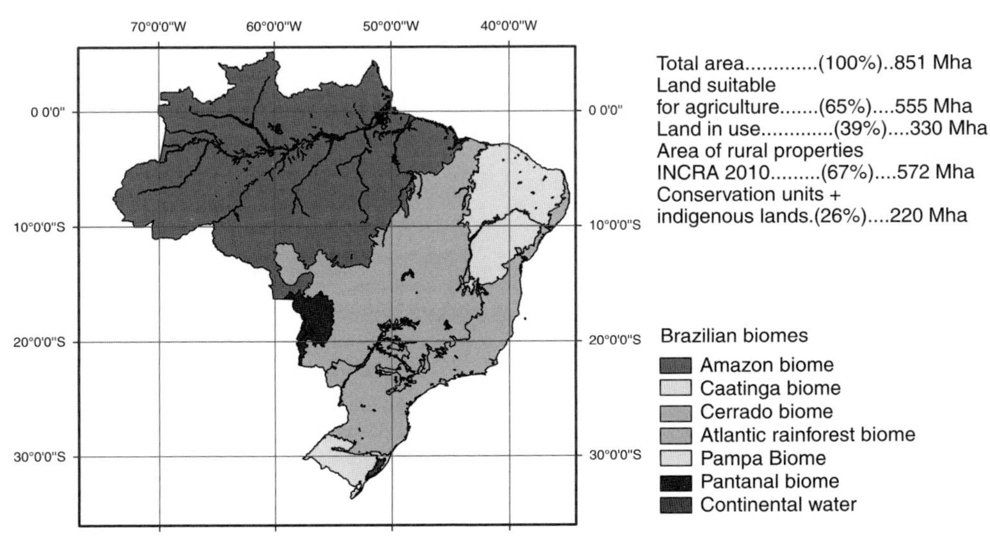

Fig. 3.1. Current land uses and biome regions in Brazil (EMBRAPA, Brasília, DF, Brazil, 2013, pers. comm.).

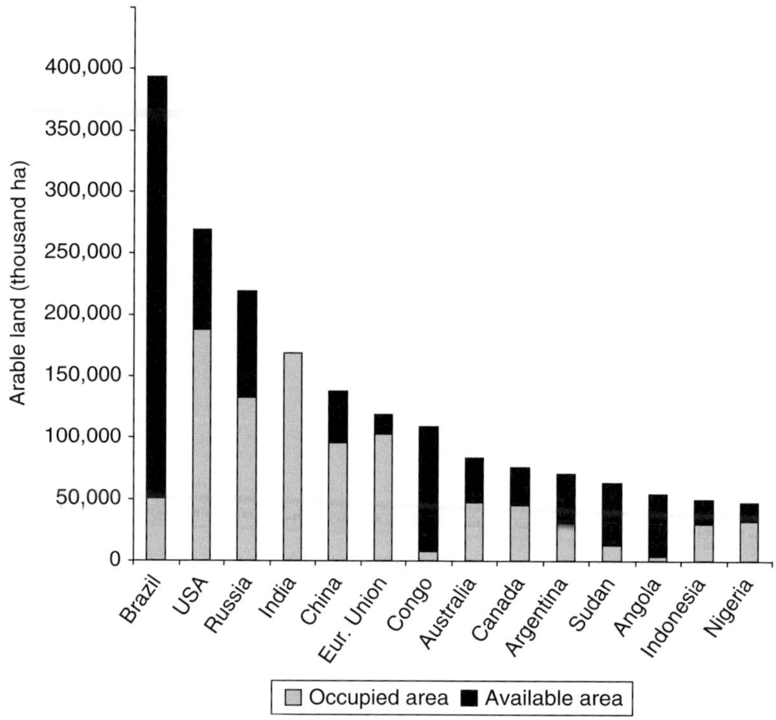

Fig. 3.2. Land area in Brazil vis-a-vis other countries of the world for potential agricultural use (FAO, cited by EMBRAPA, Brasília, DF, Brazil, 2013, pers. comm.).

The traditional or conventional soil management system (ploughing) was commonly used during the 1970s until the mid-1900s. In the 1980s, technical data from Agricultural Research Institute (IAPAR) showed that the NTS should not be merely a new alternative soil management method, but rather evolve into a system integrated to different practices that should develop in an orderly, interrelated and dependent fashion.

According to Gazzoni (Soybean Brazilian Center – Embrapa, Londrina, PR, 2013, pers. comm.), in 1992 the total area with NT in Brazil was around 1.3 Mha (4% of the total grains area). In 2012, there were around 26–28 Mha of NT (70–75% of the total grains area in Brazil). At the present time, crop yield improvement for different crops in NTS is 44% for rice, 72% for maize, 48% for soybean and 64% for wheat.

Compared from 1970s until now, the NT practice in Brazil not only increased the yields of different crops but also reduced the use of external inputs. As compared to 1992 (20 years ago), to produce 25 kg of grains used around 1 l diesel, and now in 2012 with just 1 l we can produce around 105–175 kg of grain. To produce 1 t of soybean, 20 years ago, around 70 l diesel was used, and now just 9 l is enough to produce 1 t of soybean grain. With this increase and better efficiency of NT in Brazil, the diesel consumption decreased 66%. In 2012, the NT area saved around 1.34 billion l of diesel. Thus, there was a mitigation of 3.59 billion kg of CO_2.

3.5 Advancing with the No-Till System in Brazil

The understanding of how crop residues influence nutrient cycling and soil chemical properties, and the integration of residue management into different cropping systems is the key to develop and maintain soil fertility. Continuous monocropping with cereals and less diversified cropping systems such as maize–maize, rice–rice, cotton–wheat, soybean–wheat, soybean–maize systems in Brazil and in other regions of the world has increased the incidence of pests and diseases, including the enhancement of the nematode population, weed infestation and soil degradation, resulting in decline in crop yields. On the other hand, the NTS in association with cover crops, in an appropriate rotation system, comprising other soil conserving practices, such as terracing, grassed waterways, vegetate terraces etc. (conservation agriculture – CA) has minimized the soil degradation process, promoted favourable changes in the soil attributes, chemical, physical and biological, and also decreased the use of external chemical inputs.

The results obtained in Paraná, south region and other parts of Brazil prove that the use of cover crops, as part of the productive system, is economically viable and ecologically sustainable, as it leads to greater water storage in the soil profile, reducing the loss of water by evaporation. This not only increased the greater productivity of cotton, soybean, maize, rice, sunflower, sorghum and wheat in various rotations, but also conserved, maintained and/or recovered soil fertility. In addition, the system promoted economy in the use of N fertilizers (leguminous plants), achieved better weed control by the mulch effects, led to greater biological activity and biodiversity in the soil, decreasing insect-pests and disease occurrence, and represents a very promising way to manage soils towards sustainability.

Generally, the principles and concepts of the CA system comprise a holistic approach, which can be adapted for different farming systems based on agroecological zones and by harmonious integration of different components, such as cover-crop species, crop rotation, NT, terracing and intercropping systems. The main aims of CA are to empower farmers to make more sustainable use of their land in ways that improve their incomes and welfare, and lead to acquiring the knowledge and skills to operate systems that save labour, promote soil-water retention, enhance soil fertility and improve crop yields (Merten *et al.*, 1994; Calegari *et al.*, 1995, 2008; Wildner, 2000; Rheinheimer *et al.*, 2006).

Oliveira (1994) evaluated the response of cotton during 4 years to various winter cover-crop systems and N fertilization in a Eutrophic Red Latosol (Eutrorthox) in north Paraná State, Brazil. The results showed that a Neq (nitrogen equivalent) of 160, 90 and 106 kg ha^{-1} of N was achieved by white lupin, black oat + lupin and radish production systems, respectively. The nitrogen mineralization and uptake by cotton depended on the various cover crops. The lowest cotton yield (including seeds) was obtained under fallow conditions varying from 2700 to 3000 kg ha^{-1} (0 and 120 kg ha^{-1} of mineral N, respectively): there was significant nitrogen response to the mineral N until 60 kg ha^{-1} in all other treatments. The highest cotton yield was obtained following lupin (almost 3300 kg ha^{-1} without mineral N); and when this was supplemented with 60 kg ha^{-1} of mineral N, 3500 kg ha^{-1} yield was attained. These results suggest that it is feasible to reduce the fertilizer N demand for cotton by using rotations with cover crops. Similar results were also reported by Costa *et al.* (1993) with cotton in rotation with cover crops. Also in Paraná, Muzilli (1978) and Muzilli *et al.* (1983) found that the Neq of common vetch and cerebella were 80 kg ha^{-1}, and of white lupin was greater than 90 kg ha^{-1} to the following maize crop; these results are in accordance with those reported by Derpsch *et al.* (1991) and Calegari (1998a, 2000a, b), who found more than 90 kg ha^{-1} of N from white and blue lupin to the maize crop. Calegari (2000c, 2002) also reported more than 120 kg ha^{-1} Neq from hairy vetch to the following maize crop.

In general, maize planted following leguminous cover crops shows less response to nitrogen application, conversely areas with grass fallow show higher response to applied nitrogen.

According to Pieri *et al.* (2002), the experience by Brazilian and Paraguayan farmers as well in other countries in the

Americas provides evidence to show that CA has a potential to promote a sustainable and profitable environmental approach to meet the challenge of food security and alleviate rural poverty mainly in the tropical environment with vulnerable natural resources. Nevertheless, CA is an extremely complex system, and field experiences and strategies are needed to validate farming systems in different agroecological zones to develop adaptation methods and to facilitate the dissemination process for technologies under on-farm conditions, and must be improved according to local conditions.

3.6 The Main Components for the No-Till System

3.6.1 Cover-crop species in Brazil

The practice of not ploughing the soil, and organic carbon addition by plants and maintenance of crop residues on the soil surface, preserves and attains soil organic carbon equilibrium at a higher level. Results obtained by researchers and farmers with different cover crops in NTS, conducted in different Brazilian agroecological conditions, have shown the efficiency of these systems for improving soil properties, promoting better plant–soil–water relations, and also the rotation practised including various species has contributed to improve the systems' biodiversity.

Plants used as cover crops, given their high capacity to produce high biomass and roots, with important direct and indirect effects in the soil–water–plant relations, play a fundamental role when they form an adequate part of orderly rotation systems with cash and food crops.

Despite the fact that the primary function of cover crops is to increase biomass and provide soil covering during periods when available resources are too limited for a cash crop, most cover crops used in Brazil fulfil multiple agronomic, ecological, or economic functions in concert with those performed by the main crops (Derpsch, 1986; Skora Neto and Darolt, 1996; Anderson *et al.*, 2001; Florentin *et al.*, 2001, 2010; Calegari, 2002; Calegari *et al.*, 2007). According to Bolliger *et al.* (2006), such general functions of cover crops broadly include: (i) providing additional fodder, forage, food, and secondary commercial or subsistence products for livestock and humans; (ii) directly adding or sparing N to/from the soil through symbiotic N_2 fixation from the atmosphere; (iii) converting otherwise unused resources, such as sunlight and residual soil moisture, into additional biomass and concomitantly, upon the breakdown of their residues, increasing the build-up of SOM, capturing and recycling easily leachable nutrients (nitrates, K, Ca and Mg) that would otherwise be lost beyond the rooting zone of commercial crops, ameliorating soil structure and buffering against compaction by creating additional root channels that differ from those of the main crops and by stimulating soil biological activity through, *inter alia*, the release of root exudates; (iv) improving the management of acidic soils by releasing various products that can mobilize lime movement through the soil profile, decarboxylize organic anions, function in ligand exchange and add basic cations to the soil; (v) facilitating weed management by competing against or smothering weeds that would otherwise become noxious in the main crop cycle; and (vi) breaking the cycle of certain insect-pests and diseases that could otherwise build-up in continuous monocropping systems.

The use of cover-crop species is widespread in all main production regions, from the south to the Savannah area of Brazil, and they provide mulch for NT cash crops, are used as intercrops in perennial crops (coffee, rubber tree, citrus and others perennial fruit), horticultural crops (potatoes, carrots, tomatoes, onion, garlic, cabbage, etc.), and also the cover crops can be used as animal fodder. According to Derpsch and Calegari (1985), Calegari (1990, 2009) and Calegari *et al.* (1993), several cover-crop options are available relative to crop rotation systems in Brazil, such as:

- **Winter species:** black oat (*Avena strigosa* Schreb), radish (*Raphanus sativus* L.), vetches (*Vicia sativa* L. and *Vicia villosa* L.), lupin (*Lupinus* spp.), rye (*Secale cereale* L.), ryegrass (*Lollium multiflorum* L.), triticale (*X-triticosecale*), sweet pea (*Lathyrus sativus* L.), clovers (*Trifolium* spp.), sweet clover (*Melilotus* sp.), lucerne (*Medicago sativa* L.), serradella (*Ornithopus sativus* L.), chickpea (*Cicer arietinum* L.).

- **Summer species:** pigeon pea (*Cajanus cajan* L.), sunnhemp (*Crotalaria juncea* L.), crotalarias (*spectabilis, ochroleuca, breviflora, mucronata*), buckwheat (*Fagopirum esculentum*), cowpea (*Vigna unguiculata* L.), green gram (*Vigna radiata* L.), lablab (*Dolichos lablab* L.), siratro (*Macroptilium atropurpureum* L.), stylo (*Stylosanthes* spp.), butterfly-pea, blue-pea (*Clitoria ternatea* L.), jack bean (*Canavalia ensiformis* L.), brave bean of Ceará (*Canavalia brasiliensis* L.), pear millet (*Pennisetum americanum* L.; *Pennisetum glaucum* L.), finger millet (*Eleusine coracana* L.), annual foxtail (*Setaria italica* L.), velvetbean (*Mucuna* sp.), *Centrosema* sp., *Desmodium* sp., tropical kudzu (*Pueraria phaseoloides* L.), *Stylosanthes* sp., *Tephrosia* sp., *Calopogonium mucunoides* L., *Neonotonia wightii* L.etc.; *Brachiaria* sp. has also been used as a soil cover and occasionally killed by herbicides to sow on it cash crops.

Also the effects of mixed crops (oat + vetch, oat + radish, oat + lupin, or also oat + radish + vetch, oat + lupin + radish + vetch, pearl millet + crotalaria, pigeon pea + pearl millet, etc.) or a cocktail of three, five or more species has been studied by Calegari (2010) at IAPAR for more than 20 years; these improve soil physical properties (increase soil aggregate stability indices, enhance soil water infiltration levels, etc.), chemical effects (higher levels of N, P, K, Ca, Mg and organic matter in soil surface, so by nutrient recycling and/or N fixation legume, decrease in toxic aluminium, etc.) (Miyazawa *et al.*, 1994; Tiecher *et al.*, 2012b; Vinther, 2004) and also biological effects (improving soil organisms and reduction of phytonematode population), beyond allelopathic effects by root exudates and also by plant tissues that qualitatively and quantitatively affect weed population (Skora Neto, 2001; Teasdale *et al.*, 2007; Skora Neto and Calegari, 2010). The developments in soil-water management by systematic work in watersheds have contributed to improve not only the whole agriculture, but also the socio-economic conditions of farmers in Paraná. In general the legume crops present a high potential to fix nitrogen through the symbiosis between roots and bacteria (rhizobia) (Table 3.1), and also they have a large capacity to recycle nitrogen and other nutrients that were leached to deeper layers.

Cover crops have been used in Brazil for almost a hundred years, and IAC (Campinas Agricultural Research Institute) developed many promising studies of different cropping systems (Miyasaka and Okamoto, 1992). Over 100 different species and varieties of cover crop were screened, tested and evaluated in on-farm trials throughout southern Brazil in the 1980s (Derpsch, 2003), and many different cover crops are being used by both large- and small-scale farmers in southern Brazil (Calegari, 1998c; Calegari and Alexander, 1998), including black oats, vetches (both *V. villosa* and *V. sativa* L.), oilseed radish, ryegrass, rye (*Secale cereale* L.) and white or blue lupins (*Lupinus albus* L. and *L. angustifolius* L.); in the Savannah areas of Brazil, the main cover crops used are pearl millet, crotalaria, pigeon pea, brachiaria, sorghum, stylosanthes, etc. With the diffusion of NTS, it is estimated that cover crops are grown over more than 3 Mha in Paraná, Santa Catarina and Rio Grande do Sul.

In south Brazil, and also in states such as Minas Gerais and São Paulo, winter cover crops are mainly used, while in the Savannah area the most common species used are summer cover crops. Some of the major cover crops are used in Brazil, together with their main advantages/functions and drawbacks, although we would like to draw attention to the fact that cover crops are also commonly grown in mixtures ('cocktails') rather than alone by Brazilian farmers (Calegari, 2009, 2010).

Table 3.1. Biological N fixation by some legume species (Adapted from Calegari *et al.*, 1993).

Legume species	N (kg ha^{-1} year^{-1})
Lucerne (*Medicago sativa*)	127–333
Groundnut (*Arachis hypogaea*)	33–297
Calopo (*Calopogonium mucunoides*)	64–450
Cowpea (*Vigna unguiculata* sin. *Vigna sinensis*)	73–240
Centrosema (*Centrosema pubescens*)	93–398
Crotalaria (*Crotalaria juncea* L.)	150–165
Tropical Kudzu (*Pueraria phaseoloides*)	100
Desmodium sp.	70
Peas (*Pisum sativum*)	81–148
Common vetch (*Vicia sativa*)	90
Hairy vetch (*Vicia villosa*)	110–184
Stylo (*Stylosanthes* sp.)	30–196
Faba beans (*Vicia faba*)	88–157
Jack bean (*Canavalia ensiformis*)	57–190
Galactia striata	181
Chickpea (*Cicer arietinum*)	41–270
Pigeon pea (*Cajanus cajan*)	41–90
Cyamopsis psoraloides	37–196
Lens culinaris	35–77
Lespedeza stipulacea	193
Leucaena (*Leucena leucocephala*)	400–600
Black mucuna (*Stizolobium aterrimum*)	157
Perennial soybean (*Neonotonia wightii* Lacrey) (syn. *Glycine wightii* Verdc.)	160–450
Siratro (*Macroptilium atropurpureum*)	70–181
Soybean (*Glycine max*)	17–369
Lupins (*Lupinus* sp.)	128
White clover (*Trifolium repens*)	128–268
Sweet clover (*Melilotus alba*)	9–140
Egyptium clover (*Trifolium alexandrinum*)	62–235
Red clover (*Trifolium pratense*)	17–191
Subterraneum clover (*Trifolium subterraneum*)	21–207
Trigonela (*Trigonella fænum-græcum*)	44
Vigna sp.	63–345

Some details about the main cover crops used in different cropping and farming systems in Brazil are shown in Table 3.2.

When different cover crop (summer and winter) species are grown in the field and are managed, it can promote nutrient recycling and nitrogen fixing (Tables 3.3 and 3.4).

3.6.2 Crop rotation

The continuous use of monocropping can modify the soil environment through selective nutrient uptake by exploiting similar root depth, creation of favourable conditions for stimulating the growth of specific microorganism species, effects of root exudates on soil pH and other soil characteristics. Also the monocropping systems have increased the occurrence of insect-pest and diseases and also some weed species. These effects interfere with soil–water–plant relations and also adversely affect soil fertility, leading to decline in crop yields. Conversely, in a rotation system, the most suitable crop sequences are those that comprise plants that have different growing habits for water and nutrient needs. For example, leafy horticulture crops need more nitrogen, conversely horticultural root crops and those with rhizomes need more potassium, and leguminous plants normally acquire more

Table 3.2. Some of the major cover crops grown in Brazil (Adapted from Bolliger *et al.*, 2006).

		Species	Soil and climatic requirements	Days to flowering	Dry matter (t ha^{-1} $year^{-1}$)	Advantages and limitations
Winter	Non-legumes	*Avena strigosa* (Schreb.)	S–C; LF–MF	100–145	2–11	AF; WC; decrease soil root diseases (*Fusarium* spp. etc.); FASM
		Avena sativa (L.)	S–C; LF–MF	80–145	3–9	AF; WC; decrease soil root diseases (*Fusarium* spp. etc.); FASM
		Raphanus sativus ssp. *oleiferus* Metzg.	S–L; A–	90–110	3–9	High-nutrient recycling capacity; BP; WC; FASM
		Secale cereale (L.)	S–C; LF; A+; Wlog–; DT	100–120	4–8	BP; WC; controls some soil diseases
		Lollium multiflorum (L.)	S–C	120–150	2–6	AF;WC
	Legumes	*Lupinus albus* (L.)	S–C; MF; Wlog–	120–140	3.5–5	AF; HF; BNF; BP; sensitive to diseases (*Fusarium* spp. etc.)
		Pisum arvense (L.)	S–C; A–	100–130	2.5–7	AF; FEG; BNF; sensitive to aphids and some diseases
		Lupinus angustifolius (L.)	S–C; A+; Wlog–	120–140	3–6	AF; HF; BNF; BP; sensitive to diseases (*Fusarium* spp. etc.); FASM
		Vicia sativa (L.)	S–C; HF; A–; Wlog–	120–150	3–5	AF; BNF
		Vicia villosa Roth.	S–C; LF; A+; Wlog–	140–180	3–5	AF; BNF; WC
Summer	Non-legumes	*Brachiaria* spp.	S–C;A+	n.a.	>4	AF; BP; high biomass; SOM
		Helianthus annuus (L.)	S–C; A+; LF; DT	70–120	4–8	FEG, high nutrient recycling; WC
		Panicum maximum (L.)	S–C; WD; DT; A+; Wlog–	n.a.	>20	FEG; AF; BP; SOM
		Paspalum notatum Flugge	S; DT; CT	n.a.	3–8	AF; SOM
		Pennisetum americanum (Schum.)	S; A+; LF; DT	90–120	3.5–21	AF; BP; SOM; WC; FASM
		Fagopirum esculentum (Moench)	S–L–C; L/M; Wlog+; A+; DT	45–60	3–6	AF; HF; GC; WC; FEG
		Setaria italica (L.)	S–C; WD; MF; DT	45–60	2.5–8.5	AF; FEG; FASM; high-seed production
		Sorghum bicolor (L.) Moench	S–C; WD; MF; DT	60–110	3.5–18.5	AF; BP; SOM
	Legumes	*Cajanus cajan* (L.) (dwarf variety)	S–L; LF; Wlog–	70–85	2–6.5	AF; NC; high-seed production

Continued

Table 3.2. Continued.

Species	Soil and climatic requirements	Days to flowering	Dry matter (t ha^{-1} year^{-1})	Advantages and limitations
Cajanus cajan (L.) Millsp.	S–C; LF; Wlog–	140–180	3–7.5	AF; BP; BNF + nutrient recycling, NC
Calopogonium mucunoides Desv.	S–L–C	n.a.	4–10	WC; GC
Canavalia ensiformis (L.) DC.	S–C; LF; DT	100–120	5–6	WC (allelopathic effects against *Cyperus* spp. and *Cynodon dactylon*)
Crotalaria juncea (L.)	S–L–C; MF	70–120	3–8.5	BNF; WC; NC; efficient in nutrient cycling
Crotalaria spectabilis (L.)	S–L–C; MF	80–120	4–7.5	BNF; WC; NC; efficient in nutrient cycling
Crotalaria ochroleuca (L.)	S–L–C; MF	70–120	4–9.0	BNF; WC; NC; efficient in nutrient cycling
Macroptilium atropurpureum (DC.) Urb.	S–C; WD; A+; MF; DT	n.a.	3–6.5	AF; SOM; WC
Mucuna pruriens (L.) DC.	S–C; LF	130–150	2–5	FEG; GC, BNF; NC
M. pruriens (L.) DC. (dwarf varieties)	S–C; LF	80–100	2–4	NC; FASM; rain during harvesting period can damage the seeds
Pueraria phaseloides (L.)	L; WD; Wlog–; DT	n.a.	3.5–8	AF; GC
Stylosanthes spp.	S–L–C; A+; LF; DT	n.a.	n.a	AF; BP; SOM
Vigna radiata (L.)	S–L–C; DT; WL–	60–80	3.5–6.5	AF; HF; high seed production
Vigna unguiculata (L.)	S–L–C; L/MF; A+; WL–	70–110	2.5–5.7	AF; HF

n.a., Data not available; S, light-textured (sandy) soil; L, medium-textured (loamy) soil; C, heavy-textured (clayey) soil; LF/MF/HF, low/medium/high fertility; WD, well-drained soil; Wlog–/+, intolerant/tolerant of water logging; A–/+, intolerant/tolerant of soil acidity; DT, drought tolerant; AF, animal forage; HF, human food; BNF, high-N fixation; GC, produces good cover; WC, weed suppression; BP, biological ploughing; SOM, good SOM builder; FASM, facilitates acid soil management; FEG, fast early growth; NC, nematode control.

Table 3.3. Chemical compositions of some summer cover crops; also used as animal feed at the flowering stage (Adapted from Calegari, 1995a).

	Nutrient contents									
	% of dry matter					ppm				
Cover crop species	N	P	K	Ca	Mg	C	Cu	Zn	Mn	C/N ratio
Crotalaria juncea	2.50	0.19	1.20	2.31	0.47	45.25	14	44	179	18.10
Crotalaria spectabilis	2.17	0.09	1.59	0.49	0.37	50.83	8	23	126	23.42
Cajanus cajan	2.61	0.14	2.61	1.79	0.45	56.30	7	22	87	21.57
Canavalia ensiformis	3.19	0.15	5.62	1.35	0.63	50.15	9	62	254	15.72
Canavalia brasiliensis	2.49	0.13	1.68	0.20	0.16	51.24	4	14	17	20.57
Mucuna pruriens (grey)	2.50	0.15	1.40	1.20	0.27	52.30	16	28	183	21.12
M. pruriens (black)	2.49	0.13	1.40	1.17	0.27	52.15	14	29	174	21.06
M. pruriens (dwarf)	3.10	0.19	4.49	2.14	0.65	50.83	9	85	179	16.39
Vigna radiata	2.09	0.21	4.94	1.48	0.75	52.47	10	78	127	25.10
Vigna unguiculata	2.62	0.20	2.82	0.93	0.28	45.42	–	–	–	17.33
Indigofera sp.	2.17	0.14	1.54	1.20	0.32	40.36	13	24	53	18.60
Calopogonium mucunoides	2.16	0.12	1.56	1.40	0.29	46.73	9	15	172	21.63
Pueraria phaseoloides	3.68	0.29	2.14	1.30	0.41	54.10	11	27	155	14.70
Glycine wightii	2.60	0.23	2.39	0.99	0.35	45.03	8	32	102	17.31
Centrosema pubescens	2.34	0.23	1.19	0.66	0.45	47.60	10	32	67	20.34

Table 3.4. Chemical composition of winter cover crops (Adapted from Calegari *et al.*, 1993).

	Nutrient content										
	%						ppm				
Cover crop species	N	P	K	Ca	Mg	C	Zn	Cu	Mn	Protein (%)	C/N Ratio
Hairy vetch	3.82	0.30	2.03	0.78	0.27	37.87	26	9	61	23.87	10.05
Common vetch	2.87	0.23	2.88	1.05	0.41	37.1	24	9	87	17.94	12.9
Ornithopus sativus	1.79	0.14	3.55	1.10	0.45	40.14	59	13	97	11.18	22.43
Radish	2.68	0.17	2.80	1.54	0.76	38.58	49	8	84	16.75	14.45
White lupin	3.20	0.09	2.66	0.46	0.38	47.49	57	12	330	20.0	14.84
Yellow lupin	2.94	0.16	2.50	0.59	0.39	42.25	66	14	359	18.37	14.37
Blue lupin	3.19	0.19	2.29	1.20	0.49	37.83	24	13	230	19.93	11.86
Sweet blue lupin	2.28	0.10	1.75	0.59	0.42	37.87	32	16	147	14.25	16.61
Field pea (IAPAR-83)	2.09	0.12	1.50	0.70	0.20	39.77	8	22	52	13.06	19.02
Wheat	0.77	0.06	1.15	0.22	0.10	40.38	–	–	–	4.81	52.71
Sweet pea	2.23	0.10	2.90	0.39	0.19	41.91	22	11	52	13.93	18.79
Black oat	1.93	0.28	2.15	0.43	0.21	39.69	11	7	102	12.06	20.76
White oat	0.81	0.052	2.40	0.24	0.17	38.52	9	6	138	5.06	47.55
Rye grass	1.34	0.067	2.60	0.41	0.22	59.22	23	9	214	8.37	44.20
Rye	1.22	0.075	1.40	0.18	0.14	44.59	15	6	53	7.62	36.54
Sunflower	1.80	0.15	2.40	1.55	0.62	39.95	31	18	96	11.25	22.19
Corn spurrey	2.13	0.22	3.45	0.52	0.77	41.92	44	11	136	13.31	12.78

phosphorus from the soil. Therefore to attain positive soil equilibrium, it is not recommended to repeat the same crop or plants from the same species or family every season, with similar characteristics, but look for a proper crop rotation.

Crop residues on the soil surface provided by the plant biomass or main- and cover-crop roots in the soil tend to improve selected soil properties (Jat *et al.*, 2012). In this sense, residues help improve soil structure by increasing aggregate stability (by the cementing action of the organic matter, polysaccharides and fungal hypha) (Calegari and Pavan, 1995), increasing water retention capacity, increasing water infiltration rates, greater soil porosity, greater aeration, less water evaporation and decreasing soil bulk density due the effects of organic matter addition (Basch *et al.*, 2012).

Continuous adding of organic residues to the soil along with continuous NT contributes to increase in soil fauna and their diversity (Table 3.5).

These results obtained in north Paraná, south Brazil show that soil disturbance decreases the earthworm population. Also, as organic residues accumulate at the soil surface, favourable conditions for biology and earthworm population is a good soil quality indicator. The inclusion of a cover crop after soybean enhanced the number of arthropods, and NT was more favourable than conventional tillage (ConvT) (Derpsch *et al.*, 1986, 1991; Bolliger *et al.*, 2006).

Continuous ploughing without crop or organic residue mulch on the soil surface under the conventional system cause a greater fluctuation in temperature as well as moisture, and this leads to a decline in activity and population of soil organisms, especially microorganisms. Given the greater concentration of residues and their effects on the surface, NTS tends to facilitate an increase in biological life in the soil.

From the research on bean production, in the Savannah zone of Brazil, Goiás State, Costa (1999) concluded that NTS generally in the first and second years increased the population of *Rhizoctonia solani* and *Fusarium solani* in the soil, but later on the cumulative effects of rotation with *Avena strigosa* Schreb and *Brachiaria plantaginea* and annual addition of crop residues increased soil biota number and biodiversity, leading to reduction in the soil fungi population. These results clearly showed the advantages of crop rotations and the use of cover-crop species for better soil health, as also mentioned in several results and experiences achieved by Primavesi (1982) in different Brazilian regions. Results obtained by Santos *et al.* (1990, 2000) indicated that crop rotation, including cover crops such as vetches, black oat, sorghum, soybean and maize, was efficient in reducing the incidence of root diseases in NT wheat and maize in Rio Grande do Sul, while Ribeiro *et al.* (2005) stated that among a surveyed group of smallholder farmers in Paraná, those farmers growing tobacco faced the most serious challenges in respect to insect-pests and diseases, and hence were also those that rotated crops most frequently. Yorinori (1996) observed a reduction of *Diaporthe phaseolorum* ssp. *meridionalis* dispersion in soybean when pearl millet was grown as NT cover crop, while black oat has been noted to decrease root rot diseases, such as *Fusarium* species, and pigeon pea or sunnhemp have been found successful in controlling some nematode species (Calegari, 1998c). Results obtained by Viedma (1997) showed that when vetches are mixed with oat in NT rotation

Table 3.5. Soil fauna under no-tillage as compared to that under minimum and no- tillage (Derpsch *et al.*, 1986).

	Conventional tillage	Minimum tillage	No-tillage
Earthworms m^{-2}			
March 1979	5.8	7.5	13.0
Earthworms m^{-2}			
November 1981	3.2	5.2	27.6
Arthropods 300 cm^{-3}			
Soybean/wheat	7.0	–	33.0
Soybean/cover crop	23.0	–	192.0

relying only on wheat and oat, they nearly completely eliminated the incidence of *Helminthosporium* and *Drechslera* species.

The effects of crop residues on the soil surface stimulate the growth of microflora and microfauna, and increase their biodiversity and the antagonistic organisms are able to reduce the population of phytoparasitic nematodes. The use of mixed cover crops, e.g. pearl millet + cowpea, can decrease the population of different nematode species. Results provide evidence to support the hypothesis that the management of soil organic matter in the long term can improve plant resistance to insect-pests. This is confirmed by more recent studies on the relationships between the soil biota that are on and in the soil ecosystem and suggests that the biological activity in the soil is probably much more important than what is recognized by determining individual responses of plants to the stresses caused by insects (Blouin *et al.*, 2005). As suggested by Altieri *et al.* (2007), these results have increased the understanding on the role of biodiversity in agriculture and the close relationship between the biota found on and under the soil surface, and its basis to develop ecological strategies that combines a greater crop diversification and increase soil quality.

Nowadays, an intensive and frequently irrational use of inputs such as chemical fertilizers, pesticides and also the crop sequences (monocropping), with less crop diversification contributes to increase in the insect-pest, disease and nematode populations, which results in declining crop yield, and an increased use of pesticides decreases the biodiversity. This situation has become serious and a suitable diagnosis of soil characteristics and cropping system management should be considered to promote NT with appropriate cover crops, crop rotations and crop diversification, enhancing the natural enemy population base and better soil–water–plant relationship.

Fortunately, the experiences by farmers show that with time, NT leads to better conservation and improvement of all soil characteristics, thereby reducing fertilizer use, less labour use, increased crop productivity and, consequently, greater profits from the production systems.

3.6.3 Weed management

One of the major tools in Brazilian integrated weed management under NT is the use of cover crops and crop rotations. The different cover-crop species properly rotated with other crops are important in weed management as they compete with weeds during their development, and their mulch can also suppress weed emergence. Several winter and summer cover crops have been shown to suppress weeds through their fast growth pattern. Weed biomass reductions of 22–96% have been observed by using summer cover crops depending on the plant species in southern Brazil. A similar result has been observed in the Savannah region (Cerrado) where it was possible to eliminate the use of a selective maize herbicide. Many weeds in field conditions can be properly controlled by the effects of soil mulch by residues, as a result shadow or allelopathic effect, or perhaps both together. Normally the effects are strongly linked with the amount and quality of the mulch produced and remaining on the soil surface.

Adegas (1998) described a study of an integrated weed management (IWM) programme on 58 farms in Paraná, observing that after 3 years, if optimal recommendations were followed, weed control costs decreased on average by 35% with herbicide application reductions of 25%. Ruedell (1995) also reported results of an IPW programme in Rio Grande do Sul, where, over an average of 34 sites, there was a reduction of 42% in weed control costs assuming farmers follow optimal weed management practices.

According to Skora Neto (1998), the main reasons for decrease in weed infestation over time are due to reduction in weed seed banks, and for example, results showed an exponential reduction in weed populations when weeds were controlled before seed-set and not allowed to produce seeds.

Almeida and Rodrigues (1985) and Almeida *et al.* (1983, 1984) showed that cover crops such as black oats, oilseed radish and hairy vetch can be effective in reducing weed population in the NTS and consequently reducing the amount of herbicide needed. According to them there is a linear correlation between the amount of biomass produced by cover crops and their effectiveness in suppressing weeds. These effects on weeds may not only be through competition for light, but also the allelopathic effects achieved by plant exudates (Altieri and Doll, 1978; Altieri, 1995; Teasdale *et al.*, 2007). The effects of some plant species to control different species of weeds are well known, which are mainly due to the mulching effect and also because of physical and chemical (allelopathic) effects of the roots and residues of sorghum, pear millet, mucuna, crotalaria, pigeon pea, etc., affecting the growth and number of certain weed species (Skora Neto and Darolt, 1996).

Research studies by Kliewer *et al.* (1998) working on clay soils at Colonia Iguazu (CETAPAR), Paraguay showed the benefits of cover crops and crop rotations in reducing weed populations. They evaluated the effects of different residues of winter species applied on the soil surface on weed dry mass in soybean (next crop) planted under NTS (Table 3.6). Comparing all winter treatments, the fallow showed the highest weed biomass, as a result herbicide was applied for weed control in the soybean crop; otherwise crop residues of species such as oats (black and white oat), wheat and rye had strong effects on weed control (shadow and allelopathic effects), decreasing weed population, and soybean was raised without the application of herbicide. Farmers who make a good use of cover crops and crop rotations have also made similar observations.

Table 3.6. Effects of fallow and residues of winter species on weed dry mass under no-tillage (Kliewer *et al.*, 1998).

Treatments	Weed dry mass (t ha^{-1})
Winter fallow	7.39 A
Radish	4.26 B
White lupin	3.73 BC
Field pea cv. Iapar-78	2.26 CD
Triticale	1.82 D
Sunflower	1.65 DE
Rye	0.76 DE
White oat cv. IAC-7	0.72 DE
Wheat	0.50 DE
Black oat cv. Iapar 61	0.09 E

LSD (*P*=0.05): 1.810 kg ha^{-1}; F (treatment) = 12.98

In summary, results by farmers and researchers have shown that using adequate integrated strategy and suitable cover-crop species, successful weed management in NT can be achieved with low levels of inputs. Many of these farmers from Cerrado and other areas have reduced the use of herbicides in their fields, and consequently have achieved lower production costs and reduced effect on environmental quality. Therefore, the real farm conditions (on the ground) in Brazilian diversified cropping and farming systems, however, is often more varied, and the great majority of the farmers, especially smallholders in southern Brazil, still struggle with weed control challenges and on many occasions rely on high-herbicides use. These farmers need a strategy that uses cover-crop species in suitable crop rotations to promote an efficient weed control in different Brazilian cropping systems.

3.7 Brazilian Agricultural Regions

3.7.1 CA in Savannah Region (Cerrado)

The Savannah region is generally characterized by well-defined dry and rainy seasons, high temperatures, soils low in clay and witness rapid decomposition of organic matter. Much of the Cerrado, Savannah region of Brazil (central plateau between 10 and 20°S latitude) and western central Brazilian region form an agricultural frontier with large and mechanized farms. This contrasts with southern Brazil where there is a variety of farm sizes and levels of mechanization. The seasonality of rainfall in that region often does not allow continuous cropping without irrigation.

It is common for farmers to establish fast-growing, drought-tolerant cover crops immediately after harvest of the main crop. Their aim is to grow a cover crop to produce some biomass on the residual stored soil moisture under the mulch layer. The most common cover crop is millet, but other drought-tolerant cereals or pasture and forage species are also used. Some innovative farmers plant millet at the beginning of the rainy season, rather than at the end, desiccating the millet with glyphosate 45–80 days later and planting soybean into the millet residues. Another progressive option is to use continuous NT with sequences of cover crops that remain alive throughout the 3–5 month dry season. These crop types can regrow rapidly after the first rains during the following rainy season, or after sporadic dry-season rain, and thereby ensure a permanent soil cover. This may include soybean, rain-fed rice, maize or common beans, which are grown during the rainy season and followed by a second crop of fast-growing cereals or cover crops (millet, crotalarias – *spectabilis, ochroleuca* – pigeon pea, maize, sorghum, finger millet, sunnhemp, etc.) and intercropped with forages. Recent estimates show that the area under pearl millet (*Pennisetum glaucum*) just in Cerrado and western central Brazil was greater than 4 Mha. One of the main reasons for the use of pearl millet is to reduce soil nematode population, mainly *Pratylenchus brachiurus, Meloidogyne incognita* and *Meloidogyne javanica*. A mixed cover crop (cocktail) also has been commonly used in the last 15–18 years, including: millet + *Crotalaria (spectabilis, ochroleuca, breviflora*, etc.), millet + pigeon pea, sudangrass + crotalarias, etc. (Calegari, 2010).

According to Scopel *et al.* (2005), the Cerrado region covers around 200 Mha in the mid-altitude (1000 m) savannahs of central Brazil. It is mainly constituted of large plateaux called 'chapadas'. The climate is tropical humid with good mean rainfall (from 100 to more than 2000 mm) concentrated in 8 months between September and April, and high temperatures (25°C in average) during the whole year. Since the 1970s, 'chapadas' have started to be colonized for agricultural purposes. After initial liming, the well-structured oxisols are very favourable for intensive, mechanized grain production. On the other hand, the margins of the 'chapadas' and the uneven, sloping zones between 'chapadas' (Valverde *et al.*, 2004) are made up of chemically poor soils, with few exceptions. Under hot and wet conditions, organic matter stocks that make up most of the fertility of these soils can decrease fast under severe water erosion and/or inefficient biomass production, leaving in many cases a negative organic carbon balance in the soil. On the other hand, when appropriate soil management is followed the soil organic matter content is enhanced and better soil properties lead to increased crop yield (Resck *et al.*, 1999).

In the Savannah area, many farmers are practising successfully cover crops, and a very common system in Savannah comprises of one commercial crop (soybean, maize, common beans, rain-fed rice) grown during the rainy season followed by a second crop of fast-growing cereals or cover crops (pearl millet, maize, sorghum, sudangrass, finger millet or sunnhemp (*Crotalaria spectabilis* L., *Crotalaria ochroleuca* L.)) and in selected cases cover crops are intercropped with forage species (*Brachiaria* and *Cajanus* spp., *Pannicum maximum* var. Tanzania, *Cynodon dactylon* var. Tifton, various varieties of *Paspalum notatum* and legumes such as *Stylosanthes* sp., *Calopogonium mucunoides, Arachis pintoi* etc.) at the end of the rainy season, the latter staying throughout the dry season after the cereal has been harvested (Séguy *et al.*, 1996; Scopel *et al.*, 2004). Results obtained by Séguy *et al.* (2001) in this region under irrigation or in wetter areas (>1500 mm rainfall year^{-1}), show that the total above- and below-ground annual dry matter production increased from an average of 4–8 t ha^{-1} in systems with a single annual commercial crop to an average of around 30 t ha^{-1} in the most efficient NTS using, for example, *Brachiaria* species (*B. decumbens, B. brizantha, B. humidicola*, etc.). Some farmers in the cerrado with large livestock herds and sufficient land at their disposal leave part of their land under pasture for 3–4 years, before recommencing a 3–4-year cycle of zero-till grain cultivation, as this minimizes the reestablishment costs of the pasture and the need for selective herbicides, while allowing effective SOM build-up (Séguy *et al.*, 1996; Séguy and Bouzinac, 2001). Also some farmers use the crop–livestock–tree system, where combinations of grasses and trees (eucalyptus and others can be used) in order to improve soil attributes, enhance soil organic matter and increase the net income in a sustainable way.

In the Savannah area of Brazil, where the climatic conditions are dry with higher temperatures (tropical conditions), Séguy *et al.* (1996) reported increased soil organic matter when suitable crop rotations are followed (Table 3.7).

3.7.2 Conservation Agriculture in Brazilian Subtropical Region

Soil tillage destroys soil structure, increases soil organic matter decomposition rates and causes soil exposure to the direct impact of raindrops leading to accelerated erosion with loss of soil, water and nutrients. Conversely, NTS reduces soil and water loss. A long-term experiment at the Agricultural Research Institute of Paraná (IAPAR), Ponta Grossa, Brazil, showed the effects of different soil management systems by animal traction on soil loss (Araújo *et al.*, 1993) (Table 3.8).

The soil loss (average of 4 years evaluation, under mean annual rainfall of 967.5 mm) was reduced by more than 90% in NTS when compared with the use of mouldboard plough, and more than 120 times less than when the bare soil is ploughed. Also in Brazil, Castro *et al.* (1993) studied soil and water loss from a field having 6% slope where soybean was planted in rotation with black oat by adopting different soil management practices. The results revealed that NT treatment had advantages over other soil management systems, mainly due to improved soil properties.

In south Brazil in Rio Grande do Sul State (Rheinheimer *et al.*, 2000a), the soil acidity is a challenge and it needs to be

Table 3.7. Organic matter content in the soil after 6 years (1986–1992) of cropping under different ploughing systems and crop rotations in the Brazilian savannah (Séguy *et al.*, 1996).

Soil management and crop rotation	Soil depth (cm)	Soil organic matter (%)
Heavy discs; soybean monocrop	0–10	1.0
	10–20	1.0
	20–30	1.0
Ploughed by discs; soybean–maize rotation	0–10	1.5
	10–20	1.3
	20–30	1.3
No-tillage; soybean–maize rotation	0–10	3.8
	10–20	3.4
	20–30	2.0

Table 3.8. Soil losses under different tillage systems.

Treatments	Soil loss[1] (t ha^{-1} year^{-1})	Relative %	Soil loss[2] (t ha^{-1} year^{-1})	Water loss (mm year^{-1})
Ploughed bare soil	113.8	1307	9.0	109.5
Mouldboard plough	8.7	100	7.7	93.0
Chisel plough	4.3	50	3.4	35.7
No-tillage	0.8	10	1.1	13.1

[1]Araújo *et al.* (1993); [2]Castro *et al.* (1993)

managed by liming. Liming improves soil attributes and creates better conditions for crop root growth (Tormena *et al.*, 1998).

Research conducted by UFSM and UFRGS (Rheinheimer *et al.*, 2000b, c, 2002a, b, 2003a, b, 2008; Rheinheimer and Anghinoni, 2001, 2003; Conte *et al.*, 2003; Gatiboni *et al.*, 2007; Martinazzo *et al.*, 2007; Guardini *et al.*, 2012; Tiecher *et al.*, 2012a, b) showed that it is possible to reduce phosphate fertilizer addition when NTS, along with appropriate crop rotations, is adopted as a management strategy. Under NTS, the soil phosphorus dynamics are totally modified and biological effects are maximized, decreasing the adsorption of phosphate by inorganic colloids; conversely, in conventional systems phosphorus absorption will occur and the uptake of this nutrient by roots becomes more difficult. Also, there was a significant increase in the amount of total and labile organic phosphorus stored into the soil by microbial biomass. When the soil is properly managed through the use of high amounts of crop residues, it is possible to recover more than 80% of phosphorus applied as fertilizer, higher than nitrogen recovered from the soil. This breaks the paradigm that in tropical and subtropical soils, the efficiency of phosphorus chemical fertilizer applied is low. After building and achieving the sufficient phosphorus level in the soil, it is possible to achieve high grain crop yields with reduced rates of phosphorus addition to replenish that removed in the harvested yield.

In south Brazil, Sá *et al.* (2001), working at the South region centre of Paraná on a clay red Latosol (Typic Hapludox) chronosequence following 22 years of NT, reported that the soil organic carbon stock in the 0–40 cm soil depth was 19 Mg ha^{-1} higher than under conventional tillage. Sidiras and Pavan (1985) in Paraná reported significant increase in soil pH, effective cation exchange capacity, Ca, Mg, K and P and also a decrease in Al saturation near the soil surface under NT and permanent cover as compared to that under conventional systems (Skora Neto and Darolt, 1996). Similar results were obtained by Sá (1993) in south Paraná and by Calegari *et al.* (1995, 2008) in the northern and south-western regions.

Calegari (1995a), studying different crop rotation systems in southern Brazil, Paraná State, concluded that using winter legumes such as blue lupin and hairy vetch in NT led to an economy of 90 kg N ha^{-1} when compared with fallow in the conventional system. Tiecher (2011) and Tiecher *et al.* (2012a, b) in a similar trial at IAPAR Experimental Station, Pato Branco, Paraná reported that winter cover-crop species enhanced biological properties mainly in NTS, increasing organic P, P stocked in the soil microbial biomass and acid phosphatase enzyme activity. They also found that species such as black oat with high biomass production capacity and blue lupin (cv. Iapar – 24), which has the ability to take up P from low labile pool, in the no-till system, increased inorganic labile P in the soil upper layers; conversely, in conventional system when the crop residues are incorporated in the soil, the mineralized P is adsorbed.

Results obtained with winter cover crops in southern Brazil indicate that significant improvement in soil attributes and yields can be achieved if an appropriate cover crop is included in crop rotations (Bairrão *et al.*, 1988; Medeiros *et al.*, 1989; Calegari *et al.*, 1993, 1998a; Calegari 1995b, c, 2000a, c, 2002; Calegari and Alexander, 1998). After 19 years of experimentation at IAPAR, on a clayey soil (72%) in the south-western region of Paraná State, Brazil at Pato Branco Experimental Station, Calegari *et al.* (2008) concluded that the NT management sequestered 6.84 Mg ha^{-1} more organic C compared to the conventional tillage (64.6%) in the 0–10 cm soil depth, 29.4% more in the 0–20 cm soil depth, but equivalent amounts in the 20–40 cm soil depth as compared to conventional tillage. Also, the results obtained showed that when winter cover crops were used with NT, in general, greater amounts of organic C were sequestered (Fig. 3.3).

In the 0–20 and 0–40 cm layers, the NTS sequestered higher soil organic carbon (SOC) than in conventional tillage. Independent of soil management, the fallow treatment had

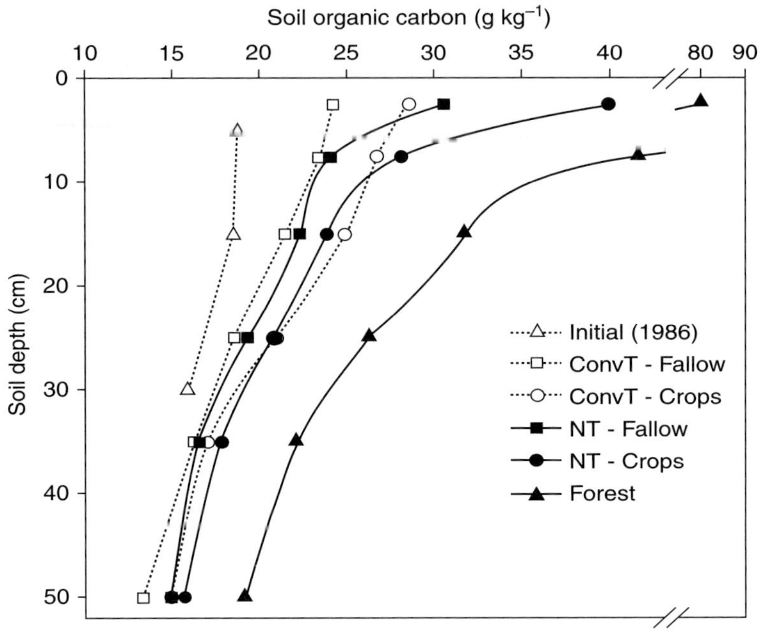

Fig. 3.3. Soil organic carbon distribution under different soil management practices and cropping systems in a Rhodic Hapludox in south-western region of Paraná State, Brazil (Calegari *et al.*, 2008). ConvT, conventional tillage; NT, no-tillage; crops: average of oat, hairy vetch, lupin, wheat, radish, rye.

the lowest SOC stocks compared to all other winter cover-crop treatments.

Continuous NT management combined with the use of winter cover crops had the greatest amount of soil organic matter in the surface soil, and this was the only cropped treatment that approached the level of SOC in undisturbed forest soil. Thus, the NTS with winter cover crops stored greater amounts of soil organic C (Bayer, 1996; Bayer *et al.*, 2000), making the CA sustainable; and this system serves as a management model for sustaining the productivity of Oxisols in tropical and subtropical regions of the world, and one to be emulated by Brazilian farmers and others who are managing similar soil types.

3.8 Influence of Cover Crops, Tillage and Residue Management on Organic Carbon, Soil Attributes and Crop Yield

Normally residues may be managed by different methods: by removal, feeding to animals, burning, incorporation, or left on the soil surface. Soil productivity is directly influenced by the fate of crop residues and the best effects are attained if they are not removed from the field. Residue incorporation effects on soil productivity are difficult to separate from the tillage effects because the incorporation is achieved through some type of tillage operation. Also, soil water content, soil temperature and porosity are influenced by the presence and redistribution of the crop residues.

Carbon and N content in crop residues along with lignin content greatly influence the decomposition rates and N availability to plants (Hargrove, 1991). Decomposition of residues with low N contents such as black oats (*Avena strigosa* Schreb.) may result in microbial immobilization of soil and fertilizer N, effectively reducing N availability to plants. Normally, residues with N concentrations below 1.5% or C:N ratios greater than 25–30 are considered to immobilize inorganic N. Despite these, residues with very similar C:N ratios can have different decomposition rates because of the differences in their chemical composition (Stott and Martin, 1989).

Studies conducted by Sidiras and Pavan (1985) and Calegari (1995a) in south Brazil, Paraná showed significant increase in pH, effective cation exchange capacity, Ca, Mg, K and P and decrease in Al saturation near the soil surface under NT, in comparison to the conventional system. Similar results were obtained by Sá (1993) in south Paraná and Calegari *et al.* (1995) in the northern region.

On-farm studies in north Paraná, Santo Antonio farm, compared the two tillage systems. The NTS yielded 34.4% and 13.7% more soybean and wheat, respectively, compared to the conventional tillage systems (Table 3.9).

In addition, the crop rotation increased yields of soybean and wheat by another 19.2% and 5.8%, respectively, in comparison to monoculture, showing that under both on-station and on-farm conditions the benefits of increased grain yield of soybean and wheat were attained.

A study conducted on a 50 ha experimental area in northern Paraná (Calegari *et al.*, 1998a) showed that an adequate NTS with soybean in crop rotation can generate good net income compared to conventional systems (Table 3.10).

These results show that NT under appropriate crop rotation by including soybean gives significantly higher benefit than conventional tillage and monocropping of soybean. These results were based on a soybean price of US$166.00 per ton grain. Therefore, it is very profitable for the farmers; and also increases the stability of the grain soybean production system.

The area under NTS continues to increase every year with many different crops being planted (soybean, maize, beans, cotton, sorghum, millet, sunflower, wheat, barley, rye, oat, lupin, rape, groundnuts and vegetable crops) with improved profitability. This has been a period of rapid improvement in agriculture in Brazil – a most exciting time!

Different agroecological zones of Brazil such as regions of Paraná, Rio Grande do

Table 3.9. Average grain yield (from 1985/86 to 1991/92) in different tillage and cropping systems during 7 years in Paraná State (Calegari *et al.*, 1995).

	Yield (kg ha^{-1})			
Treatments	Soybean	Yield (%)	Wheat	Yield (%)
No-tillage	2816	134.4	2121	113.7
Conventional tillage	2094	100.0	1864	100.0
Crop rotation[a]	3040	119.2	2200	105.8
Monoculture[a]	2550	100.0	2078	100.0

[a]Average values

Table 3.10. Economic evaluation of soybean production in an area of 50 ha in no-till crop rotation system as compared to conventional tillage in northern Paraná (Calegari *et al.*, 1998a).

Particulars	Benefits (US$)
Crop yield improvement	3960
Cheaper machine maintenance	1145
Fuel saving	731
Labour saving	2880
Fertilizer saving	186
Total benefit	8902

Sul state and also Savannah and other agricultural areas with several farming systems present a large number of species of cover crop alternatives, which had been largely used by farmers. These species grow in many regions in different cropping systems with cash crops such as maize, wheat, beans, soybean, cotton, cassava, potato, groundnut, sunflower and vegetables and are also intercropped with perennial crops such as coffee, citrus, fruit trees, grapes, etc. They also improve soil properties, and also help mobilize soil nutrients for the next crops. They have also been used for multi-purposes including as animal fodder and some species have potential as human food. The evolution of NT in Brazil has developed with the evolution of agricultural systems in the country (Fig. 3.4).

It can be observed that from 1975 until 2010, the planted area in Brazil increased 31%, while the production enhanced 228% and the crop yield increased 151%. It is important to emphasize that this has been as a result of some important advances in the use of suitable genetic material, fertilizers, machinery, pesticides and biotechnology, and one of the most important components that contributes to the increase in crop yields and the production of grains and oilseeds is of course the high adoption of NTS by Brazilian farmers.

Some results presented in Table 3.11 provide examples of the evolution of crops, beef and wood products in selected Brazilian regions.

Clearly, NT played an important role in increasing crop yields and profitability at the farm level, and also enhanced biodiversity and environmental conditions.

3.9 Strategies for Dissemination of No-Till Among Farmers in Brazil

Following pioneer NT farmers in Brazil including Mr Herbert Bartz, Mr Manoel Henrique Pereira (Nonô Pereira) and Mr Frank Dijkstra who obtained good results at their farms by implementing the NTS, many other farmers and researchers and extension personnel focused to develop, validate and spread this system to other regions and farmers from Paraná, and also to other Brazilian regions.

Interestingly, the spread of NT permanent soil cover by small farmers worldwide has generally been poor. It remains marginal outside Brazil, Paraguay and some other Latin American countries such as Bolivia,

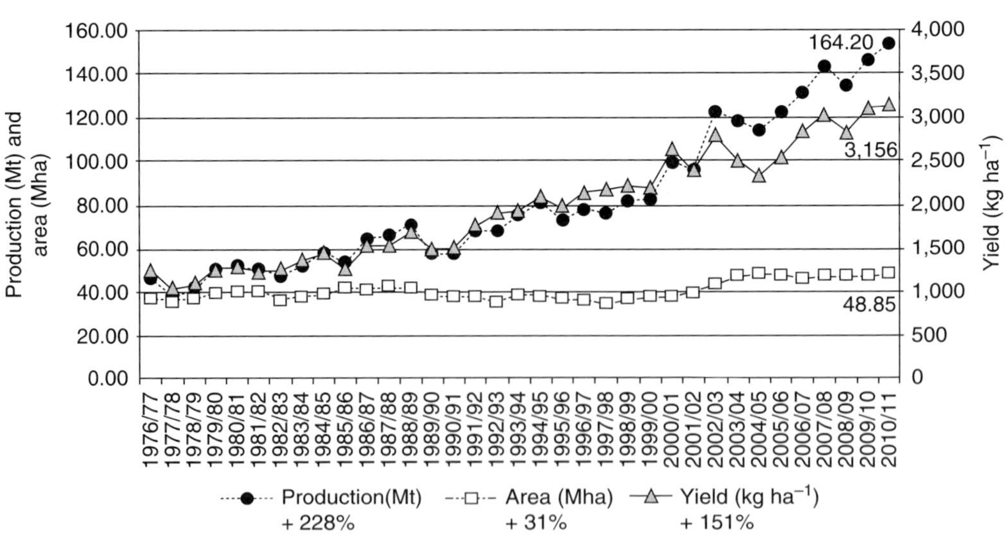

Fig. 3.4. Evolution of grains and oilseeds production (Mt), yields (kg ha^{-1}) and area (Mha) in Brazil from 1975 to 2010 (CONAB, 2010).

Table 3.11. Average yield (last 10 years) in Brazil and Savannah Biome (CONAB, 2010).

		Savannah Biome		
Products	Brazil	Ordinary farm[a]	High-tech farm	Experimental station
Maize[b]	3,507	4,546	12,000	16,000
Soybean[b]	2,613	2,846	3,900	5,000
Beans[b]	778	1,268	2,000	4,000
Eucalyptus[c]	30	40	80	120
Beef cattle[d]	60	70	90	120

[a]Yields from mid-western states (Mato Grosso, South Mato Grosso and Goiás)
[b]Average kg ha^{-1} $year^{-1}$
[c]m^3 ha^{-1} $year^{-1}$
[d]live-weight gain; ha^{-1} $year^{-1}$ (complete system)

Colombia, Uruguay and parts of Central America. The opportunity cost of labour, land and organic residues is often viewed as a stumbling block for small farmers for wide adoption of NTS. NTS have been reported to not only reduce soil erosion, but also increase crop yields and family income, and reduce drudgery by farmers (Bolliger *et al.*, 2006).

The light machinery (mini-tractors) and animal-drawn NT equipment were developed and tested on various soil types and topography by IAPAR and its equivalent in the state of Santa Catarina, EPAGRI (Table 3.12), and gradually more and more smallholder farmers started adopting zero-till technologies.

However, as the results of a recent survey in the Irati region of Paraná indicate, unlike their more commercially oriented large-scale counterparts, smallholder NT farmers, without sufficient means to buy recommended external inputs and consequently often with a high degree of risk-averseness, as well as high-opportunity costs for land, labour and crop biomass, still resort to a range of intermediate-tillage systems rather than adopting complete or 'ideal' NT models promoted by research and extension (Palmans and van Houdt, 1998; Ribeiro *et al.*, 2005). Many such farmers fall back on disc harrowing before/after certain crops in order to check weeds and pests and incorporate lime, while sometimes neglecting cover and main crop rotations that could potentially optimize the functioning of NTS. As Ribeiro *et al.* (2005) further conclude, contrary to some perhaps overly enthusiastic reports on the success of NT in Brazil, and although some very-well functioning 'ideal' smallholder NT farms do exist, numerous challenges for the resource-poor smallholder NT farmers on a general level still remain. As Calegari (2002) argues, such challenges, but also innovations and advances in terms of smallholder systems (e.g. equipment and fertility changes), need to be continuously evaluated and monitored by following testing/validation processes that involve the smallholders themselves.

NT is more than planting a crop into an undisturbed soil. The basis to make the system work is proper use of cover crops in a sound cropping sequence (crop rotation). The diagnosis of the soil-system, consideration of soil characteristics and their interactions (physical, chemical and biological aspects) determines which crops to grow in an adequate cropping system. Therefore, harmony in the use of cover-crop species and rotation systems are important components for the sustainability of the NTS.

Nevertheless, one must recognize that for small farmers the lack of equipment to cut straw and open a furrow in undisturbed soil is the main constraint in the adoption of this system. After the development of the first animal-drawn NT planter prototype Gralha Azul/IAPAR, a series of on-farm trials were established in order to assess the technical and economic efficiency and its feasibility at the farm level. The main constraints were identified and

Table 3.12. Development, testing and evaluation of light machinery under on-farm; and the dissemination and adoption process of no-till technologies by smallholder farmers in Santa Catarina State, southern Brazil (Freitas *et al.*, 1994).

1984/1985	1986	1987
Facilitation of farmer excursions to relevant research and experimental sites Formation of micro-catchment commissions	Establishment of a green manure observation unit and identification of potential cover-crop green manure systems	Establishment of the first crop through zero-till with animal traction
1988/1990	1991/1992	1993/1994
Period of testing and adapting agricultural equipment, especially equipment for zero-till with animal traction and light mechanization	Farmers started to adopt zero-till practices (5% adoption rate) Continuous research and adaptation of zero-till equipment	Increase in the area under zero-till Acquisition of equipment by individual farmers and farmer groups

some technological options were formulated and developed.

In general, for a long time most of the small farmers had no easy access to credit and information. Fortunately, in recent years this has changed, and as a result of government support (municipality, state and federal), farmers' association, public extension service and public research has focused on developing sustainable soil and water management practices through mainly supporting NT, including cover crops, rotation and options for grain, livestock, trees and sustainable production systems. Also strategies to improve local markets for inputs and outputs, agroindustries to add value to the products, as well developing locally adapted suitable soil and water conservation systems have been promoted in different agricultural regions of Brazil.

3.10 Perspectives

The cropping systems based on sound CA principles can contribute to solve some of the main constraints of the small-scale farmers and may help to suppress poorly executed soil tillage operations, ensure timely and proper crop planting, facilitate proper weed control, enhance forage production for the dry season, save labour, promote soil biology improvement (macro-, meso- and micro-soil fauna and flora – increasing biodiversity), lead to improved soil properties over time including physical, biological and chemical, ensure rational use of inputs by decreasing need for external farm inputs, help controlling wind and water erosion processes, and sequester atmospheric carbon, decrease the amount of CO_2 released to the atmosphere and mitigating the greenhouse effects.

The Old-World experience has shown that the abundance of natural resources leads individuals to their misuse. In contrast, scarce resources stimulate economic rationality and concern over predictability; in other words, responsible actions are taken for environmental preservation both in the present and the future.

The NT adoption process by smallholders of such rather complex innovations on a significant scale requires involved projects and institutions to implement approaches that are as fully participatory as possible. This includes designing technical options based on CA principles jointly with farmers from the very beginning in order to answer more closely their main constraints and/or objectives. It also includes strengthening farmers' capacity to organize themselves for at least two key reasons. First, because it is perhaps the only or the best way for farmers to gain adequate access to CA inputs including training and technical assistance. But beyond this functional reason, collective organization also opens the door to achieving non-technical innovations such as better negotiating capacity.

Some of the key technological challenges for NT evolution in the next years in southern Brazil are related to the spread of economic returns and the importance of adopting crop rotations through the use of other plant species besides soybean and maize, which will increase the permanence of straw over the soil surface and thus ensure full coverage of the soil throughout the year. This has a direct impact on mechanization inputs required, since NT seeders will have to deal with different requirements including size, geometries, quantity, spacing and depth of seeding of crops.

Furthermore, it is important to improve components of soil–tool–straw contact of NT seeders to ensure a high seeding quality, total coverage of the furrow with straw and seeding operation under conditions of large volumes of straw on the soil surface.

Beyond the NTS, we also must consider regionally some specificities of soil, water, rainfall, cropping and farming systems, in order to adjust and include components such as terracing, grassed waterways, vegetative terrace, etc., adequately tested and validated regionally under farm conditions to achieve a sustainable agriculture.

We have learnt through 40 years of farmer experience and field research in south Brazil and beyond, that NT systems combined with appropriate crop rotations are very economical and sustainable. Such robust systems ensure soil erosion control, provide higher soil-water storage, enhance soil fertility and give increased crop productivity.

We have learnt to grow cash crops in conjunction with cover crops in a sensible manner. In addition, these crops in rotation save N fertilizer, give superior weed control through the mulch effects, and give a greater biological balance in the soil, higher soil biodiversity decreasing inset-pests and disease occurrence, saving labour and fuel, and decreasing production costs and thus indeed represents a sustainable way of farming.

References

Adegas, F.S. (1998) Manejo integrado de plantas daninhas em plantio direto no Paraná. In: *Seminário nacional sobre manejo e controle de plantas daninhas em plantio direto*, 1, Palestras, Aldeia Norte, Passo Fundo, RS, Brazil, pp. 17–26.

Almeida, F.S. (1988) *A alelopatia e as plantas.* IAPAR, Circular Técnica, 53, Londrina, Brazil, p. 60.

Almeida, F.S. and Rodrigues, B.N. (1985) *Guia de herbicidas. Contribuição para o uso adequado em plantio direto e convencional.* IAPAR, Londrina, Prana, Brazil, p. 482.

Almeida, F.S., Rodrigues, B.N. and Oliveira, V.F. (1983) Influence of winter cover crop mulches on weed infestation in maize. In: *Proceedings of Symposium on Weed Problems in the Mediterranean Area.* (3: Oeiras, Portugal). EWRS, Wageningen, the Netherlands, pp. 351–358.

Almeida, F.S., Rodrigues, B.N. and Oliveira, V.F. (1984) Influence of winter crop mulches on weed infestation in maize. In: *Proceedings of the 3rd EWRS Symposium on Weed Problems in the Mediterranean Area.* Lisbon, Portugal, pp. 351–358.

Altieri, A.M., Ponti, L. and Nicholls, C.I. (2007) Melhorando o Manejo de Pragas através da Saúde do Solo: Direcionando uma Estratégia de Manejo do Habitat Solo. In: *Controle Biológico De Pragas Através Do Manejo De Agroecossistemas.* Brasília, pp. 17–31.

Altieri, M.A. (1995) *Agroecology. The science of sustainable agriculture*, 2nd edn. Westview Press, Colorado, p. 433.

Altieri, M.A. and Doll, U.D. (1978) The potential of allelopathy as a tool for weed management in crop fields. *PANS* 24(4), 495–502.

Amado, T.J.C. and Eltz, F.L.F. (2003) Plantio direto na palha – rumo à sustentabilidade agrícola nos trópicos. *Revista Ciência e Ambiente*, Santa Maria, 27, 49–66.

Anderson, S., Güendel, S., Pound, B. and Triomphe, B. (2001) *Cover Crops in Smallholder Agriculture: Lessons from Latin America.* IT Publications, London, p.136.

Araújo, A.G., Casao, Jr and Figueiredo, P.R.A. (1993) Recomendações para o dimensionamento e construção do rolo-faca. In: *Encontro Latino Americano de Plantio Direto na Pequena Propriedade*, 22–26 November 1993. Ponta Grossa, PR, Brazil, pp. 271–280.

Araújo, A.G., Yamaoka, R., Figueiredo, P.R.A. and Benassi, D.A. (1998) Equipamentos e implementos para manejo. In: *Plantio Direto. Pequena propriedade sustentável.* Instituto Agronômico do Parana (IAPAR) Circular 101, Londrina, PR, Brazil, pp. 113–127.

Ashford, D.L. and Reeves, D.W. (2003) Use of mechanical roller-crimper as an alternative kill method for cover crops. *American Journal of Alternative Agriculture* 18, 37–45.

Bairrão, J.F.M., Goelzer, L.F.D. and Bego, A. (1988) Comportamento de alternativas de inverno com vista à integração em rotação de culturas. In: *Resultados de pesquisa na safra de inverno*, 1986. Ocepar, Cascavel, Brazil, pp. 112–113.

Basch, G., Kassam, A., Friedrich, T., Santos, F.L., Gubiani, P.I., Calegari, A., Reichert, J.M. and Rheinheimer, D.S. (2012) Sustainable soil water management systems. In: Lal, R. and Stewart, B.A. (eds) *Soil Water and Agronomic Productivity*. CRC Press, Taylor & Francis Inc., Bosa Roca, USA, pp. 229–288.

Bayer, C. (1996) Dinâmica da matéria orgânica em sistemas de manejo de solos. Tese (Doutorado em Solos), Universidade Federal do Rio Grande do Sul, Porto Alegre, RS, 240 pp.

Bayer, C., Mielniczuk, J., Amado, T.C., Martin-Neto, L. and Fernandez, J.V. (2000) Organic matter storage in a clay loam Acrisol affected by tillage and cropping systems in southern Brazil. *Soil and Tillage Research* 54, 101–109.

Blouin, M., Zuily-Fodil, Y., Pham-Thi, A.-T., Laffray, D., Reversat, G., Pando, A., Tondoh, J. and Lavelle, P. (2005) Belowground organism activities affect plant aboveground phenotype, inducing plant tolerance to parasites. *Ecology Letters* 8(2), 202–208.

Bolliger, A., Magid, J., Amado, T.J.C., Skora Neto, F., Santos Ribeiro, M.F., Calegari, A., Ralisch, R. and Neergaard, A. (2006) Taking stock of the Brazilian 'zero till revolution': a review of landmark research and farmers' practice. *Advances in Agronomy* 91, 47–110.

Calegari, A. (1990) *Plantas para adubação verde de inverno no Sudoeste do Paraná*. IAPAR, Boletim técnico, 35. IAPAR, Londrina, Parana, Brazil, p. 37.

Calegari, A. (1995a) *Leguminosas para adubação verde de verão no Paraná*. IAPAR, Circular, 80, Londrina, Parana, Brazil, p. 118.

Calegari, A. (1995b) Leguminosas para adubação verde de verão no Paraná. IAPAR, Circular Técnica, 80. IAPAR, Londrina, Parana, Brazil, p. 118.

Calegari, A. (1995c) The effects of tillage and cover crops on some chemical properties of an Oxisol in south-western Paraná, Brazil. Dissertation thesis, University of Aberdeen, Department of Plant and Soil Science, Aberdeen, UK, p. 81.

Calegari, A. (1998a) The effects of winter cover crops and no-tillage on soil chemical properties and maize yield. In: *Summaries – Symposium no. 7*. August 20–26, 16th Soil World Congress of Soil Science, Montpellier, France (CD-Rom).

Calegari, A. (1998b) Espécies para cobertura do solo. In: Moacir Roberto Darolt (ed.) *Plantio direto - Pequena Propriedade Sustentável*. IAPAR, Circular, 10, Londrina, Paraná, Brazil, pp. 65–94.

Calegari, A. (1998c) Towards sustainable agriculture with a zero tillage system in south Brazil. In: Benites, J., Chuma, E., Fowler, R., Kienzle, J., Molapong, K., Manu, I., Nyagumbo, Steiner, K. and van Veenhuizen, R. (eds) *Proceedings of the International Workshop on Conservation Tillage for Sustainable Agriculture, Annexe III: Background Papers (International)*. GTZ, Eschborn, Harare, Zimbabwe, pp. 239–246.

Calegari, A. (2000a) Cover crops and crop rotation. In: *Beans Crop System Technology*. IAPAR, Research, 135. Londrina, Parana, Brazil, pp. 29–34.

Calegari, A. (2000b) Mulch provided by cover crops on no-till system. In: *Guia de Plantio Direto*. Edited by Masa S/C Ltd, São Paulo-SP, Brazil, pp. 30–37.

Calegari, A. (2000c) Crop rotation. In: *Guia de Plantio Direto*. Edited by Masa S/C Ltd, São Paulo-SP, Brazil, July, pp. 68–78.

Calegari, A. (2002) The spread and benefits of no-till agriculture in Paraná State, Brazil. In: Norman, U. (ed.) *Agroecological Innovations: Increasing Food Production With Participatory Development*. Earthscan, London, pp. 187–202.

Calegari, A. (2009) Alternative land uses and farm diversification strategies to strengthen CA. In: *Lead Papers – 4th World Congress on Conservation Agriculture on Innovations for Improving Efficiency, Equity and Environment*. M/S Print Process, 225, DSIDC Complex, Okhla Industrial Area, Phase I, New Delhi, India, pp. 117–121.

Calegari, A. (2010) Rotação de culturas/Culturas de cobertura. In: *12o. Encontro nacional de Plantio direto na Palha. Plantio direto Tecnologia que mudou a visão do produtor*. RESUMOS./ANAIS Federação brasileira Plantio direto na Palha; Editado por Lutécia Canalli, Ponta Grosa, Pr: FEBRAPDP, 2010. Foz do Iguaçu, 23 a 25 de junho de 2010, pp. 165–172.

Calegari, A. and Alexander, I. (1998) The effects of tillage and cover crops on some chemical properties of an Oxisol and summer crop yields in south-western Paraná, Brazil. *Advances in GeoEcology* 31, 1239–1246.

Calegari, A. and Pavan, M.A. (1995) Efeitos da rotação de milho com adubos verdes de inverno na agregação do solo. *Arquivo de Biologia e Tecnologia* 38(1), 45–53.

Calegari, A., Mondardo, A., Bulisani, E.A., Wildner, L. do P., Costa, M.B.B., Alcântara, P.B., Miyasaka, S. and Amado, T.J.C. (1993) *Adubação verde no sul do Brasil*, 2nd edn. AS-PTA, Rio de Janeiro, 346 pp.

Calegari, A., Ferro, M., Grzesiuk and Jacinto Jr, L. (1995) *Plantio direto e rotação de culturas. Experiência em latossolo roxo, 1985-1992*. Cooperativa dos Cafeicultores e Agropecuaristas de Maringá Ltd, Maringá, PR, Brazil, 64 pp.

Calegari, A., Darolt, M.R. and Ferro, M. (1998a) Towards sustainable agriculture with a no-tillage system. *Advances in GeoEcology* 31, 1205–1209.

Calegari, A., Costa, A. and Ralisch, R. (2007) Sustainable agriculture with no-tillage including cover crops and crop rotation. *Annals of Arid Zone* 46(1), 1–23.

Calegari, A., Hargrove, W.L., Rheinheimer, D.S., Ralisch, R., Tessier, D., Tourdonnet, S. and Guimarães, M.F. (2008) Impact of long-term no-tillage and cropping system management on soil organic carbon in an oxisol: a model for sustainability. *Agronomy Journal* 100, 1013–1019.

Casão Junior, R. and Siqueira, R. (2003) *Resultados das avaliações do desempenho de semeadoras adubadoras diretas na Costa Oeste Paranaense*. Circular no. 127, IAPAR, Londrina, Paraná, Brazil, p.134.

Casão Junior, R. and Siqueira, R. (2004) Dinâmica de semeadoras-adubadoras diretas em Guaíra-PR. Passo Fundo. In: *Revista Plantio Direto*, pp. 15–27.

Casão Junior, R., Siqueira, R., Mehta, Y.R. and Passini, J.J. (2006) *Sistema plantio direto com qualidade*. IAPAR, Londrina (PR); ITAIPU Binacional: Foz do Iguacu (PR), Brazil, 200 pp.

Casão Junior, R., Araújo, A.G. and Llanillo, R.F. (2012) No-till agriculture in southern Brazil. IAPAR/FAO, Londrina, Brazil, p. 77.

Castro, O.M., Severo, Prado, H. do, Severo, A.C.R. and Cardoso, E.J.B.N. (1993) Avaliação da atividade de microorganismos do solo em diferentes sistemas de manejo da soja. *Sci. agric. (Piracicaba, Brazil)* [online] 50(2), 212–219.

CONAB (Companhia Nacional do Abastecimento) (2010) MAPA (Ministério da Agricultura e Abastecimento). Avaliação da safra agrícola 2010. Brasília, p. 19.

Conte, E., Anghinoni, I. and Rheinheimer, D.S. (2003) Frações de fósforo acumuladas em Latossolo argiloso pela aplicação de fosfato no sistema plantio direto. *Revista Brasileira de Ciência do Solo (Impresso)*, Viçosa, 27(5), 893–900.

Costa, A., Pires, J.R. and Yamaoka, R.S. (1993) Efeito da rotação de culturas sobre o rendimento do algodoeiro. In: *Reunião Nacional do Algodão, 7., Cuiabá, Resumos... Cuiabá*: EMPAER-MT; Campina Grande: EMBRAPA-CNPA, p. 200. Embrapa Agropecuária Oeste, Campo Grande – MS, Documentos, 32.

Costa, J.L. da S. (1999) Influência de plantio direto e manejo de palhada nas podridões radiculares do feijoeiro. In: *Reunião Nacional De Pesquisa De Feijão*, 6., 1999, Salvador. Resumos expandidos. Embrapa Arroz e Feijão, Santo Antônio de Goiás, pp. 218–220. (Embrapa Arroz e Feijão. Documentos, 99.)

Darolt, M.R. (1998) *Plantio Direto – Pequena Propriedade Sustentável*. Circular, 101, IAPAR, Londrina, Brazil, p. 255.

Denardin, J.E., Kochhann, R.A., Bacaltchuk, B., Sattler, A., Denardin, N. D'a., Faganello, A. and Wiethölter, S. (2008) Sistema plantio direto: fator de potencialidade da agricultura tropical brasileira. In: Albuquerque, A.C.S. and Silva, A.G. da (eds) *Agricultura Tropical: Quatro Décadas de Inovações Tecnológicas, Institucionais e Políticas*. Embrapa Informação Tecnológica, Brasília, DF, vol. 1, cap. 1, pp. 1251–1273.

Derpsch, R. (1986) Erosion Problems in Paraná, Brazil. Research Results and Strategies for the Implementation of Efficient Soil Conservation Measures. Dissertation Thesis, Agricultural Extension and Rural Development Centre, Reading University, UK.

Derpsch, R. (2003) Why and how to use green manure cover crops in a zero tillage system: Experiences from Latin America. In: *Proceedings of a Conference sponsored by the South Dakota Zero tillage Association on 2003 Zero tillage Under Cover*. Sioux Falls Convention Center, Sioux Falls, South Dakota, pp. 5–13.

Derpsch, R. and Calegari, A. (1985) *Guia de Plantas de Inverno*. IAPAR, Documentos, 9, Londrina, Parana, Brazil, p. 96.

Derpsch, R., Sidiras, N. and Roth, C.H. (1986) Results of studies made from 1977 to 1984 to control erosion by cover crops and zero tillage techniques in Paraná, Brazil. *Soil Tillage Research* 8, 253–263.

Derpsch, R., Roth, C.H., Sidiras, N. and Köpke, U. (1991) *Controle da erosão no Paraná, Brasil: sistemas de cobertura de solo, plantio direto e preparo conservacionista do solo*. GTZ, Eschborn, Alemanha e IAPAR, Londrina, Brazil, p. 272.

EMATER – Empresa de Assistência Técnica e Extensão Rural do Paraná (1996) *Sistema de controle operacional realidade rural maquinário agrícola*. EMATER-PR, Curitiba, Brazil, p. 19.

Florentin, M.A., Peñalva, M., Calegari, A. and Derpsch, R. (2001) Abonos verdes y rotación de cultivos en siembra directa en pequeñas propiedades. MAG-GTZ, Paraguay, p. 85.

Florentin, M.A., Peñalva, M., Calegari, A. and Derpsch, R. (2010) *Green manure / cover crops and crop rotation in conservation agriculture on small farmers*. Vol. 12. FAO, Rome, p. 97.

Freitas, V.H. de (2000) *Soil management and conservation on small farms. Strategies and methods of introduction, technologies and equipment*. FAO Soils Bulletin 77, FAO, Rome.

Freitas, V.H., Gubert, R. and Bet, M. (1994) Situação da adubação verde em Santa Catarina. In: *Encontro Centro Sul Brasileiro de Adubos Verdes e Rotação de Culturas*. Anais. Chapecó, Santa Catarina, Brasil, pp. 55–63.

Gatiboni, L.C., Kaminski, J., Rheinheimer, D.S. and Flores, J.P.C. (2007) Biodisponibilidade de formas de fósforo acumuladas em solo sob sistema plantio direto. *Revista Brasileira de Ciência do Solo (Impresso)* 31, 691–699.

Guardini, R., Comin, J.J., Schmitt, D.J., Tiecher, T., Bender, M.A., Rheinheimer, D.S., Mezanni, C.P., Oliveira, B.S., Gatiboni, L.C. and Brunetto, G. (2012) Accumulation of phosphorus fractions in typic Hapludalf soil after long-term application of pig slurry and deep pig litter in a no-tillage system. *Nutrient Cycling in Agroecosystems* 93, 215–225.

Hargrove, W.L. (1991) *Cover Crops for Clean Water*. Soil and Water Conservation Society, Ankeny, Iowa, 5002 1–9764, p. 198.

IAPAR (Instituto Agronômico do Paraná) (1981) *Plantio Direto no Estado do Paraná*. Editado pelo IAPAR. Londrina, Paraná, Brazil (Circular no. 23), p. 244.

Jat, R.A., Wani, S.P. and Sahrawat, K.L. (2012) Conservation agriculture in the semi-arid tropics: prospects and problems. In: Sparks, D.L. (ed.) *Advances in Agronomy* 117, 191–273.

Kliewer, I., Casaccia, J. and Vallejos, F. (1998) Viabilidade da redução do uso de herbicidas e custos no controle de plantas daninhas nas culturas de trigo e soja no sistema plantio direto, através do emprego de adubos verdes de curto período. In: *Primeiro Seminário Nacional sobre manejo e controle de plantas daninhas em plantio direto*. Aldeia Norte, Passo Fundo RS, Brazil, pp. 120–123.

Martinazzo, R., Rheinheimer, D.S., Gatiboni, L.C., Brunetto, G. and Kaminski, J. Fósforo (2007) Microbiano do solo sob sistema plantio direto em resposta à adição de fosfato solúvel. *Revista Brasileira de Ciência do Solo (Impresso)* 31, 563–570.

Medeiros, G.B., Calegari, A. and Gaudêncio, C. (1989) Rotação de culturas. In: *Manual Técnico do Sub-programa de Manejo e Conservação do Solo*. Secretaria da Agricultura e do Abastecimento, Curitiba, pp. 189–195.

Merten, G.H., Fernandes, F.F., Machado, M., Ribeiro, M.F.S., Samaha, M.J., Benassi, D., Gomes, E.P., Siqueira, E.M. and Silva, F. (1994) Estratégias de manejo para solos de baixa aptidão agrícola da região Centro-Sul. In: Merten, G.H. (Coord.) *Manejo de Solos de Baixa Aptidão Agrícola no Centro-Sul do Paraná*. IAPAR, Circular, 84, Londrina, Parana, Brazil, p. 112.

Miyasaka, S. and Okamoto, H. (1992) Integrated production systems and organic agriculture. In: Assisi, R.L. de, Souto, S.M., Duke, Fernando, F., Almeida, D.L. and Mueller, K. (eds) *II Course on Soil Biology in Agriculture*. Seropédica: EMBRAPA CNPBS, Documents, 8, p. 41.

Miyazawa, M., Pavan, M.A. and Calegari, A. (1994) Efeitos de materiais vegetais na acidez do solo. *R. bras. Ci. Solo*., Campinas, SP, Brazil, 17, 411–416.

Muzilli, O. (1978) O manejo da fertilidade do solo: a prática de adubação verde. In: *Fundação Instituto Agronômico do Paraná, Londrina, Manual Agropecuário para o Paraná*. Londrina, Parana, Brazil, pp. 57–58.

Muzilli, O. (1981) Princípios e perspectivas de expansão. In: *Instituto Agronômico do Paraná. Plantio Direto no Estado do Paraná*. IAPAR, Londrina, Parana, Brazil, pp. 11–17.

Muzilli, O. (2006) Manejo do solo em sistema de plantio direto. In: Casão Junior, R., Siqueira, R., Mata, Y.R. and Passini, J.A. (eds) *Plantio Direto com Qualidade*. IAPAR/Itaipu Binacional, Londrina/Foz do Iguaçu.

Muzilli, O., Oliveira, E.L., Gerage, A.C. and Tornero, M.T. (1983) Adubação nitrogenada em milho no Paraná. 2. Influência da recuperação do solo com a adubação verde de inverno nas respostas à adubação nitrogenada. *Pesquisa Agropecuária Brasileira*, Brasília, DF, 18, 23–27.

Oliveira, E.L. (1994) Coberturas verdes de inverno e adubação nitrogenada em algodoeiro. *R. bras. Ci. Solo*, Campinas, SP, Brazil, 18, 235–241.

Palmans, B. and van Houdt, E. (1998) *Effet des Systemes de Culture sur la Degradation Physique du Sol et sur son Evolution Pedologique (Paraná-Brésil)*. Relatório (AOL and DAT), CNEARC, Montpellier, France, p. 140.

Pieri, C., Evers, G., Landers, J., O'Connell, P. and Terry, E. (2002) No-till farming for sustainable rural development. Agriculture & Rural development working paper. The International Bank for Reconstruction and Development Rural Development Department, Washington, DC, p. 65.

Primavesi, A. (1982) *O manejo ecológico do solo*, 4th edn. Ed. Nobel, São Paulo, Brazil, p. 541.

Resck, D.V.S., Vasconcellos, C.A., Vilela, L. and Macedo, M.C.M. (1999) Impact of conversion of Brazilian Cerrados to cropland and pasture land on soil carbon pool and dynamics. In: Lal, R., Kimble, J.M. and Stewart, B.A. (eds) *Global Climate Change and Tropical Ecosystems*. Advances in Soil Science, CRC Press, Boca Raton, Florida, pp. 169–196.

Rheinheimer, D.S. and Anghinoni, I. (2001) Distribuição do fósforo inorgânico em sistemas de manejo de solo. *Pesquisa Agropecuária Brasileira (1977. Impressa), Brasíla - DF* 36(1), 151–160.

Rheinheimer, D.S. and Anghinoni, I. (2003) Accumulation of soil organic phosphorus by soil tillage and cropping systems under subtropical conditions. *Communications in Soil Science and Plant Analysis* 34 (15–16), 2339–2354.

Rheinheimer, D.S., Kaminski, J., Santos, E.J.S., Gatiboni, L.C. and Bortoluzzi, E.C. (2000a) Alterações de atributos do solo pela calagem superficial e incorporada a partir de pastagem natural. *Revista Brasileira de Ciência do Solo*, Viçosa, MG, 24(4), 797–805.

Rheinheimer, D.S., Anghinoni, I. and Kaminski, J. (2000b) Depleção do fósforo inorgânico de diferentes frações provocada pela extração sucessiva com resina em diferentes solos e manejos. *Revista Brasileira de Ciência do Solo (Impresso), Viçosa, MG* 24(2), 345–354.

Rheinheimer, D.S., Anghinoni, I. and Conte, E. (2000c) Fósforo da biomassa microbiana em solos com diferentes teores de argila e sistemas de manejo. *Revista Brasileira de Ciência do Solo (Impresso), Viçosa, MG* 24(3), 589–597.

Rheinheimer, D.S., Conte, E. and Anghinoni, I. (2002a) Fósforo da biomassa microbiana e atividade de fosfatase ácida pela aplicação de fosfato em solo no sistema plantio direto. *Revista Brasileira de Ciência do Solo, Viçosa* 26(4), 925–930.

Rheinheimer, D.S., Anghinoni, I. and Frores, A.C. (2002b) Organic and inorganic phosphorus as characterized by phosphorus-31 nuclear magnetic resonance in subtropical soils under management systems. *Communications in Soil Science and Plant Analysis* 33(11–12), 1853–1871.

Rheinheimer, D.S., Anghinoni, I., Conte, E., Kaminski, K. and Gatiboni, L.C. (2003a) Dessorção de fósforo avaliada por extrações sucessivas em amostras de solo provenientes dos sistemas plantio direto e convencional. *Ciência Rural (UFSM. Impresso)* 33, 1053–1059.

Rheinheimer, D.S., Anghinoni, I. and Conte, E. (2003b) Sorção de fósforo em função do teor inicial e de sistemas de manejo de solos. *Revista Brasileira de Ciência do Solo (Impresso), Viçosa* 27(1), 41–49.

Rheinheimer, D.S., Tessier, D., Kaminski, J. and Bortoluzzi, E.C. (2006) Travail du sol simplifié au Brésil: un succès contrasté. *Oléoscope, Thiverval-Grignon* 87(11), 28–31.

Rheinheimer, D.S., Gatiboni, L.C. and Kaminski, J. (2008) Fatores que afetam a disponibilidade do fósforo e o manejo da adubação fosfatada em solos sob sistema plantio direto. *Ciência Rural (UFSM. Impresso)* 38, 576–586.

Ribeiro, M.F.S., Benassi, D., Triomphe, B. and Huber, H. (2005) Incorporation of zero tillage principles into family farmers' practices at Irati region, South Brazil. *III World Congress on Conservation Agriculture*, 3–7 October, Nairobi, Kenya [CD-ROM].

Ruedell, J. (1995) *Plantio direto na região de Cruz Alta*. Fundacep/Basf, Cruz Alta, RS, Brazil, p.134.

Sá, J.C.M. (1993) Manejo de fertilidade do solo no sistema plantio direto. In: Aldeia Norte, EMBRAPA-CNPT, FECOTRIGO, Fundação ABC (Org.) *Plantio direto no Brasil*. Castro: Aldeia Norte, pp. 41–47.

Sá, J.C.M., Cerri, C.C., Dick, W.A., Lal, R., Venske Filho, S.P., Piccolo, M.C. and Feigl, B.F. (2001) Organic matter dynamics and carbon sequestration rates for a tillage chronosequence in a Brazilian oxisol. *Soil Science Society of America Journal* 65, 1486–1499.

Santos, H.P., Reis, E.M. and Pereira, L.R. (1990) Rotação de culturas. Efeitos no rendimento de grãos e nas doenças do sistema radicular do trigo de 1980 a 1987. *Pesquisa Agropecuária Brasileira, Brasília, DF, Brasil* 25(11), 1627–1635.

Santos, H.P.D., Lhamby, J.C.B., Prestes, A.M. and de Lima, M.R. (2000) Efeito de manejo de solo e de rotação de culturas de inverno no rendimento e doenças de trigo. *Pesquisa Agropecuária Brasileira, Brasília, DF, Brasil* 35, 2355–2361.

Scopel, E., Findeling, A., Chavez Guerra, E. and Corbeels, M. (2005) Impact of direct sowing mulch-based cropping systems on soil carbon, soil erosion and maize yield. *Agronomy for Sustainable Development* 25, 425–432.

Scopel, E., Triomphe, B., Ribeiro, M.F.S., Seguy, L., Denardin, J.E. and Kochann, R.A. (2004) Direct seeding mulch-based cropping systems (DMC) in Latin America. In: Fischer, T., Turner, N., Angus, J., McIntyre, L., Robertsen, M., Borrell, A. and Lloyd, D. (eds) *New Directions for a Diverse Planet: Proceedings for the 4th International Crop Science Congress*, Brisbane, Australia, 26 September–1 October 2004.

SEAB (Secretaria da Agricultura do estado do Paraná/Paraná Rural) (1994) *Manual técnico do subprograma de manejo e conservação do solo*. 2nd edn. SEAB, Curitiba, PR, Brazil, p. 372.

Séguy, L. and Bouzinac, S. (2001) Direct seeding on plant covers: Sustainable cultivation of our planet's soils. In: Garcia-Torres, L., Benites, J. and Martinez-Vilela, A. (eds) *Proceedings of the First World Congress on Conservation Agriculture on Conservation Agriculture: A Worldwide Challenge*. Madrid, Spain, pp. 85–92.

Séguy, L., Bouzinac, S., Trentini, A. and Cortez, N.A. (1996) L'agriculture brésilienne du front pionniers. *Agriculture et Développement* 12, 2–61.

Séguy, L., Bouzinac, S., Maronezzi, A.C., Belot, J.L. and Martin, J. (2001) A safrinha de algodão: opção de cultura arriscada ou alternativa lucrativa dos sistemas de plantio direto nos trópicos úmidos. Boletim técnico no 37 da COODETEC CP. Cascavel, PR, Brasil.

Sidiras, N. and Pavan, M.A. (1985) Influência do sistema de manejo de solo no seu nível de fertilidade. *Rev. Bras. Ci. Solo. Campinas, SP, Brasil* 9(3), 249–254.

Skora Neto, F. (1998) Manejo de plantas daninhas. In: *Plantio Direto. Pequena propriedade sustentável.* IAPAR Circular 101, Londrina, PR, Brazil, pp. 128–158.

Skora Neto, F. (2001) Efeito da prevenção de sementes pelas plantas daninhas e da aplicação de herbicida em jato-dirigido na densidade de infestação na cultura do milho em anos sucessivos. *Planta Daninha* 19(1), 1–10.

Skora Neto, F. and Calegari, A. (2010) Sistemas de produção de palha para o manejo de plantas daninhas na cultura do feijão. In: *VII Seminário sobre pragas, doenças e plantas daninhas do feijoeiro, Anais*... 20 a 21 de Outubro de 2010, Instituto Agronômico, Campinas, SP, Documentos IAC, 95, pp. 75–84.

Skora Neto, F. and Darolt, M.R. (1996) Controle integrado de ervas no sistema de plantio direto nas pequenas propriedades. In: *I Congresso Brasileiro de Plantio Direto para uma Agricultura Sustentável*, 10–22 March 1996. Ponta Grossa, PR, Brasil, Resumos expandidos, pp. 153–154.

Sorrenson, W.J. and Montoya, L.J. (1989) *Implicações econômicas da erosão do solo e do uso de algumas práticas conservacionistas no Paraná*. IAPAR, Londrina, PR, Brazil, 104 pp. (Boletim Técnico, 21).

Stott, D.E. and Martin, J.P. (1989) Organic matter decomposition and retention in arid soils. *Arid Soil Research and Rehabilitation* 3, 115–148.

Teasdale, J.R., Brandsäeter, L.O., Calegari, A. and Skora Neto, F. (2007) Cover crops and weed management. In: Upadhyaya, M.K. and Blackshaw, R.E. (eds) *Non-chemical Weed Management: principles, concepts and technology*. Reading, UK, pp. 49–64.

Tiecher, T. (2011) Dinâmica do fósforo em solo muito argiloso sob diferentes preparos de solo e culturas de inverno. Tese (Mestrado em Ciência do Solo) – Universidade Federal de Santa Maria, Santa Maria, RS, Brazil, 82 pp.

Tiecher, T., Rheinheimer, D.S. and Calegari, A. (2012a) Soil organic phosphorus forms under different soil management systems and winter crops, in a long term experiment. *Soil and Tillage Research* 124, 57–67.

Tiecher, T., Rheinheimer, D.S., Kaminski, J. and Calegari, A. (2012b) Forms of inorganic phosphorus in soil under different long term soil tillage systems and winter crops. *Revista Brasileira de Ciência do Solo (Impresso)* 36, 271–282.

Tormena, C.A., Roloff, G.E. and Sá, J.C.M. (1998) Propriedades físicas do solo sob plantio direto influenciadas por calagem, preparo inicial e tráfego. *R. Bras. Ci. Solo, Campinas, SP, Brasil* 22, 301–309.

Valverde, S.R., Mattos, A.D.M., Jacovine, L.A.G., Silva, M.L. and Neiva, S.A. (2004) Oportunidades do mercado de CO_2. *Boletim Informativo*. SCBS, 29, 34–37.

Viedma, L.Q. de (1997) Manejo de enfermedades de cultivos extensivos en el sistema de siembra directa. In: *Curso Sobre Siembra Directa*. PROCISUR, Paraguay, pp. 203–216.

Vieira, M.J. (1991) Embasamento técnico do subprograma de manejo e conservação do solo – Paraná Rural. In: SEAB/PARANÁ RURAL. *Manual técnico do subprograma de manejo e conservação do solo*. SEAB, Curitiba, pp. 12–29.

Vinther, M.S. (2004) Hairy vetch a green manure and cover crop in conservation agriculture: N fixation, nutrient transfer and recovery of residue N. MSc Dissertation, The Royal Veterinary and Agricultural University of Denmark (KVL), Copenhagen, Denmark.

Wildner, L.P. (2000) Soil Cover. In: *Manual on Integrated Soil Management and Conservation Practices*. FAOL and Water Bulletins 8, IITA and FAO, Rome, p. 230.

Yorinori, J.T. (1996) *Cancro da haste da soja: Epidemiologia e controle*. Embrapa-Soja. Circular Técnica no. 14. Londrina, Paraná, Brazil, p. 75.

4 Conservation Agriculture on the Canadian Prairies

Guy P. Lafond,[1†] George W. Clayton[2] and D. Brian Fowler[3]
[1]*Agriculture and Agri-Food Canada, Indian Head Research Farm, Saskatchewan;*
[2]*Agriculture and Agri-Food Canada, Lethbridge Research Center, Alberta;*
[3]*University of Saskatchewan, Saskatoon, Saskatchewan, Canada;* [†]*Deceased*

4.1 Introduction

The goal of the agriculture industry is to keep increasing food production in order to meet the demands of a growing population while conserving and enhancing the land resource that produces 92 to 99% of food consumed by humans (Pimentel and Pimentel, 2000; Smil, 2000). Globally, there are 1.6 billion ha of land available for annual cropping (World Fact Book, 2009), but approximately 0.7 billion ha or 45% of arable soils worldwide are affected by one form or another of soil degradation (Lal, 2007). The present estimate is that 2–12 Mha, or 0.3–0.8% of the world's arable land, is rendered unsuitable for agricultural production annually with wind and water erosion accounting for 84% of this degradation (den Biggelaar *et al.*, 2004a). Clearly, better land stewardship is critical to meet the world's future needs for food, fibre and energy (den Biggelaar *et al.*, 2004b).

Although wind and water erosion are recognized as the major contributors to soil degradation, arable land lost to urbanization also represents another form of soil degradation. In 2008, the world reached an invisible but important milestone. More than 50% of the world's population now lives in urban areas (Parsons, 2008). In Canada, for example, 1.2 Mha of agricultural land was consumed for urban uses between 1971 and 2001 and 18% of the province of Ontario's best Class 1 farmland is now urban (Hofmann, 2001). Loss of land to urbanization is accelerating and this has important effects on watersheds, aquifers and microclimates around large urban areas.

Canada has ~3.1% of the global arable soils of which approximately 87% is located on the Canadian Prairies. Uncultivated native prairie soils contained from 0.2 to 0.7% nitrogen. By the 1940s, barely 60 years after the first plough turned over the virgin prairie sod, 15 to 40% of the N had been lost (Mitchell *et al.*, 1944). Unfortunately, this level of soil degradation continued into the 1980s with most prairie soils having lost more than 40% of their initial organic nitrogen content.

Given that wind and water erosion are the most important forms of soil degradation, an efficient prevention method is to maintain surface residues and standing stubble on the soil surface. For this reason, Conservation Agriculture (CA) on the Canadian Prairies is synonymous with no-till production systems. Achievement of this desirable state requires reduction or elimination of tillage, adoption of continuous cropping practices, elimination of summer fallow practice, proper crop fertility, appropriate

 Conservation Agriculture: Global Prospects and Challenges (eds R.A. Jat, K.L. Sahrawat and A.H. Kassam)

pest management practices, appropriate seeding equipment and crop diversity. These key factors have been thoroughly investigated in western Canada and represent the foundation for CA on the Prairies (Fowler *et al.*, 1983; Hass, 1984; Hay, 1986; Smika and Unger, 1986; Holm *et al.*, 1990; Lafond and Fowler, 1990; Lafond *et al.*, 1996, 2006).

4.2 Some Key Historical and Technological Developments Leading to the Widespread Adoption of Conservation Agriculture on the Prairies

The early establishment of Experimental Farms across Canada after 1886 provided a unique opportunity to document the impact of agricultural practices on prairie soils. As stated by Janzen (2001), 'When the ploughs began to invert the prairie sod, scientists were already there to record their effects. And from the onset, preserving the soils was a priority.' Over the next 20 years, extensive and detailed measurements of soil organic matter led Shutt (1905 in Janzen, 2001) to report on its rapid decline and raise concerns about the 'permanence', now referred to as 'sustainability', of agriculture on the Prairies. Although these shortcomings were well recognized, weed control and soil fertility technology that would permit successful shifts away from fallow-cropping practices with intensive tillage was not yet available. The continuance of these practices resulted in frequent severe wind erosion events during the next 80 years, with intense events experienced during the 1930s on the Great Plains of North America (Montgomery, 2007). However, during that 80 year period a series of technological developments and an increase in basic knowledge of soil and crop management combined to pave the way for the gradual development and widespread adoption of CA as it is practised today on the Canadian Prairies.

4.2.1 Importance and benefits of surface residues and standing stubble

In the mid- to late 1930s, studies conclusively showed that maintaining crop residues on the soil surface could improve water infiltration, reduce evaporation losses, reduce surface runoff, reduce wind and water erosion, and conserve more water because of the increased ability to trap and hold snow (Smika and Unger, 1986). These findings led to increased efforts to develop soil and crop management practices that could make better use of the potential offered by crop residues, especially for the arid and semi-arid areas. Standing stubble has since been shown to be four times more effective at protecting the soil from wind erosion than flattened or incorporated residues, thereby allowing for adequate protection from wind erosion even in years of low residue production (Lyles and Allison, 1981). More recent studies have confirmed that no-till significantly reduces sediment losses after heavy precipitation events, providing for improved water erosion control (Mostaghimi *et al.*, 1992).

4.2.2 Introduction of one-way discs and discers for seeding

An important technological development in western Canada that helped pave the way to changes in cropping practices was the introduction of one-way discs in the 1930s. These implements were heavy enough to do primary tillage and less aggressive than the conventional ploughs. As a result, they left more residues on the soil surface thereby providing greater protection against wind erosion. They were ideal as a spring preseeding tillage implement and, later, seed and fertilizer boxes installed on the one-way discers allowed for planting at the same time as the primary tillage operation. The combined operations also created an opportunity to seed into standing stubble. This increased efficiency opened the door for more extended cropping, which, in turn, reduced the intensity of summer fallow, a major contributor to soil degradation. The one-way discs later paved the way for discer seeders, which provided greater seedbed utilization reducing the toxic risks from seed-placed fertilizers (Fig. 4.1).

Efforts to curb erosion problems in the 1930s identified drying of the seedbed and poor crop establishment resulting from the pre-seeding tillage operations as a problem with stubble-cropping tillage technology. Discer seeder technology provided a solution to this problem (Fig. 4.2). Less soil moisture was lost and the seed was placed on or into moist soil at a shallower depth. The seed was immediately covered with soil and packed with harrows or a combination of harrows and packers to conserve soil moisture and ensure proper seed to soil contact. This allowed for more successful crop emergence and a greater opportunity to extend cropping. The discer also provided for the control of emerged or emerging weed seedlings. Discer seeders were used extensively until the late 1990s.

Fig. 4.1. One-way disc machine.

4.2.3 Introduction of the Noble blade, mulch tillage and air seeders

In the 1930s, the practice of strip farming was adopted as a means to address wind erosion. Charles Noble realized that the real solution to wind erosion was to adopt stubble mulch systems (The Noble Blade, 2013). This led him to adapt ideas taken from a sugarbeet farmer in California. The farmer used a flat straight blade to loosen the sugarbeet. Noble adapted the concept and created what is now known as the Noble blade. The tillage machine consisted of a heavy

Fig. 4.2. Disc seeder.

steel sub-soil blade that could cut weeds off at the roots with minimal disturbance of the soil surface. In the dry areas of the prairies, this allowed for the continued practice of summer fallow while greatly reducing the risks of wind erosion (Fig. 4.3). One can argue that this technology led to the development of heavy duty cultivators, also commonly referred to as 'deep tillers' or 'chisel ploughs'. In turn, these heavy duty cultivators were later adapted with air delivery systems for seed and fertilizer and became known as 'Air seeders'. These air seeders represented the start of what we call 'high disturbance direct seeding systems', the forerunner to no-till providing better penetration, residue clearance and depth control than discer seeders. In the 1980s, technology and production concepts evolved resulting in the development of 'low disturbance direct seeding' implements now referred to as air drills (Hood, 1990; Memory and Atkins, 1990). Air seeder technology created a fundamental change in cropping practices in Canada with earlier seeding and less passes over the field, extending the growing season by up to 3 weeks.

4.2.4 Introduction of selective and non-selective herbicides

The introduction of the selective broadleaf herbicides, 2,4-D in 1947 and MCPA in 1953, represented a huge leap forward for cereal production. This was followed

Fig. 4.3. Noble blade.

with the introduction of the selective wild oat herbicides diallate and later triallate in the early 1960s (Appleby, 2005; Timmons, 2005). These introductions allowed for more continuous cropping, especially in the moister areas of the Prairies. As a matter of interest, a study was commissioned in 1948 by the superintendent of the Experimental Farms and established at the Indian Head Research Farm in the province of Saskatchewan to determine the long-term effects of 2,4-D and later MCPA applications on wheat production under fallow and stubble cropping conditions and on the soil. The study was terminated in 1989 after 42 years. A *Pseudomonas* spp. bacterium isolated from the soil of these plots was capable of using not only 2,4-D and MCPA as its sole carbon source but also other phenoxy herbicides (Smith *et al.*, 1994). Other agronomic results of this 42 year study can be obtained from a review by Smith *et al.* (1991).

In 1962 and 1966, diquat and paraquat were registered as fast-acting, non-selective, non-translocated and non-residual herbicides (Timmons, 2005).The introduction of these herbicides allowed for more investigations into the concept of minimum-till and no-till production systems.

The 1970s and 1980s were characterized by the introduction of numerous other selective and non-selective herbicides for cereal, oilseed and pulse crops stimulating crop diversification and greater adoption of continuous cropping, especially in the drier areas of the Prairies (Appleby, 2005).

Discovery of herbicides resulted in innovative changes to cropping practices. The first studies looking at the potential for chemical summer fallow were conducted from 1949 to 1955 in Havre, Montana by Baker and Krall (1956). Their results showed that grain yields could be maintained even when tillage was completely eliminated while ensuring wind erosion control. These studies represent some of the first documented evidence that tillage was not necessary to grow a crop. Of course, no-till technology was not advanced enough to sustain the practice of chemical summer fallow with the herbicides that were available at that time.

The introduction of the non-selective herbicide glyphosate in 1971 represented a key technology for the detailed investigation of no-till or CA production systems worldwide (Appleby, 2005). However, it was not until the early to mid-1980s that economics allowed glyphosate to become associated with no-till systems. Unlike diquat and paraquat, glyphosate translocated readily into the plant, providing very good perennial weed control and overall good annual weed control when applied prior to seeding or after seeding prior to crop emergence. The practice of preharvest glyphosate applications further enhanced the tools for the control of perennial weeds. This practice is also used extensively in tillage-based systems. Since 1995, the introduction of herbicide-resistant canola to glyphosate and glufosinate made the production of canola easier and more profitable under both no-till and tillage-based cropping systems.

4.2.5 Introduction of winter wheat into Prairie cropping systems

A research and development programme was initiated by the Crop Development Centre at the University of Saskatchewan in 1972 to expand the traditional winter wheat production area in southern Alberta to the north and east into Saskatchewan and Manitoba (Fowler, 2011). Winter survival was considered the main limitation to production in this expanded region. Earlier research efforts had established that the cold hardiness genetic potential of wheat had reached a maximum that had not been improved upon for decades (Fowler *et al.*, 1983). These observations led to the conclusion that a strategy that included options complementing a breeding effort had to be explored in order effectively to address the winter survival question.

No-till seeding of small research plots began in 1974. Successes with these trials were then demonstrated in larger strip plots seeded with a Noble DK-5 high clearance hoe drill at the University of Saskatchewan in Saskatoon and Agriculture and Agri-Food Canada Research Farms at Indian Head and Melfort. Large scale commercial field testing was initiated at Clair, Saskatchewan at the same time. By 1983, most of the details for a successful no-till winter wheat production system were available on paper and by 1985 approximately 203,000 ha of no-till winter wheat were planted in western Canada. However, no-till seeding was still a new concept for farmers in the 1980s (Fowler *et al.*, 1990) and surveys conducted as part of the federal-provincial Economic Regional Development Agreement programmes and by the Western Canadian Wheat Growers Association revealed that many farmers were not employing recommended management practices for the production of winter wheat. It was obvious that successful no-till production methods would have to be clearly demonstrated if winter wheat was to become a viable cropping option for more than just a few farmers outside the traditional area of southern Alberta. In response, the winter wheat 'Conserve and Win' programme was initiated by the University of Saskatchewan and Ducks Unlimited Canada in 1991 (Fowler and Moats, 1995), with the objective of developing management packages and demonstrating production systems that would allow Saskatchewan farmers to realize the full production and conservation potential of no-till winter wheat in an integrated cropping system.

By the early 1990s, improvements in the design of seeding equipment, cheaper and more effective herbicides, a better understanding of the role of tillage in crop production systems and increased emphasis on residue management had combined to start the no-till revolution for spring-sown crops. The benefits of reduced input costs and improved soil and water conservation added momentum to this paradigm shift. The wide adoption of no-till seeding along with the large area of standing stubble available each autumn now provides the opportunity for winter wheat production in this region with minimal risk of winter-kill if cultivars with a high level of winter hardiness are grown using recommended management practices (Fowler, 1986). When combined with plant breeding improvements, no-till cropping meant that the major obstacles due to winter survival, lodging, crop residue management and rust susceptibility were no longer barriers to winter wheat production and the true potential started to be recognized. In recent years, a large group of farmers who have had long-term success have carried this momentum forward and demonstrated the many advantages associated with the inclusion of winter wheat in crop rotations.

The finding that winter wheat could overwinter and avoid winter kill when seeded into standing stubble because of the insulating effect of the trapped snow has provided new cropping opportunities and some important agronomic benefits (Fowler *et al.*, 1983). Winter wheat could take advantage of early season moisture, grow during a cooler part of the growing season and ripen earlier than other crops. This provided new opportunities to continuous cropping, especially in the drier areas of the Canadian Prairies. A basic requirement for winter wheat production was that it had to be seeded into standing stubble. Therefore, any producer seeding winter wheat was automatically introduced to no-till production practices.

4.2.6 Nitrogen management – the 'in-soil' banding concept

With the advent of inorganic fertilizers came the issue of placement, timing, form and rate. The first inorganic fertilizer used on the Prairies in the 1950s was mono-ammonium phosphate (MAP) fertilizer with the analysis of 11-48-00, which lent itself to being applied safely with the seed. Manufacturers produced granular fertilizer applicators that could be attached to existing drills and meter the product with the seed. At the same time, ammonium nitrate (AN) fertilizer (34-00-00) was also available. Ammonium nitrate was mixed with MAP creating a heterogeneous blend with an analysis of 23-23-00. This fertilizer blend (23-23-00) could be applied with the seed but with more restrictions on the total amount because of the increased potential for seedling damage. To circumvent the problem when higher amounts of AN were required, AN was broadcast on the soil surface. In the early 1970s, nitrogen fertilizer in the form of urea (48-00-00) and later anhydrous ammonia (82-00-00) were introduced. Ammonia, being a gas, had to be injected into the soil. With urea, some of it was blended with MAP to create a blended product with an analysis of 28-28-00. The presence of urea in the blend resulted in even more restrictions for seed-placed applications because of enhanced potential for damage to seedlings than the previous blend with AN (23-23-00).

Other than limited amounts of urea placed with the seed, the majority of it was broadcast on the soil surface either in the autumn or in the spring prior to seeding. This placement method quickly exposed problems that had not been observed with AN. The nitrogen responses were found to be highly variable and later shown to be due to volatilization losses. This led the Westco fertilizer company in the mid-1970s to investigate ways to circumvent this problem (Harapiak, 1990; Harapiak *et al.*, 1993). They discovered that if the urea fertilizer was placed in the soil in bands, these limitations were overcome. During this time period, producers found poor depth control with air seeding technology and shortfalls with the pneumatic air delivery systems relative to the other seeding implements in use at the time (Memory and Atkins, 1990). However, with the arrival of the fertilizer banding concept and the promising results from numerous field trials, the fertilizer industry quickly adopted the concept and there was a new use for these first-generation air seeders. This propelled the air seeding industry forward with new designs such that by the early 1980s, the precision in terms of depth control and seed metering were as good as the conventional drills of the day (Memory and Atkins, 1990). This concept also allowed the first no-till producers another option besides surface broadcasting urea. They could realize the benefits of banding the urea using narrow openers either in the autumn or spring prior to seeding without losing all the benefits of standing stubble for trapping snow.

One can argue that the concept of in-soil fertilizer bands paved the way for further improvements in air seeding technology leading to the development of the one-pass seeding and fertilizing no-till system. The fertilizing system involved placement of the fertilizer to the side and below the seed row or else mid-row banded between every second seed row.

4.2.7 Adoption of no-till – other underlying forces

A number of additional key factors must be considered in order to fully appreciate the broad-scale adoption of no-till on the Canadian Prairies. Three pivotal forces are recognized.

The most important force was the vision and determination of a select group of producers in western Canada (Table 4.1). These producers, through ingenuity and conviction, proved that it was possible to make radical changes in production practices to conserve the soil and survive economically. Jim McCutcheon from Manitoba started using no-till in 1973 and was fully converted to no-till by 1976. Jim Halford from Saskatchewan started in 1978 and rapidly assembled key technology on how to seed and fertilize using a one-pass seeding and fertilizing system while at the same time not compromising the seedbed and minimizing the dangers of seedling damage due to fertilizer toxicity.

Table 4.1. Recognized pioneers of no-till in western Canada (this does not imply that these were the only early adopters of no-till in western Canada. The list provided for western Canada could also have inadvertently missed some individuals).

Province	Producers
Alberta	Brian Hearn, Gordon Hilton, Danny Stryker, Dick Middleton, Wayne Arrison, Richard Walters, Henry Graw
Manitoba	Jim McCutcheon, Walter Klimchuk, Robert McNabb, Gordon McPhee
Saskatchewan	Jim Halford, A.S. McBain

The second force was public policy. In 1984, a report on soil conservation by the Standing Committee on Agriculture, Fisheries and Forestry presented its report to the Senate of Canada (Anonymous, 1984). The report outlined clearly the extent of soil degradation in all regions of Canada, the lack of awareness of the extent of the problem and the increasing danger of losing a large portion of agricultural production capacity unless there was a major commitment to conserving the soil. This led to the establishment of important programmes that included the National Soil Conservation Program (NSCP), the Save our Soils (SOS) programme and the Green Plan programme (Ward *et al.*, 2010). These programmes provided resources to promote CA. This, combined with increased research activities at both the federal and provincial levels established the framework necessary to make no-till a reality on the Canadian Prairies. The SOS programme made no-till seeding equipment available to producers for limited use at low cost. This provided them with first-hand experience of no-till management practices with seeding equipment, residue management and weed control. The various programmes also provided some training on no-till production practices.

The third force was knowledge transfer. The creation of the Alberta Conservation Tillage society in 1978 (Gamache, 2010) and the Manitoba-North Dakota Zero Tillage Farmers Association (http://www.mandakzerotill.org) in 1982 (Bradley, 2010) provided a forum to bring ideas, knowledge, technology and producers together on the subject of no-till. The subsequent formation of the Saskatchewan Soil Conservation Association (http://www.ssca.ca) in 1987 (McClinton and Polegi, 2010) and the Alberta Reduced Tillage Initiative (http://www.reducedtillage.ca) in 1994 (Gamache, 2010) provided even more momentum to the growing movement of soil conservation at the producer level.

In terms of technology, companies such as Haybuster in Jamestown, North Dakota, USA (http://www.dura-ind.com) provided seeding options for no-till in the mid-1970s followed by the Amazone no-till hoe drill from Germany in the early 1980s. At the same time as Amazone, the development and testing of the ConservaPak seeder was well underway and commercially introduced in the late 1980s. The ConservaPak seeder, developed by Jim Halford of Indian Head, Saskatchewan, allowed for successful seeding and fertilizing using a one-pass seeding and fertilizing no-till system. During that time, homemade triple-disc openers were fabricated and the Noble high clearance hoe press drill was being used for no-till seeding winter wheat into standing stubble. The Flexi-Coil company introduced their air drill in the late 1980s, which found favour with many producers. At the same time, development activities in seeding systems and opener designs were also underway by numerous machinery companies because the potential of no-till on the Canadian Prairies was now becoming more evident as a feasible crop production practice.The long and often difficult road taken by the no-till winter wheat development programme led by the Crop Development Centre (University of Saskatchewan) emphasized that a coordinated approach that combines programmes in agronomy, plant breeding/genetics, information transfer and market development is often required for successful crop adaptation to a new or changing environment or production system (Fowler *et al.*, 1983).

It should be noted that many private companies and farmer innovators were involved in the production of various aspects of no-till technology, such as better straw choppers and chaff spreaders for combine harvesters and better spraying technology. Their efforts contributed significantly to the success of the CA adoption movement by creating the opportunities that made it possible for no-till to succeed on the Prairies.

4.2.8 Current status of Conservation Agriculture on the Canadian Prairies

Although no-till was being practised on a very limited basis starting in the mid-1970s, the rapid phase of adoption started in the early 1990s. The rate of adoption was fastest in Saskatchewan and slowest in Manitoba (Table 4.2). Although the rate of adoption has slowed, the most recent 2011 Census of Canadian Agriculture indicates that it is still growing. If one was to use the definition provided by the Food and Agriculture Organization for conservation agriculture (FAO, 2013), the amount of adoption on the Canadian Prairies would be much higher than that reported in Table 4.2.

4.3 No-till Research Results

4.3.1 Soil physical properties

Early studies reported that changes in bulk density and penetration resistance after 5 years of no-till were not large enough to negatively impact crop production (Grant and Lafond, 1993). No-till increased macro-aggregation (>0.25 mm) and mean weight diameter of aggregates, even in coarse-textured soils (Franzluebbers and Arshad, 1996). The higher level of macro-aggregation was one reason for increased carbon sequestration observed with no-till (McConkey *et al.*, 2003). Arshad *et al.* (1999) reported that under no-till, water retention was increased with little change in soil bulk density due to a redistribution of pore size classes into more small pores and fewer large pores. They also noted better water infiltration into no-till soils. The improved soil internal structure should lead to a better carrying capacity of equipment and therefore less potential for compaction.

Table 4.2. Percentages of cultivated area using no-till as the primary soil and crop management practice on the Canadian Prairies from 1991 to 2011 (Source: 1991 to 2006 data from McClinton, 2007; data for 2011 adapted from the 2011 Census of Canadian Agriculture prepared by B. McClinton (Statistics Canada, 2013)).

Year	Saskatchewan	Alberta	Manitoba
1991	10	3	7
1996	19	10	15
2001	39	27	13
2006	60	48	21
2011	70	65	24

4.3.2 Soil chemical properties

The impact of no-till on soil chemical constituents, especially soil organic carbon (SOC) and nitrogen (SON), has been of great interest to producers and members of the research community. A positive nitrogen balance is necessary in order to maintain or increase SOC and more removal than replacement of nitrogen will lead to a decrease in SOC and SON. As a result, in the semi-arid regions of the Prairies, SOC is closely related to the amount of crop residues returned to the soil and the nitrogen content of the residues, which in turn is dependent on the type of fertility regime used (Campbell and Zentner, 1997; Campbell *et al.*, 2007a).

Janzen *et al.* (1997) showed that with erosion control, the current practices of continuous cropping and fertility management, and the use of chemical fallow rather than tillage fallow, SOC and SON can be maintained in the semi-arid to sub-humid areas of the Prairies. They also indicated that adopting no-till combined with continuous cropping would likely lead to greater increases in SOC.

In the sub-humid areas, maintenance of SOC requires continuous cropping, or the addition of manure, or the appropriate use of inorganic fertilizers to ensure optimum crop growth (Juma *et al.*, 1997). Related studies have shown that the increase in SOC is proportional to cropping frequency or the amount of residues returned to the soil, the use of a managed fertility regime, or the inclusion of legume–grass forage crops (Campbell *et al.*, 1997). More recent studies have confirmed that when no-till is combined with continuous cropping and optimum fertilizer management, SOC will increase (McConkey *et al.*, 2003; Campbell *et al.*, 2007b; Lafond *et al.*, 2011a).

4.3.3 Soil biological properties

The size of the soil microbial community is directly proportional to soil organic matter and soil microbes are the principal mediators of nutrient cycling (Hamel *et al.*, 2006). Although soil microbial biomass represents only a small proportion of total soil organic matter, it is much more dynamic. Soil microbial biomass is a better indicator of how tillage systems and cropping systems impact soil health and the soil's productive capacity (Lupwayi *et al.*, 1999; Campbell *et al.*, 2001). SOC and SON, microbial biomass carbon (MBC), light fraction carbon (LFC), light fraction organic nitrogen (LFN) and wet aggregate stability were enhanced with increased cropping frequency, fertilization and also with the inclusion of green manure crops and legume hay crops. However, LFC, LFN, MBC and potentially mineralizable N were more sensitive to changes in cropping practices than simple measures of total SOC and SON (Campbell *et al.*, 2001).

When no-till was included as a management factor, Lupwayi *et al.* (2004) noted that microbial biomass increased as well as the functional diversity and activity of microbes. They suggested that this would have a positive effect on decomposition processes of crop residues by microbes. In another study they observed that microbial biomass carbon turnover was higher with no-till than conventional tillage (ConvT). Soon and Clayton (2003) also observed higher N mineralization with no-till.

The three main factors describing the rate of crop residue decomposition are air temperature, location of residues (on the soil surface versus buried) and the nitrogen content of residues (Janzen and Kucey, 1988; Douglas and Rickman, 1992). As air temperature and nitrogen content of crop residues increase, the rate of decomposition also increases. Crop residues placed on the soil surface decompose at about two-thirds the rate of buried residues. With the increase in soil microbial activity and diversity observed under no-till, some interesting observations are noted. Residues lost nitrogen faster with tillage than no-till but overall crop yield and N uptake tended to be greater with no-till than with tillage (Soon *et al.*, 2001). As well, nitrogen mineralization was always greater with no-till even though initial immobilization of nitrogen was sometimes observed (Soon and Arshad, 2004; Lupwayi *et al.*, 2006a, b). It could be argued that the slower rate of decomposition under no-till may allow for a longer period of nutrient release thereby supplying the crop with nutrients like nitrogen over a longer period of time during the growing season and minimizing potential nitrogen losses early in the growing season from adverse climatic conditions (Lupwayi *et al.*, 2004).

4.3.4 Impact of no-till on grain yields

The positive yield benefits of no-till production systems were not always evident in the early years of no-till research (Lafond and Fowler, 1990). Some of the reasons for these results could be seeding and fertilizer application equipment limitations and lack of effective and timely weed control. Over time, more consistent improvements in no-till grain yield relative to conventional-till were observed in research trials (Lafond *et al.*, 1992, 1996, 2006). For example, in the more semi-arid areas yield benefits with no-till were not initially observed (Zentner *et al.*, 1996). It was not until more innovative approaches to stubble management were undertaken did the shift towards improved grain yields under no-till occur (Campbell *et al.*, 1992). Seeding crops into tall stubble (>30 cm) reduced water losses from evaporation and increased water use efficiencies producing higher grain yields (Cutforth and McConkey, 1997; Cutforth *et al.*, 2002, 2006). More recently, studies have quantified the long-term benefits of no-till. When 9 years of no-till were compared to 31 years of no-till, grain yields of spring wheat (*Triticum aestivum* L.) and canola (*Brassica napus* L.) were increased by 14% and 16%, respectively, with the longer period under no-till (Lafond *et al.*, 2011b). Consequently, not only was a yield improvement observed going from a tillage-based system to a no-till cropping system, but an additional yield increase was also observed as length of the no-till rotation increased.

4.3.5 Impact of no-till on economic performance

The initial economic analyses of no-till production systems were influenced by the agroecological zones where the studies were conducted. In the more semi-arid areas, the savings offered by no-till in terms of labour, fuel and oil, machinery repairs and overhead were more than offset by the increase in herbicide costs. This resulted in higher overall production costs with similar yields and economics that favoured more tillage-based systems. Improvements in stubble management practices has now shifted the balance to no-till as observed by the increasing amount of land dedicated to no-till since the mid-1990s in the semi-arid areas (Zentner *et al.*, 1996).

In the transition agroecological zones from the semi-arid to the sub-humid areas, the economic performance has favoured no-till over tillage-based production systems when combined with continuous cropping (Zentner *et al.*, 2002a). This has been demonstrated over a wide range of growing conditions.

In the sub-humid areas of the Prairies, production economics have favoured no-till (Gray *et al.*, 1996; Zentner *et al.*, 2002a). In fact the economic analyses clearly show that producers in this region will opt for diversified continuous cropping systems and no-till, regardless of the level of their risk aversion (Zentner *et al.*, 2002b).

4.3.6 Impact of no-till on energy inputs, outputs and energy use efficiency

Only a limited number of studies have quantified the impact of no-till on energy use efficiency. Energy inputs consist of the energy required to produce fertilizers, herbicides and the energy associated with fuels and lubricants for doing the various field operations. Energy outputs represent the energy from the grain harvested. The best overall strategy to increase energy use efficiency is to increase the energy from crop production by improving water conservation and water use efficiencies (Lafond *et al.*, 2011a).

Initially, energy inputs for the semi-arid Prairies were larger for no-till than conventional-till and energy output was similar between the two due to the comparable yields. As a result, energy use efficiency favoured ConvT (Zentner *et al.*, 1998). Recent research with tall stubble has shifted energy use efficiency in favour of no-till because of higher no-till grain yields due to higher crop water use efficiencies (Cutforth and McConkey, 1997; Cutforth *et al.*, 2002, 2006).

In the sub-humid areas of the Prairies, energy input was similar between no-till and ConvT and energy output was greater with no-till than conventional-till due to higher no-till grain yields (Zentner *et al.*, 2004). Energy use efficiency was either similar between tillage systems or greater for no-till than conventional-till depending on the crop rotation. Energy associated with inorganic nitrogen fertilizers accounts for the largest proportion of total energy needs in crop production. Therefore, the inclusion of grain legumes in the rotation usually results in lower energy inputs and higher efficiencies because nitrogen fertilizers are not required.

4.4 Perceived Problems Encountered with the Adoption of Conservation Agriculture on the Canadian Prairies

In the early stages of no-till, some of the hypothesized problems were due to lack of research information and this made it challenging for the initial no-till adopters. In many cases, solutions to problems were based only on circumstantial evidence and limited observations and producers had to put their faith into the overall potential that no-till production systems could offer.

4.4.1 Impact of no-till on soil temperature

The possibility of lower soil temperatures with no-till on crop germination and emergence was very much a concern to early no-till adopters. Some of the first results did indeed show that surface soil temperatures were lower under no-till (Gauer *et al.*, 1982). The sensitivity of germination and emergence to temperature, especially the temperatures observed in early spring on the Canadian Prairies, is well documented (Lafond and Fowler, 1989a, b). It was hypothesized that crop emergence would be delayed more with no-till than what was observed on ConvT. However, if crop residues were burnt or physically removed from the surface of the soil, there was no difference in soil temperature between no-till and ConvT. This was explained by the higher heat flow in the soil due to the higher soil bulk densities and soil moisture content at the soil surface under no-till. Later research showed that when emergence was directly quantified in the field, no-till systems did not delay the emergence of spring wheat relative to ConvT systems. Of interest also was the fact that spring wheat emergence under no-till was the same regardless of whether it was grown on spring wheat or field pea stubble (Lafond *et al.*, 1992). Later research showed that if a 7.5 cm strip of soil above the seed row was free of any crop residues, any difference in surface soil temperature between tillage systems disappeared resulting in similar germination and emergence times (Arshad and Azooz, 2003). This may explain why the majority of no-till seeders on the Prairies use hoe-type or tine openers rather than disc openers.

Crop emergence is a function of soil temperature, soil moisture and seeding depth. Under no-till seeding conditions, soil moisture is seldom a limiting factor to crop emergence (Lafond and Fowler, 1989b). Therefore speed of crop emergence is dictated mainly by soil temperature and planting depth. The higher surface soil moisture with no-till allows for shallower planting, which offsets some potential for lower soil temperatures, and when combined with hoe-type or tine openers, the end result is essentially no difference in crop emergence, regardless of tillage systems, which is what has been observed and documented under field conditions.

4.4.2 Crop residue decomposition and residue accumulations under no-till

An early concern with no-till was the potential for the accumulations of crop residues over time at the soil surface causing problems with the seeding operation and delaying crop emergence due to cooler soil temperatures. It was believed that tillage was necessary to accelerate decomposition, otherwise residues left standing or on the soil surface would decompose too slowly and accumulate over time resulting in impeding planting even more. These concerns were addressed by two studies. In the first study, it was shown that residue decomposition was determined in large part by their nitrogen content, regardless of species (Janzen and Kucey, 1988). In other words, if wheat, lentil or canola residues had similar nitrogen contents, their rates of decomposition were the same. The second study showed that the initial nitrogen content of the crop residues, the accumulated air heat units and residue placement (buried versus surface) after receiving a small amount of precipitation governed the rate of decomposition (Douglas and Rickman, 1992). The rate of residue decomposition at the soil surface was ~66% of the rate of buried crop residues. Hence even though the rate of decomposition was slower with residues on the soil surface, the rate was fast enough to avoid these perceived problems. Over time, it was also recognized that varying the types of residues through crop rotations combined with proper shredding or chopping and uniform spreading greatly lessened any negative impact of crop residues at the soil surface.

4.4.3 Nitrogen fertilizer management under a no-till one-pass seeding and fertilizing system

Nitrogen management proved to be challenging in the early years of no-till. The four major components to nitrogen fertilizer management, form, timing, placement, and rate and placement, created most of the early challenges. The two most common methods of placement in the early years of no-till in western Canada were seed-placed and surface broadcast. However, there was a limit to how much N fertilizer, like urea, could be applied with the seed and placing urea on the soil surface led to high losses from volatilization under certain conditions. It was not until the late 1970s and early 1980s that the technology for late autumn or prior to seeding in-soil N banding became available on a commercial scale thereby allowing amide and ammonium-based fertilizers like urea and anhydrous ammonia to be used effectively (Harapiak, 1990). Research conclusively showed that losses from urea volatilization could be almost eliminated if it was placed in the soil and covered properly (Harapiak *et al.*, 1993; Malhi *et al.*, 2001). As discussed previously, the one-pass seeding and fertilizing no-till system evolved as a result of incremental equipment innovations that provided the desired separation between seed and fertilizer (Johnston *et al.*, 1997, 2001). The one-pass seeding and fertilizing no-till system in use on the Canadian Prairies is now regarded as a highly efficient method of managing nitrogen fertilizers for achieving high nitrogen use efficiencies (Malhi *et al.*, 2001; Grant *et al.*, 2002) and it is also recognized as a best management practice for minimizing the potential for nitrous oxide emissions (Lemke and Farrell, 2008). Even farmers employing tillage in their farming operations are now purchasing no-till seeding equipment capable of seeding and fertilizing in one-pass because of the recognized efficiencies with this fertilizer management approach.

4.4.4 No-till and the long-term impact on weed densities and shifts in weed community

A major concern with early adopters of no-till was the long-term impact on weed densities, possible rapid shifts in weed communities towards more perennial type species and a greater dependence on herbicides (Lafond and Derksen, 1996; Derksen *et al.*, 2002). More recently, no-till producers have been expressing concerns about weeds resistant to herbicides increasing the vulnerability of no-till systems. Some weeds have become resistant to glyphosate in the USA and the first resistant weed (*Kochia scoparia* L.) to glyphosate on the Canadian

Prairies was reported in 2011 (Robert Blackshaw, Lethbridge, Alberta, 2012, pers. comm.).

The large anticipated change in weed communities has not yet occurred in western Canada. A number of reasons have been put forward. One reason is the increase in crop diversification that allows for a broader range of herbicide chemistries while the inclusion of diverse crop types and growth habits (spring versus winter crops, oilseeds or pulse crops versus cereals) allows for more varied selection pressure (Derksen *et al.*, 2002). Another reason involves the precise placement of fertilizer relative to the seed in the one-pass seeding and fertilizing no-till system, which increases the competitiveness of crops against weeds (O'Donovan *et al.*, 1997). A third reason is the temporal variation in weed communities as a result of temporal variability in growing season temperature and moisture observed on the Prairies. This variability represents an important source of varied selection pressure that helps guard against the dominance of particular weeds. A fourth reason is the impact of agronomic practices such as planting rates, crop rotations, crops, planting dates and herbicides all working together to reduce weed-seed recruitment in the soil seed bank and weed densities in future years (Harker and Clayton, 2003).

Another very important reason is the introduction of canola crops resistant to three specific herbicide chemistries (Beckie *et al.*, 2006). This provided new tools to combat weeds such as wild oats (*Avena fatua*) and green foxtail (*Setaria viridis*) that were showing resistance to the ACCase (Group 1) and ALS/AHAS (Group 2) groups of herbicides (Saskatchewan Agriculture and Food, 2008).

The concern that no-till producers would need to reintroduce tillage to control weeds has not yet materialized in western Canada. Some farms in Saskatchewan have been in no-till for more than 30 years and have yet to resort to tillage to control weeds. Of interest is the overall lower weed densities reported by no-till producers, which is indicative of lower soil weed seed banks (Blackshaw *et al.*, 2008). Changes in weed communities occur slowly and the temporal variability in growing season temperature and moisture along with the crop management practices utilized are the dominant factors influencing weed densities and communities and not the presence or absence of tillage. The strategy for no-till producers is to utilize a diversity of weed management tools ensuring that no one tool has a disproportionate amount of use, otherwise its effectiveness may be greatly diminished (Harker and Clayton, 2003). No-till producers have many weed management tools at their disposal to vary selection pressure and prevent any weed species from becoming dominant. The effectiveness of no-till production practices has been reflected in the overall reductions in weed densities on the Canadian Prairies (Leeson *et al.*, 2005).

4.4.5 No-till and the long-term impact on plant diseases

The first no-till producers were very uncertain about the impact of no-till on root and leaf diseases because of the crop residues left at the soil surface. However it has now been shown that the effects of environment and crop rotation are the dominant factors determining the incidence of plant diseases and the effects of tillage systems tend to be small or of no consequence (Bailey *et al.*, 2001; Turkington *et al.*, 2006). In fact no-till has been shown to reduce the severity of common root rot in cereals (Bailey *et al.*, 2001). No-till reduces many crop diseases because of its direct and beneficial effects on soil biology (Krupinski *et al.*, 2002). A healthy soil with diverse and balanced populations of soil microorganisms will provide substantial competition against root pathogens as they often use the same organic carbon substrate.

The best strategy to minimize plant diseases in no-till cropping systems is to include crop diversity. The temporal variability in growing-season climatic conditions on the Prairies also reduces the risks of certain diseases from becoming dominant. For example, in dry years, the build-up of disease inoculum will be low, shifting disease risks in the following growing season. Attention should be given to complementary disease control methods such as providing disease-resistant cultivars, disease-free seed with high vigour, use of seed treatments or foliar fungicides if warranted, balanced soil fertility, control of weeds and volunteer crops to break pathogen cycles, and careful record keeping of any recurring disease incidence (Krupinski *et al.*, 2002).

4.5 Conclusions

The rate of no-till adoption has slowed during the last 5 years but, based on the 2011 Canadian Agriculture Census (Table 4.2), the overall area under no-till is still increasing on the Canadian Prairies. This leaves the important question, foremost in the mind of agriculture stakeholders, how durable will the future of no-till be as the dominant production practice on the Canadian Prairies?

No-till represents the best solution to wind erosion and not only has it been shown to sustain soil productivity, but it can also increase it when combined with appropriate soil and crop management practices. Consequently, no-till will persist on the Canadian Prairies as the production system of choice for preventing soil degradation from wind and water erosion. Building on this strength, more importance needs to be placed on developing no-till production practices that accelerate soil organic carbon accumulation, especially in degraded soils.

In the short to medium term, more production potential can still be realized with improved no-till management practices. For example, more effort needs to be directed toward identifying the full potential offered by tall stubble and the risks associated with this approach. Tall stubble alters the microclimate at the soil surface resulting in less water lost through evaporation making more water available for transpiration resulting in higher grain yields. Research in the last 10–12 years has demonstrated that increased grain yields can be obtained by simply seeding crops between tall stubble rows. Maintaining tall stubble also reduces the amount of crop residue that needs to be processed through harvesters, reducing fuel consumption and accelerating the harvest operation.

The issue of weeds becoming resistant to herbicides such as glyphosate remains a threat to no-till production systems. More attention needs to be directed at integrated weed management to protect the effectiveness of chemical weed control and more emphasis needs to be place on developing technologies for precise application of herbicides to only the areas of the field that exceed threshold levels.

Canada's experience shows that environmental and economic sustainability are achievable in CA crop production systems. The accumulated knowledge and experience gained with no-till production systems on the Canadian Prairies provide a resource that can be shared with other areas of the world with similar soils and climatic conditions. The CA movement and experiences gained on the Canadian Prairies provide a framework to address new problems like climate change. The CA experience on the Prairies emphasizes the importance and interdependence of the different disciplines to bring about a change like no-till and the opportunities that can arise from technological advances.

References

Anonymous (1984) *Soil at Risk. Canada's Eroding Future*. Standing Committee on Agriculture, Fisheries and Forestry, The senate of Canada, Ottawa, Ontario.

Appleby, A.P. (2005) A history of weed control in the United States and Canada – a sequel. *Weed Science* 53, 762–768.

Arshad, M.A. and Azooz, R.H. (2003) In-row residue management effects on seed-zone temperature, moisture and early growth of barley and canola in a cold semi-arid region in North Western Canada. *American Journal of Alternative Agriculture* 18, 129–136.

Arshad, M.A., Franzluebbers, A.J. and Azooz, R.H. (1999) Components of surface soil structure under conventional and no tillage in north-western Canada. *Soil and Tillage Research* 53, 41–47.

Bailey, K.L., Gossen, B.D., Lafond, G.P., Watson, P.R. and Derksen, D.A. (2001) Effect of tillage and crop rotation on root and foliar diseases of wheat and pea in Saskatchewan from 1991 to 1998: Univariate and multivariate analyses. *Canadian Journal of Plant Science* 81, 789–803.

Baker, L.O. and Krall, J.L. (1956) Chemical summer fallow in Montana. *Down to Earth* 11, 21.

Beckie, H.J., Harker, K.N., Hall, L.M., Warwick, S.I., Legere, A., Sikkema, P.H., Clayton, G.W., Thomas, A.G., Leeson, J.Y., Sequin-Swartz, G. and Simard, M.J. (2006)A decade of herbicide resistant crops in Canada. *Canadian Journal of Plant Science* 86, 1243–1264.

Blackshaw, R.E., Harker, K.N., O'Donovan, J.T., Beckie, H.J. and Smith, E.G. (2008) Ongoing development of integrated weed management systems on the Canadian Prairies. *Weed Science* 56, 146–150.

Bradley, B. (2010) Manitoba-north Dakota zero tillage farmers association. In: Lindwall, C.W. and Sonntag, B. (eds) *Landscapes Transformed: The History of Conservation Tillage and Direct Seeding*. Knowledge Impact in Society, University of Saskatchewan, Saskatoon, Saskatchewan, pp. 45–49.

Campbell, C.A. and Zentner, R.P. (1997) Crop production and soil organic matter in long-term crop rotations in the semi-arid Northern Great plains of Canada. In: Paul, E.A., Paustian, K., Elliott, E.T. and Cole, C.V. (eds) *Soil Organic Matter in Temperate Agro-ecosystems: Long-term Experiments in North America*. CRC Press, Boca Raton, Florida, pp. 317–334.

Campbell, C.A., McConkey, B.G., Zentner, R.P., Selles, F. and Dyck, F.B. (1992) Benefits of wheat stubble strips for conserving snow precipitation in south-western Saskatchewan. *Journal of Soil and Water Conservation* 47, 112–115.

Campbell, C.A., Lafond, G.P., Moulin, A.P., Townley-Smith, L. and Zentner, R.P. (1997) Crop production and soil organic matter in long-term crop rotations in the sub-humid northern Great Plains of Canada. In: Paul, E.A., Paustian, K., Elliott, E.T. and Cole, C.V. (eds) *Soil Organic Matter in Temperate Agro-ecosystems: Long-term Experiments in North America*. CRC Press, Boca Raton, Florida, pp. 297–315.

Campbell, C.A., Selles, F., Lafond, G.P., Biederbeck, V.O. and Zentner, R.P. (2001) Tillage-fertilizer changes: Effect on some soil quality attributes under long-term crop rotations in a thin Black Chernozem. *Canadian Journal of Soil Science* 81, 157–165.

Campbell, C.A., VandenBygaart, A.J., Grant, B., Zentner, R.P., McConkey, B.G., Smith, W., Lemke, R. and Gregorich, E.G. (2007a) Quantifying carbon sequestration in a conventionally tilled crop rotation study in south-western Saskatchewan. *Canadian Journal of Soil Science* 87, 23–38.

Campbell, C.A., Vanden Bygaart, A.J., Zentner, R.P., McConkey, B.G., Smith, W., Lemke, R., Grant, B. and Jefferson, P.G. (2007b) Quantifying carbon sequestration in a minimum tillage crop rotation study in semiarid south-western Saskatchewan. *Canadian Journal of Soil Science* 87, 235–250.

Cutforth, H.W. and McConkey, B.G. (1997) Stubble height effects on microclimate, yield and water use efficiency of spring wheat grown in a semi-arid climate on the Canadian Prairies. *Canadian Journal of Plant Science* 77, 359–366.

Cutforth, H.W., McConkey, B.G., Ulrich, D., Miller, P.R. and Angadi, S.V. (2002) Yield and water use efficiency of pulses seeded directly into standing stubble in the semiarid Canadian Prairie. *Canadian Journal of Plant Science* 82, 681–686.

Cutforth, H.W., Angadi, S.V. and McConkey, B.G. (2006) Stubble management and microclimate, yield and water use efficiency of canola grown in the semi-arid Canadian Prairie. *Canadian Journal of Plant Science* 86, 99–107.

den Biggelaar, C., Lal, R., Wiebe, K. and Breneman, V. (2004a) The global impact of soil erosion on productivity I: absolute and relative erosion-induced yield losses. *Advances in Agronomy* 81, 1–48.

den Biggelaar, C., Lal, R., Wiebe, K., Breneman, V. and Reich, P. (2004b) The global impact of soil erosion on productivity II: effects on crop yields and production over time. *Advances in Agro*nomy 81, 49–95.

Derksen, D.A., Anderson, R.L., Blackshaw, R.E. and Maxwell, B. (2002) Weed dynamics and management strategies for cropping systems in the northern great plains. *Agronomy Journal* 94, 174–185.

Douglas, C.L. Jr and Rickman, R.W. (1992) Estimating crop residue decomposition from air temperature, initial nitrogen content and residue placement. *Soil Science Society of America Journal* 56, 272–278.

FAO (2013) What is conservation agriculture? Available at: http://www.fao.org/ag/ca/1a.html (accessed 20 January 2013).

Fowler, D.B. (1986) Snow management and winter grain cropping systems. In: Steppuhn, H. and Nicholaichuk, W. (eds) Great Plains Agriculture Council, *Proceedings of the Snow Management for Agricultural Symposium*, Swift Current, SK. Great Plains Agriculture Council, Publication No. 120, pp. 501–512.

Fowler, D.B. (2011) Wheat production in the high winter stress climate of the great plains of north America - an experiment in crop adaptation. *Crop Science* 52, 11–20.

Fowler, D.B. and Moats, L.R. (1995) Winter wheat 'Conserve and Win' demonstration and development program. In: University of Saskatchewan (ed.) *1995 Soils and Crops Workshop*, University of Saskatchewan, Saskatoon, Saskatchewan, Canada, pp. 212–226.

Fowler, D.B., Gusta, L.V., Slinkard, A.E. and Hobin, B.A. (1983) New frontiers in winter wheat production. In: *Proceedings of the Western Canada Winter Wheat Conference*. University of Saskatchewan, Saskatoon.

Fowler, D.B., Entz, M.H., Lafond, G.P. and Tompkins, D.K. (1990) Crop diversification through reduced tillage: The low input, environmentally friendly winter cereal option. In: Lafond, G.P. and Fowler, D.B. (eds) *Crop Management for Conservation*. University of Saskatchewan, Saskatoon, pp. 202–251.

Franzluebbers, A.J. and Arshad, M.A. (1996) Water-stable aggregation and organic matter in four soils under conventional and zero tillage. *Canadian Journal of Soil Science* 76, 387–393.

Gamache, P. (2010) Alberta conservation tillage society and Alberta reduced tillage initiative and Alberta reduced tillage linkages. In: Lindwall, C.W. and Sonntag, B. (eds) *Landscapes Transformed: The History of Conservation Tillage and Direct Seeding*. Knowledge Impact in Society, University of Saskatchewan, Saskatoon, Saskatchewan, pp. 67–76.

Gauer, E., Shaykewich, C.F. and Stobbe, E.H. (1982) Soil temperature and soil water under zero-tillage in Manitoba. *Canadian Journal of Soil Science* 62, 311–323.

Grant, C.A. and Lafond, G.P. (1993) The effects of tillage systems and crop sequences on soil bulk density and penetration resistance on a clay soil in southern Saskatchewan. *Canadian Journal of Soil Science* 73, 223–232.

Grant, C.A., Brown, K.R., Racz, G.J. and Bailey, L.D. (2002) Influence of source, timing and placement of nitrogen fertilization on seed yield and nitrogen accumulation in the seed of canola under reduced and conventional tillage management. *Canadian Journal of Plant Science* 82, 629–638.

Gray, R.S., Taylor, J.S. and Brown, W.J. (1996) Economic factors contributing to the adoption of reduced tillage technologies in central Saskatchewan. *Canadian Journal of Plant Science* 76, 661–668.

Hamel, C., Hanson, K., Selles, F., Cruz, A.F., Lemke, R., McConkey, B. and Zentner, R.P. (2006) Seasonal and long-term resource-related variations in soil microbial communities in wheat-based rotations of the Canadian prairies. *Soil Biology and Biochemistry* 38, 2104–2116.

Harapiak, J. (1990) Air seeders and fertilizer placement. In: Holm, F.A., Hobin, B.A. and Reed, W.B. (eds) *Air Seeding '90. Proceedings of the International Symposium on Pneumatic Seeding for Soil Conservation Systems in Dryland Areas*. University of Saskatchewan, Saskatoon, pp. 75–82.

Harapiak, J.T., Malhi, S.S., Campbell, C.A. and Nyborg, M. (1993) Fertilizer N application practices. In: Rennie, D.A. (ed.) *A Review of the Impact of Macronutrients on Crop Responses and Environmental Sustainability on the Canadian Prairies*. Canadian Society of Soil Science, Ottawa, Ontario, pp. 251–313.

Harker, K.N. and Clayton, G.W. (2003) Diverse weed management systems. In: Inderjit (ed.) *Weed Biology and Management*. Kluwer Academic Publishers, the Netherlands, pp. 251–266.

Hass, G. (1984) *The Optimum Tillage Challenge*. University of Saskatchewan Printing Services, Saskatoon, p. 280.

Hay, J. (1986) *Proceedings of the Tillage and Soil Conservation Symposium*. Indian Head Research Farm, Indian Head, Saskatchewan, p. 180.

Hofmann, N. (2001) Urban consumption of agricultural farmland in rural and small town Canada. *Analysis bulletin catalogue* no. 21-006-XIE Vol. 3, No. 2. Statistics Canada, 13 pp.

Holm, F.A., Hobin, B.A. and Reed, W.B. (1990) *Air Seeding '90. Proceedings of the International Symposium on Pneumatic Seeding for Soil Conservation Systems in Dryland Areas*. University of Saskatchewan, Saskatoon, Canada, p. 194.

Hood, N.W. (1990) Air-seeder overview – Australia: the past, the present and the future. In: Holm, F.A., Hobin, B.A. and Reed, W.B. (eds) *Air Seeding '90. Proceedings of the International Symposium on Pneumatic Seeding for Soil Conservation Systems in Dryland Areas*. University of Saskatchewan, Saskatoon, pp. 9–30.

Janzen, H.H. (2001) Soil science on the Canadian prairies – peering into the future from a century ago. *Canadian Journal of Soil Science* 81, 489–503.

Janzen, H.H. and Kucey, R.M.N. (1988) C, N, and S mineralization of crop residues as influenced by crop species and nutrient regime. *Plant and Soil* 106, 35–41.

Janzen, H.H., Johnston, A.M., Carefoot, J.M. and Lindwall, C.W. (1997) Soil organic matter dynamics in long-term experiments in southern Alberta. In: Paul, E.A., Paustian, K., Elliott, E.T. and Cole, C.V. (eds) *Soil Organic Matter in Temperate Agro-ecosystems: Long-term Experiments in North America*. CRC Press, Boca Raton, Florida, pp. 283–296.

Johnston, A.M., Lafond, G.P., Harapiak, J.T. and Head, W.K. (1997) No-till spring wheat and canola response to side banded anhydrous ammonia at seeding. *Journal of Production Agriculture* 10, 452–458.
Johnston, A.M., Lafond, G.P., Hultgreen, G.E. and Hnatowich, G.L. (2001) Spring wheat and canola response to nitrogen placement with no-till sideband openers. *Canadian Journal of Plant Science* 81, 191–198.
Juma, N.G., Izaurralde, R.C., Robertson, J.A. and McGill, W.NB. (1997) Crop yield and soil organic matter trends over 60 years in a Typic Cryoboralf at Breton, Alberta. In: Paul, E.A., Paustian, K., Elliott, E.T. and Cole, C.V. (eds) *Soil Organic Matter in Temperate Agro-ecosystems: Long-term Experiments in North America*. CRC Press, Boca Raton, Florida, pp. 273–282.
Krupinski, J.M., Bailey, K.L., McMullen, M.P., Gossen, B.D. and Turkington, K. (2002) Managing plant disease risk in diversified cropping systems. *Agronomy Journal* 94, 198–209.
Lafond, G.P. and Derksen, D.A. (1996) The long term potential of conservation tillage on the Canadian Prairies. *Canadian Journal of Plant Pathology* 18, 151–158.
Lafond, G.P. and Fowler, D.B. (1989a) Soil temperature and moisture stress effects on kernel water uptake and germination of winter wheat. *Agronomy Journal* 81, 447–450.
Lafond, G.P. and Fowler, D.B. (1989b) Soil temperature and water content, seeding depth and simulated rainfall effects on winter wheat emergence. *Agronomy Journal* 81, 609–614.
Lafond, G.P. and Fowler, D.B. (1990) Crop Management for Conservation. In: *Proceedings of the Soil Conservation Symposium*. University of Saskatchewan, Saskatoon, 297 pp.
Lafond, G.P., Loeppky, H. and Derksen, D.A. (1992) The effects of tillage systems and crop rotations on soil water conservation, seedling establishment and crop yield. *Canadian Journal Plant Science* 72, 103–115.
Lafond, G.P., Boyetchko, S.M., Brandt, S.A., Clayton, G.W. and Entz, M.H. (1996) Influence of changing tillage practises on crop production. *Canadian Journal of Plant Science* 76, 641–649.
Lafond, G.P., May, W.E., Stevenson, F.C. and Derksen, D.A. (2006) Effects of tillage systems and rotations on crop production for a thin Black Chernozem in the Canadian Prairies. *Soil and Tillage Research* 89, 232–245.
Lafond, G.P., Brandt, S.A., Clayton, G.W., Irvine, R.B. and May, W.E. (2011a) Rainfed farming systems on the Canadian Prairies. In: Tow, P. and Cooper, I. (eds) *Dryland Farming Systems*. Springer-Verlag, the Netherlands, pp. 467–510.
Lafond, G.P., Walley, F., May, W.E. and Holzapfel, C.B. (2011b) Long term impact of no-till on soil properties and crop productivity on the Canadian prairies. *Soil and Tillage Research* 117, 110–123.
Lal, R. (2007) Anthropogenic influences on world soils and implications to global food security. *Advances in Agronomy* 93, 69–93.
Leeson, J.Y., Thomas, A.G., Brenzil, C.A., Andrews, T., Brown, K. and Van Acker, R. (2005) *Weed Survey Series*. Publication 05-1. Agriculture and Agri-Food Canada, Saskatoon, Saskatchewan.
Lemke, R. and Farrell, R. (2008) Nitrous Oxide Emissions and Prairie Agriculture. *Prairie Soils and Crops* 1, 11–15. Available at: http://www.prairiesoilsandcrops.ca (accessed 20 January 2013).
Lupwayi, N.Z., Rice, W.A. and Clayton, G.W. (1999) Soil microbial biomass and carbon dioxide flux under wheat as influenced by tillage and crop rotation. *Canadian Journal of Soil Science* 79, 273–280.
Lupwayi, N.Z., Clayton, G.W., O'Donovan, J.T., Harker, K.N., Turkington, T.K. and Rice, W.A. (2004) Soil microbiological properties during decomposition of crop residues under conventional and zero tillage. *Canadian Journal of Soil Science* 84, 411–419.
Lupwayi, N.Z., Clayton, G.W., O'Donovan, J.T., Harker, K.N., Turkington, T.K. and Rice, W.A. (2006a) Decomposition of crop residues under conventional and zero tillage. *Canadian Journal of Soil Science* 84, 403–410.
Lupwayi, N.Z., Clayton, G.W., O'Donovan, J.T., Harker, K.N., Turkington, T.K. and Soon, Y.K. (2006b) Nitrogen release during decomposition of crop residues under conventional and zero tillage. *Canadian Journal of Soil Science* 86, 11–19.
Lyles, L. and Allison, B.E. (1981) Equivalent wind-erosion protection from selected crop residues. *Transactions of the American Society of Agricultural Engineers* 2, 405–407.
Malhi, S.S., Grant, C.A., Johnston, A.M., and Gill, K.S. (2001) Nitrogen fertilization management for no-till cereal production in the Canadian Great Plains: a review. *Soil and Tillage Research* 60, 101–122.
McClinton, B. (2007) Highlights from the 2006 Census. *Prairie Steward* 51(1). Available at http://www.ssca.ca (accessed 20 January 2013).
McClinton, B. and Polegi, J. (2010) Saskatchewan Soil Conservation Association. In: Lindwall, C.W. and Sonntag, B. (eds) *Landscapes Transformed: The History of Conservation Tillage and Direct Seeding*. Knowledge Impact in Society, University of Saskatchewan, Saskatoon, Saskatchewan, pp. 52–66.

McConkey, B.G., Liang, B.C., Campbell, C.A., Curtin, D., Moulin, A., Brandt, S.A. and Lafond, G.P. (2003) Crop rotation and tillage impact on carbon sequestration in Canadian prairie soils. *Soil and Tillage Research* 74, 81–90.

Memory, R. and Atkins, R. (1990) Air Seeding – The North American situation. In: Holm, F.A., Hobin, B.A. and Reed, W.B. (eds) *Air Seeding '90. Proceedings of the International Symposium on Pneumatic Seeding for Soil Conservation Systems in Dryland Areas*. University of Saskatchewan, Saskatoon, Saskatchewan, pp. 1–8.

Mitchell, J., Moss, H.C. and Clayton, J.S. (1944) *Soil Survey Report 12*. University of Saskatchewan, Saskatoon, Saskatchewan, Canada.

Montgomery, D.R. (2007) *Dirt: The Erosion of Civilizations*. University of California Press, Los Angeles, California.

Mostaghimi, S., Younos, T.M. and Tim, U.S. (1992) Crop residue effects on nitrogen yield in water and sediment runoff from two tillage systems. *Agriculture, Ecosystem and Environment* 39, 187–196.

O'Donovan, J.T., McAndrew, D.W. and Thomas, A.G. (1997) Tillage and nitrogen influence weed populations dynamics in barley (*Hordeum vulgare* L). *Weed Technology* 11, 502–509.

Parsons, G.F. (2008) Managing change: prospects, opportunities and issues in Saskatchewan's agricultural future. In: Saskatchewan Soil Conservation Association (ed.) *Proceedings of the 20th Annual Meeting and Conference of the Saskatchewan Soil Conservation Association*. Saskatchewan Soil Conservation Association, Indian Head, Saskatchewan, pp. 147–172.

Pimentel, D. and Pimentel, M. (2000) Feeding the world's population. *Bioscience* 50, 387.

Saskatchewan Agriculture and Food (2008) *2008 Guide to Crop Protection – Saskatchewan Weeds, Plant Diseases and Insects*. Saskatchewan Agriculture and Food, Regina, Saskatchewan.

Smika, D.E. and Unger, P.W. (1986) Effects of surface residues on soil water storage. *Advances in Soil Science* 5, 111–138.

Smil, V. (2000) *Feeding the World: A Challenge for the 21st Century*. MIT Press, Cambridge, Massachusetts.

Smith, A.E., Hume, L., Lafond, G.P. and Biederbeck, V.O. (1991) Review of the effects of long term 2,4-D and MCPA applications on wheat production and selected biochemical properties of a black chernozem. *Canadian Journal of Soil Science* 71, 73–87.

Smith, A.E., Mortensen, K., Aubin, A.J. and Molloy, M.M. (1994) Degradation of MCPA, 2,4-D and other phenoxyalkanoic acid herbicides using an isolated soil bacterium. *Journal Agriculture and Food Chemistry* 42, 401–405.

Soon, Y.K. and Arshad, M.A. (2004) Tillage, crop residue and crop sequence effects on nitrogen availability in a legume-based cropping system. *Canadian Journal of Soil Science* 84, 421–430.

Soon, Y.K. and Clayton, G.W. (2003) Effects of eight years of crop rotation and tillage on nitrogen availability and budget of a sandy loam soil. *Canadian Journal of Soil Science* 83, 475–481.

Soon, Y.K., Clayton, G.W. and Rice, W.A. (2001) Tillage and previous crop effects on dynamics of nitrogen in a wheat-soil system. *Agronomy Journal* 93, 842–849.

Statistics Canada (2013) 2011 Census of Agriculture. [Online] Available at: http://www.statscan.ca (accessed 20 January 2013).

The Noble Blade (2013) Alberta inventors and inventions. Available at: http://www.abheritage.ca/abinvents/inventions/inv_ag_noble_blade.htm (accessed 20 January 2013).

Timmons, F.L. (2005) A history of weed control in the United States and Canada. *Weed Science* 53, 748–761.

Turkington, T.K., Xi, K., Clayton, G.W., Burnett, P.A., Klein-Gebbinck, H.W., Lupwayi, N.Z., Harker, K.N. and O'Donovan, J.T. (2006) Impact of crop management on leaf diseases in Alberta barley fields, 1995–1997. *Canadian Journal of Plant Pathology* 28, 441–449.

Ward, B., Smith, D., Shaw, G., Haak, D. and Fredette, J. (2010) Policy and program response to land management issues. In: Lindwall, C.W. and Sonntag, B. (eds) *Landscapes Transformed: The History of Conservation Tillage and Direct Seeding*. Knowledge Impact in Society, University of Saskatchewan, Saskatoon, Saskatchewan, pp. 15–23.

World Fact Book (2009) The world fact book. Available at: http://www.cia.gov/library/publications/the-world-factbook/geos/xx.html (accessed 20 January 2013).

Zentner, R.P., McConkey, B.G., Campbell, C.A., Dyck, F.B. and Selles, F. (1996) Economics of conservation tillage in the semi-arid Prairie. *Canadian Journal of Plant Science* 76, 697–705.

Zentner, R.P., McConkey, B.G., Stumborg, M.A., Campbell, C.A. and Selles, F. (1998) Energy performance of conservation tillage management for spring wheat production in the Brown soil zone. *Canadian Journal of Plant Science* 78, 553–563.

Zentner, R.P., Wall, D.D., Nagy, C.C., Smith, E.G., Young, D.L., Miller, P.R., Campbell, C.A., McConkey, B.G., Brandt, S.A., Lafond, G.P., Johnston, A.M. and Derksen, D.A. (2002a) Economics of crop diversification and soil tillage opportunities in the Canadian Prairies. *Agronomy Journal* 94, 216–230.

Zentner, R.P., Lafond, G.P., Derksen, D.A. and Campbell, C.A. (2002b) Tillage method and crop diversification: effect on economic returns and riskiness of cropping systems in a Thin Black Chernozem of the Canadian Prairies. *Soil and Tillage Research* 67, 9–21.

Zentner, R.P., Lafond, G.P., Derksen, D.A., Nagy, C.N., Wall, D.D. and May, W.E. (2004) Effects of tillage method and crop rotations on non-renewable energy use efficiency for a thin Black Chernozem in the Canadian Prairies. *Soil and Tillage Research* 77, 125–136.

5 Conservation Agriculture in Australian Dryland Cropping

Jean-Francois (John) Rochecouste[1] and Bill (W.L.) Crabtree[2]
[1]*Conservation Agriculture Australia, Toowoomba, East Queensland, Australia;* [2]*Crabtree Agricultural Consulting, Beckenham, Western Australia*

5.1 Introduction

The Australian grains industry generates approximately 45 megatonnes (Mt) of grain annually, depending on seasonal conditions. They do this within a 200 mm to 800 mm annual rainfall zone that extends from central Queensland to Western Australia (WA). Most of this production occurs on light, low-fertility soils with limited water-holding capacity and an annual rainfall of less than 450 mm. Grain production is directly reliant on rainfall and there is a strong correlation between yield and available soil moisture in the northern Australian states, and in-crop rainfall in the southern states.

The incentive for a change in farming practices in Australia was created through three significant consequences of the traditional tillage farming system; erosion, the loss of soil moisture and delayed time of sowing. The most visible consequence of full-cut tillage was erosion from both water and wind depending on local climate patterns. In the northern cropping zones of Australia, high-intensity summer storms prior to summer cropping resulted in severe loss of topsoil and the associated loss of organic matter in the A horizon. In the southern and western cropping regions where lighter soils predominate, pre-frontal late autumn dust storms were similarly removing topsoil with severe impacts on soil fertility.

The consequence of these seasonal weather events was not immediately felt by most pioneer farmers as the negative impact on yield was a gradual process. The exception was circumstances in the sandier regions where crops were killed by sandblasting in high winds. However, the economic and emotional impact of declining yields from land degradation was a strong incentive for change. Less visible, but more evident to farmers on a seasonal basis was the loss of soil moisture from cultivation and the resulting lack of planting opportunities in the dry years.

Following visits to the USA and the UK, Australian soil conservation researchers and innovative farmers in the mid-1970s began experimenting with reduced tillage in all states. They were primarily interested in managing soil erosion from high-intensity rainfall events on hill slopes in Queensland, and managing severe wind erosion in South Australia (SA), WA and Victoria. By the late 1970s, the herbicide companies Monsanto and Imperial Chemical Industries had established a number of demonstration trials where herbicide was substituted for tillage, and crop residue was maintained as a form of soil cover to better manage the off-site impact of erosion.

The early results demonstrated both a significant reduction in erosion and a significant boost in available soil moisture.

From the early 1980s, leading farmers across Australia began experimenting with reducing the number of tillage operations to two, then to zero cultivation prior to sowing. Later farm trials showed increased planting opportunities over time and returned significant financial benefit relative to traditional multiple cultivation systems. Despite the obvious financial benefits (Table 5.1), uptake of such a new farming system by farmers at the time was relatively slow. It required a significant paradigm shift in the attitudes of farmers and support for change was limited by a range of factors. Foremost was the required change in seeding machinery and the lack of commercially available equipment. Weed control was also an issue, because without maintenance tillage in the fallow, cost-effective herbicides and sprayers were needed.

5.2 Reduced Tillage

In the early stages of reduced tillage adoption, no-till equipment was not commercially available and many farmers were already locked into conventional planters/seeders designed for a pre-seeding finely worked seedbed rather than one that would need to develop its own seedbed. The process of adoption took many years, led most often by farmers in the more marginal areas who had the most to gain from retaining soil moisture and timeliness of sowing. Adoption was faster in the drier western part of Australia, and is ongoing in eastern Australia where some farmers are still experimenting with reducing the number of tillage operations. Locally made commercial products are now supporting more rapid adoption. The cost of the herbicide glyphosate also became more competitive and over a span of 40 years reduced tillage has become the standard practice (Fig. 5.1).

5.3 Definitions of Tillage

The definitions of tillage practices have been variously described over time and it is likely that farmers' interpretations have also varied over time. This has implications for survey questions that compare today's practices with those of the past. Australian Conservation Agriculture (CA) farmer groups use the terminology below for common practices:

- Conventional (or multiple) tillage – two or more tillages before seeding.
- Reduced tillage – one pass of full-soil disturbance prior to seeding.
- Direct drilling – one-pass seeding with a full-cut or greater than 20% topsoil disturbance.
- No-tillage – knifepoint or disc seeding with 5–20% topsoil disturbance.
- Zero-tillage – disc seeding without soil throw, but note that some discs do throw soil (Crabtree, 2010).

Table 5.1. Wheat yields (t ha^{-1}) comparing farming practices over 4 years at two locations in Queensland. Relative cost benefit to growers based on current grain market price (Wylie and Moll, 1998).

Compared practice	Biloela (1989–92)	Goondiwindi (1989–92)
Conventional cultivation	2.5	1.6
Stubble mulch	3.1	1.8
Reduced tillage	3.3	2.0
Zero-till	3.4	2.2
Relative income differences in moving from conventional to zero-till in today's dollar value (AUS$212 t^{-1})[a] for a 500 ha year^{-1} crop	$95,400	$63,600

[a]Price based on multi-grade APW1 at Goondiwindi on the 28 May 2012, sourced from Graincorp (http://www.graincorp.com.au/pricing).

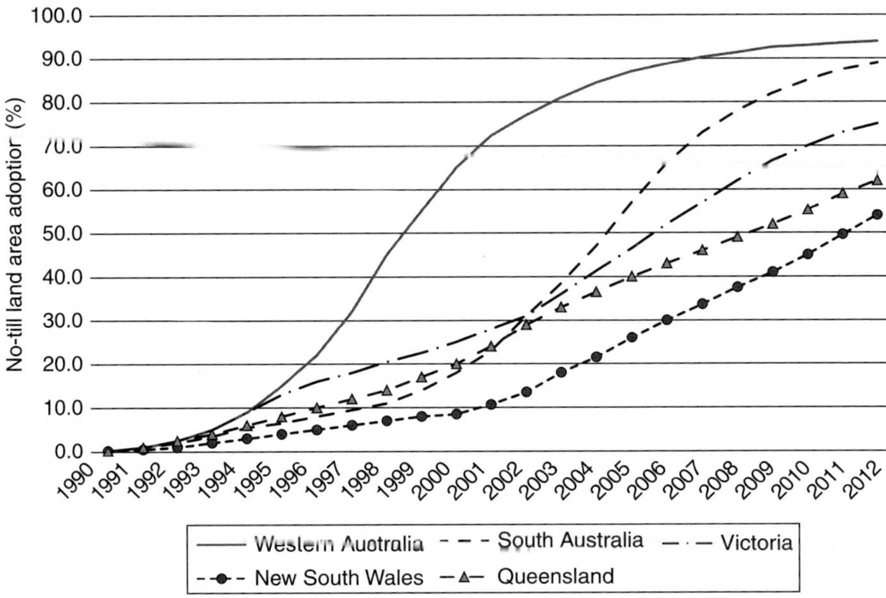

Fig. 5.1. The estimated adoption of reduced tillage farming practices in area terms by Australian states, updated to 2012 (Crabtree, 2010).

The term 'conventional' is becoming misleading as it now represents a minority practice in most cropping regions. The current trend has been for farmers to continue to reduce soil disturbance. Adoption of disc seeders has been more common in areas where livestock has been removed from the farming system and where diverse crop rotations are economically feasible. There are some regions where disc zero-till has been popular and is close to 100% adoption (Crabtree, 2010). The dominant reduced tillage system in Australia now is no-tillage and seeding with narrow (20–40 mm) knife points (Fig. 5.2) on 25–35 cm row spacing, along with press wheels (Fig. 5.3). No-tillage seeding using knife points following surface-applied pre-emergent herbicides has sufficient soil throw to cover the inter-row area and allows for safe and effective weed control. This does not work as efficiently with disc zero-till.

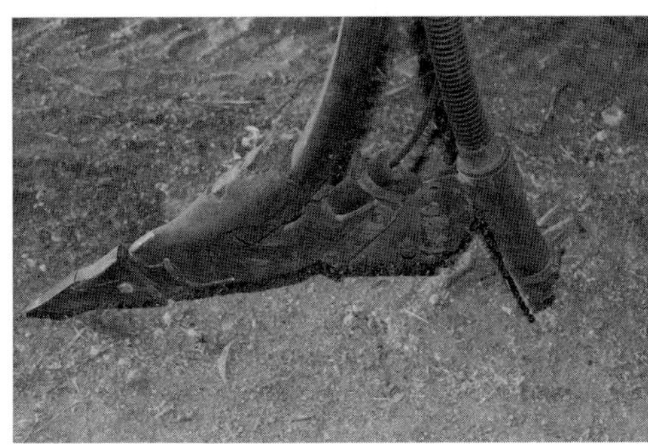

Fig. 5.2. Narrow knife point with seed slot at rear (photo courtesy of Neville Gould, Conservation Agriculture and No-till Farmers Association, Wellington, New South Wales).

5.4 Retained Stubble

Australia has seen an increasing trend to stubble retention, which represents a change in practice. In the past, one purpose of ploughing was to incorporate the plant residue left after harvest, allowing it to be broken down by soil microorganisms and facilitating the next planting (Thomas *et al.*, 2007). This involved a considerable amount of energy and often required several machinery passes to break up the plant material and mix it with the soil (Quick *et al.*, 1984). In the more arid regions of Australia which experience dry conditions for a large part of the year, there was generally insufficient topsoil moisture to allow the breakdown

Fig. 5.3. Press wheel located behind knife point (photo courtesy of Neville Gould, Conservation Agriculture and No-till Farmers Association, Wellington, New South Wales).

process to occur to an acceptable level for planting without mechanical intervention (Roper, 1985). Planting problems were more pronounced following years when high yields created biomass levels greater than 4 t dry matter ha^{-1} (Ashworth *et al.*, 2010). In the past, farmers responded to these high levels by grazing the stubble, baling it as feed or burning it with the aim of removing much of the crop residue prior to cultivation (Anderson, 2009). Although grazing and burning is still an option, many Australian farmers and agronomists see the value of leaving the stubble in place to protect the soil from high-intensity rainfall and erosion by water and wind (Silburn *et al.*, 2007). This benefit could be extended post-sowing but this required seeding equipment capable of operating successfully in these conditions, without the tines serving as a rake (Butler, 2008; Ashworth *et al.*, 2010). A range of approaches was used in adapting seeding machinery for this purpose, including many combinations of:

- Cutting crops 15–20 cm high, using harvesters with residue choppers/spreaders;
- Increasing row spacing (often to 30 cm or more);
- Using coulter cutting blades ahead of the seeding times to cut through residue;
- Deflecting residue ahead of each tine to be inter-row space;
- Distributing seeding tines between three to five machine bars, to increase the gap between adjacent tines on the same bar.

Despite the issues of seeding through crop residue, farmers and natural resource officers all considered the benefits exceeded the costs in effort and expense. Today more than 30 commercial machinery suppliers offer a large range of seeding machines and seeding machine adaptations. This includes a variety of seed-trench firming 'press wheels', which can have a major positive impact on crop emergence under marginal moisture conditions. Overload release ('stump jump') systems are universal in some areas, and individual row depth control ('parallelogram mount') mechanisms are increasingly common.

To further reduce soil disturbance, farmers have moved to disc seeding equipment, usually with individual row depth control and varying fertilizer placement systems. Some of these units are extremely heavy and capable of cutting through substantial volumes of dry residue. Most have some problems of pushing residue into the seed trench in soft moist conditions, when soil adhesion can also be a problem. An increasing number of farmers are addressing the issue of seeding through heavy residue by 'inter-row seeding' using high-precision guidance to place seed in a precise relationship to the standing stubble rows of the previous crop. Disc seeders disturb less soil, and hence encourage less weed-seed germination, but they are not as good as knife openers in producing even soil throw to incorporating pre-sowing residual herbicides. Consequently, where disc seeders are common, farmers are relying more on diversity in crop rotation as a weed management tool.

Some of the more important benefits of stubble retention in Australia's dry climate and poor soils include a reduction in surface sealing and herbicide movement into the seed furrow resulting from raindrop impact, together with improved infiltration, and reduced soil erosion (Scott *et al.*, 2010). Crop residue can also impair weed growth, return nutrients to the soil and provide some protection for emerging seedlings (Roper, 1985; Jacobson *et al.*, 1992; Unger, 1994; Malinda, 1995; Lal, 2008; Anderson, 2009).

Stubble is also a source of organic material contributing to the nutrient cycling performed by soil microorganisms and increases soil organic carbon (Table 5.2). Wheat stubble consists of approximately 40% carbon, 0.58% nitrogen, 0.05% phosphorus, 1.42% potash and 0.19% sulfur and the degradation of crop residues releases about 55–70% of the carbon to the atmosphere as CO_2 (Schomberg *et al.*, 1994; Tan, 2009). Microbial biomass takes up 5–15% of the carbon and the remaining 15–40% is partially stabilized in soil as new humus (Jenkinson, 1971).

The level of carbon returned to the soil is variable depending on the stubble type, soil characteristics, environmental conditions and management practices (Chan *et al.*, 2003; Wang and Dalal, 2006; Robertson and Thorburn, 2007; Liu *et al.*, 2009; Luo *et al.*, 2010). The 2010 Australian grain crop left a potential 56.5 Mt of residue after harvest prior to burning, grazing, or slow breakdown when retained for surface protection. This equates to 22.6 Mt of carbon, so changes in farm practices that involve residue retention became a bipartisan component of Australian government policy. This is currently expressed as part of the Commonwealth 'Caring for Our Country' initiative to support projects that help

Table 5.2. The advantages of crop stubble retention in Australian agricultural systems (Scott *et al.*, 2010).

Benefit	Description
Water erosion control	Reduced erosion by protecting the soil surface from the impact of raindrops during high-intensity storms predominantly in the north
Wind erosion control	Reduced loss of soil from the winds that cause dust storms, as wind speed is significantly decreased at the soil surface. Standing wheat stubble with rows across the wind direction reported to be the most effective to reduce wind erosion
Slows evaporation of soil moisture at the surface	Effectiveness is proportional to volume of stubble. Standing stubble more effective in resisting evaporation from wind
Increases soil moisture storage	In the higher rainfall areas, stubble cover increases net soil moisture by reducing the amount of surface run-off. In the southern lower rainfall areas stubble cover reduces evaporation to retain soil moisture
Nutrient conservation	Nutrient component of the stubble is returned to the system but with some immobilization during decomposition
Soil organic carbon (SOC) accumulation	May increase net SOC to a higher equilibrium or reduce the ongoing decline of SOC depending on other farming practices
Increased micro-fauna	Populations of several species of earthworms have increased with stubble retention when combined with reduced tillage

farmers maintain ground cover. However, in farm management terms retaining stubble (Table 5.3) can create a number of logistical and production problems that need to be considered in any policy development (Unger, 1994; Scott *et al.*, 2010).

The 2010/11 seasons were La Niña years and the fourth wettest on record for the eastern states, following similar La Niña events in 1973/74, 1955/56 and 1949/50. Wet seasons create excess stubble that becomes difficult to manage and also increases the occurrence of pest and disease carry-over. This is exemplified by yield impacts from such diseases as yellow leaf spot (*Pyrenophora tritici-repentis*) and crown rot (*Fusarium pseudograminearum*) on wheat, and can be a significant incentive for stubble removal. If problems become excessive, residue disturbance or even burning becomes a management option.

In 2007/08, an Australian Bureau of Statistics survey indicated that only 43% of crop farmers (all sectors) left their stubble intact, although it should be recognized that the percentage of farmers is not the same as the percentage of production. Another 33% tilled crop residues and 34% baled or grazed the stubble with some overlapping practices (Pink, 2009). Other surveys suggest the area of stubble burned is about 20% of the cereal area (Llewellyn *et al.*, 2009), but burning is less common in states such as WA and Queensland (Pink, 2009), except in continuous wheat areas where weed resistance is becoming a problem.

Overall, the ongoing benefits of stubble retention to stored moisture and improved soil health have seen a majority of farmers make the choice to retain crop residues after harvest, and manage the associated disadvantages as best they can. Retained residue is more acceptable than burning in terms of soil carbon impact, but the proportion of the remaining residue that degrades into the more stable humus fraction of soil carbon is both small and uncertain. This uncertainty creates a problem when we consider measuring the carbon balance of cropping soils for sequestration under the climate change policy being developed.

5.5 Controlled Traffic Farming

The impact of soil compaction by heavy farm machinery has become more apparent as larger tractors are used to operate more land per unit time (Chamen *et al.*, 1992, 2003; Batey, 2009). The effect of tractor wheels on soil compaction has resulted in crop production issues stemming from a disruption of structure (Hamza and Anderson, 2005; Kirchhof and Daniels, 2009), although not all soils are equally affected. The consequence has been reduced microbial activity, reduced water infiltration and poor root growth leading to yield limitations (Jones *et al.*, 2003; Tullberg *et al.*, 2007; Ahmad *et al.*, 2009; McKenzie *et al.*, 2009; Botta *et al.*, 2010). Controlled traffic farming (CTF) restricts the wheels of all heavy field traffic

Table 5.3. The disadvantages of crop stubble retention in Australian agricultural systems (Scott *et al.*, 2010).

Disadvantages	Description
Interference with seeding operation	Retained stubble can be a problem for machine operation at seeding, causing blockages between the tines or poor establishment by interfering with seed placement
Slow decomposition	In dry areas decomposition is slow and can interfere with future crop operations
Disease carry-over	Can be serious under the right conditions for disease development
Pest carry-over and habitat	Stubble can provide shelter that supports an increase in pest populations; more notably snails
Weed adaptation	Some weeds have adapted to high stubble loads and the stubble can interfere with foliar application of herbicides

to permanent traffic lanes to prevent damage to the whole paddock area from conventional 'random' operation.

The compacted permanent traffic lanes are laid out and managed for efficient traction, traffic and drainage, allowing the intervening crop beds to remain soft and in better condition for crop production. Because the harvester is the most difficult machine, controlled traffic operation has usually been achieved by modifying tractors to the harvester track width, using a harvester and seeder of the same width, and a sprayer which is a multiple of this width. This can provide machinery footprints in the range 12–16% of paddock area. This practice was taken up initially by farmers on heavier soil types, providing evidence of soil structural improvements, increased yields (Li *et al.*, 2007) and substantial reductions in fuel use. Farmers also report that hard permanent traffic lanes allow a wider window of operation for machinery as they do not have to wait so long for soft soils to dry out. Although farmers were initially concerned about the cost of machinery modification and tractor warranties, there has been an increasing adoption of CTF across the Australian cropping zone.

Use of compacted permanent traffic lanes resulted in greater energy efficiency than operating randomly on softer soils. The difference recorded by Tullberg *et al.* (2007) showed a 39% reduction in energy requirement from employing CTF. Gas exchange between soft soils and compacted soils are still under investigation, but preliminary results (J. Tullberg, Queensland, 2013, pers. comm.) show substantial reductions in nitrous oxide emissions from controlled traffic cropping beds. Emerging problems of CTF include the difficulty of controlling weeds in wheel tracks, and deep rutting of traffic lanes by continuous wheel passes in clay soils.

Despite the yield benefits of using CTF systems, the major barrier seems to be a false perception that machinery conversion is very expensive. Some current estimates of CTF in Australian agriculture indicate the number of farmers using some form of CTF at 15,320, which is about 21.9% of all crop farmers (ABS).[1] Given the overall energy savings, yield benefits and improved soil condition across most soil types, there is an argument for CTF to be considered an important practice in Conservation Agriculture (CA)(Yule and Chapman, 2011).

5.6 Crop Rotations

Rotating crops by growing different types of plants sequentially in the same paddock has been a long-term practice of agriculture to reduce build-up of pathogens and manage the nutrient demand of different crops (Bailey, 1996; Feller *et al.*, 2003; Korstanje and Cuenya, 2010). Legume production crops are also highly valued in rotations as a means of increasing nitrogen inputs or minimizing commercial demands for the next crop (Angus, 2001; Lindemann and Glover, 2003), but cereal crops are more profitable in the drier cropping regions. It can be financially difficult for farmers to rotate into alternative crops with poorer cash returns, despite the risk of plant disease carryover or increased weed burdens from not doing so (Godsey *et al.*, 2007; Thomas *et al.*, 2011).

Farmers will also move from one crop to another depending on market price; they will seek more profitable crop options if they are confident that they know how to grow the crop. Risk is another factor affecting the choice of crops. High input crops that are complex to grow often require a bigger outlay for greater returns, but there is also more to lose if conditions become unfavourable.

The value of legumes in supplying soil nitrogen for following crops is well-known to Australian farmers, but the economics of introducing a legume crop is not always acceptable when cereal grain prices are high but pulse crop prices low (Malcolm *et al.*, 2009). Grain seasonal prices have varied as much as 200% since 2004/05, with some legumes having similar variations though not necessarily in tandem. This has resulted in variable production volumes and a gradual decrease in the area planted to pulses over the last decade (O'Connell, 2010). Some of the more effective legume crops for fixing nitrogen are not always the most economical from a production perspective (Lindemann and Glover, 2003; Thomas *et al.*, 2011).

In a recent report, the Grains Research Development Corporation evaluated the benefit of break crops from a series of long-term trials in WA (GRDC, 2011). It indicated that the yield benefits of legume break crops were highly variable, often riskier and less profitable then cereals. The average yield benefit from the inclusion of lupin or field pea in the rotation compared to 'wheat following wheat' was in the range 0.3–0.6 t ha^{-1} in favour of a legume break crop.

These yield benefits were more evident in the higher rainfall areas, with improved water use efficiency over time attributed in part to no-till practices. Following the break crop, the following cereals still responded to nitrogen application; however, the rate of response was relatively low and more often dominated by non-nitrogen-related benefits (diseases and weed control). The break-crop benefit was also reported to last up to a third wheat crop (Seymour *et al.*, 2012). Despite the perceived value of crop rotations, especially legumes, the choice of cereals is predominantly driven by economics in many marginal areas.

5.7 Current Trends in Australian Conservation Agriculture Practices

Conservation Agriculture is said to offer a new paradigm that offers greater productivity from the same area of land using fewer resources and reducing negative impacts of agriculture on the environment (Collette *et al.*, 2011). Innovative farmers in Australia have moved beyond reducing tillage, maintaining ground cover and including crop rotations. They have sought further efficiencies in the use of resources from CTF and the application of precision agriculture. Precision agriculture is defined as farming using computers and information technology; it combines various sensors on-farm with global positioning systems to match farming practices more closely with crop needs (Bloomer and Powrie, 2011).

These innovations have not been without their challenges in the management of weeds, pest and diseases. Australia has benefited greatly from engineering innovations, research in weed control, digital sensors and satellite technology. General acceptance of the benefits of CA by farmers has encouraged industry suppliers to provide products that further support CA practices.

5.7.1 Machinery advances

Many modern no-till seeders can achieve precision seed placement in changing soil types (wet and dry); they can place the seed and fertilizer separately, ensure the crop seed is safely separated from herbicides, are capable of seeding through thick crop residues, and can ride over obstacles efficiently with less machinery damage. No-till seeders, for example, are increasingly using hydraulic systems to provide adjustable down-force control for openers and press-wheels, together with overload protection.

5.7.2 Herbicide resistance

Research into weed control has been critical to the development of CA. Fallow weed control usually depends on glyphosate and some populations of annual ryegrass (*Lolium rigidum*) have become glyphosate resistant (Fig. 5.4), an issue that first emerged in 1996 in Victoria. Later, glyphosate resistance also occurred in awnless barnyard grass (*Echinochloa colona*), liverseed grass (*Urochloa panicoides*) and windmill grass (*Chloris truncata*) in New South Wales (NSW). The first recording of broadleaf resistance was in fleabane (*Conyza bonariensis*) in Queensland. The most recent occurrence of resistance was in great brome (*Bromus diandrus*) in 2011 in SA (C. Preston, 31 May 2012, University of Adelaide, pers. comm.). Resistance problems

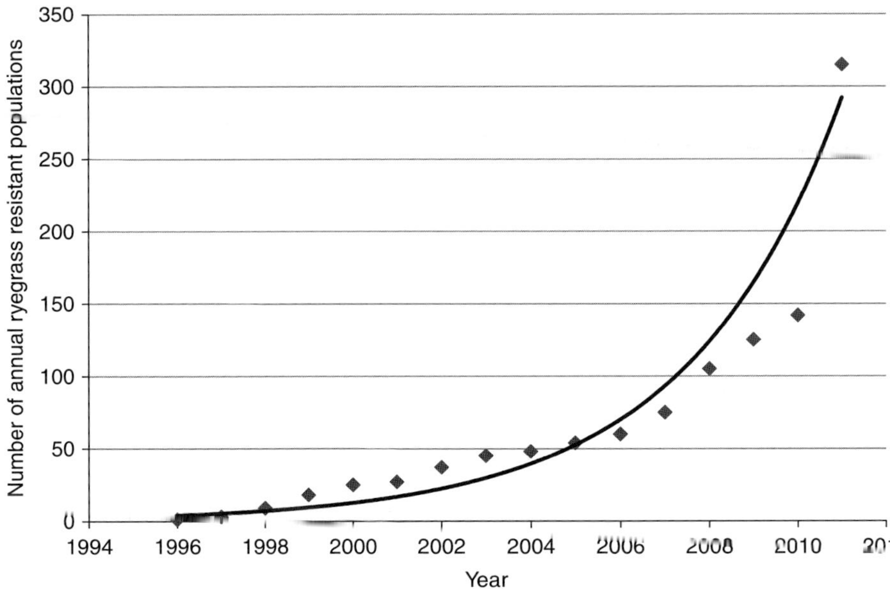

Fig. 5.4. The cumulative increase in the number of ryegrass resistant populations over time (Source: C. Preston, 31 May 2012, University of Adelaide, pers. comm.).

have not been limited to broad-acre cropping but are also evident in horticulture, industrial weed control areas, railway lines and roadsides.

To counter the increasing threat of resistant weeds, research has focused on rotating herbicides from different chemical groups, managing the postharvest weed seed bank with windrow burning where the harvested chaff is stacked in rows and burnt.

Reducing tillage limits moisture loss from evaporation but not from weeds, so when cultivation is not an option, weed control relies heavily on herbicides. Glyphosate has been an inexpensive and effective broad-spectrum knockdown herbicide, but its continuous use for fallow weed control has created an increasing problem of herbicide resistance (Code and Donaldson, 1996; Peltzer *et al.*, 2009; VanGessel *et al.*, 2009). An integrated weed management strategy to slow the development of resistance requires the addition of other herbicides and a range of agronomic strategies, such as rotation and harvest adaptations to reduce the weed seed bank. These all threaten to increase the cost of weed control (Beckie, 2011).

Technology that uses optical sensors to detect weeds (Fig. 5.5), along with on-off solenoids on the spray line would limit herbicide delivery to weed infestation areas instead of spraying the whole paddock (Hilton, 2000). The increasing use of this 'weed-seeker' technology is aimed at reducing the volume of herbicides, thus allowing a broader range of herbicides at reduced cost (M. Burgis, 15 June 2009, Conservation Farmers Inc., pers. comm.). The difference in application may not be obvious when wet years produce high weed populations, but becomes more significant in drier years with non-uniform establishment. Although expensive, this technology can provide substantial resource savings in the fallow weed control required to reduce soil moisture loss (Fig. 5.3).

5.7.3 Precision agriculture

The label precision agriculture was first applied when the combination of harvester

Area to be sprayed	264 ha
Water rate	80 l ha^{-1}
Actual usage	4.5% of volume
Actual area sprayed	11.88 ha
Actual cost of chemical	AUS$583.30
Chemical cost normal spray	AUS$12,962.40
Actual cost saving	**AUS$12,379.10**

Fig. 5.5. Demonstration by Crop Optic Australia at a farmer field day on how the optic sensor identifies a weed and activates the spray solenoid over that area (picture on left). The data are based on a case example for a grain and cotton farmer on the Darling Downs (M. Burgis, 15 June 2009, Conservation Farmers Inc., pers. comm.).

yield monitoring and satellite-based global positioning systems (GPS) allowed the economic production of paddock yield maps, but 'PA' is now a generic term covering a wide range of satellite and sensor-based technologies. The most widely adopted of these is 'GPS Autosteer' (self-steering) for farm equipment.

The world's first commercial satellite-based auto-steer using Real Time Kinematic (RTK) GPS correction for precise tractor steering, the 'Beeline Navigator', was developed by an Australian in the early 1990s. Guidance equipment of this type is now manufactured by several international organizations, and is a built-in option or standard unit in many tractors and harvesters. Inexpensive units claim pass-to-pass (repeatable only in the short term, not year-to-year) accuracies of ±10–30 cm, but more sophisticated units provide '2 cm' precision (±2 cm 67%; ± 4cm 95% of time). This development was originally driven by early controlled traffic adopters, but benefits, such as increased productivity with the elimination of overlap, and reduction in operator fatigue are large, often quantifiable, and easily justified by farmers managing increasingly large areas.

Accurate digital GPS position monitoring and recording now provides a platform for a large number of precision agriculture applications where data from various proximal sensors (Table 5.4) and other spatial information (e.g. satellite images) can be combined to provide resource efficiencies of the key farm inputs: labour, fuel, fertilizer and chemicals. The input cost benefits are balanced by the cost of establishing a digital network system on-farm and the human factor of having to learn how to use the system efficiently.

'Site-specific management' – the matching of seed, fertilizer and crop chemical inputs to crop requirements or soil characteristics of each paddock zone – became possible with the development of GPS-based harvest yield monitors and variable rate applicators. It has also become cheaper as more monitoring capability is standard equipment built into harvesters and applicators. Many Australian grain growers have now used yield mapping to provide useful management information, but the next step – 'zone management' using variable rate technology – requires complex assessment of soil/crop response characteristics and their interaction with climate probabilities.

Table 5.4. Some examples of sensors and related information for farmers.

Farm asset	Sensor	Data	Information
Soil texture	Electromagnetic induction	This is a non-contact method of measuring electrical conductivity involving inducing a magnetic field into the soil and measuring the electrical current response field	Texture and depth of topsoil
Soil moisture	Various: reflectometry, microwave or radio frequency via probe	Moisture curves	Current soil moisture trend
Soil pH and nutrient	Electrochemical sensors	Various	Field pH and nutrient status
Crop vigour/weed presence	Optical and radiometric	Crop vigour (relative)	Areas of poor growth, nutrient, disease, insect damage or presence of weeds
Yield monitors	Flow meters	Grain yield (relative)	Harvestable yield based on management
Variable seeding	Ground speed sensor	Seed volume	Plant population
Variable rate fertilizing	Flow meters/ chlorophyll sensors	Fertilizer output	Fertilizer volume

Scientific enthusiasm and investment in this technology has not been matched by practical adoption, which has been slow.

More recent development of crop condition sensing equipment shows greater promise of rapid application, particularly when early problem detection (e.g. nutrient deficiency) can enable timely and effective management response (e.g. foliar nutrient application). Development of systems to integrate proximal and remote sensor outputs to deploy farm operation more accurately has also interested farmers managing increasingly expensive and limited resources (Rochecouste, 2009). The aim is to optimize economic performance and avoid wasteful uniform applications by limiting inputs (e.g. fertilizers and chemicals) to 'what is needed, where it is needed' (Whitlock, 2006; Butler, 2008). This use of precision agriculture continues the trend towards increasing efficiency in the use of limited resources (Cook and Bramley, 2000; Shoup *et al.*, 2004).

Most farmers and agronomists have taken up some aspect of digital technology as part of their management, and the trend is increasing. Continuously Operating Reference Stations (CORS) are being built and gradually covering much of rural Australia. CORS is a network of permanent Global Navigation Satellite System (GNSS) tracking stations, which provide the RTK correction signals necessary for precise satellite positioning for industry and agriculture (Janssen *et al.*, 2011). CORS installation in Victoria is complete with 100% coverage. New South Wales is more than 50% complete, Queensland has coverage but mostly in the south-east and WA and SA have limited coverage. This technology will be an integral component of a more resource-efficient, productive and sustainable mechanized farming future.

5.7.4 Inter-row seeding

As the practice of retaining crop residues increased to protect soils, farmers noticed that crops sown between standing stubble rows performed better. Leaving the stubble standing after harvest reduces the problems of tine planter residue blockage and disc planters 'hairpinning' through failure to cutting through flat, wet stubble on soft soils. Precision auto-steer made it possible to routinely place an alternate row between existing rows of standing stubble, perhaps after some increase in row spacing, and adjustment of sprayer nozzle positioning. Inter-row planting provides a more consistent soil cover and associated weed-control benefits (Roberts and Leonard, 2008), and is simply achieved by use of an offset hitch to displace the seeder frame relative to the previous year's planting. Yield improvement of legumes sown within cereal stubble has also been reported, attributed mainly to reduced lodging and improvements in harvest efficiency (Roberts, 2008).

5.7.5 Cover cropping

Planting cover crops helps protect the soil from erosion. Cover crops add organic matter and immobilize soluble nutrients that would otherwise be lost down the soil profile. A cover crop is generally not grown primarily for harvest but returned to the soil as a green manure input. If the cover crop is a legume, there is an additional nitrogen input. The benefits are well recognized, but dryland farmers are concerned that cover-crop moisture requirements will compromise moisture availability for the following economic crop. Cropping windows on the lighter soils in the south and west are also short. Positive evidence about the impact of cover crops continues to accumulate, but it is still not common, except in those areas with reliable rainfall in the off-season.

5.7.6 Recycled organics

The nutrient value of animal industry waste as an alternative fertilizer and means of improving long-term soil structure has been a point of discussion among farmers. This applies particularly to those cropping zones where, a number of intensive livestock enterprises producing animal waste is conveniently located. Farmers have started purchasing and applying this waste, and in most cases seen a yield increase, primarily due to the nutrient content, released over several years. The cost of manure is comparable to traditional inorganic fertilizers, but manures are generally less predictable in their NPK nutrient value, and transport/application costs are significant. Uncomposted product has high water content, and raw manure can also tie up nitrogen for some period of time. Despite these issues a significant increase in its use occurred in 2008/09 when global fertilizer prices rose sharply.

In addition to the nutrient benefit, some farmers have also reported better long-term water-use efficiency from increased organic matter. This could be attributable to improved water-holding capacity (WHC) where agriculture is dominated by sandy soils, as outlined by the Western Australian Waste Authority (WAWA, 2010). Other benefits attributed to recycled organic amendments include increased water infiltration and improved soil structure. Although the linkage between water-use efficiency and WHC is well researched, it is unclear if adding a range of unspecified animal manure to fine-textured, low-fertility soils in an arid climate will lead to long-term improvement in WHC.

Use of urban sewerage is being trialled in some areas but there is concern about the likely build-up of heavy metal contaminants. Grain farms are also generally distant from major urban areas, making transportation costs prohibitive, so while the practice is favoured by many farmers, logistics limit its use to certain areas within easy transport reach of waste outlets.

5.7.7 Ecosystem services

Ecosystem services are defined as the public benefit of maintaining land in good condition, and payment for ecosystem services has often been advocated. Public benefit could include changes in land characteristics that improve soil and water quality, increase biodiversity or sequester carbon.

As a compensation for adopting land practices that reduce externalities, the proposal is that farmers be paid by governments on behalf of taxpayers or by private organizations looking to demonstrate their corporate social responsibility. This is still being explored in Australian policy terms.

Some environmental services are already being delivered by conservation farmers in the form of reduced erosion and improved soil biodiversity from retained stubble leading to improved water quality. Extending this scheme could include an annual performance-based cash flow to farmers to support revegetation on non-cropping marginal land (biodiversity refuges, carbon sequestration), maintaining or establishing natural vegetation along riparian areas (hydrological services), protecting established natural habitats (biodiversity), and the use of cover cropping in the rotation when economic crops are not available (soil carbon sequestration, soil biodiversity).

5.8 Policy Impacts on Conservation Agriculture in Australia

The Australian Government has three rural policy programmes directed at farmers that are likely to impact on conservation farmers.

1. 'Care for Our Country' is a two-billion Australian dollar spending initiative to improve Australia's environmental assets, which includes a multi-year budget of AUS$15 million for sustainable farm practices. The target involves improving land management practices of 42,000 farmers across 70 Mha, and includes initiatives to reduce tillage, maintain ground cover and build-up soil organic matter.
2. The Carbon Farming Initiative (CFI) was announced by the government in August 2010 with the aim of giving farmers, forest growers and land-holders access to domestic voluntary and international carbon markets by providing a framework to remove carbon dioxide from the atmosphere and to avoid the emission of greenhouse gases (GHG). The CFI is supported legislatively by the Carbon Credit (Carbon Farming Initiative) Act 2011, and is a market-based instrument to encourage farmers to become a net sink of carbon.
3. As part of the Clean Energy Future plan, the government included within the Tax Act a provision entitled The Conservation Tillage Refundable Tax Offset 3.1 Schedule 2 to the Clean Energy (Consequential Amendments) Bill 2011. This amends the Income Tax Assessment Act 1997 (Cth Australia s 67-23 (24)) to provide a Refundable Tax Offset (RTO) for certain new depreciating assets used in conservation tillage farming practices. The new law entitles the taxpayer to an RTO of 15% of the cost of an eligible asset. This would include tine machines fitted with minimum tillage points to achieve minimum soil disturbance, disc openers and suitable hybrid machines.

These rural policy programmes offer some form of incentive to reduce tillage, retain on-farm biomass, increase soil organic carbon, or to support new methodologies to reduce on-farm GHG emissions. Farmers applying CA practices have some opportunities to benefit from these policies.

5.8.1 Carbon sequestration using no-till in an Australian context

The concept that no-till practices will lead to significant carbon sequestration does not seem very likely in Australian dryland farming, where low rainfall limits biomass production and high temperatures accelerate the loss of soil organic matter. Soil carbon sequestration faces the same 'additionality' and 'permanence' tests as other sequestration mechanisms participating in carbon offset trading. The potential role of increasing soil organic carbon (SOC) in Australia has been reviewed by Sanderman *et al.* (2010). Grain cropping covers approximately 23 Mha of production (GRDC, 2012) dominated by light-textured soils. Cultivated soils lose organic carbon at variable rates depending on the clay content and annual rainfall (Swift, 2001). In a range of clay soils, losses of organic carbon averaged 0.6% year^{-1}

(Dalal and Chan, 2001). The limited rainfall and high summer temperatures of the cropping region limits the opportunity to significantly increase the organic carbon content of these soils (Chan *et al.*, 2008; Baldock *et al.*, 2009).

Under these conditions, reduced tillage practices have limited capacity to increase soil organic content, and in most situations they can only mitigate the ongoing loss (Wang *et al.*, 2010). This would mean that many of the cropping soils would show only marginal increases in SOC over time (Luo *et al.*, 2010; Chan *et al.*, 2011). Such small changes are unlikely to find sufficient offset units across the average grain farm to interest traders and would require some form of pooling to create the necessary economies of scale (Renwick *et al.*, 2003).

This is further complicated by the error margins associated with measuring SOC that emanate from variations in bulk density (Throop *et al.*, 2012) when sampling occurs to fixed depth, rather than equivalent mass across heterogeneous soil types (Sanderman *et al.*, 2010). Sanderman and Baldock (2010) also argue that predicted stock change data from agricultural trials may not truly reflect sequestration when the state of the soil carbon at the beginning of the trial is unknown; that is, when there is no comparable baseline at the start of the field trials. Thus current International Panel for Climate Change (IPCC) accounting methodologies developed from trial results may not show the true value of the carbon storage based on the management activities (Sanderman and Baldock, 2010).

This uncertainty is likely to affect confidence in the market allocation of carbon credit units for offsetting a unit of emission using soil carbon sequestration. Nevertheless, CA significantly reduces the loss of SOC to the atmosphere, and in certain seasons does create a carbon sink. Although it may not fit the mainstream carbon market, this should perhaps still be considered as a market-based instrument to encourage the benefits attributable to CA through reduced emissions and positive effects on the soil carbon balance.

5.8.2 Carbon market options

The role of agriculture in carbon trading has been reviewed in Australia by CSIRO (Walcott *et al.*, 2009). Current carbon markets in Australia are mostly voluntary and involve predominantly offsets derived from designated carbon sinks – usually forest plantations – with variable project methodologies (Ribon and Scott, 2007; Hassall, 2010). The operation of these markets using offset units from agricultural practices is still evolving. This is in part due to the uncertainties perceived by farmers that relate to contract terms in the offset market; that is, what sort of monitoring is involved and how long would the payment last? (Sanderman *et al.*, 2010).

The extent of reduction and the means of measuring emission performance from a farm practice in Australia are still unclear (Sanderman *et al.*, 2010). The determination of an accepted methodology for international markets is currently determined by agreements within the United Nations Framework Convention on Climate Change (Hodgkinson and Garner, 2008). In Australia, research is underway into methodologies that can produce an Australian Carbon Credit Unit. The Domestic Offsets Integrity Committee has endorsed four land-based methodologies (capture and combustion of landfill gas, destruction of methane generated from manure in piggeries, environmental plantings and savannah burning) and these have been approved by the Parliamentary Secretary for Climate Change and Energy Efficiency. As farming provides a critical service in terms of food production, it is important that emission reduction should not be at the cost of our food production. This would be likely to shift unintended consequences of food shortages to other nations to meet local national emission reduction targets.

At present, market options for a carbon credit unit based on CA practices are limited by not having a methodology, due in part to the complex biophysical processes of both the carbon and nitrogen cycles in seasonal agricultural practices.

5.8.3 Climate change consideration on future production

Following the IPCC Fourth Assessment Report: Climate Change 2007, the IPCC Working Group I noted in its executive summary that a 0.6°C increase was observed across the Australian continent. They also noted that southern Australia, which holds a significant portion of the cropping belt, is becoming drier. In 2010, the Australian Parliament's House of Representatives Standing Committee on Primary Industries and Resources held an inquiry into the role of government in assisting Australian farmers adapt to the impacts of climate change. The Conservation Agriculture Alliance of Australia and New Zealand (CAAANZ) made a submission on behalf of its members. A farmer representative informed the committee that conservation farmers had already been adapting to climate changes by deploying technology such as zero-till, CTF and retaining crop residues to conserve moisture. The committee was further advised that although gradual changes can be managed with adaptation strategies, of more concern to farmers is an increase in the timing of temperature extremes, and in the pattern as well as the level of precipitation. CAAANZ alliance members sought support not only in research for adaptive strategies but also requested that it be coupled with suitable extension programmes.

The average rainfall in the grain production areas lies between 200 and 800 mm year^{-1}, but this can fluctuate with drought and flood years depending on the various climatic patterns of the Indian and Pacific oceans. While the production areas are familiar with drought and flood years, they are nevertheless economically vulnerable to future climate change impacts on rainfall, evaporation and temperature (Crimp *et al.*, 2008; Howden *et al.*, 2010). Of particular concern to crop yield in the short-term outlook are reductions in net rainfall and the timing of that rainfall, with the possibility of a trend to increases in rainfall intensity going to run-off and limiting infiltration (Stephens and Lyons, 1998; van Herwaarden *et al.*, 1998; Hope and Ganter, 2009).

Potential changes in rainfall will vary across regions but overall the trend is towards reduced rainfall across the cereal belt (–30% to +20%). Increasing temperatures in the range 0 to 4°C will also impact on evaporation. Cropping as a farming enterprise generally yields better profit than livestock production, but it is also more economically vulnerable to climate risk in dry years due to grain yield sensitivity to moisture loss (van Herwaarden *et al.*, 1998; Day *et al.*, 2010). Predicted meteorological changes increase the risk conditions of reduced rainfall and reduced crop production, which is likely to have a significant impact on the future of farm profit if those risks are realized (Stephens and Lyons, 1998). Research may provide future solutions but that is purely speculative at this point. Successful adaptation therefore relies on the capacity of farmers to manage their production vulnerability through better farm management.

Conservation Agriculture has played a key role in the marginal grain production areas to manage the risk of drought over the last 30 years (Armstrong *et al.*, 2003; Turner and Asseng, 2005; Thomas *et al.*, 2007, 2011). The compelling benefits of CA in increasing crop yield by managing soil moisture and fertility have allowed farmers to meet the economic realities of increases in production costs and a reduction in the relative price of grains (Turner, 2004; Mullen, 2007). These gains are being further challenged by the risks associated with climate change (Howden *et al.*, 2010). Available soil moisture will be a key driving factor for farmers in managing future risk in Australia (Acuna and Wade, 2005; Branson, 2011).

Note

[1] Australian Bureau of Statistics ARMS Survey: Broad-acre crop farmers include those who planted cereals, canola, lupins, sugarcane and cotton (excludes fruit and vegetables).

References

Acuna, T.L.B. and Wade, L.J. (2005) Root penetration ability of wheat through thin wax-layers under drought and well-watered conditions. *Australian Journal of Agricultural Research* 56(11), 1235–1244.

Ahmad, N., Hassan, F.U. and Belford, R.K. (2009) Effect of soil compaction in the sub-humid cropping environment in Pakistan on uptake of NPK and grain yield in wheat (*Triticum aestivum*) I. Compaction. *Field Crops Research* 110(1), 54–60.

Anderson, G. (2009) *The Impact of tillage practices and crop residue (stubble) retention in the cropping system of Western Australia*. Bulletin number 4786, Western Australian Government.

Angus, J.F. (2001) Nitrogen supply and demand in Australian agriculture. *Australian Journal of Experimental Agriculture* 41(3), 277–288.

Armstrong, R.D., Millar, G., Halpin, N.V., Reid, D.J. and Standley, J. (2003) Using zero tillage, fertilisers and legume rotations to maintain productivity and soil fertility in opportunity cropping systems on a shallow Vertosol. *Australian Journal of Experimental Agriculture* 43(2), 141–153.

Ashworth, J., Desbiolles, J. and Tola, E. (2010) *Disc Seeding in Zero-till Farming Systems: A review of technology and paddock issues*. Western Australia No-till Farmers Association, Perth.

Bailey, K.L. (1996) Diseases under conservation tillage systems. *Canadian Journal of Plant Science* 76(4), 635–639.

Baldock, J., Grundy, M., Wilson, P., Jacquier, D., Griffin, T., Chapman, G., Hall, J., Machmet, D., Crawford, D., Hill, J. and Kidd, J. (2009) *Identification of Areas within Australia with the Potential to Enhance Soil Carbon Content*. CSIRO, Glen Osmond, South Australia.

Batey, T. (2009) Soil compaction and soil management – a review. *Soil Use and Management* 25(4), 335–345.

Beckie, H.J. (2011) Herbicide-resistant weed management: focus on glyphosate. *Pest Management Science* 67(9), 1037–1048.

Bloomer, D. and Powrie, J. (2011) Precision Agriculture: Doing the right thing in the right place at the right time. In: Bloomer, D. and Powrie, J. (eds) *A Guide to Smart Farming*. Landwise, Feilding, New Zealand, p. 4.

Botta, G.F., Tolon-Becerra, A., Lastra-Bravo, X. and Tourn, M. (2010) Tillage and traffic effects (planters and tractors) on soil compaction and soybean (*Glycine max* L.) yields in Argentinean pampas. *Soil & Tillage Research* 110(1), 167–174.

Branson, M. (2011) Using Conservation Agriculture to improve water use efficiency in wheat crops on the Branson farm in South Australia. In: *5th World Congress of Conservation Agriculture incorporating 3rd Farming Systems Design Conference - Resilient Food Systems for a changing world*. Brisbane, p. 547.

Butler, G. (ed.) (2008) *Conservation Agriculture: Moving Beyond Adoption*. SANTFA, Clare, South Australia.

Chamen, T., Alakukku, L., Pires, S., Sommer, C., Spoor, G., Tijink, F. and Weisskopf, P. (2003) Prevention strategies for field traffic-induced subsoil compaction: a review Part 2. Equipment and field practices. *Soil & Tillage Research* 73(1–2), 161–174.

Chamen, W.C.T., Vermeulen, G.D., Campbell, D.J. and Sommer, C. (1992) Reduction of traffic-induced soil compaction - a synthesis. *Soil & Tillage Research* 24(4), 303–318.

Chan, K.Y., Heenan, D.P. and So, H.B. (2003) Sequestration of carbon and changes in soil quality under conservation tillage on light-textured soils in Australia: a review. *Australian Journal of Experimental Agriculture* 43(4), 325–334.

Chan, K.Y., Gowie, A., Kelly, G., Singh, B. and Slavich, P. (2008) *Scoping Paper: Soil Organic Carbon Sequestration Potential for Agriculture in NSW*. NSW DPI Research & Science, NSW Department of Primary Industries, Sydney.

Chan, K.Y., Conyers, M.K., Li, G.D., Helyar, K.R., Poile, G., Oates, A. and Barchia, I.M. (2011) Soil carbon dynamics under different cropping and pasture management in temperate Australia: Results of three long-term experiments. *Soil Research* 49(4), 320–328.

Code, G.R. and Donaldson, T.W. (1996) Effect of cultivation, sowing methods and herbicides on wild radish populations in wheat crops. *Australian Journal of Experimental Agriculture* 36(4), 437–442.

Collette, L., Hodgkin, T., Kassam, A., Kenmore, P., Lipper, L.S., Nolte, C., Stamoulis, K. and Steduto, P. (2011) *Save and Grow - A policymaker's guide to the sustainable intensification of smallholder crop production*. Food and Agriculture Organization of the United Nations, Rome.

Cook, S. and Bramley, R. (2000) Precision agriculture: Using paddock information to make cropping systems internationally competitive. Bureau of Rural Sciences Publications. Available at: http://www.daff.gov.au/brs/publications/all_brs_publications (accessed 7 January 2011).

Crabtree, B. (2010) *Search for Sustainability in Dryland Agriculture*. Crabtree Agricultural Consulting, Beckenham, West Australia.

Crimp, S., Howden, M., Power, B., Wan, E. and De Voil, P. (2008) Global climate change impacts on Australia's wheat crops. In: GCC Review (ed.), Australian Government, Canberra.

Dalal, R.C. and Chan, K.Y. (2001) Soil organic matter in rainfed cropping systems of the Australian cereal belt. *Australian Journal of Soil Research* 39(3), 435–464.

Day, P., Cribb, J., Burgi, A. and Stanley, M. (2010) *Responding to Climate Change - Eyre Peninsula Research Findings 2010*. Eyre Peninsula Natural Resources Management Board Port Lincoln, South Australia.

Feller, C.L., Thuries, L.J.M., Manlay, R.J., Robin, P. and Frossard, E. (2003) 'The principles of rational agriculture' by Albrecht Daniel Thaer (1752-1828). An approach to the sustainability of cropping systems at the beginning of the 19th century. *Journal of Plant Nutrition and Soil Science* 166(6), 687–698.

Godsey, C.B., Vitale, J., Damicone, J.P., Sholar, J.R., Nickels, J. and Baker, J. (2007) Rotational effects in Oklahoma peanut production: Prospects for peanut rotations in the post-quota era. *Agronomy Journal* 99(5), 1238–1244.

GRDC (2011) *Break Crop Nenefit Factsheet - Western Region: Why make the Break*. Grain Research Development Corporation, Canberra, Australia.

GRDC (2012) *Grain Yearbook 2012*. Greenmount Press, Toowoomba.

Hamza, M.A. and Anderson, W.K. (2005) Soil compaction in cropping systems - A review of the nature, causes and possible solutions. *Soil & Tillage Research* 82(2), 121–145.

Hassall, G. (2010) *The Implication of Greenhouse Mitigation Policies on the Demand for Agricultural land - Research Report*. Australian Farm Institute, Surry Hills, Australia.

Hilton, P.J. (2000) Laser induced fluorescence for discrimination of crops and weeds. In: Gonglewski, J.D., Vorontsov, M.A. and Gruneisen, M.T. (eds) *High-Resolution Wavefront Control: Methods, Devices, and Applications*. Spie-Int Soc Optical Engineering, Bellingham, vol. 4124, pp. 223–231.

Hodgkinson, D. and Garner, R. (2008) *Global Climate Change: Australian Law and Policy*, 1st edn, LexisNexis Butterworths, Sydney New South Wales.

Hope, P. and Ganter, C. (2009) Recent and projected rainfal trends in south-west Australia and the associated shifts in weather systems. In: Jubb, I., Holper, P. and Cai, W. (eds) *Managing Climate Change*. CSIRO Publishing, Collingwood, Victoria.

Howden, S.M., Gifford, R.G. and Heinke, H. (2010) Grains in adapting agriculture to climate change. In: Stokes, C.J. and Howden, S.M. (eds) *Adapting Agriculture to Climate Change*. CSIRO Publishing, Melbourne, pp. 22–48.

Jacobson, C., Keith, K. and Kamel, T. (1992) *Understanding Soil Ecosystem Relationships*. Queensland Department of Primary Industries, Brisbane.

Janssen, V., Haasdyk, J. and McElroy, S. (2011) CORSnet-NSW Network RTK: Same look and Feel ... only Better. Paper presented to Association of Public Authority Surveyors 16th Annual Conference, Bathurst, New South Wales.

Jenkinson, D.S. (1971) Studies on the decomposition of C 14 labelled organic matter in soil. *Soil Science*, 64–70.

Jones, R.J.A., Spoor, G. and Thomasson, A.J. (2003) Vulnerability of subsoils in Europe to compaction: a preliminary analysis. *Soil & Tillage Research* 73(1–2), 131–143.

Kirchhof, G. and Daniels, I. (2009) Changing tillage management practices and their impact on soil structural properties in north-western New South Wales, Australia. *ACIAR Technical Reports Series*, no. 71, pp. 60–69.

Korstanje, M.A. and Cuenya, P. (2010) Ancient agriculture and domestic activities: a contextual approach studying silica phytoliths and other microfossils in soils. *Environmental Archaeology* 15(1), pp. 43–63.

Lal, R. (2008) Crop Residues and Soil Carbon. Available at: http://www.fao.org/ag/ca/Carbon%20Offset%20 Consultation/CARBONMEETING/3FULLPAPERSBYCONSULTATIONSPEAKERS/PAPERLAL.pdf (accessed on 31 May 2010.

Li, Y.X., Tullberg, J.N. and Freebairn, D.M. (2007) Wheel traffic and tillage effects on runoff and crop yield. *Soil & Tillage Research* 97(2), 282–292.

Lindemann, W.C. and Glover, C.R. (2003) *Nitrogen Fixation by Legumes*. Cooperative Extension Service - College of Agriculture and Home Economics, New Mexico State University.

Liu, D.L., Chan, K.Y. and Conyers, M. (2009) Simulation of soil organic carbon under different tillage and stubble management practices using the Rothamsted carbon model. *Soil & Tillage Research* 104, 65–73.

Llewellyn, R., Demden, F. and Gobbett, D. (2009) *Adoption of No-till and Conservation Farming Practices in Australian Grain Growing Regions: current status and trends*. CSIRO, Glen Osmond, South Australia.

Luo, Z.K., Wang, E.L. and Sun, O.J. (2010) Soil carbon change and its responses to agricultural practices in Australian agro-ecosystems: A review and synthesis. *Geoderma* 155(3–4), 211–223.

Malcolm, B., Sale, P., Leury, B. and Barlow, S. (2009) *Agriculture in Australia*, 2nd edn. Oxford University Press, Melbourne.

Malinda, D.K. (1995) Factors in conservation farming that reduce erosion. *Australian Journal of Experimental Agriculture* 35(7), 969–978.

McKenzie, B.M., Kuhner, S., MacKenzie, K., Peth, S. and Horn, R. (2009) Soil compaction by uniaxial loading and the survival of the earthworm *Aporrectodea caliginosa*. *Soil & Tillage Research* 104(2), 320–323.

Mullen, J. (2007) Productivity growth and the returns from public investment in R&D in Australian broadacre agriculture. *Australian Journal of Agricultural and Resource Economics* 51(4), 359–384.

O'Connell, L. (2010) *Grain Yearbook 2010 - Report to Industry*. Toowoomba.

Peltzer, S.C., Hashem, A., Osten, V.A., Gupta, M.L., Diggle, A.J., Riethmuller, G.P., Douglas, A., Moore, J.M. and Koetz, E.A. (2009) Weed management in wide-row cropping systems: a review of current practices and risks for Australian farming systems. *Crop & Pasture Science* 60(5), 395–406.

Pink, B. (2009) *Land Management and Farming In Australia 2007-08*. Australian Government, Hobart, Tasmania.

Quick, G.R., Andrews, A.S. and Erbach, D.C. (1984) *Opportunities to Reduce Energy Consumption in Tillage Operations in Australia*. Agriculture Engineering Branch, Department of Agriculture, New South Wales.

Renwick, A., Ball, A.S. and Pretty, J. (2003) Economic, biological and policy constraints on the adoption of carbon farming in temperate regions. In: Swingland, I.R. (ed.) *Capturing Carbon and Conserving Biodiversity*. Earthscan Publications, London.

Ribon, L. and Scott, H. (2007) *Carbon Offset Providers in Australia 2007*. RMIT Univesity, Melbourne.

Roberts, M. (2008) Multiple Benefits from inter-row sowing. In: Butler, G. (ed.) *Conservation Agriculture - Moving Beyond Adoption*. South Australian No-till Farmers Association, Clare, South Australia, pp. 34–36.

Roberts, M. and Leonard, E. (2008) Inter-row seeding part of a systems package. In: Butler, G. (ed.) *Conservation Agriculture - Moving Beyond Adoption*. South Australian No-till Farmers Association, Clare, South Australia, pp. 37–38.

Robertson, F.A. and Thorburn, P.J. (2007) Management of sugarcane harvest residues: consequences for soil carbon and nitrogen. *Australian Journal of Soil Research* 45(1), 13–23.

Rochecouste, J.-F.G. (2009) Integrating proximal and remote sensor technologies to improve production efficiency in a low emission cropping system. Ppaper presented to 7th National CTF Conference Australia - Hi-Tech Low Emissions Cropping Systems, Canberra, 7–8 September 2009.

Roper, M.M. (1985) Straw decomposition and nitrogenase activity (C_2H_2 reduction): effects of soil moisture and temperature. *Soil Biology & Biochemistry* 17(1), 65–71.

Sanderman, J. and Baldock, J.A. (2010) Accounting for soil carbon sequestration in national inventories: a soil scientist's perspective. *Environmental Research Letters* 5, 3.

Sanderman, J., Farquharson, R. and Baldock, J. (2010) *Soil Carbon Sequestration Potential: A review for Australian agriculture*. CSIRO Land and Water, Canberra. Available at: http://www.csiro.au/files/files/pwiv.pdf.

Schomberg, H.H., Ford, P.B. and Hargrove, W.L. (1994) Influence of crop residues on nutrient cycling and soil chemical properties. In: Unger, P.W. (ed.) *Managing Agricultural Residues*. Lewis Publishers, Boca Raton, Florida, pp. 99–121.

Scott, B.J., Eberbach, P.L., Evans, J. and Wade, L.J. (2010) *Stubble Retention in Cropping Systems in Southern Australia: Benefits and Challenges*. EH Graham Centre Monograph 1, Industry & Investment NSW, Orange.

Seymour, M., Kirkegaard, J.A., Peoples, M.B., White, P.F. and French, R.J. (2012) Break-crop benefits to wheat in Western Australia – insights from over three decades of research. *Crop and Pasture Science* 63(1), 1–16.

Shoup, D.W., Lee, W. and Harrison, T. (2004) Precision technologies for precision management. In: Peart, R.M. and Shoup, D.W. (eds) *Agricultural Systems Management - Optimizing Efficiency and Performance*. Marcel Dekker Inc., New York.

Silburn, D.M., Freebairn, D.M. and Rattray, D.J. (2007) Tillage and the environment in sub-tropical Australia - Tradeoffs and challenges. *Soil & Tillage Research* 97(2), 306–317.

Stephens, D.J. and Lyons, T.J. (1998) Rainfall-yield relationships across the Australian wheatbelt. *Australian Journal of Agricultural Research* 49(2), 211–223.

Swift, R.S. (2001) Sequestration of carbon by Soil. *Soil Science* 166(11), 858–871.

Tan, K.H. (2009) *Environmental Soil Science*, 3rd edn. CRC Press, Boca Raton, Florida.
Thomas, G.A., Titmarsh, G.W., Freebairn, D.M. and Radford, B.J. (2007) No-tillage and conservation farming practices in grain growing areas of Queensland - a review of 40 years of development. *Australian Journal of Experimental Agriculture* 47(8), 887–898.
Thomas, G.A., Dalal, R.C., Weston, E.J., King, A.J., Holmes, C.J., Orange, D.N. and Lehane, K.J. (2011) Crop rotations for sustainable grain production on a vertisol in the semi-arid subtropics. *Journal of Sustainable Agriculture* 35(1), 2–26.
Throop, H.L., Archer, S.R., Monger, H.C. and Waltman, S. (2012) When bulk density methods matter: Implications for estimating soil organic carbon pools in rocky soils. *Journal of Arid Environments* 77, 66–71.
Tullberg, J.N., Yule, D.F. and McGarry, D. (2007) Controlled traffic farming - From research to adoption in Australia. *Soil & Tillage Research* 97(2), 272–281.
Turner, N.C. (2004) Sustainable production of crops and pastures under drought in a Mediterranean environment. *Annals of Applied Biology* 144(2), 139–147.
Turner, N.C. and Asseng, S. (2005) Productivity, sustainability, and rainfall-use efficiency in Australian rainfed Mediterranean agricultural systems. *Australian Journal of Agricultural Research* 56(11), 1123–1136.
Unger, P.W. (1994) *Managing Agricultural Residues*. Lewis Publishers, Bushland, Texas.
van Herwaarden, A.F., Farquhar, G.D., Angus, J.F., Richards, R.A. and Howe, G.N. (1998) 'Haying-off', the negative grain yield response of dryland wheat to nitrogen fertiliser - I. Biomass, grain yield, and water use. *Australian Journal of Agricultural Research* 49(7), 1067–1081.
VanGessel, M.J., Scott, B.A., Johnson, Q.R. and White-Hansen, S.E. (2009) Influence of glyphosate-resistant horseweed (*Conyza canadensis*) growth stage on response to glyphosate applications. *Weed Technology* 23(1), 49–53.
Walcott, J., Bruce, S. and Sims, J. (2009) *Soil Carbon for Carbon Sequestration and Trading: a Review of Issues for Agriculture and Forestry*. Commonwealth of Australia, Canberra.
Wang, E.L., Zhongkui, L. and Smith, C.J. (2010) Potential change of soil carbon in Australia agro-ecosystems as affected by conservation management: data synthesis and modelling. Paper presented to 19th World Congress of Soil Science - Soil Solutions for a Changing World, Brisbane.
Wang, W.J. and Dalal, R.C. (2006) Carbon inventory for a cereal cropping system under contrasting tillage, nitrogen fertilisation and stubble management practices. *Soil & Tillage Research* 91(1–2), 68–74.
WAWA (Waste Authority) (2010) *Recycled Organics and Water Use Efficiency - Information Paper*. WAWA, The Government of Western Australia.
Whitlock, A. (2006) Precision farming - Supporting innovation. In: Horn, R., Fleige, H., Peth, S. and Peng, X.H. (eds) *Soil Management for Sustainability. Advances in GeoEcology* 38, pp. 242–250.
Wylie, P. and Moll, J. (1998) *Opportunity Cropping*, 2nd edn. Conservation Farmers Incorporated, Toowoomba, Queensland.
Yule, D. and Chapman, W. (2011) Controlled Traffic Farming - more productivity, sustainability and resilience. Paper presented to 5th World Congress of Conservation Agriculture incorporating 3rd Farming Systems Design Conference, 26–29 September 2011, Brisbane, Australia.

6 Conservation Agriculture in Europe

Theodor Friedrich,[1] Amir Kassam[1,2] and Sandra Corsi[1,3]
[1]*Plant Production and Protection Division, Food and Agriculture Organization of the United Nations, Rome;* [2]*School of Agriculture, Policy and Development, University of Reading, UK;* [3]*University of Teramo, Italy*

6.1 Introduction

This chapter tries to provide a snapshot of the Conservation Agriculture (CA) development and adoption in Europe as far as reported. It is based on reports from countries with at least one organization dedicated to CA and member of the European Conservation Agriculture Federation (ECAF) or in some other way connected to the global community of practice on CA. However, it is quite possible that there is also some CA adoption taking place in countries that were not reached and have not reported for this chapter.

Europe is considered to be a developing continent in terms of the adoption of CA. Only Africa, with about 1 Mha under CA corresponding to 1% of the arable land in the reporting countries has a smaller area under CA/no-till than Europe (including Russia) with 6 Mha corresponding to about 3% of the cropland. According to Basch (2005):

> European and national administrations are still not fully convinced that the concept of CA is the most promising one to meet the requirements of an environmentally friendly farming, capable to meet the needs of the farmers to lower production costs and increase farm income, and to meet the consumer demands for enough and affordable quality food with a minimum impact on natural, non-renewable resources. The reliance of CA on the use of herbicides and the alleged increased input of herbicides and other chemicals for disease and pest control are the main constraints to the full acceptance of CA as a sustainable crop production concept.

The global proliferation of negative environmental events, such as soil degradation and erosion, increasing humus decomposition through intensive soil cultivation and the associated release of CO_2 into the atmosphere, decreasing biodiversity through the removal of plant residues from the ground surface and also the political context (cadastral maps of erosion) make a change from conventional agriculture (ConvA) to CA essential in the future. All recent studies as well as field observations show that European soils are threatened by erosion, compaction and loss of organic matter in moist areas as well as in dry zones. Water pollution with nitrates, phosphorus and pesticides is widespread over Europe. In addition the economic viability of farming is declining, for different reasons:

1. It is highly dependent on fossil fuel for agricultural machinery and for the manufacture of nitrogen fertilizer, on protein for concentrated livestock production, and on inorganic fertilizers such as phosphates.

2. Norms and regulations on the environment and animal welfare frequently result in economic handicaps, on the basis that intensive production usually results in increased pollution.

The reasons for adoption of CA across Europe vary. In the wetter and cooler northern and western parts, characterized by low intensity rainfall, the main drivers behind CA adoption are cost reduction, the capability of finishing field work in shorter time-windows to respond to unreliable climatic conditions, and pollution reduction. In the hotter and drier south-western parts, also characterized by heavier rainstorms, soil and water conservation have been the main drivers for CA adoption (Soane *et al.*, 2012).

6.1.1 History of Conservation Agriculture in Europe: beginnings and expansion over the years in different regions and cropping/production systems

The history of CA varies in Europe from country to country. It is mostly characterized by consideration of different levels of reduced tillage leading to a general confusion and, only in exceptional cases, to conclusive development and promotion of a full CA system as defined by FAO (FAO, 2012a), which in fact has been adopted by only few pioneer farmers throughout Europe. An important milestone for CA in Europe, resulting from the developments in different European countries, was the foundation of ECAF in 1999, which, together with the UN Food and Agriculture Organization, held the first World Congress on Conservation Agriculture in 2001 in Madrid, initiating a series of such congresses (2003 Brazil, 2005 Kenya, 2009 India, 2011 Australia, 2014 Canada) and promoting CA also at European policy levels.

The adoption of no-tillage technologies was very rapid in Finland. The area of no-till (NT) in Finland increased rapidly from 1998 to reach 8–12% of the total area of cereals and oilseed crops by 2005 and 13% by 2008 (Soane *et al.*, 2012). This corresponds, according to FINCA (the Finnish CA Association), to 200,000 ha in 2008. In this way Finland has advanced to be one of Europe's leading NT countries. The reason for this quick adoption was that the process was farmer-driven: those farmers who believed in the NT system and made it work communicated their experiences to their peers. The extension service and research organizations as well as the agribusiness sector took interest in this development only later. FINCA has played a major role in spreading NT in Finland.

The situation is completely different in Denmark: in the 1960s and 1970s some Danish farmers tried to practise NT, but they discontinued mainly because of problems with perennial weeds. In the 1980s some farmers again used burning of straw before NT direct seeding. The burning of straw in the fields was prohibited in 1987 and so NT stopped again. In 1999 the Danish association for CA was established (FRDK). Since then the number of farmers that practise no-ploughing has increased considerably. The system that is used is harrowing before seeding, and some of the farmers who now have practised the no-ploughing system for some years, including the vice-president of the board of FRDK, moved further and practise complete NT.

A different situation evolved in Ireland, where the initial impulse came from the commercial sector beginning early in 2000 with an information and awareness campaign targeted at the farming community about the benefits of conservation tillage. The technique was called 'ECOtillage', which was based on shallow cultivation with soil disturbance limited to depths under 10 cm. Early pioneers of the system were commercial growers who mainly practised monoculture winter wheat systems. In 2003 an organization called 'CA Ireland' (CAIR) was established by a group of farmers with a common interest in raising education and awareness about CA among crop producers. CAIR became an affiliated member of ECAF in 2004 and continues to be funded solely by farmer member subscription. CAIR has organized field events on members' farms where some of the problems growers were experiencing were discussed.

Other farmers began using reduced tillage and by 2005 there were approximately 100 farmers practising some form of conservation tillage on approximately 11,000 ha. Yet, adoption of CA in Ireland is still nearly non-existent.

In the UK in 1989, a ground-breaking farm-scale whole rotational experiment began at Long Ashton Research Station (south-west England), which led to the initiation of a network of similar research farms around the country each specializing in different aspects of crop production. This 14-year project (known as the 'Less Intensive Farming and Environment' (LIFE) Project) provided strategic and applied information to underpin the development of economically viable, ecologically and environmentally sound and sustainable arable crop production systems. Such systems targeted the stepwise replacement of off-farm inputs by the integration of natural regulation, on-farm alternatives and management skills in order to maintain species and landscape diversity, minimize pollution and losses, provide a safe and wholesome food supply and to sustain income (Jordan *et al.*, 1997). The LIFE experiment demonstrated that input costs could be reduced and even accepting a small reduction in yield, greater margins could be achieved by the farmer. A pinch-point in autumn-dominated arable crops was the clash between late-harvest and early crop establishment, a key restriction being the use of the plough and the subsequent follow-up cultivations required to make a seedbed. In 1991 an organization called 'Linking Environment and Farming' (LEAF) was set up in the UK to promote the integrated approach pioneered by the LIFE Project. In 1996 the 'Integrated Arable Crop Production Alliance' (IACPA) was formed with the aim to pool the knowledge of the experts conducting the experimental work. In 1998, IACPA produced a report (MAFF, 1998), which concluded that non-plough cropping systems reduced energy inputs, reduced nitrogen losses, improved soil physical properties, allowed different weed control strategies to be used, reduced the risk of soil erosion, increased beneficial flora and fauna and most importantly required 36 less working days at a busy time on a 1000 acre arable farm. Unsurprisingly, farmers seized this opportunity and a rapid and substantial switch to minimum tillage followed. Between 1999 and 2005 the amount of land ploughed in the UK dropped from over 90% to less than 50%, whilst minimum tillage increased from less than 10% to over 40% (Lane *et al.*, 2006). Contextually farmers sought expert advice to overcome problems in using a range of new machinery, with different crop rotations on different soil types. In 1999 the organization 'UK Soil Management Initiative' (SMI) was established to provide this expertise and allow knowledge exchange on CA on a Europe-wide basis through the co-foundation of ECAF. Despite SMI efforts, as yet, adoption of the complete CA system in the UK is still low.

In Switzerland interest in CA resulted from erosion problems. The country is characterized by sloping and undulating areas as well as a cool and wet climate with annual precipitation of 1000 mm and more. Therefore, soil erosion is a major concern in arable farming. In addition, axle-loads of farm machinery have increased significantly during the last decade resulting in pronounced soil compaction and decreased soil quality. In particular with maize, where the surface remains uncovered during a relatively long juvenile crop stage, soil erosion has been observed regularly on fields cropped with intensive soil tillage. Therefore, one of the first attempts to reduce tillage intensity was reported in maize in the 1980s (Sturny and Meerstetter, 1990). In the late 1980s and early 1990s, a cropping system of maize with strip-band tillage was developed at the Swiss Federal Research Station in Zürich-Reckenholz in collaboration with commercial contractors. Strips of 25 cm were tilled with adapted rotary harrows and maize was planted with attached planters into these bands (Ammon *et al.*, 1990). The area between the rows remained undisturbed. This method has been successfully practised by farmers, mainly on temporary leys of red or white clover and Italian ryegrass harvested as silage prior to planting maize on an estimated actual maize area of 5%.

The tilling systems in Germany are divided, by intensity, into ploughing, conservation tillage with loosening of the soil, conservation tillage without loosening and direct drilling (NT) (KTBL, 1993). Misunderstandings frequently occur in segregating between conservation tillage and direct drilling. In the definition, which is recognized in the German-speaking countries and internationally, direct drilling (NT) is defined as a form of cultivation without any soil disturbance and tillage since the previous harvest, while conservation tillage follows the internationally accepted definition of minimum 30% soil cover remaining after tillage. The first research activities on conservation tillage and NT took place in Germany from 1970 to 1980 at locations in Braunschweig, Göttingen and Gießen. Within this, the prime comparison in the trials was between mulch sowing and conventional ploughing. It was not until the beginning of the 1980s that there were technical developments in sowing technology that enabled seed placement into an undisturbed soil and that the agro-chemical industry developed products to enable these new cultivation methods to be established in practice. This was the time to put conservation tillage, in the form of research and development projects, into practical use. As statistical data on direct sowing for Germany are lacking, estimates are based on surveys carried out by market research institutes (Kleffmann Group) and figures from the subsidy programmes of the German federal states. In 2001 mulch and direct sowing was only applied to just under one-third of the area used for winter oilseed rape. By 2012 this share had grown to 53%. For winter wheat the figures were 56% of the 3.26 Mha of areas under cultivation, for maize just under one-third of a total of 2.52 Mha (Lezovic, 2011).

Long-term experiments in France with different minimum tillage techniques (including NT) were initiated by INRA and ITCF in 1970 mainly with cereals (Boisgontier *et al.*, 1994). In 1999 the 'Association pour la Promotion d'une Agriculture Durable' (APAD) was founded and in 2008 it decided to focus on CA according to the more strict definition of FAO specifying the three principles of CA as minimum soil disturbance, permanent soil cover and crop rotations. In the same year the 'Institut de l'Agriculture Durable' (IAD) was founded with the 'Compagnie Européenne d'Intelligence Stratégique' (CEIS), a partnership with private companies and a cooperative. IAD created a set of indicators of sustainability on farm, with a central role given to soil and ecosystem management by farmers, and a strategy for conversion, with proposals for policy, based on the Payment for Ecosystem Services (PES), as developed by the United Nations Millennium Ecosystem Assessment scheme and FAO. Since 2008, IAD has been organizing a yearly international conference in Paris on sustainable agriculture, with key leading international experts in sustainability and CA.

The history of CA in Spain also began in the mid-1970s, in the southern part of the country. In the 'Haza del Monte' farm in Seville, a soybean crop trial under NT was performed in order to advance the sowing time and to try to harvest a second crop. The success of the study encouraged other researchers to conduct another trial in 'El Encín' in Central Spain where the starting point was an agreement between the Technical School of Agricultural Engineers (ETSIA) of the Polytechnic University of Madrid and the National Research Institute for Agriculture and Food Technology (INIA) (Fernández-Quintanilla, 1997). The results were promising: NT not only did not impact on winter wheat yields, but also reduced energy consumption by 80% (Juste *et al.*, 1981). These trials, which began in 1982 and still continue today, were extended to other Spanish regions, and were performed by the Agricultural Research Service of Andalusia and the School of Agricultural and Forestry Engineering of the University of Cordoba in the 'Tomejil' farm (Carmona, Sevilla), the Technical and Farm Management Institute in Navarra and the technical departments of companies of the agriculture sector in Castille Leon (Fernández-Quintanilla, 1997). Based on these experiments, González *et al.* (2010)

and González-Sánchez *et al.* (2010) reported that CA leads to higher yields than conventional tillage (ConvT). A milestone in the introduction of CA in Spain came in 1986 with the First Symposium on Minimum Tillage in Arable Crops. Since that time, research studies have multiplied and spread to other geographical areas. In February 1995, a group of farmers, technicians and scientists, many of them participants of the above-mentioned projects, founded the 'Spanish Association of CA Living Soil' (AEACSV, in Spanish). Thanks to the development of European projects, such as LIFE 99ENV/E/308 (LIFE, 1999) and LIFE 96ENV/E/338 (LIFE, 1996), and the support of private manufacturers of plant protection products and machinery, a number of activities that required technical-scientific knowledge were conducted with a high degree of regularity. Another important event was the 1st World Congress on CA, held in Madrid in 2001, with the support of ECAF, FAO, the European Commission LIFE Unit and the Spanish Ministries for Agriculture and Environment.

In Portugal, the Mediterranean climate and soil conditions only allow a rather extensive agricultural land use under rainfed conditions, with the exception of the north-western districts where the share of land under irrigation reaches almost 50%. Despite an average total annual rainfall of between 450 and 800 mm in most of the territory, precipitation can vary greatly from year to year (250–1200 mm $year^{-1}$ for the south of Portugal) and its distribution between autumn, winter and spring can be very erratic. In general, and with the exception of the humic Cambisols (north-west), soils are very low in organic matter (mostly around 1%) and very shallow (Alves, 1989). Water retention capacity and thus water availability for the crops is very low, limiting the yield potential of most crops grown under rainfed conditions. On the other hand, waterlogging during the rainy season can be a very severe problem for winter crops. The low organic matter content and low pH are responsible for the poor structure of the majority of the soils with the known consequences of soil compaction, surface sealing, low infiltration rates, surface runoff and soil erosion. The root causes of the severe soil degradation problems are found in the intensive soil tillage, practised since the introduction of widespread mechanization, and the removal of all crop residues as feed for ruminants leading to soil loss mainly through water erosion and soil organic matter (SOM) decline. It was the low SOM content of Portuguese soils that made Azevedo and Fernandez (1972, 1973, 1974/75) start to study the effects of minimum soil disturbance on the evolution of SOM. Based on these first experimental results an extensive research programme on the study of the effects of different tillage systems and crop rotations was initiated at the University of Évora in 1984 (Basch, 1988). This was the beginning of a series of research projects and studies on the agronomic, environmental and economic impacts of CA-based soil management systems. In the late 1980s the first dissemination and demonstration activities followed, but despite an apparent interest there was no notable uptake of CA by the farming community.

The situation changed after the foundation of the 'Portuguese Association for Conservation Tillage' (APOSOLO) in 1999, which became a foundation member of ECAF. As a result of the recognition of the need for soil conservation both at European (see Soil Thematic Strategy, Van-Camp *et al.*, 2004) and national levels and through the voice of APOSOLO, the first agri-environmental measures were proposed and implemented in 2001 in Portugal. All these measures, however, were limited to an eligible area of 200 ha per farm. Based on an inquiry among its members and service providers, APOSOLO's first estimation for the area under NT in 2002 was 6400 ha and for strip till around 3600 ha. The first official numbers available on CA were provided by the Portuguese Ministry of Agriculture in 2005 and shown a 240% increase of the area under direct drilling/strip tillage of annual crops from 2004 to 2005 and increase of 107% of the area under cover crops in perennials in the same time frame.

In Italy, in the early 1980s and 1990s CA started spreading as a result of the need to reduce production costs; the potential agronomic and environmental benefits of CA production systems with crop diversification were not yet regarded as a priority. The rate of CA adoption has however remained relatively low over a long time. In order to encourage its adoption and discourage tillage-based forms of agriculture, appropriate agricultural development policies would be needed. The Common Agricultural Policy (CAP) instead aimed at providing incentives for high yields rather than for ecosystem services from the agricultural sector. This was one reason why CA uptake was particularly slow; the other and most important reason was that the lower yields obtained under reduced tillage systems discouraged adoption and lead to the misconception that high yields could not be achieved under such systems. The main causes that lead to low yields were: (i) the lack of knowledge and experience of farmers, contractors and extensionists on the right implementation of CA systems; (ii) the over-simplification and the faulty application of the technique (e.g. NT in the absence of crop residues or in uninterrupted monocrop systems); and (iii) its introduction in unsuitable conditions (e.g. in marginal lands, on eroded and compacted soils) without remedial measures. Only in the 1990s did the adoption of CA start to increase thanks to the foundation of the 'Associazione Italiana per la Gestione Agronomica e Conservativa del Suolo' (AIGACoS). Since its foundation in 1998 in Osimo (Ancona, Marche) AIGACoS played an important role in disseminating scientific results achieved on durum wheat-, maize-, soybean-based cropping systems and convincing farmers that, through the correct implementation of CA systems, high yields can be achieved. The term 'Agricoltura BLU' was coined in 2002 by AIGACoS to refer to CA systems and highlight the relevance of water (hence 'blue') for agriculture and the role of the latter in the provision of ecosystem services.

In Russia the idea of reduced tillage has some history behind it: the idea of farming without tillage was proposed for the first time at the end of 19th century by I.E. Ovsinsky (Karabayev *et al.*, 2000), who consolidated scientific and practical works of outstanding Russian scientists, among them V.V. Dokuchaev (Dobrovol'ski, 1983) and P.A Kostychev (Mishustin, 1955). Unfortunately those developments were far ahead of their time. In the 1930s, N.M. Tulaykov (Vorontsova, 2007) worked out the theory of surface tillage for arid lands of the Volga region. Non-inversion tillage methods were introduced and work on conservation tillage continued in the 1960s and 1970s under A.I. Baraev (Baraev, 1983). However, only in 1998 the programme 'The grain production improvement in Samara region using water and resource saving technology' picked up reduced tillage systems again. In 2004 the Presidium of the State Council with a session 'On the role of modern technologies in sustainable development of the agro-industry in the Russian Federation' recognized the importance of water- and resource-saving technologies and the necessity of new technologies resulting in executive orders for implementation.

6.1.2 Current status and dynamics

Despite some history on CA development, the overall adoption levels of CA in Europe remain low and development is rather slow, again with large differences between countries (Tables 6.1 and 6.2).

In Western Europe, Spain is the leading country in terms of NT adoption. According to AEAC/SV, 650,000 ha of annual crops and 893,000 ha of perennial trees in most cases in combination with cover crops are under NT in Spain. The main annual crops under NT are wheat, barley and, to a lesser extent, maize and sunflowers. The main perennial systems under NT are plantations and orchards for olives, apples, oranges and almonds. In total it is reported that CA in annual crops is applied on about 10% of arable land in Spain. CA finds increasing interest in Spain from both farmers and official institutions. The evidence is reflected in the increasing area that is cultivated

Table 6.1. Conservation Agriculture adoption in annual crops in some European countries as reported by FAO-AQUASTAT (country contributions) (FAO, 2012b).

Country	CA area ('000 ha)	% of arable land	Arable land area ('000 ha)	Area under no-till ('000 ha)
Finland	160.00	7	2,199.00	200.00
France	200.00	1	18,442.00	200.00
Germany	5.00	0	11,792.00	354.00
Hungary	8.00	0	4,611.00	8.00
Ireland	0.10	0	1,120.00	0.10
Italy	80.00	1	8,293.00	380.00
Netherlands	0.50	0	916.00	0.50
Portugal	80.00	4	1,988.00	80.00
Slovakia	10.00	1	1,416.00	350.00
Spain	650.00	5	13,739.00	650.00
Switzerland	16.30	4	409.00	16.30
UK	150.00	3	5,761.00	250.00
Ukraine	600.00	2	32,537.00	600.00
Russia	4,500.00	4	123,491.00	15,000.00
Total	6,459.90	3	226,714.00	18,088.90

Table 6.2. Conservation Agriculture adoption in perennial crops in selected European countries (as reported by country authors).

Country	CA area ('000 ha)
Italy	500
Portugal	30
Slovakia	10
Spain	893

under this farming system, as well as by the increasing financial support given by governmental agencies, primarily through regional rural development programmes (Table 6.3) and energy saving programmes.

Table 6.4 shows official data from the Spanish Government regarding the yearly evolution of CA both in arable and perennial crops. AEAC/SV believe that NT is underestimated in the official data and estimate the actual area being around 700,000 ha for 2012. However, the trend is upwards for CA in recent years. Effective equipment is available to farmers everywhere across the country, but because skilled technicians are not as widespread many failures in CA come from the wrong implementation of the system: CA is sometimes perceived as just avoiding ploughing and not as a holistic agricultural approach.

Among the more advanced countries in Europe in terms of adoption of CA/NT farming is France. APAD estimates that NT is practised on about 200,000 ha in this country. Some farmers have developed superior NT systems with green manure cover crops and crop rotations, which are working very well. The 2008 IAD International Conference on Sustainable Agriculture under the patronage of the president of France and the following launching of the IAD Charter for Sustainable Agriculture was aiming at raising the political profile of CA in France.

Surprisingly, one of the smaller European countries, Finland, has 160,000 ha of CA adoption (out of 200,000 ha NT, part of which is not permanent) and is one of the leading CA-adopting countries in Europe. This contrasts very much with the situation in the UK, where, despite the extended history, CA development has been slow and fairly recent. In the UK moisture conservation is less of an issue than managing soil water: soil moisture limits direct-drilling and NT unless overall management changes are made to the farming system. In North-Western Europe, autumn sown crops go into a semi-dormant period over winter, which may cause poor rooting and hence stunted growth and poor yields of later sown crops. Managing crop

Table 6.3. Agri-environmental measures in Spain in 2006. Investment in Conservation Agriculture (Adapted from MAGRAMA, 2012a).

	Number of farmers	%	Area (ha)	%	Public support (€1,000)	%
Total agri-environmental measures	98,302	100	3,034,511	100	201,996	100
CA measures	17,613	17.9	144,403	4.6	27,133	13.4
Woody crops	16,943		141,190		26,959	
Arable crops	670		3,213		174	

Table 6.4. Conservation Agriculture adoption in Spain (adapted from MAGRAMA, 2012b).

	2011	%	2010	%	2009	%
Woody crops						
Total	4,932,002	100	4,986,046	100	5,043,896	100
Cover crops	1,178,297	23.89	1,218,726	24.4	1,066,182	21.1
Arable crops						
Total	7,378,280	100	7,100,060	100	7,341,709	100
No-tillage	510,773	6.9	428,638	6.0	274,628	3.7

residue is one of the keys to success with CA/NT in the UK's wetter climate. Experience has shown the beneficial combined effect of maintaining crop residues on the soil surface (that encourages earthworm activity) and leaving the harvested plant intact (that maintains 'communication' between the soil horizons) to aid drainage and soil aeration. Additionally, in this system weed and volunteer seeds left on the soil surface are easier to control, and finally surface cover protects the soil and soil structure from extreme rainfall and potential erosion. With this understanding the area under CA has over the last years increased in the UK to about 150,000 ha.

Even slower is the development in Ireland. A CAIR (CA Ireland)-organized visit to a NT farm in the UK in 2008 prompted one member to purchase a second-hand triple-disc drill and, having spent 7 years doing minimum tillage, he started NT in 2009. Yields on this farm have improved and, due to significant savings on machinery and fuel combined with reduced inputs, annual profits have increased. Since 2010 at least five other drills have been purchased that are designed for direct drilling crops. The area of direct seeded crops is now in excess of 200 ha.

In Portugal APOSOLO estimates the total area under CA in 2006 (APOSOLO, 2006) at around 80,000 ha for annual crops sown under NT or strip-till and around 30,000 ha of cover crops in perennials. However, an abrupt change in the Portuguese agricultural policy as a result of the change of the government in 2005, together with the decoupling of the support for agriculture and the consequent extensification of land use made the area under arable crops (mainly cereals) decrease by 30% on average both nationally and in the Alentejo, the bread basket in Portugal (INE, 2011). This contributed to a reduction of the area of arable crops grown under CA to only 4% of the total in 2009 (INE, 2011). The agricultural census still cites the use of 'reduced' tillage practices on 20% of the area under arable crops at the national level. With regard to the establishment of 'vegetative cover' in the inter-row space (which includes the maintenance of spontaneous vegetation) the agricultural census of 2009 (INE, 2011) does not provide the area where this CA practice is applied, but only a figure of 10% of all farms growing perennial crops using this technique.

Despite the relevance of CA for Italian agriculture, no direct data on its adoption are available, as CA is not monitored through the official agriculture census and often farmers allegedly implement CA systems on an irregular base. However, a survey of manufacturers of NT machinery shows that more than a thousand seeders have been sold (two-thirds of these in the north of Italy). The survey of contractors also shows that every year each sod-seeder is used on an average of 300 ha. Based on the coupled analysis of these surveys, it is presumed that the surface under NT systems is approximately 380,000 ha for cereal crops and 500,000 ha for orchards, for a potential of 900,000 ha, provided all this area adheres to the CA concept. In general terms, the potential for CA in Italy is particularly high for cereal-based systems (and more specifically for durum wheat, winter wheat, barley, maize), rapeseed, sunflower, soybean, fodder crops, horticulture systems and orchards (especially vineyards and olive orchards). However, there are no reliable data available on how much of this area is actually under a permanent NT system.

No-till systems without any soil disturbance (CA, contrary to high disturbance or temporary NT) are becoming more frequently used on Swiss fields mainly due to the improved availability of NT equipment, as a result of rising concerns by farmers, extension specialists and researchers on soil protection and cost efficiency as well as increased experience with this modern cropping system by the stakeholders. In consequence, NT has been established as a recognized and defined cropping system. The area cultivated with NT practices increased constantly, reaching 16,000 ha or nearly 5% of the arable land in 2011 (survey of SWISS NO-TILL, http://www.no-till.ch). In some parts of Switzerland the proportion of NT fields has reached 10% (Schneider *et al.*, 2010).

In Germany there is still major confusion about the concepts and most of the research efforts go towards reduced tillage rather than NT systems. For that reason the adoption of CA is probably only around 5000 ha. However, there are outstanding farmers practising NT in the country, one of them having been awarded the Environmental Award of the State of Saxony in 2006. In the regions endangered by erosion, such as the Ambergau (Lower Saxony), large farms use mulch sowing methods fairly often. For instance up to 70% of sugarbeet is grown with mulch sowing (including NT and minimum tillage) using straw and/or the remains of cover crops. In 2011, 59% of the farms with an area of between 200 and 500 ha applied methods without the use of a plough for winter cereals. For farms with over 1000 ha the figures were 70% (winter cereals) and 61% (winter oilseed rape) for the use of methods without a plough (Voßhenrich *et al.*, 2005). In addition to the size of the farms, significant regional differences may also be seen in the application of methods without the use of the plough. Direct sowing (NT) and mulch sowing are seen more frequently in eastern Germany, where the annual precipitation is less than 500 mm. Mulch sowing with loosening is done in regions with high precipitation and where soil conservation is necessary due to the hilly landscape. The strongest use of the plough is found in Bavaria and Schleswig-Holstein with 75 and 67% of the winter wheat area, respectively, and in the western federal states, which are also marked by high annual amounts of precipitation of up to over 800 mm. The adoption of conservation tillage and possibly direct drilling is not explained in Germany by cost savings and the combating of erosion alone, but is also a result of the improved load-bearing capacity of the soil when driving with high loads, such as harvesting and transport machines. Therefore the greatest development can be seen in maize where the area using mulch or direct drilling methods has doubled in the last 6 years alone. With increased fertilizer and fuel prices, erosion problems in some regions and regular droughts in others, interest in NT farming is growing steadily and adoption and consistency with CA over the years is increasing.

Much larger numbers in NT adoption are expected in the near future from Eastern European countries (Fig. 6.1). However, since in most of these countries the NT farmers are not organized, the data that are available are even less reliable. In Slovakia the economic situation urging farmers to reduce the cost, as well as impact of climate change requiring soil moisture-saving technologies is driving farmers towards the adoption of reduced tillage and specifically NT technologies. The adoption of NT increased from a total of 37,000 ha in 2008 to 350,000 ha in 2011 of a total area of 1,416,000 ha of arable land. However, since there are no official data and the area is deduced from the existing capacity of NT equipment in the country, it remains unclear how much of this area is actually complying with CA. The area of CA in perennial crops in the same time period (2008–2011) has increased from 7000 ha to 10,000 ha.

Ukraine is a country where estimates on the adoption of NT also vary greatly depending on the source of information. Estimates vary from less than 30,000 ha to more than 1 Mha. Official government statistics on NT state an adoption of 250,000 ha. Unfortunately, NT systems conforming to the definition of CA have not progressed as much as some people might wish. According to AgroSoyuz (a large cooperative farm in Dnipropetrovsk), there are about 1.1 Mha of direct seeding technology being practised in Ukraine. However, most of that direct seeding is done with very high disturbance tools, leaving practically the entire soil surface disturbed after seeding. For this reason this form of seeding does not comply with the CA definition and can only be classified as reduced tillage or mulch tillage. AgroSoyuz has estimated the CA area in Ukraine as 600,000 ha in 2011.

In Russia NT is often referred under the umbrella term 'Resource Saving Technology'. However, also here the database on actual CA adoption is not very reliable. Several machine manufacturers have exported NT machines to Russia in significant numbers. With the National Foundation for development of CA (NFDCA), Russia also has an organization promoting CA and

Fig. 6.1. Conservation Agriculture in Eastern Europe: no-till planting immediately following the combine harvester (Photo: Theodor Friedrich).

is part of ECAF. NFDCA estimates the total area under reduced tillage in Russia as 15 Mha, of which 4,500,000 ha are supposed to be CA.

Yet in many countries the general trend towards reduced tillage agriculture has not yet resulted in significant uptake of CA. For example in Denmark 12–15% of the arable land is harrowed before seeding and no plough is used, but only on less than 0.1% of the arable land is NT practised.

6.1.3 Prospects for Conservation Agriculture in Europe

Compared to other world regions such as the Americas or Asia, CA development in Europe has been particularly slow, with some few exceptions, for example Finland. There is a number of reasons for this slow adoption in Europe. One of these is the moderate climate, which does not cause too many catastrophes requiring urgent action. Another reason is that agricultural policies in the European Union (including direct payments to farmers and subsidies for certain commodities) take the pressure off farmers for extreme cost savings and discourage the adoption of diversified crop rotations. In addition to this, there are interest groups opposed to the introduction of CA, which results for example in difficulties for European farmers to buy good quality NT direct seeders with low soil disturbance and high residue handling capacity. Most of the European farmers practising CA have directly imported CA equipment from overseas or have had contact with small import agents. However, also in the EU, the environmental pressure is increasing and a new European CAP is being prepared, which most likely will be more favourable to CA.

Yet, in France, for example, prospects for adoption are still poor and, despite some very positive experiences, development is slow. One problem is, as in many other countries, the confusion between concepts and the belief that reducing tillage might be a gradual pathway towards CA. Unfortunately this is in most cases not true and farmers face many problems with this approach, which force them to revert to the plough and not to adopt CA.

Soil type and water availability are the major yield-determining factors and also influence the attraction for farmers to switch to CA. Based on the two above-mentioned variables, the Italian territory below 800 m above sea level (i.e. approximately 77% of the total surface area) has been divided into three vocational classes for maize and wheat production under CA (high, medium, low), showing than 30% of the Italian territory is highly suitable or easy to adapt for CA, 39% of it is challenging and in 8% agriculture in general is challenging. In poorly-drained asphictic soils the application of CA techniques can be difficult and it is challenging to obtain similar yields as in tillage-based systems. However, in heavy soils in semi-humid and humid areas, positive results can be achieved if drainage problems are addressed adequately. The best comparative advantage is achieved in heavy soils in dry areas.

Overall there is no conclusive picture for the future prospects of CA in Europe. Climate change with increased incidences of drought and more intensive rainfall, resulting in increased erosion problems, could favour adoption, yet, wetter soil conditions in some parts of Europe could be a challenge for CA. Rising fuel prices and an increasing attention of EU legislation on soils might further favour adoption, while the ongoing uncertainty about carbon sequestration and emission reductions under CA will not encourage farmers or policy makers to promote adoption (ECAF, 2012; Soane *et al.*, 2012).

6.2 Research Results Reported in Europe

As in other parts of the world research has not really been the engine for successful adoption of CA in Europe. In many countries research results, mainly focusing on comparing different tillage treatments but not really concentrating on optimizing

CA-based systems, have contributed to more confusion than clarity. Obviously, as shown in the different adoption levels, there are also differences in CA research between European countries.

No-tillage research in Spain started in 1982 and is still a major theme for Spanish researchers. On the clay soils of southern Spain NT was found to be advantageous in terms of energy consumption and moisture conservation, as compared to both conventional or minimum tillage techniques (Giráldez and Gonzáles, 1994). In 1996, a network of academics and technicians joined in a Thematic Network within the 'Creating a Thematic Network on Conservation Tillage' programme (AGF96-1613-E) to promote CA (Hernanz *et al.*, 1996).

In Portugal several research and also extension projects were run after 1984 on reduced tillage systems. At the very beginning agronomic and environmental aspects dominated the research interest, later economic and other increasingly specific studies followed.

In Ireland the semi-state Agriculture and Food Development Authority, Teagasc, began their minimum tillage research trials in autumn 2000. Experiments have been conducted on machinery and fuel costs as well as different aspects of agronomy from 2000 to date. No formal state-funded research has been conducted specifically on CA although third-level students have carried out unpublished dissertations on various aspects of the system as part of their studies.

In Germany most research has been done on comparing conservation tillage with ploughing. Experience with direct drilling (NT) and CA is sadly restricted to a few individual farms that have consistently practised CA over the long term. There are still challenges in the areas of equipment, plant protection and in the optimal form for the transition.

In Switzerland research on NT systems has been carried out in the framework of field experiments where different tillage systems have been compared at the Swiss federal research stations at Changins, Zürich and Tänikon, at the Swiss Federal Institute of Technology in Zürich, and at the Bernese Soil Conservation Service at Zollikofen. Compared with other tillage systems, crop yields and other basic parameters varied across experiments and years but tended to be more positive in treatments with soil tillage than in NT (Table 6.5). However, the principles of CA have been respected only in the Oberacker field trial at Zollikofen (Berne). In addition, in most experiments the plant protection measures and crop rotations were chosen according to the national guidelines, which are based and optimized in cropping systems with intensive soil tillage with mouldboard ploughs. Despite a systematic disadvantage of NT compared with other systems, the performance of NT systems seems to be robust and stable even under the cool and humid conditions of Central Europe. Research has also been carried out to optimize NT systems. One key element of any CA system has been the availability of adequate seeders; therefore, different NT planters for maize have been evaluated over 3 years (Streit *et al.*, 2005). Experiments have been carried out on strategies for herbicide replacement for organic CA (Hiltbrunner *et al.*, 2007) with combinations of cover crops and knife rollers to suppress weeds (Stadler *et al.*, 2009). Experiments have been carried out on methods to reduce mycotoxin content in cereals related to residue mulch (Vogelgsang *et al.*, 2011). The outcome of several projects has been summarized in a leaflet for farmers and extension specialists (Blum *et al.*, 2011).

6.2.1 Effect on soil quality (physical, chemical, hydrological and biological)

In general, soil organic matter levels and aggregate stability increase in soils that have been subject to CA (Jat *et al.*, 2012). The increased earthworm activity and undisturbed root channels result in a vertical structuring of the soil, improving water infiltration and aeration. Penetration resistance and bulk density tend to increase, resulting together with the higher aggregate stability in higher mechanical strength and

Table 6.5. Experiments/Projects in Switzerland where treatments with no-tillage have been studied and their evaluation with regard to the principles of Conservation Agriculture.

Site/Experiment	Tillage systems/ treatments[a]	Crop yield	Crops/crop rotations[b]	Plant protection	Basics of CA respected	Tillage system and transition period prior to the experiments	Reference
Agroscope ART, Tänikon: 'Hausweid'	NT, different MT treatments, P	Minimum tillage > no-tillage > plough, SM: P > MT > NT	GM-WW-SB-WW-SM-SW-SR-WW-SM	Standard	No	No transition period, but 14 years experiment	Anken *et al.*, 2004
Agroscope ART, Tänikon: 9 year experiment 'Langwies'	NT, different MT treatments, P	Minimum tillage > no-tillage > plough, SM: P > MT > NT	GM-WW-SB-WW-SM-SW-SR-WW-SM	Standard, slugs in maize	No	No transition period, but 9 years experiment	Anken *et al.*, 1997
Agroscope ART, Tänikon: experiment with repetition over 3 years	NT (hoe opener, disc opener), MT, P	3% higher in P, no difference between NT (hoe opener) and MT	3 single year experiments, WW (previous crop=SM)	Standard	No	Mouldboard plough, prior to each field experiment	Anken *et al.*, 1999
Agroscope ART, Tänikon: 3 years experiment 'Grund'	NT, P	NT<P	SM-WW-SM	Standard	No	2 years of pasture	Anken, 2003
ETH, Zollikofen, Schafisheim (4 year experiment)	NT, MT, P	GM: NT<tillage systems WW: no difference	GM-WW-OR-WW	Standard	No	Oat (NT) prior to the start of the project	Rieger, 2001 Rieger *et al.*, 2008
Oberacker, Zollikofen	NT, P	In both tillage systems equal or slightly superior in NT	WW-SP-WR-SM-WB-SB (including cover crops where possible)	CA based	Yes	No transition period, but 17 years experiment	Sturny *et al.*, 2007
Numerous non-scientific tillage trials at different agricultural colleges	NT, various MT systems, P	In NT systems lower than in systems with tillage	Mainly SM (Strickhof), SB (e.g. Strickhof, every year), single year experiments	Standard	No	Mouldboard plough	Bopp *et al.*, 2011

[a]NT, no-tillage; MT, minimum tillage/surface tillage; P, plough

[b]SM, silage maize; GM, grain maize; SB, sugarbeet; WW, winter wheat; SW, spring wheat; OR, oilseed rape; SP, spring peas; WR, winter rye; WB, winter barley; SR, silage rape

trafficability of soils. However, conclusive changes in soil structure cannot be expected in less than 3 years, which makes short-time experiments meaningless. Bulk density levels, while initially increasing, can after 6 years decrease again and reach values even below those of ploughed land (Soane *et al.*, 2012).

The results obtained in Switzerland so far show continuous NT of long duration to be an alternative to traditional plough tillage: NT is ready to be put into agronomical practice, it leads to a biologically active soil of stable structure and thus of high load capacity, reduces the risk of soil erosion, the number of vehicle crossings and the consumption of fuel and presents an overall more favourable life cycle assessment (Schaller *et al.*, 2006).

Measurements in Ireland with a shear vane showed significant differences in soil strength between plough and reduced cultivation treatments. Shear vane measurements to 40 and 120 mm showed that the shear strength at these depths was substantially higher on the reduced cultivation areas (Fortune *et al.*, 2003). Resistance increased very rapidly from 8 to 18 cm in minimum tillage (Fortune *et al.*, 2005). In general the soil profile becomes more homogeneous without the clear distinction of horizons.

In the UK maintaining crop residues is key to the management systems. CA systems start each year with the production and distribution of residue from the previous year's crop. Farmers report that the increase in crop residues at the soil surface create, over time, a higher level of soil organic matter (SOM) and rich soil life in this critical zone, making operations easier particularly in dry conditions. Two Research Studies on zero-tilled land have shown significant increases in soil organic matter. Longhurst (2010) showed 20 times more earthworms in three fields of Denchworth series clay compared to ploughed comparisons nearby, giving rise to greater water infiltration and recorded organic matter levels of over 30% in the top 20 cm of a no-tilled silty loam soil compared to less than 5% in the ploughed comparisons. Allton (2006), using soil taken from the site of the Soil and Water Protection Project (SOWAP) (Lane *et al.*, 2006), which consisted of a series of farm-scale erosion plots comparing plough-based tillage with NT, subjected them to rainfall simulation in laboratory conditions. The no-tilled plots showed reduced erosion and analysis indicated this was due to increased biological function in the soil. A further reason for concern is the management of soil compaction. A number of larger farmers are now using controlled traffic systems in conjunction with direct-drilling. But success or failure will also depend on SOM and its distribution within the soil profile. Good levels of SOM in the top 100 mm will act as a buffer against all kinds of extremes: compaction, drought, waterlogging, nutrient deficiencies, pests and so on.

Even considering the higher mechanical strength of NT soils, soil compaction under European climatic conditions, with frequently moist soils and with equipment masses of modern harvesting machines reaching 60 t, cannot be avoided. For a continuous NT system to be successful under those conditions, strict compaction management, for example with controlled traffic systems using permanent tramlines, is essential (Soane *et al.*, 2012).

In Portugal the suitability of soils under Mediterranean conditions for agricultural land use is frequently limited due to the aforementioned constraints of reduced effective soil depth, generally low to very low SOM and cation exchange capacity and consequently reduced water-holding capacity and structural stability. Several studies on the effects of the reduction of tillage intensity clearly indicate that the poor soil physical conditions, namely aggregate stability (Teixeira *et al.*, 2000), soil porosity and water-holding capacity (Carvalho and Basch, 1995) can be considerably improved through the shift from traditional plough tillage to NT (Tebrügge *et al.*, 1997). Structural stability and a much higher machine-bearing capacity of the soil have also been pointed out by Barros *et al.* (2008) as important benefits of NT to allow the best timing for field operations under wet soil conditions during the Mediterranean winter rainfall season.

Many studies in Spain confirm that CA has positive effects on soil quality. In a long-term study in southern Spain, Melero *et al.* (2008) reported that NT, as core component of CA, was the most effective technique for the improvement of the biochemical quality in the soil under a rainfed system. In northern Spain, Imaz *et al.* (2010) used a multivariate analysis for selecting 11 soil quality indicators (physical, chemical and biological), concluding that NT on Mediterranean semi-arid cropland has positive effects on soil quality. Ordóñez *et al.* (2007) studied for 21 years a wheat–sunflower–legume rotation where nitrogen, available phosphorus and potassium contents were found greater in CA. The changes in organic matter content were detected at progressively deeper layers in the soil profile.

Long-term studies carried out in Italy show a positive influence of the absence of soil disturbance in terms of higher chemical fertility, and more specifically of higher total SOM (Piovanelli *et al.*, 2006) and higher nitrogen content (Mazzoncini *et al.*, 2011). The positive influence of CA is also evident in terms of biological fertility with a greater amount of microbial biomass in different types of undisturbed soils under different climates (Gardi *et al.*, 2002; Piovanelli *et al.*, 2006; Marzaioli *et al.*, 2010). The effects of CA on soil physical characteristics (i.e. better structure and higher porosity) are well known as well and described in research studies by Basso *et al.* (2011), De Vita *et al.* (2007) and Pisante and Basso (2000).

6.2.2 Carbon sequestration and greenhouse gas emissions

There is increasing interest in using agricultural soils as a carbon sink, and evidence from literature shows that the implementation of CA can help increase soil organic carbon (SOC) and restore a degraded agro-ecosystem to a sustainable and productive state. However, SOC sequestration is generally non-linear over time (Freibauer *et al.*, 2004) and the effectiveness of conversion of tillage-based agriculture to CA depends on many variables: for example, the soil carbon sink strength increases most rapidly soon after a carbon-enhancing change in land management has been implemented, and reduces with time as the stable SOC stock approaches a new equilibrium (Smith, 2004). Even though some authors report significant increase in microbial activity soon after transition to CA, fuller advantages of CA in terms of soil health can usually be seen only in the medium- to longer-term run, when CA practices and soil biological processes become well established within the farming system. To provide an idea of the time scale, Smith (2004) reports that the period for European agricultural soils to reach a new steady-state level, after a carbon-enhancing land-use change has been introduced, is approximately 100 years.

In a comparison of reduced tillage and a conventional plough-based system over 8 years Hackett *et al.* (2010) reported that minimum tillage resulted in a significant increase in SOC compared to ploughing, 1.83% compared with 1.56% in the 0–15 cm soil horizon ($p<0.001$), but there was no significant difference between systems below 15 cm. When a carbon sequestration rate of 0.77 t ha^{-1} year^{-1}, as proposed by McConkey *et al.* (2000), is adopted for NT and a conservative estimate that 30% of cropped land in Ireland is suitable for CA, the potential for reducing CO_2 emissions via carbon sequestration is approximately 417,000 t. Geraghty (2008) used published research data for diesel consumption on Irish tillage farms to estimate that the adoption of CA/NT would reduce CO_2-related fuel emissions by 12,000 t on 100,000 ha of cropland.

In the UK a Department of Environment, Food and Rural Affairs (Defra) Scientific Report compiled by Bhogal *et al.* (2008) on the carbon content of arable soils in England concluded the following:

1. Increases in SOC measured have been accentuated in the top 10–15 cm. In deeper samples differences between tillage systems diminish.
2. The best estimate of the C storage potential of NT under English and Welsh conditions is 310 (+180) kg C ha^{-1} year^{-1}, based on

measurements at six study sites. This equates to 0.35% of the typical carbon content of an arable soil in England and Wales.
3. Reduced tillage is estimated to have half the C storage potential of NT at 160 kg C ha^{-1} $year^{-1}$.
4. These estimated C storage potentials can only be regarded as the initial rate of increase (<20 years). Annual rates of SOC accumulation decline (eventually to zero) as a new equilibrium is reached after more than 100 years.
5. SOC accumulation is finite and reversible. SOC levels will only remain elevated if the practice is continued. Carbon stocks are depleted again if land is ploughed every 3 to 4 years, and the reduction is much faster than the sequestration.

In France, according to the EU-funded SoCo project, organic matter levels increased by 1% in 10 years and C sequestration amounted to 1–4 t ha^{-1} $year^{-1}$ (SoCo, 2009).

In Italy, long-term experiments comparing NT with tillage show that after 15 years SOM in tilled soils was approximately 1% in the topsoil layer, while in NT systems it was approximately 2% (R. Santilocchi, 2010, unpublished data). Other studies show that crop residues left on the soil surface significantly increase the content in SOC (Bonari *et al.*, 1996; Borin *et al.*, 1997a; Masciandaro *et al.*, 1998; Mazzoncini *et al.*, 2001, 2004), the biodiversity and the resilience of the agroecosystem, soil structure and help prevent soil erosion (Pagliai *et al.*, 1989, 1995; Campiglia, 1999; Pisante, 2007; Colecchia *et al.*, 2009; Stagnari *et al.*, 2009).

A study on the potential of NT for carbon sequestration on agricultural land in the south of Portugal was subject of a research project between 2003 and 2008. Before that, it was already clear that the absence of soil tillage for crop establishment alone was able to invert the decline of SOM (Carvalho and Basch, 1995; Tebrügge *et al.*, 1997) on the extremely depleted Mediterranean soils. Yet, this research project confirmed the huge contribution that the amount and management of crop residues can play for the carbon sequestration potential of soils under sub-humid to semi-arid Mediterranean conditions (Basch *et al.*, 2010). In fact, some of the highest carbon sequestration rates across northern, western and south-western Europe have been reported for Spain and Portugal (Soane *et al.*, 2012).

Recent studies in Spain confirm that CA is a key element for soil-carbon sequestration. Alvaro-Fuentes and Cantero-Martinez (2010) did an estimation of the C mitigation potential of tillage reduction in Mediterranean climate and rainfed crops in Spain. A review of eight studies on arable crops showed that SOC sequestration would be 2.18 and 0.72 Tg C $year^{-1}$, representing 17.4% and 5.8% of the total CO_2 equivalent emissions generated in 2006 from the agricultural sector in Spain. González-Sánchez *et al.* (2012) reviewed 29 studies on arable and woody crops. Based on the research conducted and the data of agricultural area in Spain dedicated to CA, authors concluded that about 2 Gg C $year^{-1}$ would be fixed by CA. On the other hand the authors found that minimum tillage practices may increase CO_2 emissions in relation to ConvT, so every effort concerning carbon sequestration in arable crops should be made in favour of NT.

6.2.3 Crop yields under Conservation Agriculture

Overall it has been found that crop yields under CA are within a band of 5% around conventional crop yields, with the weather having a strong influence. Under drier climatic conditions CA yields tend to be higher than conventional yields (Soane *et al.*, 2012). In cases where yield reductions were observed the most common reasons were soil compactions, residue or weed management problems. Over years yields under CA appear to increase due to the build-up of soil structure and N availability in the soil, and even under very difficult conditions, such as in Finland, eventually any initial yield reductions disappear after few years (Soane *et al.*, 2012).

Swiss research showed after a 7-year conversion period, slightly higher plant yields of comparable quality were obtained in NT, due to more soil water being preserved and continually delivered to plant roots, as well as to a higher N-efficiency.

Since 2000 yields in Ireland have remained steady and have been comparable to plough-based systems across a wide range of crops (Forristal and Murphy, 2009).

Similarly in France yields under CA are maintained or improved. In some cases irregular yields were observed during the first years after conversion, but only in summer crops.

Also in a long-term study in Spain, Ordóñez *et al.* (2007) reported that the mean yields were not statistically significantly different as a whole. Wheat resulted in lesser mean yields in NT than in ConvT, estimated at 92% for NT. In dry years, sunflower in NT yielded higher figures, confirming the better water balance under NT.

In Denmark results were not so conclusive, mainly due to the fact that no true low-disturbance NT and CA has been applied (Figs 6.2, 6.3). The Danish association for CA (FRDK) and the national advisory service are working on a trial set up for a realistic evaluation on CA in Denmark. FRDK believes that cover crops are very important in making a good soil structure and a good rhizosphere and therefore requires that the trials fulfil the criteria for CA: NT, cover crops and crop rotation.

Similarly in the UK yields were found to be reduced by between 1 and 7% comparing alternative tillage systems relative to ploughing, with all other factors remaining constant. However, this work did not include a plot following the CA definition, neither did it take into account, for example, the timeliness of the operations, which is easier to achieve under CA (Ogilvy, 2000). Trials at the Focus on Farming Project, which included a seven split-field comparison, showed that wheat drilled in mid-September yielded around 1.0 t ha^{-1} more than that sown in late October (Leake, 1995). All of 12 commercial farm case-studies featured in the SMI Crop Establishment Guide (SMI/Defra, 2001) showed substantial reductions in work days, often over 50%, and where these studies were combined with local rainfall data, very often there would have been insufficient work days available for all crops to be established in good time where ploughing was used as the principle cultivation, resulting in yield penalties.

On the other hand under similarly mostly humid conditions Switzerland has made remarkable progress in terms of research, development and adoption of NT practices. Research performed in Switzerland

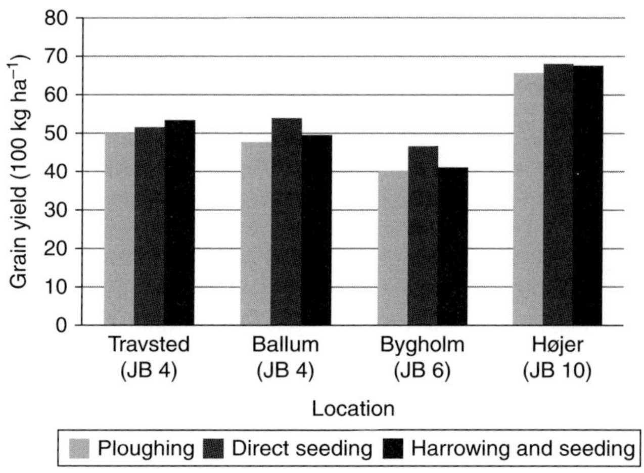

Fig. 6.2. Winter wheat yield: Danish government trials from 1981 to 1986, four locations (Rasmussen, 1988).

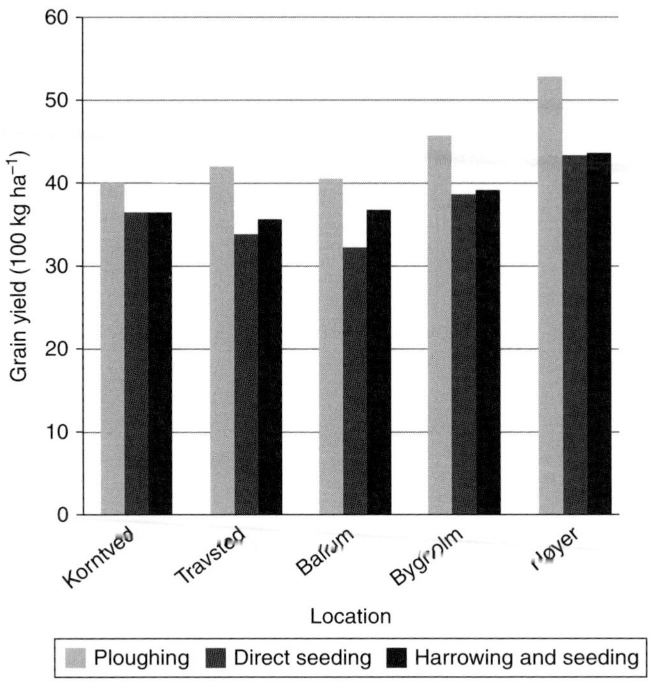

Fig. 6.3. Spring barley yield: Danish government trials from 1981 to 1986, five locations (Rasmussen, 1988).

over more than 10 years has shown equal or better yields under NT in a variety of crop rotations.

In Italy cereal-based cropping systems (especially wheat) are among the first systems to have transitioned to CA. A number of experiments on durum wheat comparing NT, minimum tillage and plough tillage have been conducted. On heavy soils in hilly areas in central Italy (Umbria and Marche) no significant differences are found either in yield (Bonciarelli, 1985; Archetti *et al.*, 1989) or in the grain quality (Antonelli *et al.*, 2001, 2003; Seddaiu *et al.*, 2003). For this reason and the lower costs in the implementation of the cropping system, farmers in the neighbourhood of the experimental fields have shown their interest in NT and today thousands of hectares are grown under NT systems. In the south of Italy, short term (Basso *et al.*, 1996; Pisante *et al.*, 2001; De Vita *et al.*, 2007) and long term (Basso *et al.*, 2010) experiments on durum wheat comparing NT, minimum tillage and plough tillage have highlighted the role of NT techniques in overcoming dry spells without causing any relevant physiological stress to the plants. The experiments also highlight that timeliness and the choice of adequate seeders for the type of soil are very important variables that strongly influence yields: improperly adjusted seeders can leave the seed-furrow partially open and cause the irregular emergence of plants; in addition, lightweight planters and drills cannot penetrate hard soils, resulting in poor seed-to-soil contact.

In Portugal the very first results obtained in the 1980s showed that yield levels under different tillage systems, including NT, were very similar, despite the lack of experience, simple NT equipment available and mostly unfavourable soils with regard to their structural condition. In particular autumn and winter sown crops, but also spring crops under irrigation, provided always similar or even higher yields than those obtained under reduced or ConvT. However, sunflower sown in spring and grown under rainfed conditions frequently yielded less than when grown conventionally. This was attributed to the higher root penetration resistance under NT and dry soil conditions prevailing in spring (Basch *et al.*, 1998). In long-term trials comparing different soil management systems it became evident that the NT treatment more and more outperformed reduced and ConvT systems in terms of grain yields (Fig. 6.4) (Carvalho, 2003).

6.2.4 Runoff, infiltration, soil water content; soil conservation

A study conducted in Spain by the Institute for Nature Conservation ICONA (1991) estimated the direct costs of erosion amounted to €280 million annually, due to the loss of agricultural production, damaged dams and flood damage among other factors. It further estimated the cost of actions taken against erosion and recovery would require €3000 million over a period of between 15 and 20 years. Indeed, soil conservation is an urgent environmental need for Spain, as soil erosion is a key factor in Mediterranean environments (García-Ruiz, 2008). Quantifying the effect of historical soil management on soil erosion over a 250-year period in south Spain, Vanwalleghem *et al.* (2011) estimated soil loss mean rates between 29 and 47 t ha^{-1} year^{-1}. However, authors found considerable historical variation and two origins: between 8 and 124 t ha^{-1} year^{-1} for water and between 3 and 42 t ha^{-1} year^{-1} for tillage. Undoubtedly, CA is a good solution to prevent soil degradation in Spain, as it can reduce erosion and runoff by on average 90–95% and 40–60%, respectively, compared to ConvT (Ordóñez *et al.*, 2001; López and Arrúe, 2005; Márquez *et al.*, 2008).

In Ireland cultivation in the autumn followed by heavy rainfall has led to crop establishment problems and yield losses especially on silt soils. Evidence on farms by growers shows that after a few years of not ploughing there is better drainage and reduced ponding in fields after heavy rainfall events, reduced leaf curling in cereals during dry periods and more resilience to traffic in soils in non-plough systems.

About 14% of the arable land in Germany suffers from a long-term average soil erosion of more than 3 t ha^{-1} year^{-1}. The use of good agricultural practices is mandated in paragraph 17 of the German Federal Soil Protection Act. One of the fundamentals of good agricultural practice is, among

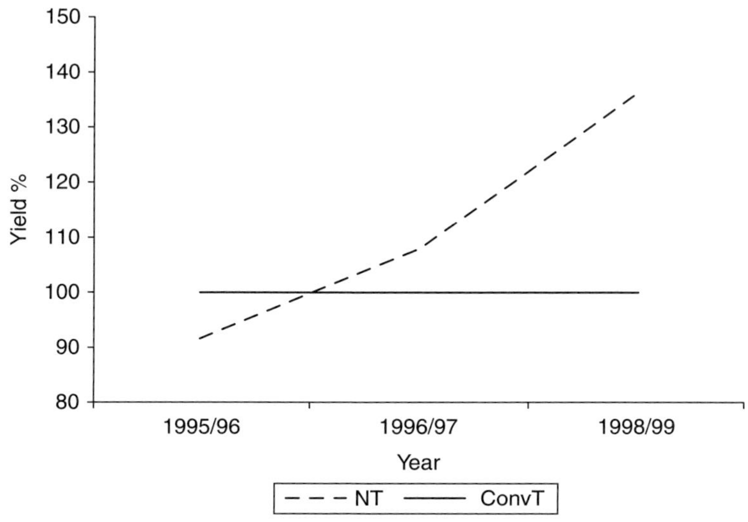

Fig. 6.4. Relative wheat grain yield (percentage) after the adoption of no-till in an experimental field trial (adapted from Carvalho, 2003).

other things, that the soil structure is maintained or improved and that soil compaction and soil erosion be avoided wherever possible. In the light of this, farming methods that protect the soil are being increasingly applied. This applies for conservation tillage methods in conjunction with cover crops and/or straw manure and subsequent mulch drilling, but in particular for CA (Table 6.6).

In France the most noticeable benefit of NT is erosion reduction: with covered soils and NT, erosion is totally absent, even in fragile soils like loams or sand, while in minimum tillage erosion is still observed, as in all tilled soils. Likewise runoff is also completely suppressed and water infiltration and water storage are visibly better. In dry spring conditions, cereals have been reported as suffering less, maize suffers less from drought if rain-fed, and on irrigated maize farmers have been able to save one or two passes of irrigation.

According to research in Portugal, improved soil cohesion, pore continuity and aggregate stability, and the protection of the soil surface from the direct impact of rain drops are the most important conditions to improve water infiltration into the soil and to reduce surface runoff (Basch *et al.*, 2012a). The concentration of the rainfall during the winter months with often high rainfall intensities makes the Mediterranean region especially prone to severe runoff and erosion events. Only the absence of soil disturbance and effective soil cover during the rainfall season are measures capable to prevent this root cause of soil degradation. Several studies both under rainfed and irrigated condition give evidence of the effectiveness of CA soil management practices on the reduction of surface runoff, but especially the reduction of sediment yield. In small scale trials, Basch (1988) and Basch and Carvalho (1998) found a clear positive correlation between soil tillage intensity and the amount of surface runoff and eroded soil.

In most, but not all cases, soils under CA tend to have significantly higher water infiltration capacity than ploughed soils as a result of the better aggregate stability and vertical network of soil pore structure. However, compaction management in NT is an essential element, particularly where the aggregate stability is not improving fast enough. Over time, infiltration appears to further improve under NT. Water retention depends very much on the climatic conditions. In the humid north-western countries there is little difference in water retention and yield between NT and tilled soils, but in the dryer south-western countries NT soils appear to have a better water availability resulting in higher yields during dry years (Soane *et al.*, 2012).

6.2.5 Climate change mitigation and adaptation

Agriculture in Europe emits 9.2% of the total European greenhouse gases (GHG). CA can therefore play an important role and help reducing GHG emissions attributable to the use of fossil fuels (direct emissions) and help sequester carbon to the soil by reducing its mineralization rate and increasing the quantity of the fresh organic matter returned to the soil.

Table 6.6. Effects of varying tillage methods on various ecological and economic aspects (Brunotte, 2002).

Problem	Conventional tillage With/without seedbed preparation		Mulch seeding With/without seedbed preparation		No-till, direct seeding
Silting	–	–	o	x	xx
Erosion	–	o	o	x	xx
Compaction	–	x	x	xx	xx
Nitrate leaching	–	–	o	o	x
Cost	o	x	x	xx	xx

Problem solving: xx, very good; x, good; o, satisfactory; –, unsatisfactory.

This is particularly true for the Italian agricultural sector, which contributes 6.7% to the total national GHG emissions and that to comply with the Kyoto Protocol would have needed to reduce its CO_2 emissions by 6.5% compared to 1990. But in fact CO_2 emissions have increased since 1990.

Besides the carbon sequestration in soils discussed in section 6.2.2, it is important to also consider other GHGs and their emissions from agricultural land. Nitrous oxide emissions depend very much on the aeration status of the soil, i.e. are related to soil moisture and compaction. For this reason, nitrous oxide emissions in NT soils tend to be higher than under ploughing in wet, badly drained and compacted soils. Yet, even under those conditions, the emissions seem to increase only over the first 10 years and to decline after 20 years. On well-aerated soils, however, there are no significant differences in nitrous oxide emissions between NT and ploughed soils (Soane *et al.*, 2012). CO_2 emissions resulting from the use of fossil fuels can be significantly reduced in CA systems, compared to ploughing. In the European literature fuel savings between 50 and 84% are reported, resulting in the corresponding savings in CO_2 emissions (Soane *et al.*, 2012). CO_2 emissions from soils are also reduced in most, yet not all cases. Under hot moist conditions with high amounts of decaying crop residues, CA soils can emit more CO_2 than ploughed soils (Soane *et al.*, 2012). The amount of carbon sequestered in the soil under CA, despite the inconclusive results for the CO_2 fluxes, varies widely but in general carbon sequestration can be obtained by applying good CA (Corsi *et al.*, 2012). On balance the potential of agriculture to contribute to climate change mitigation depends very much on optimizing each component, i.e. carbon sequestration as well as the reduction of emissions, particularly from nitrous oxides, which in some cases can be a challenge (Soane *et al.*, 2012).

As reported in sections 6.1 and 6.2.3, CA is more easily accepted in the southern European regions, suffering increasingly from drought spells and it is here, where also the yield benefits of CA, particularly in dry years, appear attractive. With this CA appears to be a good choice not only for climate change mitigation, but also for adaptation. This is also true for other weather extremes, such as excessive rainfall, where the increased water infiltration on CA soils (see section 6.2.4) would reduce the danger of erosion and flash-floods.

6.2.6 Off-site environmental benefits

CA, despite its relatively low level of adoption in Europe, is already showing off-site environmental benefits.

The EU SoCo project (2009) report a reduction in nitrate pollution in waters by 50% in France as a result of CA. Additionally it can be observed that surface water in watersheds is clear, and off-site erosion is suppressed, particularly on loamy soils frequent in south-west and north France.

A research project in Portugal on soil and water quality affected by agrochemicals under different soil tillage systems showed that the dissipation of herbicides in the soil was clearly accelerated under NT when compared to plough tillage. This was attributed to the combined effect of the retention of the applied herbicides in the residues and the higher surface SOM content under NT leading to a faster decomposition of the chemicals through a higher microbiological activity in the presence of more SOM (Borin *et al.*, 1997b). In addition, the off-site transport of the herbicides Isoproturon and its metabolite Monomethyl-Isoproturon, under rainfall conditions, and Atrazine and Metolachlor under irrigated conditions, was clearly reduced under NT compared to ConvT (Basch *et al.*, 1995).

Similar results regarding the off-site transport of herbicides from NT and ploughed fields have been reported from other European countries. The importance of erosion, however, varies. The reduced water erosion under CA is a strong driver for CA in southern Europe, while in northern Europe the importance of erosion as driver for CA adoption depends very much on the soil type and climatic conditions, which influence also whether wind- or water erosion is the major issue (Soane *et al.*, 2012).

Regarding the water eutrophication from phosphates, the total amount of P reaching surface waters seems to be significantly reduced under NT. This is particularly the case for phosphates bound to soil particles. However, the soluble fraction of P, resulting from mobilizing organic acids and decaying weeds or cover crops, for example, is much higher under CA than in ploughed soils, which can lead to an increased off-site transport of this soluble fraction with water runoff. A coping strategy for this problem is the reduction of surface runoff under NT due to the better water infiltration (Soane *et al.*, 2012). For nitrate leaching the results are not yet conclusive, depending very much on the specific management practices, soil and climatic conditions.

6.2.7 Insect-pest and disease dynamics

Several studies show that there is no conclusive trend on pest and disease dynamics in relation to tillage. Other factors, such as crop rotation and climate seem to have more dominant influence. In Ireland aphid numbers and BYDV (barley yellow dwarf virus) incidence are generally lower after reduced cultivation. Where straw was incorporated on barley treatments aphid numbers were reduced by 68% and virus by 56% in reduced cultivation treatments and grain yield was 1 t ha^{-1} higher on reduced cultivation plots (Fortune *et al.*, 2005). Research results have also been inconclusive regarding slug numbers under reduced cultivation treatments. There is some evidence that slug activity is far greater in heavy residue conditions with increases in leaf damage to young plants but there was no noticeable increase in seed hollowing or reduction in plant population, indicating eventually increased activity of predators. Fortune *et al.* (2003) reported that in wheat, take-all (*Gaeumannomyces graminis*), eyespot (*Oculimacula yallundae* and *Oculinamacula acuformis*) and sharp eyespot (*Ceratobasidium cereale*) levels have been lower in reduced cultivation treatments, but in barley there were higher levels of net blotch (*Pyrenophora teres*) and rhynchosporium (*Rhychosporium secalis*) infection, which could also be confirmed in farmers' fields.

Also in the UK the principal problem associated with reduced tillage is the damage caused to germinating cereal seedlings through seed hollowing by slugs. There is a perception amongst farmers that the presence of residues on the surface and the use of disc drills in minimum tillage systems pre-dispose crops to attack. Increasing the drilling depth from 20 mm to 40 mm reduced this problem from 26% to 9% through excluding slug–seed access and by reducing the germination time by placing seed in more moist conditions (Glen *et al.*, 1990). There is also a perception that the presence of high levels of residue at the soil surface provides a source of inoculum to infect subsequent crops. The extent to which infectious agents can survive and infect depends on a host of factors, not least the specific disease, but also the prevailing weather conditions (Jordan and Hutcheon, 1999). A literature review of cereal diseases, take-all, sharp eyespot and ear blight, and oilseed rape pathogens, dark and light leaf spot, downy mildew, stem canker and stem rot indicated that disease levels were not observably different (Leake, 2001). Residue management was a key topic that SMI found necessary to deal with in considerable detail during the transition from a plough-based system to adoption of a CA system. With good residue management pests such as slugs in time become less of a problem due to the build-up/improved balance of 'beneficials' such as ground beetles whose numbers can be depleted by cultivations.

In a study over 3 years in north Portugal comparing maize fields established under conventional and NT using integrated pest management, no differences in terms of pest occurrence were found, with the exception of rodents, which caused some problems under NT (Xavier *et al.*, 2005). However, depending on soil type and crop species, Mota *et al.* (1997) observed a higher level of lesion nematodes under NT compared to ConvT, resulting in a reduced plant growth and dry matter production of winter cereals at the end of the winter period.

6.2.8 Nutrient use efficiency

Nutrient dynamics change under NT in CA systems. P and K can become more stratified, particularly close to the soil surface as a result of the decomposition of crop residues. However, under European conditions that does not have any negative effect on crop yield. In the case of N, an increase of N-fertilizer requirements can be observed in some cases during the adoption phase, for a number of reasons, such as denitrification losses in unstructured soils, which disappear once the soil structure improves under CA, reduced mineralization, N requirements for the build-up of SOM, to name only some. In the long term the fertilizer requirements in CA systems are reduced, due to greater biologically fixed nitrogen, increased nutrient conservation and improved efficiency, resulting in cost savings and higher profitability (Soane *et al.*, 2012).

In Swiss research in the analysed cropping systems only about 60% of the standard amounts of N-fertilizer were applied under CA. In the coming years the systems will be tested further and optimized with regard to environmental sustainability and energy consumption by introducing more legume crops, applying ammonium-based N-fertilizer, and by reducing the application of glyphosate in NT and the tillage intensity in conventional plough tillage.

A 2-year study by Hooker *et al.* (2006) found that the mean soil solution NO_3 concentrations were between 38 and 70% lower when a cover crop (catch crop) was used, and total N load lost over the winter was between 18 and 83% lower with the highest reductions achieved where a cover crop was used in conjunction with reduced cultivation. It was concluded that cover crops are important to reduce nitrate losses from spring cereal systems in countries with mild winters like Ireland. In the field, farmers have been experimenting with reduced nitrogen application rates.

In the UK potassium, phosphorus, magnesium and calcium are usually supplied to maintain the recommended plant nutrient contents in soil; but they should also correspond to the actual demands of the rotation, with the nutrient supply balanced with nutrient removal. Phosphate loss is mainly due to movement of particulate matter from soil to watercourses through soil erosion, and is much reduced by non-inversion tillage and even more by NT (Brown *et al.*, 1996). Whilst more precise fertilization to meet crop needs may be achieved with inorganic sources, more efficient exploitation of the organic nitrogen contribution from incorporated residues and cover crops may offer opportunities to reduce the amount of applied nitrogen thereby minimizing waste. Improved nutrient management that takes account of crop rotation is likely to decrease the requirement for off-farm inputs that challenge sustainability through their effects on the environment. In France, for example, a reduction in fertilizer use was observed with CA farms, mainly due to the use of legume cover crops, more diverse crop rotations and a higher environmental awareness of farmers.

Due to the high nutrient leaching potential of typical winter rainfall regions, as in Portugal, nutrient efficiency of the more mobile nutrients is rather low. In this context the content of SOM plays a decisive role both due to its high ion exchange capacity but also as a source of nutrients. The response curves found to different levels of nitrogen fertilization on the same soil but with different levels of SOM, achieved through long-term differentiated soil tillage, illustrate convincingly the importance of CA-based soil management for an enhanced nutrient cycling and use efficiency (Fig. 6.5).

In Spain N availability has been indicated as one of the most critical aspects of CA. Rodríguez-Lizana *et al.* (2010) evaluated the straw decomposition releases of N, P and K in a pea–wheat–sunflower rotation and concluded that in Spain's climate, the release of nutrients from the crop residue, mainly N, is not significant. However, long-term studies on CA-based wheat–sunflower–legume rotations show the effect of the crop rotation in enhancing nutrient content to a depth of 13 cm (Ordóñez *et al.*, 2007). Moreover, in farming systems greatly affected by erosion processes (such as plantation crops), decomposition of plant residues

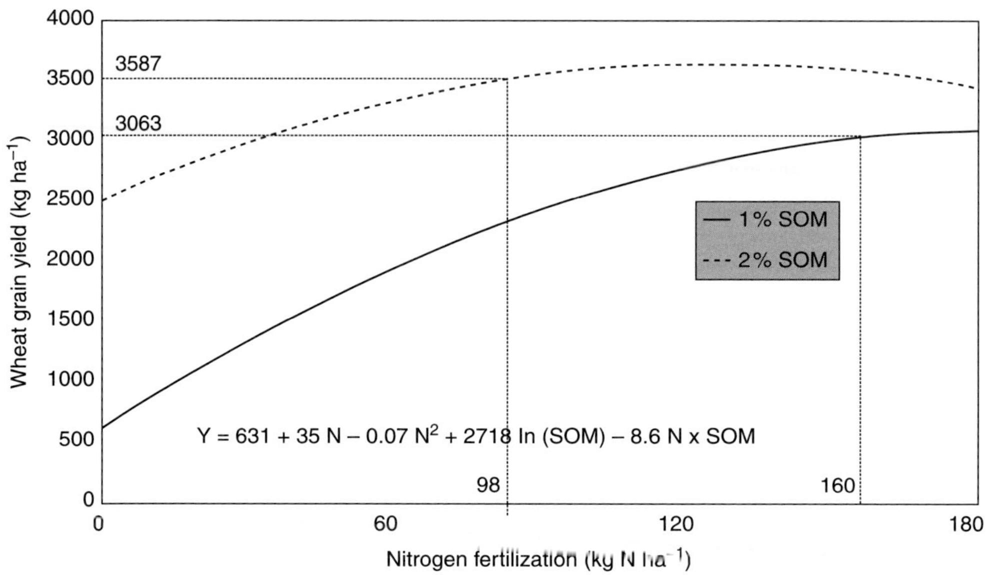

Fig. 6.5. Wheat grain yield response to N-fertilization under different levels of SOM under water-limited Mediterranean conditions (adapted from Carvalho *et al.*, 2010).

has proved to compensate for nutrients carried by the runoff flow. During one season, Ordóñez *et al.* (2009) assessed the mineralization and nutrient release from cover crop residues from different grass species used in olive groves: for N, P and K, respectively, *Brachypodium distachyon* released 81.6, 7.3 and 78.2 kg ha^{-1}, *Eruca vesicaria* 24.3, 3.4 and 33.4 kg ha^{-1} and *Sinapis alba* 21.5, 3.5 and 8.6 kg ha^{-1}. Also, Ordóñez-Fernández *et al.* (2007) evaluated the effect of an ongoing cover of olive prunings over a period of 6 years observing major improvements in soil fertility.

6.2.9 Input use efficiency

In general a significant reduction of input use is reported in Europe as a result of better input use efficiency with CA, amounting up to 70% savings in fuel, 30% in fertilizers, 50% in chemicals and 50% in time (SoCo, 2009).

Analysing the economic performance of a 650 ha farm in the south of Portugal with 350 ha of arable crops, before and after shifting from ConvT to NT, Freixial and Carvalho (2010) found a reduction of fuel and labour costs of 60% and 40%, respectively.

The LIFE+Agricarbon project in Spain is delivering positive results on input use efficiency (Table 6.7). Results show reduced fuel consumption in more than 45% in all crops studied, and energy use reductions between 13% and 25% (Márquez *et al.*, 2011). Additionally, other parameters also improve, for example energy efficiency (EE), defined as the ratio of the heat energy contained in the final product and that required to develop the product, and energy productivity (EP), defined as the amount of product produced (g ha^{-1}) per unit of energy supplied (MJ ha^{-1}).

6.2.10 Biodiversity

Earthworm numbers as the most visible effect of reduced tillage increased significantly in Irish Teagasc CA trials with and without straw (Fortune *et al.*, 2003). In an unpublished study by Russell (2011), earthworm

Table 6.7. Energy used in inputs (MJxha^{-1}) and its efficiency and productivity (gxha^{-1}) for selected crops in Spain (Márquez *et al.*, 2011).

				Energy consumed (MJ ha^{-1})							
					Indirect energy						
Field	Crop	Tillage system	Energy produced	Direct energy	Machinery	Seeds	Fertilizers	Agri-chemicals	Total	EE	EP
1	Wheat	NT	14,950	1,257	316	2,940	8,918	997	14,428	1.04	80
		ConvT	11,200	2,805	704	2,940	9,642	406	16,497	0.68	50
	Sunflower	NT	18,904	1,094	275	84	1,688	1,179	4,320	4.38	310
		ConvT	20,989	2,853	716	84	1,892	9	5,554	3.78	270
2	Wheat	NT	21,313	1,199	301	3,454	16,317	302	21,573	0.99	80
		ConvT	18,750	1,625	408	3,454	18,291	346	24,124	0.78	60
	Sunflower	NT	13,358	1,152	289	84	2,451	1,299	5,275	2.53	180
		ConvT	12,913	1,983	498	84	2,748	493	5,806	2.22	160
	Legume	NT	18,696	898	226	2,357	348	339	4,168	4.49	390
		ConvT	15,960	1,562	392	2,726	431	388	5,499	2.90	250
3	Wheat	NT	45,750	1,175	295	3,234	11,240	701	16,645	2.75	220
		ConvT	43,875	2,824	709	3,234	12,880	681	20,328	2.16	170
	Sunflower	NT	10,230	1,013	254	84	0	298	1,649	6.20	450
		ConvT	9,619	2,255	566	84	0	16	2,921	3.29	240
	Legume	NT	5,016	1,070	269	1,704	0	3,705	6,748	0.74	70
		ConvT	11,799	2,905	730	1,704	0	1,960	7,299	1.62	140

NT, no-tillage; ConvT, conventional tillage; EE, energy efficiency; EP, energy productivity (g ha^{-1}).

numbers were 25% higher in CA versus a plough-based system in the same soil type at the same location. There was a threefold difference in earthworm biomass in favour of CA. This was due to a greater proportion of larger deep-burrowing earthworm species identified in samples from CA fields. Fortune *et al.* (2005) concluded that the increase in worm biomass in minimum tillage was relatively greater than the increase in numbers over a 3-year trial period, indicating an increase in worm size.

In France the number of earthworms increased fivefold (SoCo, 2009) as the most visible impact. A few studies have been made on micro-arthropods showing more diversity and density in NT. Beetles are reported in several studies and assessments, as a big difference between tilled fields and NT. Wildlife such as hares, partridges and several bird species is reported to increase in NT fields. Larks have been reported in NT maize fields, as in tilled fields they have difficulty in establishing their nests.

In the UK a number of studies have shown benefits to biodiversity and wildlife through the adoption of CA. Changes in soil fauna, both micro and macro, are positive as are the numbers of ground-dwelling invertebrates. The Fisher Alpha diversity index of species assemblages showed the zero-tilled system to be significantly higher, indicating a more stable ecosystem (Longhurst, 2010). Studies of birds visiting split field plots in winter showed a very high preference for zero-tilled stubbles sown with winter wheat over their ploughed comparison particularly through the late winter period. The absence of food during this period is well known to be a major contributor to the decline in farmland birds in modern times. Tillage systems that retain resources close to the soil surface are more likely to be attractive to foraging birds. A comprehensive review of the evidence, funded by SMI, is provided by Holland (2004), who reviewed an extensive body of Europe-wide experiences on the biodiversity impacts of reduced cultivation.

There is a broad consensus in the scientific community that the intensification of agriculture has led to significant reductions in the numbers of animal species, both above and below the soil surface. This also applies to the same extent to arable weed vegetation. Research in Germany indicates that, in addition to the use of pesticides and the reduction in the crop varieties, intensive cultivation with the plough is essentially responsible for this (Emmerling *et al.*, 2003). In Saxony a significant increase in the size of the earthworm population came about as result of long-term conservation tillage (Krück *et al.*, 2001).

In Portugal there were a few biodiversity impact studies carried out on the abundance of earthworms under different soil tillage systems. Carvalho and Basch (1995) found a much higher number of biopores down to a depth of 35 cm of a Vertisol under 6 years of NT compared to ConvT. Other, unpublished data originating from a Luvisol site show an almost threefold number of earthworms after 5 years of NT when compared to ConvT (112 versus 39 individuals m^{-2}), and an increase from 100, 122 and 136 earthworms m^{-2} after leaving 0, 2000 and 4000 kg straw ha^{-1} on the soil surface over a period of 3 years (Basch 1999 and 2008, unpublished data).

In Spain agricultural systems with abundant crop residues on the soil provide food and shelter for many animal species during critical periods of their life cycle. Hence, with CA large numbers of, for example, species of birds, small mammals, reptiles and worms thrive. Also, CA allows the development of a living structure on the ground; more stratified, richer and diverse organisms such as microorganisms, nematodes, earthworms and insects. In a study of earthworms, CA reached 200 individuals m^{-2} in the upper 20 cm of soil, compared to just 30 individuals in ConvA (Cantero and Ojeda, 2004), meaning 600 kg biomass ha^{-1}, almost 700% more than in conventional farming. Espejo-Pérez *et al.* (2006) had similar conclusions in a study that compared in four farms, up to 40 cm deep, the earthworm population in CA and ConvA.

Overall the soil life in quantity as well as in diversity increases significantly under

CA, reflected in higher enzymatic and respiratory activity, a wider range of species, including fungi, and a higher count of individual representatives of the mesofauna and macrofauna, such as beetles and earthworms. This applies across Europe to all climatic zones (Soane *et al.*, 2012).

6.2.11 Economic return

Forristal and Murphy (2009) calculated that in a 100 ha winter wheat unit in Ireland, the adoption of a minimum tillage system could save €53 ha^{-1} compared to ploughing. A saving of €66 ha^{-1} year^{-1} was estimated on a 400 ha unit amounting to €26,400. From a labour perspective the adoption of minimum tillage was attractive as it could reduce the labour required to establish a crop from 2.14 to 1.01 h ha^{-1}. They also found that on a 400 ha unit a two-person team could effectively replace a five-person team where minimum-tillage was used for crop establishment. Heaney (2012) conducted an unpublished study on winter oilseed rape establishment on three farms and calculated that the yield required to cover production costs was 2.3, 2.8 and 2.9 t ha^{-1} for CA (autocast), minimum tillage and plough-based systems, respectively.

Investigations into NT technologies in Germany started in 1966 (Bäumer, 1979). Intensive and long-term research in Germany by Bäumer, Czeratzki, Kahnt and later Teebrügge and Böhrensen, concluded that NT is a viable cultivation system. According to Tebrügge and Böhrnsen (1997), NT is a very profitable cultivation system compared to ConvT because of the lower machinery costs and lower operating costs. No-tillage decreases the purchase costs, the tractor power requirement, the fuel consumption, the amount of required labour as well as the variable and fixed costs. Since the same crop yields can be achieved by NT compared to plough tillage, on average the profit will be greater with NT systems. Despite these facts and opportunities, adoption of NT farming in Germany is still very low.

In France the cost reduction under CA with maintained or improved yields was on average €300 ha^{-1} (SoCo, 2009).

Several studies on the economics of the use of different soil management systems have been conducted in Portugal. Basch *et al.* (1997) found a reduction of total costs for traction of €91 ha^{-1} when changing from the traditional system to NT. Similarly, Marques and Basch (2002), calculating the wheat productivity necessary to obtain a break-even net margin on a 100 ha farm, obtained 1340 kg of grain ha^{-1} for the NT system, against 1773 kg ha^{-1} for the traditional system. From the studies for his PhD thesis on the technical and economic evaluation of tillage systems, Marques (2009) concludes that different tillage systems did not significantly affect yields but that the total production costs on a 300 ha farm using NT soil management were around €115 ha^{-1} less than under ConvT, which corresponded to cost reductions of around 20%. According to the same author, additional savings can be expected in the medium and long term through the reduction of fertilizer inputs through the improvement of SOM and overall soil fertility.

In Spain González-Sánchez *et al.* (2010) stated €235 ha^{-1} extra benefit for NT farms in comparison to farms using conventional soil tillage in a wheat–sunflower crop rotation in southern Spain. Fuel cost for farmers in Spain is increasing steadily, having risen from 50 cents per litre to almost €1 in the last few years. In a study in the Vega of Carmona area, Perea and Gil-Ribes (2006) compared NT to ConvT in a wheat–sunflower rotation, and concluded that NT could save 70 l ha^{-1} of fuel.

In general the cost reduction and time and labour saving under CA are the strongest reasons for adoption. The reduced production costs would even make up for eventual yield reductions and for farms above 100 ha in Finland a yield reduction of 10–15 % is still economically acceptable for the farmers. Overall the profitability of CA appears to be higher than conventional farming (Soane *et al.*, 2012).

6.3 Challenges Encountered in Scaling-up Conservation Agriculture in Europe

Regarding yields, weeds, pests and diseases as crucial themes showed in an empirical study with 95% confidence that in balance there are many more benefits than drawbacks when shifting to CA (E.J. González-Sánchez, University of Cordoba, Spain, 2012, unpublished data). If CA is so good, why is adoption still low in Europe?

There are several reasons for this, such as the poor government support when compared to other agricultural systems. Only a few agri-environmental measures under Pillar 2 of CAP support CA, and where farmers find those subsidies, sometimes the schedule of asking for grants is antagonistic with agronomical practices. As an example, farmers were informed in January 2008 of practices to be done November 2007. So no migration from conventional farming to CA was really supported. Unfortunately, CA is not perceived by government officials as being capable of establishing really sustainable agriculture, which science has demonstrated and continues to do so in Europe and in other countries with similar environment such as Canada.

The second reason is the strong 'agricultural establishment', as undoubtedly, CA goes against plough manufacturers and related companies. This means that there are seldom companies interested in creating strong links to the CA community. Actually, sometimes farmers receive contradictory messages. CA works or CA does not work, depending on who is visiting them and what they want to sell. As a pioneer stated, CA works wherever you can do agriculture, you just have to understand your field and adapt CA to it.

6.3.1 Residue management and supply

With the exception of forage crops, the retention of crop residues on the soil surface is a common feature in CA and has to be dealt with during planting. Under European conditions residue levels for cereal straw, for example, are commonly between 3.5 and 10 t ha^{-1}, with extremes also exceeding this (Soane *et al.*, 2012). While residues serve for soil and moisture conservation resulting in higher yields in south-western Europe, in other regions they might delay the warming and drying of soils during spring planting resulting in yield penalties (Soane *et al.*, 2012). Other reported problems are to establish a good seed–soil contact during seeding, without pushing crop residues into the seed furrow, which is a challenge under moist conditions. Different residue management practices, such as chopping or high stubble are applicable under different conditions, but in any case an even distribution of the residues is important. The wrong residue management strategy under NT can result in yield penalties up to 16% (Soane *et al.*, 2012).

In Germany the main driver for increasing residue retention in the field was the improved capacity of harvest equipment to chop and evenly spread the residues.

While residues have an important role in CA in view of carbon, weed and pest management, there are also other competing uses, which will have to be balanced. Ireland, for example, has an annual market for cereal straw at harvest time, which is used in the mushroom industry, for animal bedding or in some cases for feed. Straw and crop residue is looked on as an extra source of income as well as for supplying bioenergy plants. In France with the more experienced CA farmers residue management is no more an issue, avoiding thick layers of residue where possible by keeping long stubble, or using appropriate NT seeders and planters with a good residue handling capacity. Small seeds like rape can even be seeded by gravity just before or during cereal harvest, and in this case straws are no more a problem either (Figs 6.6, 6.7).

Whereas in central Europe the huge amount of crop residues may pose some problems for the establishment of the following crop, the contrary is frequently the case in Portuguese rainfed production systems. Low total biomass production, straw removal for fodder and even subsequent stubble grazing often leave the soil almost bare even under a NT system. These practices reduce drastically the beneficial effects of the NT system as one of the main principles of CA, i.e. permanent soil cover is missing. The importance of residue management and the maintenance of crop residues for the build-up of SOM was clearly evidenced by the results of a recently terminated research project (Basch *et al.*, 2012b). Although double cropping is frequent in the north-western parts of Portugal, the option for the establishment of cover crops in the dry summer in the rest of the territory is limited to irrigated conditions, where summer crops are the main crop.

6.3.2 Non-availability of suitable implements and inputs

The availability of suitable implements, particularly in the wetter parts of Europe with more challenging residue handling conditions, has been the main impediment for spreading CA and in fact, where it could be overcome, it has resulted in a relatively faster adoption. For example one manufacturer of NT seeders in Finland took interest in NT very early and claims to have sold almost a thousand NT seeding machines up to 2007, having about 50% of the market share in the country. About ten NT seeder manufacturers from around the world have been able to place their NT machines in the Finnish market and four of them are made in Finland, lifting Finland despite very difficult conditions to one of the fastest CA adopting countries in Europe.

Spain, another country leading in CA adoption in Europe, is not a high-yield country compared to central Europe, residue handling is not a big problem for NT seeding with the exception of irrigated maize. Over 20 machinery companies were identified as

Fig. 6.6. Conservation Agriculture/no-till planting of maize into flowering turnip rape (*Brassica rapa*) (Photo: Peter Hofer).

Fig. 6.7. ... the result – maize growing in the mulch of the turnip rape (Photo: Wolfgang G. Sturny).

suppliers for CA equipment in the latest FIMA Machinery Fair in Zaragoza in 2012.

In Ireland on the other hand, there is a dearth of suitable implements for CA systems in the market in contrast with the abundance of trailed or powered cultivation equipment available. Machinery manufacturers and their sales staff seem to understand little about soil properties and dynamics and are unaware or ignore the importance of minimal soil disturbance, one of the guiding principles of CA. There is also an emphasis on tine or knife coulter drills, which farmers seem to like because they are able to break up compacted layers in the soil. This has sometimes led to establishment issues in wet soil conditions as seed has fallen through the fissure cracks created. Disc drills are thought by many farmers to be unsuitable for wet soil and trashy conditions. Despite these reservations disc drills have worked very well on farms practising CA.

In Germany good CA equipment is still a major challenge, although more recently farmers are sourcing equipment, partly from outside the country. Important elements to facilitate CA adoption have been the increased attention of harvest equipment manufacturers on management and spreading of crop residues. However, the challenge is still to match equipment and requirements of new diversified crop rotations.

In the coastal regions of the centre and north of Portugal average farm size is very small. There, NT drilling equipment would have to be shared by several farms to be an economically viable option or to be run either by service providers or cooperatives. This is not the case in the Alentejo region where large estates predominate.

Besides equipment, in France the availability of maize varieties to perform under NT conditions is still a challenge due to the different N dynamics and lower soil temperatures during germination in NT soils.

This is one of the main reasons for the initial yield reductions in maize sometimes observed in the early years of CA adoption. Availability of cover crop seeds and particularly of special mixes of cover crops is also a challenge for CA farmers in Europe.

In France farmers adopting CA have been the ones able to take risks, to have their own mind, even against extension advisors and public opinion. However, this kind of progressive farmer is usually in the minority.

6.3.3 Tillage mindset

In many European countries the cultural entrenchment of ploughing and preparing a clean seedbed for sowing is a strong reason to maintain this tradition. This mindset together with the lacking professional agronomic skills make even the attempt to try different production methods very difficult. In southern Portugal, where cropping systems and soil and climate conditions plead for the adoption of soil- and water-conservation production techniques (Basch and Carvalho, 1994), many of the farms are still run not by the landowners but by employees with long-term empirical experience but with limited professional skills. In addition, the average age of Portuguese farmers increased by 4 years in the last decade and around 50% are more than 65 years old. Around 62% of the farmers have no or only very basic (4 years) educational level (INE, 2011).

There is still a lack of understanding or belief in the capacity of the soil biosphere to improve and restore itself when left uncultivated with soil surface protected with crop residues. This is a major factor in not realizing the many practical benefits of CA in a shorter time frame. The requirement to cultivate is supported and reinforced by the machinery trade in promotions and advertising. Consequently there is an almost romantic notion about the benefits of ploughing, ripping or sub-soiling throughout the tillage sector among farmers and professionals and in society amongst the wider public.

An increasing number of farmers in Germany are prepared to change tillage practices. The change began first in the drier regions of Germany, triggered also by economic aspects and market liberalization.

6.3.4 Skill requirement

CA is sometime referred to as a 'knowledge and management intensive' system. In any case, it requires new knowledge and skills, since it is fundamentally different from conventional farming. As in other parts of the world, successful CA development in Europe happened when pioneer farmers became organized, exchanged their experience and advanced the knowledge of the entire group. In France the successful CA farmers have been organized in farmers' groups, similar to the developments in South America. With good training by experienced experts and colleagues, and operating in local groups, farmers can find a way to minimize the risk of change to CA and improve their chances of success.

In Italy, AIGACoS, since its foundation, has provided information and technology transfer and institutional support to farmers that adopt CA. Since 2000 it has collaborated with several important seed, fertilizer, pesticide and GPS companies in the organization of open field days called 'Agricoltura Blu in campo'. During these events technical support to farmers is provided and different machines can be seen in action and compared. Over the years the number of visitors has increased from hundreds to thousands. Since 2010 AIGACoS has also started closely collaborating with regional authorities in the promotion of CA.

In Spain the success of CA depends largely on the degree of adaptation of the techniques to the particular conditions of the area and the crop. The great variability in Spain in terms of soil and climate characteristics of each region does not allow CA to provide a single valid prescription to correctly apply each of the practices. Despite the extensive literature on the basic principles of

CA, it is necessary to know site-specific farmer needs; as there is no exact knowledge on the part of government and public research about CA, farmers' experiences are an important factor.

On the education side, it is difficult to find subjects on CA in Spanish universities. Therefore it is difficult to find skilled specialists supporting farmers in field. Many technicians are trained by private companies selling CA products, with the respective bias, which is not necessarily leading to the 'best' CA. Research on CA in Spain depends mostly on the awareness of some scientists, as there are no specific research calls devoted to CA.

6.3.5 Weed infestation

One of the initial observations after adoption of minimum tillage systems in Ireland was that while overall weed numbers declined, specific weed species became more abundant. Grass weeds such as sterile brome (*Bromus sterilis*) and annual meadow grass (*Poa annua*) predominated while broadleaved weed included cleavers (*Gallium aparine*) and plants whose seed is wind dispersed like groundsel (*Senecio vulgaris*) and willowherb species (*Epilobium montanum* and *Chamerion angustifolium*). Control of grass weeds became an issue for many farmers and fallow stale seedbeds were used to control weeds and volunteers before autumn crops were sown. One constraint in Ireland, for example, was the practice of growing monoculture winter wheat or winter barley. There was increased herbicide use, particularly graminicides, in these situations. Forristal and Murphy (2009) reported that additional herbicide costs in minimum tillage could amount to an additional €33 to €67 ha^{-1} and negate the machinery cost benefits of the system. But as market prices for oilseed rape improved, backed by an interim government subsidy for growing biofuel crops, many farmers started growing rape in rotation. In recent years strengthening prices in addition to merchant contracts for field beans has added another crop to viable rotations. The majority of farmers now gain adequate control of different weed species using a combination of rotation and herbicide application.

In Denmark pesticide legislation is very restricted and the farmers are not allowed to use many pesticides that are permitted in the rest of Europe. This means that it can be very difficult for Danish farmers to handle weed infestation and diseases, increasing the perceived risk for farmers to adopt CA. In the UK weed control can become a problem where the whole concept of CA (rotations and residue management etc.) is not fully understood. The increase in the prevalence of grass weeds is considered the biggest impediment to the widespread adoption of reduced tillage systems. The build-up of herbicide-resistant black grass (*Alopecurus myosuroides*) across large areas has led to increased costs and reduced yields. However, the black grass problem is not specific to minimum tillage farms in the UK: whilst ploughing will bury freshly shed black grass seed below the germination zone, it will also bring up previously shed seed into the germination zone, as black grass takes a minimum of 3 years to lose 90% of its viability. On the other hand, UK farms applying good quality CA with low disturbance NT, a diversified weed management and a good residue management do not have any black grass problems (Sims and Ellis-Jones, 2011).

In Portugal, under Mediterranean rainfed conditions, weeds are by far the most severe pest problem for 'arable' crop production. Fortunately it is mainly annual weeds that have to be controlled, as summer drought allows the survival of perennial weeds only on areas with deficient drainage. In general the weed dynamics under NT systems change, with different types of weeds becoming more predominant, compared to tillage-based systems. In humid regions of Europe perennial weeds and grasses can create major problems, while annual weeds are reduced. Serious weed problems can arise in long-term monocropping under NT or, worse, minimum tillage. Therefore important elements in weed management of CA systems are, besides NT, surface mulch and crop rotations (Soane *et al.*, 2012).

6.3.6 Yield reduction

Experience in France, as in other parts of Europe, showed that yield reductions were only seen when mistakes were made, particularly during early adoption, or on degraded soils, before the system has restored a good enough soil condition to enable a crop to grow without the help of tillage. In general these mistakes can be avoided or minimized by better learning practices. Routinely after several years, on successful NT farms, there is no yield reduction.

In Germany yield reductions were observed on badly drained or badly structured soils, which would require special attention to specific crop rotations or complementary measure to overcome the initial problems until a stable soil structure and internal drainage in soils would have been established.

6.3.7 Insect-pest and disease challenges

In general the presence of mulch seems to increase problems with slugs in CA, which can be overcome with molluscicides. Yet, this increases production costs and affects the beneficial fauna, which in the long term seems to be effective against slugs. In fact, observations in France seem to show that a strategy against slugs might be to avoid anti-slug treatments, which damage beetles and natural enemies of slugs, and rather wait for the populations of predators to be restored. They might then be able to control slugs. A similar approach can be taken with mice. Some farmers are looking for the restoration of the whole ecosystem including the field margins management to provide habitat for foxes, stoats, weasels and birds.

While other pests do not create particular problems, a suitable response strategy against diseases, which is applied by the advanced NT farmers in France, is to use a mix of varieties (normally four in one field) for cereals. They have different sensitivity to diseases, and thus the mix is more robust, and eventually needs fewer fungicide treatments. Likewise the association of crops, like rape + white clover, or cereals + clover, are used. This impacts weeds as well due to a better soil cover. It is supposed that better balance of diverse species will result in fewer problems with pests and diseases, and some indications on farms seem to show this, but this is not documented precisely and is still questioned. At least the situation is not worse than in conventional farming in France, where despite intensive tillage the use of pesticides is still prevalent.

As for insect pests, some, such as springtails (*Onychiurus* spp.) in sugarbeet, seem to be reduced by mulch, others, such as the European corn borer (*Ostrinia nubilalis*) in maize, seem to increase, but only in monocropping (Soane *et al.*, 2012). Similar effects can be observed with crop diseases, which in general do not differ with tillage treatments but, particularly in the presence of residue mulch, depend very much on the crop rotations (Soane *et al.*, 2012).

6.3.8 Lack of enabling government policies

With very few exceptions, shown in section 6.4, there are no specific policies in European countries to support CA. This is particularly true for the low-adoption countries like Denmark or Ireland where there is a reluctance to publicly promote CA adoption at different levels within the official institutions.

In general the CAP as actually applied in the European Union is not providing any incentives for the adoption of CA. On the contrary, since it has been formulated considering ConvA as the standard method, there are even disincentives for farmers to adopt CA. Subsidies derived from EU are for European farmers such an important part of their income and hence compliance with EU regulations has a high priority for farmers, even if those work against good practices, such as diverse and healthy crop rotations. With the newly proposed CAP reforms the EU is attempting to address some of these issues, but not to a satisfactory level.

A recent report published by Teagasc in Ireland (Teagasc, 2012), developed to give guidance for research and development in the Irish tillage sector for the period 2014–2020, repeatedly identified significant weaknesses in current crop production systems as being production costs including land and machinery, and the increasing cost of diesel, fertilizer and plant-protection products. Despite this, the report never mentioned the positive contribution CA could make to addressing many of these core problems, neither was it recommended that further research into CA systems was warranted to meet environmental policy objectives.

6.4 Government Support and Policy Towards Conservation Agriculture

Switzerland is one of the few countries in Europe with policies in support of CA. The instruments include penalties as well as positive incentives. Farmers cropping erosion-prone areas are obliged to maintain soil fertility in the long term, due to the federal law relating to the protection of the environment and the implementation of the preventative principle (Soils Report, 2009). In case of repeated reports of soil erosion damage at the same site, this will be considered as a management failure. Farmers can be prosecuted in accordance to the guidelines of the requirements of the proof of ecological performance (PEP) resulting in a reduction in their direct payments received. After all it is in the interest of farmers to avoid repeated soil losses by using appropriate soil conservation techniques. Erosion control is being implemented by the cantons. In order to respond to erosion alerts, the Canton of Berne elaborated an enforcement scheme in collaboration with agricultural control organizations in 2005. A situation assessment is being conducted (identification of the erosion problems, crop rotation, soil tillage, etc.) as well as an appropriate, site-specific 5-year action plan elaborated together with the affected farmers in order to prevent further soil losses. If the action plan is kept, but soil erosion damage still occurs, then the farmer will not be affected by direct payment cuts. A key element of the action plan includes CA cropping techniques. In accordance with the Bernese ordinance on preservation of natural resources and the cultural landscape (LKV, 1997), farmers in regions particularly susceptible to soil erosion, compaction and nitrate leaching are directed to implement these CA production systems. To date, the enforcement scheme has been applied in about 30 cases.

Very few other countries in Europe promote CA with national policies, and if done, it is mostly limited to certain provinces or regions within the countries. In Italy the Rural Development Programme (Piano di Sviluppo Rurale, PSR) of each region implements the EU Regulation and establishes regional strategies and interventions in agriculture, agribusiness, forestry and rural development matters. Veneto was the first region in Italy and Europe that in 2010 included CA management as part of Measure 214 – Sub-Measure ‘Eco-compatible management of agricultural lands’ in its PSR 2010–2013. More recently (Forristal and Murphy, 2009), Lombardia has modified its PSRs to include this measure. Emilia Romagna, Puglia and Basilicata have initiated an audit to amend and supplement the measure in their PSRs. This encompasses agroenvironmental payments for farmers transitioning from tillage-based systems to CA. Because CA systems are knowledge-intensive, in the first years of adoption lower yields could be observed due to lack of experience, and to make up for this the above mentioned measure would provide subsidies.

It should be noted that minimum tillage (MT) cannot be recommended under any circumstance: a review of the scientific literature conducted by Corsi *et al.* (2012) shows that yields and environmental benefits under MT are lower relative to both tillage-based systems and CA. However, the payment for the adoption of MT in the case of Lombardia is regarded as a first step toward more sustainable systems. Subsidies introduced to compensate for short-term economic losses and encourage the uptake of sustainable agronomic management systems should be coupled with the introduction of a label system to certify SOC preservation and accumulation, and prize the societal value for the soil carbon sequestered and for the lower GHG emissions from agricultural soils. Regional extension services will have to play an important role to monitor the correct implementation of the techniques, provide technical support to adopters and plan long-term policies.

In 2001 Portugal introduced compensatory payments for NT and strip-till in row crops with additional payments being granted for complementary measures such as the establishment of cover crops, maintenance of all (stubble and straw) crop residues and the non-grazing of cereal stubbles. In the same period an agri-environmental measure was also launched for permanent irrigated crops (with the exception of olives).

In Germany the state of Saxony has supported some agricultural-environmental measures, which are demonstrated in the level of reduced tillage adoption, including very good CA in that state. In the years from 1995 to 2005/2006 the area under conservation tillage supported by subsidies from the environmentally sustainable farming programme increased from under 5% to around 34% of the arable land in Saxony, with the share of areas not using the plough estimated as being at least 50%, and in some regions of Saxony up to 100%.

6.4.1 Research support

There is still relatively little public support to CA research in Europe, and it is mainly focused on minimum tillage and on comparison trials, rather than on optimizing the performance of CA systems. In Ireland some research was undertaken comparing minimum tillage with plough-based cereal production from 2000 to 2008. Equally in France scientific references for CA practically do not exist in the public sector, because most research programmes are either oriented towards fundamental research, while the private sector is not interested in this still small market. In Denmark some research support is given through the National Advisory system and through the government research institutes.

6.4.2 Incentives in the form of subsidy on implements

In the mid-2000s, and for several years, the Spanish Ministry of Agriculture, according to agricultural organizations, cooperatives and regional governments, developed a plan in order to achieve the renovation of agricultural machinery. The plan subsidized up to 30% of the cost of a new NT seeder. The Institute for Energy Diversification and Saving of Energy (IDAE, in Spanish), which is part of the Ministry for Industry, offered a subsidy of up to 40% of the cost of a NT seeder. Both plans were a very good opportunity for helping farmers to invest in CA equipment. However, the best promotion in Spain for CA has been the Rural Development Programmes, under the CAP. The huge increase of CA in woody crops was thanks to an intelligent investment in favour of cover-crop use in olive groves in hilly areas. In the southern region of Andalusia, the measure involved up to 158,462 ha in 6 years (2000–2006). Every farmer received €132 ha^{-1}. It was estimated by AEAC SV that at the end of the period, there were 450,000 ha covered in the region; so the imitative ratio was very good: for every hectare with subsidy there were two hectares adopting without subsidy. Given the success of CA in woody crops, Andalusia started to fund NT under the same programmes for Rural Development in 2007. After 2008 and due to the big economic crisis, many regional governments have cut down all these subsidies, as a part must be co-financed at regional level.

6.4.3 Promotional campaigns/training

In Ireland the Department of Agriculture, Food and the Marine (DAFM) have co-funded a series of agri-environment schemes with the EU during the past 20 years. These initiatives were mainly suitable for extensive livestock producers. Since 2008 measures such as minimum tillage and the use of cover crops over winter were supported but uptake was relatively low. The current DAFM development plan for agriculture, called Food Harvest 2020, places little emphasis on the crop production sector and no reference is made to CA. Interestingly the Department of Foreign Affairs (DFA) overseas section 'Irish Aid' recently started supporting CA development projects in their programme countries in sub-Saharan Africa. Since the end of the ECOtillage in 2005, promotion of CA has largely been carried out by CAIR. A quarterly newsletter was distributed to members and relevant government agencies up to the end of 2010 and three farmer meetings were held per year but CAIR activity has also dwindled in recent years.

The British government's approach to agriculture during the period since the CAP reform has been to allow market forces to prevail whilst setting a framework for environmental protection, often as a result of EU-wide directives. However, both the government and levy bodies have been proactive in providing resources for knowledge exchange through the funding of printed guidance documents and field demonstration programmes such as 'Soil2Crop' and 'Sow2Succeed'.

Over the past decade SMI has gathered a substantial body of evidence regarding reduced tillage cultivation systems. Much of this information is published in the SMI Guides, which include 'A Guide to Managing Crop Establishment' (SMI/Defra, 2001), 'Target on Establishment' (Vaderstad/SMI, 2004) and 'Visual Soil Assessment' (Vaderstad/SMI, 2006), along with numerous papers published in scientific journals, the scientific press and many articles in farming magazines.

In Switzerland key elements supporting the relatively rapid uptake of CA were the founding of the Swiss Soil Conservation Association (SNT) in 1995 and an increasing support of the regional soil conservation services by starting field demonstrations and initiating incentive programmes on NT systems.

6.5. Efforts/Policies Required for Scaling-up Conservation Agriculture

In France since 2011, APAD has focused on strict zero-tillage CA, according to the definition of FAO for CA and the guidance of the international CA community. It now has 100 leading NT farmers as active members, and is growing fast by creating local subsidiaries. Its final objective is the conversion to CA of most of the 300,000 professional farmers producing most of the annual crops. The strategy is no more approaching directly individual farmers, but a strategy supported by two pillars:

1. Political advocacy towards citizens, communities and their representatives, i.e. policy makers. At European level, addressing CAP, as well as addressing its national and local implementations; locally, for example, water agencies and operators of water management and quality are potential partners of choice because CA is able to solve the issue of water pollution without compromising farming economy on their territories.
2. Promotion to all partners of agriculture: engaging into cooperative partnerships with diverse farmers' groups as well as all kind of organizations of farming community or involved companies.

As recognized by the European Commission in the policy report 'The implementation of the Soil Thematic Strategy and ongoing activities' (EC, 2012) CA plays an important role to protect soils. In Italy CA is gradually spreading and, where properly implemented, it has proven to reduce significantly soil degradation and help improve chemical and biological soil fertility, while reducing GHG emissions from fossil fuels and reduce those ascribed to the mineralization of organic carbon (Pisante, 2007).

In Portugal there were several agri-environmental measures implemented by the Ministry of Agriculture in 2002 to promote not only NT and strip-till systems, but also the maintenance of crop residues, or at least the stubble, or the establishment of cover crops. This, in fact, boosted the uptake of CA systems, including cover crops in perennials, until 2006, when these measures were cancelled. Later, in 2008, CA systems were again included in agri-environmental schemes, however only if a farmer adopted the 'integrated production system' on the whole farm. The bureaucracy around this certification scheme made farmers practising CA reluctant to apply for any support with regard to CA.

As reported by SMI in the UK there is a massive amount of interest in CA and any meetings that are organized always draw large audiences. As yet it has not been possible despite a number of attempts to set up an organization where farmers can pool their ideas and resources to take the concept forward. Funding has always been the key reason for failure. As a concept it is not a system that is going to endear itself to large machinery, chemical or fertilizer manufacturers, as the system in-time requires fewer artificial inputs of any kind, so it will be up to the specialist drill manufacturers (many of whom are small companies with limited funds), the smaller seed and crop nutrient suppliers and motivated groups of farmers to invest time and money into carrying out research and disseminating the information to those that are prepared to contribute financially towards it.

Spain is in a good position for scaling-up CA practices. There are successful stories with experienced farmers across the country to support the system and also with the help of a network of 11 regional CA associations and a national one. In Spain efforts in training farmers and technicians in CA are undertaken by the private sector. The National and Regional associations for CA have regular courses and field days, normally supported economically by R&D projects or funded directly by the industry. Nowadays, most Spanish farmers know about CA; however, more specific courses and field days, reaching local farmer interest issues, are required.

Everywhere in Europe where the CAP plays a decisive role in the decision making process of farmers with regard to both what and how they grow, there should be a clear time-limited support for the adoption of sustainable production methods to cover the risks inherent to each change of production methods, and an initial support to facilitate the access to necessary new equipment. A transition period of two 5-year periods has long been practised in Switzerland, with differentiated levels of support, which could be a valuable investment towards the adoption of sustainable production systems.

6.5.1 Active research

One of the drawbacks in private and public research work carried out since 2000 in Ireland was the tendency to do comparative trials between minimum tillage/reduced cultivations and ploughing while using different treatments implemented at the same times in either system. This was neither fair to the plough nor the minimum tillage treatments. More timely operations and specific agronomy practices are necessary to achieve optimum results in any system. Based on CA developments and practice overseas it may in future be more appropriate to use CA fields on farms and evaluate and monitor performance based on farmer practices with a more flexible research model.

In Portugal today there seems to be a core of pioneer farmers who implemented the CA system with or without the support from the agri-environmental measures and managed to overcome sometimes severe difficulties without blaming the system itself but the missing solutions available. There is no doubt that active research or experimentation has to go along with the process of adoption of a locally new production process. The support of governmental (universities, research institutes, extension services) or non-governmental institutions working in the dissemination and extension of sustainable farming practices seems vital for the adoption of new systems, especially in an environment where farmer-driven innovation efforts are somehow hampered by CAPs that sustain the maintenance of unsustainable farming practices. It may be due to the perceived benefits or due to the bigger farm size that CA adoption in Portugal is highest in the Alentejo region. But it may also be the consequence of the technical and research support delivered now over 25 years by the research team from the University of Évora and the dissemination work realized by APOSOLO located as well in the district capital of the Alentejo.

No-tillage systems are accepted as an environmentally sound farming system. However, there are aspects that should be developed in order to improve the surface cropped by NT such as reduction of pesticides used, improved nutrient efficiency, more efficient and lighter machinery (Fig. 6.8).

Fig. 6.8. Winter wheat – using one-third of the seed density – precision planting directly into an established cover crop composed of eight species. The green manure plants die back in winter and provide a protection against soil erosion, pesticide runoff and nitrate leaching, among others (Photo: Wolfgang G. Sturny).

6.5.2 Identifying suitable cover crops/augmenting residue supply

Research on cover crops has only been carried out on single species and despite positive results is viewed as an unnecessary cost by the majority of farmers and advisers. International practice is to use cover-crop mixes that include a range of different species that have multiple benefits. In Ireland, cover-crop seed is quite expensive because, due to low demand, merchants have to order small quantities and ultimately pass the cost on to the farmer. It is likely that farmers who want cover-crop mixes will order their requirements directly from the UK or mainland Europe in future years.

In Germany over the past 5 years attention to cover crops by farmers has significantly increased particularly as a complement to direct drilling (NT). The development was initiated by growing cover crops for sugarbeet in areas subject to erosion (e.g. *Phacelia* spp.). Today the numerous advantages of quantities of cover crops have been recognized and adjusted mixtures of cover crops are recognized for different purposes. Mixtures of cover crops are selectively used in order to regulate the water management system, to practise active soil protection through coverage of the soil, to make a contribution to the nutritional and humus balance in the soil and to increase the load-bearing capacity and its ability to support wheeled traffic. This positive development is a result of the research and development of commercial seed companies. State-supported trials with catch crops are currently taking place in Saxony and will become part of the official extension advice given (https://publikationen.sachsen.de/bdb/artikel/14650).

6.5.3 Developing and providing suitable machinery

Much of the seeding equipment used in Ireland is also popular with farmers in the UK and hence easily available: farmers like to have confidence in a reliable backup service when replacement parts are required. Often farmers are used as sales agents but in practice success is dependent on the knowledge of the individual farmer agent about the CA system and their understanding of sustainable soil management. Generally speaking these knowledge levels are quite poor and in some cases this has led to mishaps with crop establishment, weed proliferation and other setbacks, which result in CA getting a bad name.

In Spain the equipment and machinery quality is not a major problem, but its price surely is. High costs drive farmers to service-providers for seeding. Sometimes it is a good approach, as at early stages farmers can make tests at an affordable price; but if a famer is going to finally shift towards CA, one of his certain needs would be a NT seeder on site. A major requirement would be to re-start incentives for the purchase of machinery to make CA equipment again affordable.

6.5.4 Developing effective integrated weed management techniques

In France weed infestation is usually seen as an argument against NT systems. However, experience has shown that not disturbing the soil imposes some delay for weed seeds to germinate and emerge. Covering the soil with high levels of thick biomass makes it difficult or impossible for weeds to develop. As a result only a few of them can produce seeds for the next generation. If in addition the farmers use herbicides properly in combination with other techniques such as soil cover and crop rotations, the weed pressure is reduced over time. On nearly all successful CA farms, after 3 to 7 years, less and less annual weeds are observed. The same applies to perennial weeds, as long as a good weed management strategy, including the use of herbicides when necessary, is applied. The secret of success is to get good cover crops and crops: in every place in a field where the cover is poor, there is a concern with weeds developing. On the contrary in minimum tillage, with repeated surface tillage annual grass weeds can become dominant forcing farmers to revert back to the plough. This is one of the reasons for the misconception of weeds being a particular problem in CA, but it refers in reality to reduced tillage systems which are not CA.

In Portugal studies revealed that the delay of autumn seeding until the emergence of the first wave of weeds, remaining on top of the soil under NT, is decisive for the successful pre-emergence weed control (Calado *et al.*, 2010). They further confirm that an efficient pre-emergence control of weeds is able to reduce late re-infestation considerably under NT when compared to ConvT where weed seeds are buried or brought from deeper soil layers to the topsoil, from where they germinate during the growing season (Barros *et al.*, 2008; Calado *et al.*, 2010). Another important finding of these studies was the fact that the improved soil-bearing capacity under NT makes post-emergence weed control possible under almost all soil moisture conditions allowing the correct timing and thus the reduction of herbicide doses (Barros *et al.*, 2008).

Weed and pest management in Europe is a key issue as agri-chemicals are needed for agricultural production. In Europe, products are controlled by Regulation (EC) No 1107/2009 of the European Parliament and the Council of 21 October 2009 concerning the placing of plant protection products on the market. Safe products and a safe use are both important. CA can help to make herbicide use safer and to even reduce it (Fig. 6.9).

6.5.5 Developing effective integrated insect-pest and disease management techniques

The recent introduction of the Sustainable Use Directive (SUD) governing the efficient use of pesticides will place greater emphasis on integrated pest management (IPM) practices at farm level in Ireland. This presents an opportunity for highlighting CA, the guiding principles of which are consistent with good IPM practice. However, due to the absence of formal research in CA/NT systems in Ireland, farmers will continue to rely on contact with their peers or personal advisers or agronomists about specific weed, pest or disease problems that arise. Many management techniques developed result

Fig. 6.9. The use of a knife roller to manage cover crops before no-till direct seeding can replace herbicides for weed management (Photo: Wolfgang G. Sturny).

from trial and error and informal testing on farm. Little of this information is accurately recorded or quantified but is spread by word of mouth. Due to the favourable temperate climate, weeds and diseases will require regular attention in all crop production systems.

6.5.6 Technology dissemination through training/field days/media

Despite the required proof of ecological performance (PEP) in Switzerland, additional measures are necessary to improve and stabilize soil structure, reduce erosion and maintain soil fertility in the long term. No-till agriculture can make a substantial contribution in this respect. Adaptations in crop rotation, including cover crops, seeding techniques and nitrogen fertilizers can help to optimize cropping. Farmers have been made aware of cropping systems that conserve the soil since 1996, and have received financial support during the transition phase (Schwarz *et al.*, 2007). Today about 7% of the cropland in the Canton of Berne is under NT. Knowledge transfer preferably takes place in successful show-and-tell events among those interested in application of these systems. The farmer-to-farmer approach (Fry, 2009) helps to bridge the gap between agricultural and environmental institutions and measures by:

1. Establishing an accompanying group with all relevant actor groups to induce a learning process.
2. Developing short films in collaboration with these actor groups since film is an ideal means to record farmer knowhow, which is usually spread verbally. Fundamental elements of nonverbal communication are transported by pictorial language. These allow a high degree of identification.
3. Triggering discussions within farmer networks as well as among policy makers. A consolidated view indicates that farmers can take up arguments much more easily from successful colleagues (same profession, same culture and same language).

The Canton of Berne's 'Soil Support Programme' launched by farmers and soil experts pursues a comprehensive and sustainable problem-solving approach to soil protection at the interface of water and air. It is based on voluntary participation and allows for financial incentives for implementation of different measures related to cropping systems that protect the soil (mulch-till, strip-till or NT; offset ploughing in organic farming), soil development and cropping measures (crop rotation, soil cover over winter, undersown crops, abandonment of herbicides, manure composting) and ammonia-reducing techniques for the application of liquid manure (umbilical application system, soil-conserving drive gear such as low-pressure tyres or rubber tracks). This catalogue of measures is part of the programme concept, which, together with educational and extension components, constitutes an overall farmer-to-farmer approach, along with impact monitoring that includes plant protection and emission measurements. Roughly one-sixth of the 12,000 farms participate. Eighty per cent of the costs of this CHF60 million Soil Support Programme is being assumed by the Federal Office for Agriculture, the remaining 20% by the Canton of Berne. Following completion of the 6-year programme in 2015, these measures should be economically feasible without additional incentives and can be pursued further.

In general, however, there is a distinct lack of practical knowledge about sustainable soil management at extension level among both public and private agricultural information providers. As a result advisers, consultants and commercial representatives are reluctant to promote CA practices as they have neither confidence in their own knowledge levels nor do they possess adequate understanding, enthusiasm for or belief in the appropriateness of CA under European conditions.

In Germany insufficient advice is given to farms that wish to change to CA. Farmers who have an interest in the application of direct drilling are left completely alone. As a reaction, interest groups have formed. Assistance and support for example is provided by the German Conservation Tillage association (GKB) throughout Germany (http://www.gkb-ev.de) and regionally, for example, through the Saxony-based Society for Conservation Tillage/No-Tillage (KBD) (http://www.kbd-sachsen.de). Through the GKB the necessary interfaces are also generated and maintained with organizations and farmers working within Europe (http://www.ecaf.org).

6.6 Concluding Remarks

> The age-old practice of turning the soil before planting a new crop is a leading cause of farmland degradation. Tillage is a root cause of agricultural land degradation – one of the most serious environmental problems world wide – which poses a threat to food production and rural livelihoods.
>
> Huggins and Reganold, 2008

With increasing awareness that sustainability of agricultural production is a must if sustainable development at national and global level is to be achieved, CA/NT systems will continue to grow worldwide. But for sustained growth to take place the main barriers to NT adoption need to be overcome:

- Mindset (tradition, prejudice);
- Knowledge on how to do it (knowhow);
- Availability of adequate machines;
- Availability of adequate inputs;
- Adequate policies to promote adoption.

These barriers must be overcome by politicians, public administrators, farmers, input supply industry, researchers, extension agents and university professors. With adequate policies to promote CA/NT, it is possible to obtain what is called the triple bottom line, economic, social and environmental sustainability, while at the same time improving soil health and increasing production (Friedrich and Kassam, 2009; Friedrich *et al.*, 2009).

Farmers, researchers and extensionists need to reflect on the benefits of NT farming systems (SoCo, 2009):

- 96% less erosion;
- 66% reduction in fuel consumption;

- Reduced CO_2 emissions;
- Enhanced water quality;
- Higher biological activity;
- Increased soil fertility;
- Enhanced production stability and yields;
- Incorporation of degraded areas into production;
- Lower production costs.

Recognizing the multiple benefits of NT farming over reduced and ConvT-based farming systems should foster research and development efforts in order to overcome the bottlenecks of the system and help extensionists in diffusing the technology so that farmers can have a sound basis for practical application.

The wide recognition of CA as a truly sustainable farming system should ensure the growth of this technology to areas where adoption is still low as soon as the barriers for its adoption have been overcome. The widespread adoption of NT systems (Derpsch and Friedrich, 2009; Kassam *et al.*, 2009) shows that this way of farming cannot any longer be considered a temporary fashion. Instead, this farming system has established itself as a technology that can no longer be ignored by politicians, scientists, universities, extension workers, farmers as well as machine manufacturers and other agriculture-related industries.

The EU is about to 'redefine' its CAP, adapting goals and farmers' support to changing realities. One of the major outcomes of this adaptation is expected to be the greening of the 1st Pillar payments to farmers (Direct Payments), including measures aiming at an increase of the non-producing or set-aside area in order to enhance the environmental performance of farming. Besides conflicting with another important goal of the CAP reform, which is the contribution of EU agriculture to global food security, the obligatory implementation of the so-called Ecological Focus Areas (minimum of 7% of the farmland) in countries like Portugal, where agricultural production of many commodities is far from achieving self-sufficiency, seems to completely ignore the reality of the extensive farming systems in the Mediterranean countries. Furthermore, albeit a vague obligation for the respect of a minimum of crop diversity, this greening action does not care about how the rest of the farmland is managed. Especially in southern European countries with high water-erosion risk and extremely low levels of SOM, real greening would mean incentivizing or even obliging farmers to adopt soil and water conservation farming practices, such as the principles of CA, on the largest possible area.

CA is also a pertinent agricultural system for Spain. Its multiple environmental benefits have been demonstrated for Spain's climatic conditions and soils. Farmers know about CA, but demand more and updated information. Since CA is in economic terms performing better than ConvA, the low adoption compared with American countries makes us think that it is not due to agronomic reasons. Tillage has 2000 years of history; even farmers are known as 'tillagers' in the Spanish language. Not only in Spain, but also in the rest of Europe, CAP 2020 will have a major influence on the next agricultural model. Would it be closer to CA? We still do not know, but we have certainly made some progress since 1995.

In keeping with experiences in the early years of adoption in a number of countries in South America, it would appear that development of CA in low adoption countries will have to be mainly driven by farmers. Formal research and extension is already many years behind the experiences gathered by pioneering CA farmers here and decades behind developments in other countries, particularly those outside the EU. Due to austerity measures it is also unlikely that public funding will be made available to encourage the promotion or adoption of the guiding principles of CA through agri-environmental schemes.

It has proven extremely difficult to secure funding to support CA education and awareness campaigns. Within the agriculture industry there is little appetite to facilitate the development of CA. Oil companies, who for years have been the major sponsors of ploughing championships, would hardly benefit from a 70% reduction in diesel use to establish crops. Machinery companies, other than purveyors of specific NT seeding equipment, would experience a significant loss in revenue due to falling sales of tractors, cultivation equipment and associated parts. And finally input suppliers are unlikely to invest energy or financial resources promoting a system that over time is likely to lead to reduced fertilizer, pesticide and other input sales.

Practical experience at farm level has shown that many farmers have successfully adopted minimum tillage while a few are enjoying enhanced benefits with CA/NT systems. It is these pioneering farmers who will provide the impetus for greater adoption of CA but there has to be a dramatic overhaul of technical support and extension for this to happen. Rather than a conventional top-down model of information transfer a more facilitative approach needs to be introduced. A model that puts farmers at the centre of research and extension efforts is required, one that promotes farmer to farmer dissemination of experiences while encouraging an aptitude for problem solving. As stated above it is most unlikely that the commercial sector will fund and support this approach. Government departments or agencies are unlikely to divert financial assistance towards CA promotion notwithstanding the fact that the resultant benefits of adoption are precisely in agreement with desired agricultural and environmental objectives.

CA has been farmer led and farmer driven in other parts of the world, so what makes us think it should it be any different in Europe?

Overall in Europe, much of the misleading results from short term research or incomplete implementation of CA is still discouraging adoption. While adoption seems to be more acceptable in the dryer regions of Europe, there are still challenges in the wetter and cooler parts, particularly with residue and weed management (Soane *et al.*, 2012). Those challenges require specific responses and high quality of CA implementation, including the use of good equipment and diversified crop rotations. While not being impossible, in fact those approaches are applied by successful CA farmers even in cool and moist parts of Europe, they are more difficult and challenging for the majority of farmers than plough-based agriculture.

No-tillage and CA have initially been developed as farming methods to reduce erosion. It has been proven that with CA the erosion rates can be brought to levels below the rate of soil formation, which makes the system in the long term sustainable. A review of human history and the fate of human civilizations through the millennia of human development on earth have shown that the survival of civilizations has directly been linked to the way they treated their soils. Each decline of a civilization was accompanied with significant soil erosion events, which still today can be geomorphologically proven (Montgomery, 2007). With this the adoption of NT and CA is becoming a question of the long-term survival of human civilization in the way we know it today.

Acknowledgements

This chapter has been compiled with the active contribution of the following authors each them compiling a complete country report for their respective countries, which were then incorporated into the Europe chapter. The contributing authors were as follows.

Denmark: Bente Andersen, FRDK (bea@plantekonsulenten.dk)

France: Benoit Lavier, APAD (benoit.lavier21@free.fr); Gérard Rass, APAD (gerard.rass@wanadoo.fr); François Sarreau, IAD (sarreaujfr@wanadoo.fr); Eric Schmid, CEIS (eschmidt@ceis-strat.com); Jean-Konrad Schreiber, IAD (konrad.schreiber@worldonline.fr)

Germany: PD Dr Joachim Brunotte, TI Institute of Agricultural Technology and Biosystems Engineering, Braunschweig (joachim.brunotte@vti.bund.de); Dr agr. Jana Epperlein, German Association for Conservation Tillage (GKB), Neuenhagen/

Berlin (jana.epperlein@gkb-ev.de); PD Dr Heiner Voßhenrich, TI Institute of Agricultural Technology and Biosystems Engineering, Braunschweig (hans.voss-henrich@vti.bund.de)

Ireland: Gerry Bird, Conservation Agriculture Ireland (CAIR) (info@geraghty-consulting.ie); John Geraghty, Department of Life Sciences, Waterford Institute of Technology, CAIR (info@geraghtyconsulting.ie)

Italy: Giovanni Cafiero, PhD, University of Teramo (gcafiero@unite.it); Prof Michele Pisante, University of Teramo, Italian Association for an Agronomical and Conservative Land Management (AIGACoS) (mpisante@unite.it); Fabio Stagnari, PhD, University of Teramo (fstagnari@unite.it)

Portugal: Prof Gottlieb Basch, University of Évora, Institute of Mediterranean Agricultural and Environmental Sciences; European Conservation Agriculture Federation (ECAF); Portuguese Association of Conservation Tillage (APOSOLO) (gb@uevora.pt)

Russia: Ludmilla Orlova, Russian no-till federation (priemnaya@eurotechnika.ru)

Slovakia: Rastislav Bušo, PhD, Slovak no-till club (buso@vurv.sk); Roman Hašana, PhD, Slovak no-till club (hasana@vurv.sk)

Spain: Ing. Manuel R. Gómez-Ariza, Spanish Association for Conservation Agriculture. Living Soils (AEAC.SV) (mgomez@aeac-sv.org), Prof Emilio J. González-Sánchez, University of Córdoba, Spain and AEAC.SV (egonzalez@agriculturadeconservacion.org); Francisco Márquez-García, PhD, University of Córdoba, Spain and AEAC.SV (fmarquez@agriculturadeconservacion.org); Ing. Oscar Veroz-González, AEAC.SV (overoz@aeac-sv.org)

Switzerland: Dr Bernhard Streit, Bern University of Applied Sciences, School of Agricultural, Forest and Food Sciences HAFL, Zollikofen (bernhard.streit@bfh.ch); Dr Wolfgang G. Sturny, Swiss-No-till (sturny@no-till.ch)

UK: Dr V.W.L. Victor Jordan, FRAgS, FIAgrE, SMI (vwljordan@btinternet.com); Dr Alastair R. Leake, SMI (aleake@gwct.org.uk)

The chapter authors acknowledge the good and detailed contributions received from all the contributing authors, without which this chapter would not have been possible.

References

Allton, K.E. (2006) Interactions between soil microbial communities, erodibility and tillage practices. PhD thesis, School of Applied Sciences, The National Soil Resources Institute, Cranfield University.

Alvaro-Fuentes, J. and Cantero-Martinez, C. (2010) Potential to mitigate anthropogenic CO_2 emissions by tillage reduction in dryland soils of Spain. *Spanish Journal of Agricultural Research* 8(4), 1271–1276.

Alves, J.A. (1989) Fertilidade de alguns solos e problemas dela decorrentes. Ministério da Agricultura Pescas e Alimentação Instituto Nacional de Investigação Agrária, Oeiras, Lisboa, Portugal.

Ammon, H.-U., Bohren, C. and Anken, T. (1990) Breitbandfrässaat von Mais in Wiesen und Zwischenfrüchte. *Landwirtschaft Schweiz* 3(3), 121–124 (in German).

Anken, T. (2003) Pflanzenentwicklung, Stickstoffdynamik und Nitratauswaschung gepflügter und direktgesäter Parzellen. Dissertation Thesis, ETH Zürich No. 15393. Zürich, Switzerland.

Anken, T., Heusser, J., Weisskopf, P., Zihlmann, U., Forrer, H.R., Högger, C., Scherrer, C., Mozafar, A. and Sturny, W.G. (1997) Bodenbearbeitungssysteme – Direktsaat stellt höchste Anforderungen. *FAT-Bericht 501*, Tänikon, Switzerland (in German).

Anken, T., Irla, E., Ammann, H., Heusser, J. and Scherrer, C. (1999) Bodenbearbeitung und Bestellung, Winterweizen eignet sich bestens für minimale Bestellverfahren. *FAT-Bericht 534*, Tänikon, Switzerland (in German).

Anken, T., Weisskopf, P., Zihlmann, U., Forrer, H., Jansa, J. and Perhacova, K. (2004) Long-term tillage system effects under moist cool conditions in Switzerland. *Soil and Tillage Research* 78, 171–183.

Antonelli, M., Petrini, A. and Santilocchi, R. (2001) Conservation tillage of durum wheat in Central Italy. In: Garcia-Torres, L., Benites, J. and Martnez-Vilela, A. (eds) *Conservation Agriculture, A worldwide Challenge. Proceedings of the 1st World Congress on Conservation Agriculture Volume I; Keynote Contributions*. ECAF, Madrid, Spain, pp. 51–54.

Antonelli, M., Guzzini, A. and Santilocchi, R. (2003) Grano duro tiene la resa riducendo le lavorazioni. *Terra e Vita* 39, 69–72.

APOSOLO (Associação Portuguesa de Mobilização de Conservação do Solo) (2006) *Conservar a Terra 6*. Santarém, April 2006. Available at: http://www.aposolo.pt/admin//ficheiros/PUBLICAO160.pdf (accessed 31 July 2012).

Archetti, R., Bonciarelli, F. and Farina, G. (1989) Results of trials carried out in 1981-1987 in central Italy. *Rivista di Ingegneria Agraria* 1, 43–49.

Azevedo, A.L. and Fernandes, M.L.V. (1972) Evolução do teor em matéria orgânica de barros castanho-avermelhados sujeitos à um sistema de mobilização mínima. I. Carbono orgânico. *Anais do Instituto Superior de Agronomia* XXXIII, 181–213.

Azevedo, A.L. and Fernandes, M.L.V. (1973) Evolução do teor em matéria orgânica de barros castanho-avermelhados sujeitos à um sistema de mobilização mínima. II. Azoto total. *Anais do Instituto Superior de Agronomia* XXXIV, 115–137.

Azevedo, A.L. and Fernandes, M.L.V. (1974/75) Evolução do teor em matéria orgânica de barros castanho-avermelhados sujeitos à um sistema de mobilização mínima. III. Razão C/N. *Anais do Instituto Superior de Agronomia* XXXV, 125–145.

Baraev, A.I. (1983) Spring wheat in north Kazakhstan (translated from Yarovaya pshenitsa v Severnom Kazakhstane), New Delhi: Published for the US Dept of Agriculture and the National Science Foundation, Washington, DC, Amerind Pub. Co., Springfield, Virginia. Available from NTIS.

Barros, J.F.C., Basch, G. and Carvalho, M. (2008) Effect of reduced doses of a post-emergence graminicide to control *Avena sterilis* L. and *Lolium rigidum* G. in no-till wheat under Mediterranean environment. *Crop Protection* 27(6), 1031–1037.

Basch, G. (1988) Alternativen zum traditionellen Landnutzungssystem im Alentejo/Portugal unter besonderer Berücksichtigung der Bodenbearbeitung. *Göttinger Beiträge zur Land- und Forstwirtschaft in den Tropen und Subtropen* 31, p. 188.

Basch, G. (2005) Europe: the developing continent regarding conservation agriculture CA. In: AAPRESID (ed.) *Proceedings of the XIII Congreso de AAPRESID El Futuro y los Cambios de Paradigmas Rosario Argentina*. Rosario, Argentina, pp. 341–346.

Basch, G. and Carvalho, M. (1994) Conditions and feasibility of no-tillage in Portugal. In: Tebrügge, F. and Böhrnsen, A. (eds) *Proceedings of the EC-Workshop I: Experience with the applicability of no-tillage crop production in the West-European Countries*. Wissenschaftlicher Fachverlag Giessen, Germany, pp. 93–104.

Basch, G. and Carvalho, M. (1998) Effect of soil tillage on runoff and erosion under dryland and irrigated conditions on Mediterranean soils. *Geoökodynamik* XIX(3–4), 257–268.

Basch, G., Carvalho, M., Düring, R.-A. and Martins, R. (1995) Displacement of herbicides under different tillage systems. In: Tebrügge, F. and Böhrnsen, A. (eds) *Proceedings of the EC-Workshop II: Experience With the Applicability of No-Tillage Crop Production in the West-European Countries*, Silsoe, Wissenschaftlicher Fachverlag/Giessen, Germany, pp. 25–38.

Basch, G., Carvalho, M. and Marques, F. (1997) Economical considerations on no-tillage crop production in Portugal. In: Tebrügge, F. and Böhrnsen, A. (eds) *Proceedings of the EC-Workshop IV: Experience With the Applicability of No-Tillage Crop Production in the West-European Countries*, Boigneville, Wissenschaftlicher Fachverlag Giessen, Germany, pp. 17–24.

Basch, G., Mendes, J.P., Carvalho, M., Marques, F. and Santos, M.J. (1998) Influence of tillage system on water regime in irrigated and rainfed sunflower production. In: Pereira, L.S. and Gowing, J.W. (eds) *Water and the Environment - Innovation Issues in Irrigation and Drainage*. E and FN Spon, London, pp. 381–389.

Basch, G., Carvalho, M., Barros, J.F.C. and Calado, J.M.G. (2010) The importance of crop residue management for carbon sequestration under no-till. In: ECAF (ed.) *Proceedings of the European Congress on Conservation Agriculture: Towards Agro-Environmental Climate and Energetic Sustainability*, Madrid, Spain, pp. 241–248.

Basch, G., Kassam, A., Friedrich, T., Santos, F.L., Gubiani, P.I., Calegari, A., Reichert, J.M. and Dos Santos, D.R. (2012a) Sustainable soil water management systems. In: Lal, R. and Stewart, B.A. (eds) *Soil Water and Agronomic Productivity Advances in Soil Science*. CRC Press, pp. 229–288.

Basch, G., Calado, J., Barros, J. and Carvalho, M. (2012b) Impact of soil tillage and land use on soil organic carbon decline under Mediterranean conditions. In: ISTRO (ed.) *Proceedings of the 19th ISTRO Conference*, Montevideo/Uruguay. Available at: http://iworx5.webxtra.net/~istroorg/p_publications_frame.htm (accessed 1 July 2013).

Basso, B., Cammarano, D., Troccoli, A., Chen, D. and Ritchie, J.T. (2010) Long-term wheat response to nitrogen in a rainfed Mediterranean environment: Field data and simulation analysis. *European Journal of Agronomy* 33(2), 132–138.

Basso, B., Sartori, L., Bertocco, M., Cammarano, D., Martin, E.E. and Grace, R.P. (2011) Economic and environmental evaluation of site-specific tillage in a maize crop in NE Italy. *European Journal of Agronomy* 35, 83–92.

Basso, F., Pisante, M. and Basso, B. (1996) Influenza dei residui colturali e delle lavorazioni sull'umidità del terreno sull'accrescimento e produzione del favino da seme e frumento duro. *Rivista di Agronomia* 30, 0. 3, 212–221.

Bäumer, K. (1979) First experiences with direct drilling in Germany. *Netherland Journal of Agricultural Science*. Papers on zero-tillage 18(4), 283–292.

Bhogal, A., Chambers, B., Whitmore, A.P. and Powlson, D.S. (2008) The potential to increase carbon storage in agricultural soils. Defra Report, London.

Blum, A., Chervet, A., Forrer, H.R., Vogelgsang, S. and Schmid, F. (2011) Fusarien in Getreide. *Merkblatt 2.5.25*. Agridea, Lindau, Germany (in German and French).

Boisgontier, D., Bartholomy, P. and Lescar, L. (1994) Feasibility of minimum tillage practices in France. In: Tebrügge, F. and Böhrnsen, A. (eds) *Proceedings of the EC-Workshop-I - Experience with the Applicability of No-Tillage Crop Production in the West-European Countries*. Giessen, Wissenschaftlicher Fachverlag Giessen, pp. 81–91.

Bonari, E., Mazzoncini, M., Ginanni, M. and Menini, S. (1996) Influenza delle tecniche di lavorazione del terreno sull'erosione idrica dei terreni argillosi della collina Toscana. *Rivista di Agronomia* 2–3, 277–287.

Bonciarelli, F. (1985) Vecchi e nuovi sistemi di lavorazione del terreno. *Macchine e Motori Agricoli* 8, 41–48.

Bonciarelli, F., Archetti, R., Farina, G. and Battistelli, A. (1986) Effetto di nuovi sistemi di lavorazione su alcune proprietà chimiche e meccaniche del terreno. *Rivista di Agronomia* 2–3, 172–177.

Bopp, M., Carrel, K., Bertschi, M. and Rüsch, A. (2011) Strickhof Versuchsbericht 2011 (in German). Unpublished report, Strickhof, Switzerland.

Borin, M., Menini, C. and Sartori, L. (1997a) Effects of tillage systems on energy and carbon balance in North-Eastern Italy. *Soil and Tillage Research* 40, 209–226.

Borin, M., Sartori, L., Guipponi, C., Mazzoncini, M., Düring, R.-A. and Basch, G. (1997b) *Effects of Tillage Systems on Herbicide Dissipation - an experimental approach at field scale*. Unipress Padova, Italy.

Brown, L., Donaldson, G.V., Jordan, V.W.L. and Thornes, J.B. (1996) Effects and interactions of rotation, cultivation and agrochemical input levels on soil erosion and nutrient emissions. *Aspects of Applied Biology 47, Rotations and Cropping Systems*, 409–412.

Brunotte, J. (2002) Recommendations for acting out good agricultural practice: Reducing soil erosion, promoting soil life. *Agricultural research Völkenrode* 256, 79–86.

Calado, J.M.G., Basch, G. and Carvalho, M. (2010) Weed management in no-till winter wheat (*Triticum aestivum* L.). *Crop Protection* 29(1), 1–6.

Campiglia, E. (1999) Colture di copertura utilizzate in agroecosistemi mediterranei. Nota I: modificazioni dell'ambiente colturale. *Rivista di Agronomia* 33, 90–103.

Cantero, C. and Ojeda, L. (2004) Efectos sobre la población de lombrices de las técnicas de laboreo del suelo en zonas de secano semi-árido, Agricultura. *Revista Agropecuaria* 73, 866, 724–728.

Carvalho, M. (2003) Contribuição da sementeira directa para o aumento da sustentabilidade dos sistemas de culturas arvenses. In: Barros, V.C. and Ramos, J.B. (eds) *Agricultura Sustentável - Ciclo de Seminários*. INIAP-EAN, Oeiras, Portugal, pp. 59–73.

Carvalho, M. and Basch, G. (1995) Effects of traditional and no-tillage on physical and chemical properties of a Vertisol. In: Tebrügge, F. and Böhrnsen, A. (eds) *Proceedings of the EC-Workshop II: Experience with the Applicability of No-Tillage Crop Production in the West-European Countries*. Silsoe, Wissenschaftlicher Fachverlag Giessen, Germany, pp. 17–23.

Carvalho, M., Basch, G., Barros, J., Calado, J., Freixial, R., Santos, F. and Brandão, M. (2010) Strategies to improve soil organic matter under Mediterranean conditions and its consequences on the wheat response to nitrogen fertilization. In: ECAF (eds) *Proceedings of the European Congress on Conservation Agriculture: Towards Agro-Environmental Climate and Energetic Sustainability*, Madrid, Spain, pp. 303–308.

Colecchia, S., Pisante, M., Gallo, A., Farina, R., Russo, M., Cattivelli, L. and Troccoli, A. (2009) L'erosione del suolo si combatte anche con le giuste lavorazioni. *L'Informatore Agrario* 39, 52–55.

Corsi, S., Friedrich, T., Kassam, A., Pisante, M. and Sà De Moraes, J. (2012) Soil organic carbon accumulation and carbon budget in conservation agriculture: a review of evidence. *FAO Integrated Crop Management* Vol. 16, FAO, Rome, Italy.

De Vita, P., Di Paolo, E., Fecondo, G., Di Fonzo, N. and Pisante, M. (2007) Effect of no-tillage and conventional tillage systems on durum wheat yield grain quality and soil moisture content in southern Italy. *Soil and Tillage Research* 92, 69–78.

Derpsch, R. and Friedrich, T. (2009) Global overview of conservation agriculture adoption. In: ICAR (ed.) *Proceedings of the 4th World Congress on Conservation Agriculture: Innovations for Improving Efficiency Equity and Environment.* ICAR, New Delhi, India, pp. 429–438.

Dobrovol'ski, G.V. (1983) The Role of V.V. Dokuchaev's 'Russian Chernozem' in the Formation and Development of Soil Science. *Moscow State University Soil Science Bulletin* 38, 3–8.

EC (European Commission) (2012) The implementation of the soil thematic strategy and ongoing activities; Report from the Commission to the European Parliament, the Council, the European Economic and Social Committee and the Committee of the Regions. COM (2012) 46 final, European Commission, Brussels, Belgium.

ECAF (2012) *Making sustainable agriculture real in CAP 2020, the role of conservation agriculture.* ECAF, Brussels, Belgium.

Emmerling, C., Fortune, T., Kennedy, T., Mitchell, B. and Dunne, B. (2003) Reduced cultivations – agronomic and environmental aspects. In: Teagasc (ed.) *Proceedings of the National Tillage Conference*, Teagasc, Carlow, Ireland, pp. 70–82.

Espejo-Pérez, A., Márquez, F. and Rodríguez-Lizana, A. (2006) Aumentos de la biodiversidad en suelos de olivar. *Vida Rural* 236, 46–48.

FAO (2012a) What is CA? Conservation Agriculture Website of FAO. Available at: http://www.fao.org/ag/ca/1a.html (accessed December 2012).

FAO (2012b) CA Adoption Worldwide, FAO AQUASTAT database. Available at: http://www.fao.org/ag/ca/6c.html (accessed December 2012).

Fernández-Quintanilla, C. (1997) Historia y evolución de los sistemas de laboreo. El laboreo de conservación. In: García Torres, L. and González Fernández, P. (eds) *Agricultura de Conservación: Fundamentos Agronómicos Medioambientales y Económicos.* Asociación Española Laboreo de Conservación/Suelos Vivos, Córdoba, Spain, pp. 1–12.

Forristal, D. and Murphy, K. (2009) Can we reduce costs and increase profits with min till? In: Teagasc (ed.) *Proceedings of the National Tillage Conference 2009*, Teagasc, Carlow, Ireland, pp. 48–67.

Fortune, T., Kennedy, T., Mitchell, B. and Dunne, B. (2003) Reduced cultivations - agronomic and environmental aspects. In: Teagasc (ed.) *Proceedings of the National Tillage Conference 2003*, Teagasc, Carlow, Ireland, pp. 70–82.

Fortune, T., Kennedy, T., Mitchell, B., Dunne, B., Murphy, K., Connery, J.J. and Grace, J. (2005) Reduced cultivations - update from Oak Park experiments. In: Teagasc (ed.) *Proceedings of the National Tillage Conference 2005*, Teagasc, Carlow, Ireland, pp. 18–34.

Freibauer, A., Rounsevell, M., Smith, P. and Verhagen, A. (2004) Carbon sequestration in the agricultural soils of Europe. *Geoderma* 122, 1–23.

Freixial, R. and Carvalho, M. (2010) Aspectos prácticos fundamentales en la implantación de la Agricultura de Conservacion/Siembra Directa en el sur de Portugal. In: ECAF (eds) *Proceedings of the European Congress on Conservation Agriculture: Towards Agro-Environmental Climate and Energetic Sustainability*, Madrid, Spain, pp. 361–369.

Friedrich, T. and Kassam, A.H. (2009) Adoption of conservation agriculture technologies: constraints and opportunities. In: ICAR (ed.) *Proceedings of the 4th World Congress on Conservation Agriculture: Innovations for Improving Efficiency Equity and Environment.* ICAR, New Delhi, India, pp. 257–264.

Friedrich, T., Kassam, A.H. and Taher, F. (2009) Adoption of conservation agriculture and the role of policy and institutional support. In: Suleimenov, M. *et al.* (eds) *Proceedings of the International Consultation Conference on No-till with Soil Cover and Crop Rotation: a Basis for Policy Support to Conservation Agriculture for Sustainable Production Intensification.* Astana-Shortandy, Kazakhstan; Shortandy 2009, ISBN 9965-407-55-X (Russian with English).

Fry, P. (2009) Von Bauern für Bauern, Erfolgsgeschichten für eine schonende Bodennutzung. DVD, Agridea, Lindau, Switzerland.

García-Ruiz, J.M. (2008) The effects of land uses on soil erosion in Spain: a review. *Catena* 81, 1–11.

Gardi, C., Tomaselli, M., Parisi V., Petraglia, A. and Santini, C. (2002) Soil quality indicators and biodiversity in northern Italian permanent grasslands. *European Journal of Soil Biology* 38, 103–110.

Geraghty, J. (2008) Sustainable crop production and climate change - reducing emissions in the Irish arable sector. In: Institute of International and European Affairs (ed.) *Proceedings of the Conference The Greening of Irish Agriculture*, Institute of International and European Affairs, Dublin Castle, Ireland, pp. 20–21.

Giráldez, J.V. and González, P. (1994) No-tillage in clay soils under Mediterranean climate: Physical aspects. In: Tebrügge, F. and Böhrnsen, A. (eds) *Proceedings of the EC-Workshop-I – Experience with the Applicability of No-Tillage Crop Production in the West-European Countries*. Wissenschaftlicher Fachverlag Giessen, Germany, pp. 111–117.

Glen, D.M., Milsom, N.F. and Wiltshire, C.W. (1990) Effect of seed depth on slug damage to winter wheat. *Annals of Applied Biology* 117, 693–701.

González, P., Ordónez, R., Perea, F. and Giráldez, J.V. (2010) Estudio comparativo de las cosechas recogidas a lo largo de 26 campañas en un ensayo con distintos manejos del suelo. In: ECAF (eds) *Proceedings of the European Congress on Conservation Agriculture. Towards Agro-Environmental Climatic and Energetic Sustainability*, Madrid, Spain, pp. 433–440.

González-Sánchez, E., Pérez-García, J.J., Gómez-Ariza, M., Márquez-García, F. and Veroz-González, O. (2010) Sistemas agrarios sostenibles económicamente: el caso de la siembra directa. *Vida Rural* 312, 24–27.

González-Sánchez, E.J., Ordóñez-Fernández, R., Carbonell-Bojollo, R., Veroz-González, O. and Gil-Ribes, J.A. (2012) Meta-analysis on atmospheric carbon capture in Spain through the use of conservation agricultura CA. *Soil and Tillage Research* 122, 52–60.

Hackett, R., Spink, J., Mitchell, B. and Creamer, C. (2010) Impact of management practices on soil organic carbon levels under Irish conditions. In: Teagasc (ed.) *Proceedings of the National Tillage Conference 2006*, Teagasc, Carlow, Ireland, pp. 77–91.

Heaney, C. (2012) A comparison of winter oilseed rape sown using different establishment methods. BSc (Hons) thesis in Land Management. Department of Life Sciences, Waterford Institute of Technology, Waterford, Ireland.

Hernanz, J.L., Arrúe, J.L., Cantero, C., Sombrero, A., Giráldez, J.V., González, P., Gil Ribes, J.A., San Martín, C., Navarrete, L., López-Fando, C., Moreno, F. and Sánchez-Girón, V. (1996) Creación de una red temática sobre laboreo de conservación. Plan Nacional I+D, Programa Nacional de Ciencias Agrarias, AGF96-1613-E, Spain, 1996–1997.

Hiltbrunner, J., Jeanneret, P., Liedgens, M., Stamp, P. and Streit, B. (2007) Response of weed communities to legume living mulches in winter wheat. *Journal of Agronomy and Crop Science* 193, 93–102.

Holland, J.M. (2004) The environmental consequences of adopting Conservation tillage in Europe; reviewing the evidence. *Agriculture, Ecosystem and Environment* 103, 1–25.

Hooker, K., Coxon, C., Hackett, R., Kirwan, L., O'Keeffe, E. and Richards, K. (2006) Evaluation of cover crop and reduced cultivation for reducing nitrate leaching in Ireland. *Journal of Environmental Quality* 37, 138–145.

Huggins, D.R. and Reganold, J.P. (2008) No-till: the quiet revolution. *Scientific American* July, 70–77.

ICONA (1991) *Plan Nacional de lucha contra la erosión*. Ministerio de Agricultura Pesca y Alimentación. Instituto Nacional para la Conservación de la Naturaleza, Madrid, Spain.

Imaz, M.J., Virto, I., Bescansa, P., Enrique, A., Fernandez-Ugalde, O. and Karlen, D.L. (2010) Soil quality indicator response to tillage and residue management on semi-arid Mediterranean cropland. *Soil and Tillage Research* 107, 17–25.

INE (Instituto Nacional de Estatística) (2011) *Recenseamento Agrícola 2009 - Análise dos principiais resultados*. Lisbon, Portugal, 185 pp.

Jat, R.A., Wani, S.P. and Sahrawat, K.L. (2012) Conservation agriculture in the semi-arid tropics: prospects and problems. In: Sparks, D.L. (ed.) *Advances in Agronomy*, 117, 191–273.

Jordan, V.W.L. and Hutcheon, J.A. (1999) Disease management in less-intensive, integrated wheat systems. In: Lucas, J.A., Bowyer, P. and Anderson, H.M. (eds) *Septoria on Cereals: a Study of Pathosystems*. CAB International, Wallingford, UK, pp. 263–273.

Jordan, V.W.L., Hutcheon, J.A., Donaldson, G.V. and Farmer, D.P. (1997) Research into development of integrated farming systems for less-intensive arable crop production: experimental progress (1989-1994) and commercial implementation. *Agriculture Ecosystems and Environment* 65, 141–148.

Juste, F., Sanchez-Giron, V. and Hernanz, J.L. (1981) Estudio comparativo de la siembra directa con el cultivo tradicional de los cereales. In: Asociación Nacional de Ingenieros Agrónomos (eds) *Proceedings 13 Conferencia Internacional de Mecanización Agraria,* FIMA 81, D.L. 1981, Zaragoza, Spain, pp. 133–145.

Karabayev, M., Satybaldin, A., Benites, J.R., Friedrich, T., Pala, M. and Payne, T. (eds) (2000) *Conservation Tillage: a viable option for sustainable agriculture in Eurasia*; Proceedings of an international workshop, Shortandy – Astana, Republic of Kazakhstan, 19–24 September 1999. FAO/CIMMYT/ICARDA, Almaty/Aleppo, 2000.

Kassam, A.H., Friedrich, T., Shaxson, F. and Pretty, J. (2009) The spread of conservation agriculture: justification sustainability and uptake. *International Journal of Agricultural Sustainability* 7(4), 292–320.

Krück, S., Nitzsche, O., Schmidt, W. and Uhlig, U. (2001) Influence of tilling on soil life and soil structure. *Notice from the German Society for Soil Science* 96(2), 747–748.

KTBL (Kuratorium für Technik und Bauwesen in der Landwirtschaft) (1993) Definition und Einordnung von Verfahren der Bodenbearbeitung und Bestellung. *Landtechnik* 48(1/2), 50–53.

Lane, M., Kibblewhite, M. and Montanarella, L. (2006) *Conservation Agriculture in Europe - An Approach to Sustainable Crop Production by Protecting Soil and Water?* SOWAP, Rome, Italy.

Leake, A.R. (1995) Focus on farming practice – an integrated approach to solving crop protection problems in conventional and organic agriculture. In: McKinlay, R.G. and Atkinson, D. (eds) *BCPC Symposium Proceedings No.63 Integrated Crop Protection: Towards Sustainability?* Brighton, UK.

Leake, A.R. (2001) Integrated pest management for conservation agriculture. In: Garcia-Torres, L. Benites, J. and Martnez-Vilela, A. (eds) *Proceedings of the 1st World Congress on Conservation Agriculture on Conservation Agriculture – A Worldwide Challenge*. Vol. 1, *Keynote Contributions*. ECAF, Cordoba, Spain, pp. 534

Lezovic, G. (2011) Where we are now? *Landwirtschaft ohne Pflug* 09/2011.

LIFE (1996) Demonstration actions and technology transfer for soil erosion reduction, LIFE96 ENV/E/000338. Available at: http://ec.europa.eu/environment/life/project/Projects/index.cfm?fuseaction=search.dspPage&n_proj_id=1120 (accessed December 2012).

LIFE (1999) Co-ordination of activities and technology transfer actions to reduce water contamination, erosion and emissions of CO_2 from agricultural land in Europe (1999-2003), LIFE 99ENV/E/308. Available at: http://www.ecaf.org/index.php?option=com_content&task=view&id=48&Itemid=47 (accessed December 2012).

LKV (Verordnung über die Erhaltung der Lebensgrundlagen und der Kulturlandschaft) (1997) *Verordnung über die Erhaltung der Lebensgrundlagen und der Kulturlandschaft* vom 5. November 1997. BSG 910.112 (in German), Bern, Switzerland.

Longhurst, K. (2010) Investigating the conservation implications of using zero-tillage in the agricultural systems in the UK. MSc thesis, University College, London.

López, M.V. and Arrúe, J.L. (2005) Soil tillage and wind erosion in fallow lands of central Aragon (Spain): an overview. In: Faz-Cano, A., Ortiz, R. and Mermut, A.R. (eds) Sustainable Use and Management of Soils – Arid and Semiarid Regions. *Advances in GeoEcology* 36, 93–102.

MAFF (1998) *Integrated Farming – Agricultural Research into Practice; a Report from the Integrated Arable Crop Production Alliance*. PB 3618, Crown Print, London.

MAGRAMA (Ministerio de Agricultura Alimentación y Medio Ambiente) (2012a) Programas de desarrollo rural 2000-2006. Available at: http://www.magrama.gob.es/es/desarrollo-rural/temas/programas-ue/periodo-de-programacion-2000-2006/default.aspx (accessed July 2012).

MAGRAMA (Ministerio de Agricultura Alimentación y Medio Ambiente) (2012b) Análisis de las técnicas de mantenimiento del suelo y métodos de siembra en España 2011. Available at: http://www.magrama.gob.es/es/estadistica/temas/estadisticas-agrarias/CUBIERTAS2011rev1mama_tcm7-188433.pdf (accessed August 2012).

Marques, F. (2009) Avaliação técnica e económica de sistemas de mobilização do solo. PhD thesis, Universidade de Évora, Portugal, 309 pp.

Marques, F. and Basch, G. (2002) Comparação da viabilidade económica de quatro sistemas de mobilização do solo. In: Basch, G. and Teixeira, F. (eds) *Proceedings of 1st Congresso Nacional de Mobilização de Conservação do Solo*, APOSOLO, Évora, Portugal, pp. 283–298.

Márquez, F., Giráldez, J.V., Repullo, M., Ordóñez, R., Espejo, A.J. and Rodríguez, A. (2008) Eficiencia de las cubiertas vegetales como método de conservación de suelo y agua en olivar. In: Instituto Geológico y Minero de España (eds) *Simposio del Agua en Andalucía*. IGME, Madrid, Spain, pp. 631–641.

Márquez, F., González-Sánchez, E.J., Agüera, J., Blanco, G. and Gil-Ribes, J.A. (2011) Conservation agriculture and precision agriculture as a method to reduce energy consumption in agricultural systems. In: CIGR (eds) *Proceedings of the 11th International Congress on Agricultural Mechanization and Energy in Agriculture*, Istanbul, Turkey, pp. 277–282.

Marzaioli, R., D'ascoli, R., De Pascale, R.A. and Rutigliano, F.A. (2010) Soil quality in a Mediterranean area of southern Italy as related to different land use types. *Applied Soil Ecology* 44, 205–212.

Masciandaro, G., Ceccanti, B. and Gallardo Lancho, J.F. (1998) Organic matter properties in cultivated versus set-aside arable soils. *Agriculture Ecosystems and Environment* 67, 267–274.

Mazzoncini, M., Crocè, L., Bàrberi, P., Menini, S. and Bonari, E. (2001) Crop management systems to conserve soil fertility after long-term set-aside in southern Italy. In: Rees, R.M., Ball, B.C., Campbell, C.D.

and Watson, C.A. (eds) *Sustainable Management of Soil Organic Matter*. CAB International, Wallingford, UK, pp. 163–172.

Mazzoncini, M., Di Bene, C., Coli, A. and Bonari, E. (2004) Gestione degli Agroecosistemi e Mitigazione dell'Effetto Serra. *L'Informatore Agrario* 16, 37–41.

Mazzoncini, M., Sapkota, T.B., Barberi, P., Antichi, D. and Risaliti, R. (2011) Long-term effect of tillage nitrogen fertilization and cover crops on soil organic carbon and total nitrogen content. *Soil and Tillage Research* 114, 165–174.

McConkey, B., Chang Liang, B., Padbury, G. and Lindwall, W. (2000) Carbon sequestration and direct seeding. In: Saskatchewan Soil Conservation Association (eds) Proceedings of Direct Seeding 'Sustainable Farming in the new Millennium' 12th Annual Meeting of the Saskatchewan Soil Conservation Association, SSCA, Saskatoon, Canada. Available at: http://www.ssca.ca/conference/2000proceedings/McConkey.html (accessed December 2012).

Melero, S., Vanderlinden, K., Ruiza, J.C. and Madejon, E. (2008) Long-term effect on soil biochemical status of a Vertisol under conservation tillage system in semi-arid Mediterranean conditions. *European Journal of Soil Biology* 44, 437–442.

Mishustin, Y.N. (1955) Soil microbiology and its current problems, translation of '*Pochvennaya mikrobiologiya i yeye ocherednyye zadachi*.' Trudy Instituta Mikrobiologii, *Akademiya Nauk SSSR* 1(1), 155–175.

Montgomery, D.R. (2007) *Dirt, the Erosion of Civilizations*. University of California Press, Berkeley, Los Angeles and London, 285 pp.

Mota, M.M., Carvalho, M., Basch, G., Mcgawley, E.C. and Murcho, D.F. (1997) Soil tillage and plant effects on nematode communities in southern Portugal. *Journal of Nematology* 29(4), Abstracts 595.

Ogilvy, S.E. (2000) *LINK Integrated Farming Systems, Final Project Report*, January 2000. LINK CSA2163, UK.

Ordóñez, R., González, P., Perea, F., Llanos, I. and Giráldez, J.V. (2001) The protective role of stubble cover in dry farming conservation agriculture in south-western Spain. In: Garcia-Torres, L., Benites, J. and Martinez-Vilela, A. (eds) *Proceedings of the 1st World Congress on Conservation Agriculture*, Vol. II, Cordoba, Spain, pp. 435–439.

Ordóñez, R., González, P., Giráldez, J.V. and Perea, F. (2007) Soil properties and crop yields after 21 years of direct drilling trials in southern Spain. *Soil and Tillage Research* 94, 47–54.

Ordóñez, R., Carbonell, R., Repullo, M.A., Alcántara, C. and Rodríguez-Lizana, A. (2009) Nutrients released in the decomposition of the residue of different types of plan covers in olive groves. In: CIEC (eds) *Proceedings of the 18th Symposium of the International Scientific Centre of Fertilizers*, Rome, Italy, p. 29.

Ordóñez-Fernández, R., González-Fernández, P. and Pastor Muñoz-Cobo, M. (2007) Cubiertas inertes: los restos de poda como protección y mejora de las propiedades del suelo. In: Rodríguez-Lizana, A., Ordóñez-Fernández, R. and Gil-Ribes, J. (eds) *Cubiertas Vegetales en Olivar*. Consejería de Agricultura y Pesca, Junta de Andalucía, Spain, pp. 159–168.

Pagliai, M., Pezzarossa, B., Mazzoncini, M. and Bonari, E. (1989) Effect of tillage on porosity and microstructure of a loam soil. *Soil Technology* 2, 345–358.

Pagliai, M., Raglione, M., Panini, T., Maletta, M. and La Marca, M. (1995) The structure of two alluvial soils in Italy after 10 years of conventional and minimum tillage. *Soil and Tillage Research* 34, 209–223.

Perea, F. and Gil-Ribes, J.A. (2006) Consumo de Gasoil agrícola y tiempos de trabajo de la maquinaria agrícola. *Agricultura de Conservación* 3, 23–26.

Piovanelli, C., Gamba, C., Brandi, G., Simoncini, S. and Batistoni, E. (2006) Tillage choices affect biochemical properties in the soil profile. *Soil and Tillage Research* 90, 84–92.

Pisante, M. (2007) *Agricoltura Blu*. La via italiana dell'agricoltura conservativa. Principi tecnologie e metodi per una produzione sostenibile. IlSole24Ore-Edagricole, Bologna, Italia, p. 317, ISBN-978-88-506-5253-2.

Pisante, M. and Basso, F. (2000) Influence of tillage systems on yield and quality of durum wheat in southern Italy. In: Royo, C., Nachit, M.M., Di Fonzo, N. and Araus, J.L. (eds) *Durum Wheat Improvement in the Mediterranean Region: New Challenges*. Zaragoza, Spain, pp. 549-554, ISBN/ISSN: 2-85352-212-1.

Pisante, M., Fecondo, G. and D'Eercole, M. (2001) Conservation agriculture on *durum* wheat through no-tillage. In: Garcia-Torres, L., Benites, J. and Martinez-Vilela, A. (eds) *Proceedings of the 1st World Congress on Conservation Agriculture*, Vol. II, Cordoba, Spain, pp. 623–626, ISBN/ISSN: 84-932237-2-7.

Rasmussen, K.J. (1988) Pløjning, direkte såning og reduceret jordbearbejdning til korn (Ploughing, direct seeding and harrowing before seeding in cereals). *Tidsskrift for Planteavl* 92, 233–248.

Rieger, S., Richner, W., Streit, B., Frossard, E. and Liedgens, M. (2008) Growth, yield, and yield components of winter wheat and the effects of tillage intensity, preceding crops, and N fertilization. *European Journal of Agronomy* 28, 405–411.

Rieger, S.B. (2001) Impacts of tillage systems and crop rotation on crop development, yield, and nitrogen efficiency. Thesis, ETH Zürich No. 14124, Zürich, Switzerland.

Rodríguez-Lizana, A., Carbonell, R., González, P. and Ordóñez, R. (2010) N P and K released by the field decomposition of residues of a pea-wheat-sunflower rotation. *Nutrient Cycling in Agroecosystems* 87(2), 199–208.

Russell, T. (2011) A study of earthworm populations in no-tillage and plough based systems. BSc (Hons) thesis, Department of Life Sciences, Waterford Institute of Technology, Waterford, Ireland.

Schaller, B., Nemecek, T., Streit, B., Zihlmann, U., Chervet, A. and Sturny, W.G. (2006) Vergleichsökobilanz bei Direktsaat und Pflug. *Agrarforschung* 13(11/12), 482–487.

Schneider, F., Ledermann, T., Fry, P. and Rist, S. (2010) Soil Conservation in Swiss agriculture – approaching abstract and symbolic meanings in farmers' life-worlds. *Land Use Policy* 27(1), 332–339.

Schwarz, R., Chervet, A., Hofer, P., Sturny, W.G. and Zuber, M. (2007) Le canton de Berne favorise les techniques culturales qui préservent les ressources naturelles. *Revue Suisse Agricole* 39(3), 117–122 (in French).

Seddaiu, G., Iezzi, G. and Roggero, P.P. (2003) Riduzione delle lavorazioni e della concimazione azotata nell'avvicendamento biennale frumento duro-girasole nella collina marchigiana. In: S.I.A. (eds) *Atti XXXV Convegno della S.I.A. 'Obiettivo qualità integrale: il ruolo della ricerca agronomica'*, Napoli, Italy, pp. 23–24.

Sims, B.G. and Ellis-Jones, J. (2011) Conservation agriculture for sustainable cropping and environmental protection. *Agriculture for Development (UK)* 14, 17–20. Available at: https://docs.google.com/file/d/0BwyIPGne8KZ-S2Jna0FYM3NhVWs/edit (accessed December 2012).

SMI/Defra (2001) *A Guide to Managing Crop Establishment*. Defra, London.

Smith, P. (2004) Carbon sequestration in croplands: the potential in Europe and the global context. *European Journal of Agronomy* 20, 229–236.

Soane, B.D., Ball, B.C., Arvidsson, J., Basch, G., Moreno, F. and Roger-Estrade, J. (2012) No-till in northern western and south-western Europe: a review of problems and opportunities for crop production and the environment. *Soil and Tillage Research* 118, 66–87.

SoCo (2009) Final report on the project 'Sustainable Agriculture and Soil Conservation (SoCo)'; European Commission Directorate-General for Agriculture and Rural Development Luxemburg 2009; EU23820EN, Luxemburg.

Soils Report (2009) Bodenbericht 2009, VOL, Volkswirtschaftsdirektion des Kantons Bern, Bern, Switzerland, 127 pp. (in German, English summary).

Stadler, M., Dorn, B., Zihlmann, U., Scherrer, C., Jossi, W. and Streit, B. (2009) Verschiedene Gründüngerpflanzen - Anbaueignung und Unkrautunterdrückung im Direktsaatsystem vor Winterweizen. In: Mayer, J., Alfoldi, T., Leiber, F., Dubois, D., Fried, P., Heckendorn, F., Hillmann, E., Klocke, P., Lüscher, A. and Riedel, S. (eds) *Proceedings of the 10th Scientific Conference on Organic Agriculture*, vol.1, 11–13 February 2009, Zürich, Switzerland.

Stagnari, F., Ramazzotti, S. and Pisante, M. (2009) Conservation agriculture: a different approach for crop production through sustainable soil and water management: a review. In: Lichtfouse, E. (ed.) *Agronomy for Sustainable Development*. Organic Farming Pest Control and Remediation of Soil Pollutants Sustainable Agriculture Reviews 1, Springer Science and Business Media B.V., pp. 55–83. DOI 10.1007/978-1-4020-9654-9.

Streit, B., Sturny, W.G. and Lauper, H. (2005) Maisdirektsaat: Fünf Maschinen im Vergleich. *Schweizer Landtechnik* 5/2005, 28–31 (in German and French).

Sturny, W.G. and Meerstetter, A. (1990) Mulchsaat von Mais in Gründüngungsbestände. *FAT-Bericht 376,* Tänikon, Switzerland (in German).

Sturny, W.G., Chervet, A., Maurer-Troxler, C., Ramseier, L., Müller, M., Schafflützel, R., Richner, W., Streit, B., Weisskopf, P. and Zihlmann, U. (2007) Comparison of no-tillage and conventional plough tillage system – a synthesis. *Agrarforschung* 14(8), 350–357 (in German and French).

Teagasc (2012) Tillage Sector Development Plan 2012. A report compiled by the Tillage Crop Stakeholder Consultative Group, November 2012, Dublin, Ireland.

Tebrügge, F. and Böhrnsen, A. (1997) Crop yields and economic aspects of no-tillage compared to plough tillage: Results of long-term soil tillage field experiments in Germany. In: Tebrügge, F. and Böhrnsen, A. (eds) *Proceedings of the EC-Workshop-IV – Boigneville; Experience with the Applicability of No-Tillage Crop Production In the West-European Countries*. Wissenschaftlicher Fachverlag, Giessen, Germany, pp. 25–43.

Tebrügge, F., Borin, M., Basch, G. and Mazzoncini, M. (1997) Effects of tillage system on physical chemical and biological soil characteristics. In: Borin, M., Sartori, L., Giupponi, C., Mazzonicini, M., Düring, R.-A. and Basch, G. (eds) *Effects of Tillage Systems on Herbicide Dissipation - an experimental approach at field scale*. Unipress, Padova, Italy, pp. 41–47.

Teixeira, F., Basch, G. and Carvalho, M.J. (2000) Tillage effects on splash detachment overland flow and interril erosion. In: Riley, T.W. and Desbiolles, J.M.A. (eds) *Proceedings of the 4th International Conference on Soil Dynamics*. Adelaide, School of Advanced Manufacturing & Mechanical Engineering, University of South Australia, Mawson Lakes Campus, Mawson Lakes, South Australia, pp. 307–314.

Vaderstad/SMI (2004) Target on Establishment. Guideline-brochure. Available at: http://www.ecaf.org/docs/smi/Target%20on%20Establishment.pdf (accessed December 2012).

Vaderstad/SMI (2006) *Visual Soil Assessment*. Guideline brochure, Mollington, Chester, UK.

Van-Camp, L., Bujarrabal, B., Gentile, A.-R., Jones, R.J.A., Montanarella, L., Olazabal, C. and Selvaradjou, S.-K. (2004) Reports of the Technical Working Groups Established under the Thematic Strategy for Soil Protection. EUR 21319 EN/1, Office for Official Publications of the European Communities, Luxembourg, 872 pp.

Vanwalleghem, T., Infante Amate, J., González de Molina, J., Soto Fernández, D. and Gómez, J.A. (2011) Quantifying the effect of historical soil management on soil erosion rates in Mediterranean olive orchards. *Agriculture Ecosystems and Environment* 142, 341–351.

Vogelgsang, S., Hecker, A., Musa-Steenblock, T., Dorn, B. and Forrer, H.-R. (2011) On-farm experiments over five years in a grain maize - winter wheat rotation: effect of maize residue treatments on *Fusarium graminearum* infection and deoxynivalenol contamination in wheat. *Mycotoxin Research* 27, 81–96.

Vorontsova, T. (2007) Betriebswirtschaftliche Analyse des Einsatzes moderner Agrartechnik in der Körnerfrüchteproduktion in Russland. Dissertation zur Erlangung des Grades eines Doktors der Agrarwissenschaften vorgelegt der Fakultät Agrarwissenschaften Aus dem Institut für Landwirtschaftliche Betriebslehre Universität Hohenheim. Available at: http://opus.ub.uni-hohenheim.de/volltexte/2007/211/pdf/dissertation.pdf (accessed December 2012).

Voßhenrich, H.H., Korte, K., Ortmeier, B. and Brunotte, J. (2005) Survey on the status of tilling without the use of the plough for winter oilseed rape. *UFOP writings* 26, 2005.

Xavier, M.A., Nogueira, A., Bras, A. and Basch, G. (2005) Estudo das pragas da cultura do milho forrageiro em função do sistema de mobilização do solo. In: Escola Superior Agrária de Coimbra (eds) *A Produção Integrada e a Qualidade e Segurança Alimentar, Proceedings of 'VII Encontro Nacional de Protecção Integrada'*, Vol. I. Edições IPC, Coimbra, Portugal, pp. 327–334.

Annexes

Institutions working on Conservation Agriculture in Europe

Denmark

Aarhus University, research centre Foulum and research centre Flakkebjerg

France

IAD (Institut de l'Agriculture Durable)
University of Rennes (Daniel Cluzeau, research on earthworms)
Extension services of GDA (Groupes de Developpement Agricoles): some individual technicians experiment on how to improve CA in close partnership with farmers

Germany

Limited research on CA by state research institutes, for example in Saxony

Ireland

Teagasc Research Centres and Education Colleges

Italy

Università Politecnica delle Marche, Dipartimento di Scienze Ambientali e delle Produzioni Vegetali, Via Brecce Bianche, 60131 Ancona/Italy
Prof. Rodolfo Santilocchi, r.santilocchi@univpm.it
Agronomy and Crop Sciences Research and Education Center, Department of Food Science, University of Teramo, Via C.R. Lerici, 1, 64023 Mosciano S. Angelo (TE), Italy
Prof. Michele Pisante, mpisante@unite.it

Portugal

ICAAM (Instituto de Ciências Agrárias e Ambientais Mediterrânicas), University of Évora

Spain

University of Cordoba

7 Conservation Agriculture in South-east Asia

Pascal Lienhard,[1,2,3] Stéphane Boulakia,[1,2,4] Jean-Claude Legoupil,[1,2,3] Olivier Gilard[5] and Lucien Séguy[6]

[1]Centre for International Cooperation in Agricultural Research and Development (France), Conservation Agriculture and Systems Engineering Research Unit, Montpellier, France; [2]Conservation Agriculture Network in South-East Asia, Vientiane, Lao PDR; [3]National Agriculture and Forestry Research Institute, Vientiane, Lao PDR; [4]Support Project for the Development of Cambodian Agriculture – Ministry of Agriculture, Forestry and Fisheries/General Directorate of Agriculture, Phnom Penh, Cambodia; [5]French Development Agency, Vientiane, Lao PDR; [6]Agroecoriz, France

7.1 Introduction

7.1.1 History of conservation agriculture in South-east Asia

Taking stock of conservation tillage history in South-east Asia

The history of Conservation Agriculture (CA) in South-east Asia (SEA), taken as any cropping systems integrating the three principles of minimal soil disturbance, permanent soil cover and crop rotations (FAO, 2007), is hardly dissociable from the history of conservation tillage (CT), defined by Lal (1989) as any tillage system that reduces loss of soil or water relative to conventional tillage (ConvT), for the following reasons:

1. Both approaches look for similar objectives (e.g. soil erosion control, soil fertility improvement), promote to some extent similar tools (e.g. use of cover crops, soil mulching, reduction of soil tillage intensity) and are facing similar constraints regarding their broad adoption (e.g. opportunity cost of land and labour, field protection against communal grazing, management skills).
2. Most of the institutions involved in CA farming systems' design and promotion in SEA also promote other CT technologies, as complementary approaches towards more sustainable agricultural practices.
3. There is undoubtly a blurred area between CT and CA.

As stated by Harrington and Erenstein (2005), estimates of CA adoption are difficult since the extent to which all CA principles in a particular region have been met are often hard to determine and may rapidly change. In addition, many CT farming systems (e.g. maize monocropping with residues' management in Laos) have been designed and promoted as transient farming systems towards a full CA package in a step-by-step implementation approach with farmers (Jullien *et al.*, 2008a; Lestrelin *et al.*, 2012a).

A long history of conservation tillage in sloping areas in South-east Asia

South-east Asian countries have a long experience in CT, notably in sloping areas that are not only the most diverse, fragile and threatened ecosystems, but also the most geographically extensive. Stibig *et al.* (2007) estimate that upland areas cover 40–90% of the total land areas of each of the countries of the region.[1]

In recent decades, agrarian landscapes and livelihoods in the uplands of SEA have undergone dramatic changes. Farming households have had to adapt to the mounting influence of global drivers such as demographic changes, market forces and government policies, that have led to the rapid expansion and intensification of agriculture (Castella, 2012). The necessity to buffer the negative consequences of these land use changes (e.g. deforestation, land degradation) has rapidly emerged. The experiments and the promotion of soil and water conservation practices started in the early 1970s (Garrity, 1996) and included various technical packages including contour hedgerow systems, agroforestry practices, natural vegetative strips, managed fallows, strip tillage etc., among which two are worth mentioning for their similarities with CA principles and the problems they face for their broad diffusion.

Contour hedgerow systems were developed in the Philippines in the early 1970s and are based on the principle of growing field and permanent crops in 3 m to 5 m-wide bands between double contoured hedgerows of nitrogen-fixing trees. These leguminous trees are regularly pruned and the cuttings are placed in alley-ways to serve as organic fertilizers (MBRLC, 2004). Contour hedgerows have been widely promoted during the 1980s and 1990s in several SEA countries (e.g. Indonesia, Myanmar, Thailand, Vietnam and the Philippines) to reduce soil erosion and maintain soil fertility. They were the first experiments in SEA showing an interest in soil mulching. Two main constraints have been identified for their broad diffusion (Garrity, 1996): the tendency for the perennial pruned-tree hedgerows to compete for growth resources and hence reduce yields of associated annual crops planted in adjacent rows, and the enormous amount of labour needed to prune and maintain the hedgerows. The diffusion of contour hedgerow systems has also certainly been hindered by the increasing pressure on land to increase and sustain agricultural production (Lal, 2005), smallholders being more and more confronted with the opportunity cost of growing hedgerows where staple or cash crops may be grown. Competition between main and relay crops, labour requirement and penibility, and above all, the opportunity costs of land and labour are similar constraints experienced for CA diffusion in SEA, where smallholder farmers prevail, for which the main challenge remains how to make the best use of limited resources (land, labour, capital).

Much agronomic work has also been done in SEA countries in the 1990s to improve the benefits of fallowing through the establishment and management of leguminous species during fallow period of less than 2 years. Experiments were based on the use of herbaceous, fast-growing legume cover crops (von Uexkull and Mutert, 1995), shrubby legumes (Roder and Maniphone, 1998) or forage legumes (Garrity, 1996). All studies have pointed out the benefits of using legume species in short-term managed fallows to accelerate soil fertility regeneration, weed suppression and/or provide a possible source of other economic benefits. The main constraints highlighted in these studies for the greater diffusion of the use of cover crops were: field protection from communal grazing, protection from dry season fires and a dependable seed supply, which are also common constraints in CA diffusion.

A recent history of Conservation Agriculture in South-east Asia

Conservation Agriculture history in SEA is much more recent, with less than a decade of on-field experiments. The first projects including a CA component were located in continental SEA (Cambodia, Laos and Vietnam) and have been supported by the French Development Agency (AFD) with technical support from the French Agricultural Research Centre for International Development (CIRAD) (Table 7.1).

Table 7.1. History of Conservation Agriculture experiments in South-east Asia.

Cambodia	Kampong Cham	2004–2008	PHF (rubber for smallholder project), crop diversification component	Rubber	AFD	MAFF, CIRAD
	Kampong Cham, Battambang	2008–2013	PADAC (Support Project for the Developt of Cambodian Agriculture)	Maize, cassava, soybean	AFD	MAFF, CIRAD
	Battambang	2010–2014	SANREM CRSP (Sustainable Agriculture and Natural Resources Management Collaborative Research Support Program)	Maize, cassava	USAID	US universities, CIRAD
Laos	Sayabouri	2001–2002	PRODESSA (Project for the Development of the South of Sayabouri Province), agricultural research component	Maize	AFD	NAFRI, CIRAD
	Sayabouri, Xieng Khouang	2003–2008	PRONAE (National Program in Agroecology), PASS (Development project for the South of Sayabouri Province)	Maize, upland rice, forages	AFD	NAFRI, CIRAD
	National	2007–2011	PROSA (Sector-based program in Agroecology)	Scaling-up strategy	AFD	MAF, CIRAD
Philippines	Claveria	2010–2014	SANREM CRSP (Sustainable Agriculture and Natural Resources Management Collaborative Research Support Program)	Maize	USAID	US universities, CIRAD
Thailand	Sakhon Nakhon	Since 2005	Soil biology laboratory	Cover crops, upland rice	Thai Gov	KU, CIRAD
Vietnam	Bac Kan	1998–2004	SAM (Mountainous Agrarian Systems Project)	Rice, maize, cassava, forages	French Gov	VAAS, CIRAD, IRD, IRRI
	Pleiku	1999–2004	ADP (Agricultural Diversification Project)	Rubber	WB	NIR, CIRAD
	Phu To, Son la, Yen Bai	2008–2012	ADAM (Support to CA Extension in Mountainous Areas of Vietnam)	Maize, tea	AFD	NOMAFSI, CIRAD
	Son la, Yen Bai	2009–2013	Improved Market Engagement for Sustainable Upland Production Systems in the North West Highlands of Vietnam	Maize	ACIAR	NOMAFSI, UQ

China	Yunnan	Since 2003	CA experiments in the prefecture of Wenshan	Maize, tea	Chinese Gov	YAAS
Regional (6 countries)		Since 2009	CANSEA (Conservation Agriculture Network in South East Asia)	Research and training issues	AFD	Secretariat (CIRAD)
Regional (4 countries)		2009–2013	PAMPA (Transversal Program in Agroecology)	Impact studies	AFD	CIRAD, IRD

Institutions acronyms: ACIAR, Australian Center for International Agricultural Research; AFD, French Development Agency; CIRAD, French Agricultural Research Centre for International Development; Gov, Government; IRD, French Research Institute for Development; KU, Kasetsart University; MAF(F), Ministry of Agriculture and Forestry (and Fisheries); NAFRI, National Agriculture and Forestry Research Institute; NIR, National Institute of Rubber; NOMAFSI, Northern Mountainous Agricultural and Forestry Science Institute; UQ, University of Queensland; USAID, United States Aid; VAAS, Vietnamese Academy of Agricultural Science; WB, World Bank; YAAS, Yunnan Academy of Agricultural Science.

Conservation Agriculture technologies developed in subtropical (Brazil) and temperate (the USA, Australia, Argentina) areas are mainly based on former crop residue management (Lal, 1989) and are unlikely sufficient to preserve overall soil fertility in tropical areas since crop residue mineralization is faster in hot and wet regions. Hence, research has been underway since the late 1990s by CIRAD and its national partners aimed at creating CA farming systems for tropical areas, based on the production of additional biomass prior to, with and/or in succession of, main crops. Farming systems were built taking stock notably of the Brazilian experience on large-scale agriculture (Séguy *et al.*, 2006; Séguy and Bouzinac, 2008) and the Malagasy experience on small-scale farms (Husson *et al.*, 2006). All projects were built using the DATE (Diagnosis, Design, Assessment, Training and Extension) approach (Fig. 7.1). A large diversity of systems has been designed and tested with small-scale farmers (Table 7.2).

More recently, new institutions (US and Australian universities, ICRAF) supported by other donors (USAID, AusAid, ACIAR) have similarly initiated work on CA in continental (Cambodia, Vietnam) and insular (the Philippines) SEA. Harrington and Erenstein (2005) reported that CA has been adopted to varying degrees in Japan, Malaysia, Korea, Taiwan and Sri Lanka, but no documents on CA are available.

7.2 Current Status

Conservation Agriculture history in SEA is recent (less than a decade), which mostly explain that CA development is today mainly limited to the research sector and barely to the extension area.

The main successful stories of CA systems adoption are found for maize cropping due to the great expansion of this crop in the region over the last decade (Lestrelin and Castella, 2011). Maize cultivation under zero tillage with former crop residue management and/or relay association with a legume (beans, forage or shrubby legumes) are the most popular CA systems. After 6 years of research and 4 years of extension support, adoption estimates of maize-based CA systems in the south of Sayabouri province (northern Laos) were of 2500 ha implemented by 1800 smallholders in 2008

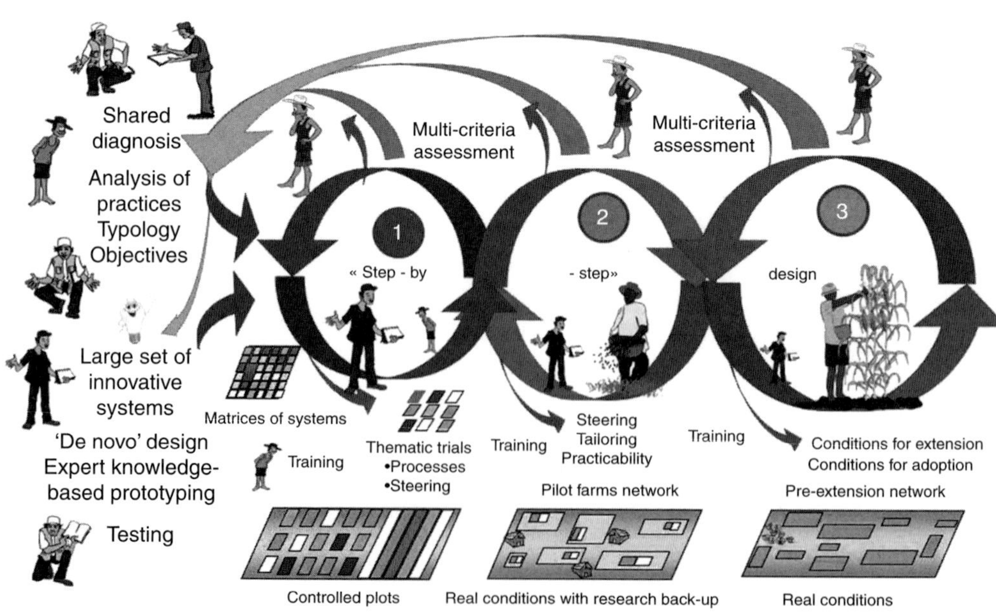

Fig. 7.1. The DATE (Diagnosis, Design, Assessment, Training and Extension) approach.

Table 7.2. Main Conservation Agriculture cropping systems tested in South-east Asia (step-by-step approach).

	Main CA farming systems	Countries
Soil cover	❖ Maize monocropping with mini-terrasses and mulching	V
	❖ **Maize monocropping under no-till with crop residue management**	**L**, V
Diversification of monocropping systems	❖ Maize with:	
	○ **Pulses** [soybean, **rice bean, mung bean, cow pea**, black bean, **faba bean**]	C, **L**, P, T, **V, Y**
	○ **Forages [*Stylosanthes sp.***, finger millet]	C, L, V
	○ Legume tree [*Cajanus sp., Crotalaria sp.*]	C, L, V
	○ Annual cover crops [*Mucuna sp., Dolichos l., Canavalia e.*]	C, L, P, V
	○ Perennial cover crops [*Arachis p., Centrosema p., Desmodium u., Macroptilium a.*]	C, L, P, V
	❖ Rice with legume cover crops [*Stylosanthes g., Centrosema p., Sesbania sp.*]	L, T
	❖ **Cassava** with forages [***Stylosanthes g.***, *Brachiaria r.*]	**C**, L, P, V
	❖ Perennial crops with permanent soil cover [Tea, fruit trees, rubber+coffee with *Arachis pintoi;* rubber with stylo]	C, V, Y
Rotational systems	❖ Annual sequence	
	○ Millet + *Stylosanthes g.* – soybean or maize	C
	○ Maize – Maize + relay crop	P, V
	○ Maize + cow pea – rice + cow pea (alternative bands)	P
	❖ 2-year rotation	
	○ **Maize/rice bean**	**L**
	○ **Maize + stylo/cassava + stylo**	**C**
	○ Rice/finger m. + pigeon p. – black beans	L
	○ Rice or maize + *Stylosanthes g. / Stylosanthes g.*	L, P, V
	❖ 3-year rotation	
	○ Rice/maize + ruzi grass/rice bean	L
	○ Rice/maize + forage grass/soybean – oat	L
	○ Rice/rice or maize + legume forage or tree/legume	L, V, T
	❖ Long term rotation	
	○ Rice + forage grass/grass // grass/soybean + forage grass	L

Countries: C, Cambodia; L, Laos; P, Philippines; T, Thailand; V, Vietnam; Y, Yunnan Province
In bold: most successful stories on CA adoption with smallholders

(Slaats and Lestrelin, 2009) and of 5000 ha in 2011 (Panyasiri *et al.*, 2011) with however only a limited (~10%) and highly variable percentage of these surfaces implemented in association or rotation with a legume (Slaats and Lestrelin, 2009). Maize associated or intercropped with legume crops is also the main system tested and promoted in northern Vietnam (Tuan and Doanh, 2008; Hauswirth *et al.*, 2011; Nicetic *et al.*, 2011), Yunnan province (Tao *et al.*, 2008) and the Philippines (Mercado *et al.*, 2011). In Cambodia, CA systems tested and promoted with farmers are mainly based on maize and/or cassava associated with stylo legume (*Stylosanthes guianensis* CIAT 184) (Boulakia *et al.*, 2008, 2012a), with about 500 ha of experiments conducted and evaluated with farmers (Chabierski *et al.*, 2011).

7.3 Prospects for Conservation Agriculture in South-East Asia

Conservation Agriculture is often deemed knowledge- and capital-intensive, hence hardly compatible with smallholder farming. For Giller *et al.* (2009), dissemination should concentrate on 'socio-ecological niches' where CA is the most likely to be adopted by smallholders. Soil erosion issues, good access to farm inputs and markets and the presence of smallholders with sufficient land, labour and capital would constitute key criteria for identifying these niches. Lestrelin and Castella (2011) have shown in the Laos context and maize-production areas, that it was more relevant to identify the key moments for intervention along specific agroecological transition pathways. They identified four agroecological zones, representing windows of opportunity for CA diffusion, according to a gradient of land productivity and of integration/access to market (Fig. 7.2).

These zones correspond to successive stages in a historical pattern of land use intensification that can be found in all SEA countries.

Areas under **Stage 1** (productive lands covered by extensive subsistence-oriented agriculture) can be found in various SEA mountainous regions such as the remote areas of northern Laos, Thailand, Vietnam and Myanmar. Shifting cultivation and slash-and-burn farming prevail, despite having been banned in many countries to prevent further deforestation. Prospect for CA in these areas is linked to its ability to maintain/improve labour productivity, propose better weed and pest control, and increase biomass production for both soil improvement and animal feeding, in a context of low access to market and to external inputs. The potential for CA diffusion in these areas will remain low unless strong and long-term support in research and extension, and specific incentives (e.g. seeds, seedlings, land titling) are given to promote the transition from shifting to permanent agriculture.

Areas under **Stage 2** (productive lands engaged in a process of commoditization and intensification of agriculture) are undoubtedly increasing in SEA alongside national efforts made to open new roads and tracks in mountainous areas and improve access to market (ADB, 2011). Possible land grabbing by private entrepreneurs and soil degradation are new social and environmental issues. Lestrelin and Castella (2011) have shown for Laos context that CA was, during that early stage of land use intensification, an attractive option for smallholders willing to engage in market-oriented agriculture with a limited increase in production costs.

Areas under **Stage 3** (degrading lands covered by intensive market-oriented monoculture) are found in rainfed plains (e.g. Thailand, Cambodia, coastal areas), smooth hills, and increasingly on steep slopes in mountainous areas (Stibig *et al.*, 2007). Environmental issues are major (deforestation, soil erosion, biodiversity losses, soil

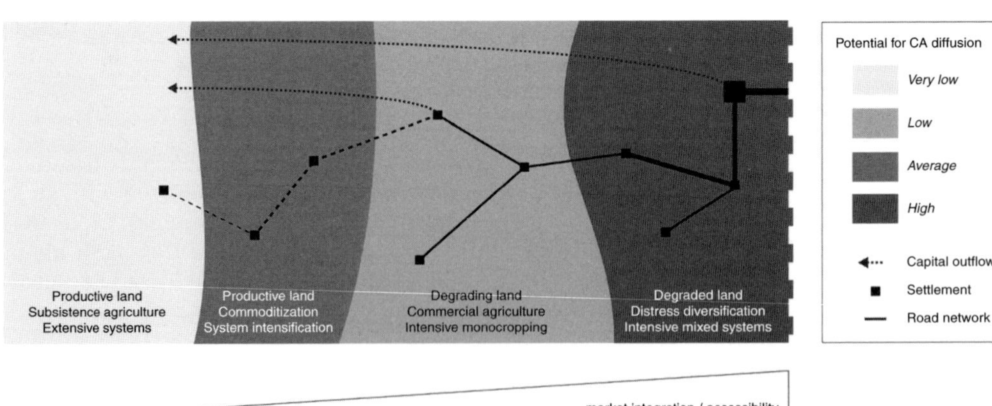

Fig. 7.2. Agroecological transition stages and potential for adoption of Conservation Agriculture in maize-prone areas, Laos context (Lestrelin and Castella, 2011).

and water pollution) and these areas are certainly where CA efforts (research, sensitization, supports and incentives) should be concentrated.

Vast areas under **Stage 4** (degraded land, distress diversification, intensive mixed systems) can be found in several SEA areas that have been long engaged in intensive agriculture (Cambodia, Indonesia, Laos, the Philippines, Vietnam and Yunnan). The improvement of soil fertility and the limitation of population migration towards cities are key agricultural and social issues. In Laos, Lestrelin and Castella (2011) have shown that CA dissemination efforts at that stage have had stronger impacts than at any other stage of the agro-ecological transition. Similar results might be expected in other SEA regions.

An ultimate stage (not presented in Fig. 7.2) would be the over-degraded areas and/or naturally poor land where the low fertility of soils strongly limits agricultural activities. This situation includes notably the bare hills of northern Vietnam and Yunnan, a major part of the acid and sandy rainfed lowlands in Cambodia, Laos, Thailand and Vietnam, and the acid savannah grasslands in Indonesia and Laos. The design of farming systems combining the use of organic and mineral fertilizers, and CA principles to restore these heavily degraded areas is a key challenge regarding the need to increase and sustain agricultural production (De Schutter, 2011). Preliminary research results (Boulakia *et al.*, 2012a; Lienhard *et al.*, 2013) have shown that sustainable CA farming systems were possible but required more investment (to rebuild fertility) and financial support to allow a broad diffusion of CA systems as compared to transition phases 2, 3 and 4.

7.4 Research Results Reported in South-east Asia

Conservation Agriculture systems are here evaluated according to the main objectives assigned to agricultural food systems (De Schutter, 2011), which include the need: (i) to increase agricultural production (to respond to future needs); (ii) to increase farmers income (to reduce poverty) and notably smallholders income; and (iii) to sustain the resources supporting agricultural activities.

There are limited published works on CA in SEA. Most of the information comes from grey literature (reports, technical leaflets and communications to congress) and is accessible on the following websites:

CIRAD: http://www.agroecologie.cirad.fr
CANSEA: http://www.cansea.org.vn
CA and Agroforestry: http://www.conservationagricultureandagroforestry.org
ORCATAD: http://orcatad.nafri.org.la

7.4.1 Conservation Agriculture systems' impact on soil productivity

Similar to other regions (Kassam *et al.*, 2009; Derpsh *et al.*, 2010), the level of integration of CA principles appears as the main factor affecting soil productivity changes under CA in SEA.

Similar or limited increase in grain yields under a partial Conservation Agriculture package

Five years of experimentation with maize monocropping under residue management versus deep ploughing on steep slopes showed no significant difference in maize yields (Tran Quoc *et al.*, 2008). However, increased crop yields are one of the reasons (8% of answers) given by farmers for expanding cultivated areas under CA (Lestrelin *et al.*, 2012b).

Affholder *et al.* (2009) reported no significant difference between maize with mulch (remaining from previous crop plus imported) and maize under traditional burning, after 2 years of experimentation. However, significant differences (up to 40%) have been reported in other studies in relation to soil erosion reduction (Tuan and Doanh, 2008; Nicetic *et al.*, 2011).

Significant higher grain yields under a full Conservation Agriculture package

Grain yield benefits related to the cultivation of a legume prior to a cereal (rice, maize) have long been demonstrated in managed fallows (von Uexkull and Mutert, 1995; Garrity, 1996; Roder and Maniphone, 1998). Similar results are reported for CA permanent rotational systems with significant production gain (up to 30% increase) observed for maize cultivated on mulch of *Mucuna* sp. (Tuan *et al.*, 2002), rice bean (Tran Quoc *et al.*, 2008; Slaats and Lestrelin, 2009), cow pea and black bean (Nicetic *et al.*, 2011), pigeon pea (Jullion *et al.*, 2008b) or *Stylosanthes guianensis* (Chabierski *et al.*, 2011). Chabierski *et al.* (2011) reported higher tuber yields (up to 65% increase) of cassava on stylo mulch with reduced tillage (chisel) on planting rows as compared to cassava monocropping under full disc ploughing.

7.4.2 Conservation Agriculture systems' impact on economic returns at field and farm level

In the absence of government subsidies for the agricultural sector and/or payment for environmental services, clear economic benefits must be apparent for smallholders to induce a change from ConvT to CA. The effect of CA on economic returns, calculated as value of production minus operational costs per unit area vary according to its effect on the main grain or tuber yield of crops and implementing costs, but also according to the economic valorization of the additional biomass produced. Various situations are described in SEA case studies.

Reduced production costs x increased yields: the win-win combination

Five-year experiments of a 2-year rotation of maize with rice bean compared to maize monocropping under tillage showed a significant increase (from 20 to 50% depending on the year) in economic returns due to reduced production costs for land preparation and weed control, associated with increased maize yields (Tran Quoc *et al.*, 2008).

Reduced production costs x similar yields

In Laos, the fee-for-service for ploughing with discs under ConvT is higher than the cost for rolling and herbicide spraying under CA (Tran Quoc *et al.*, 2008; Slaats and Lestrelin, 2009; Lienhard *et al.*, 2013). In addition, land preparation costs are significantly greater under the conventional system when herbicide use is required before sowing to supplement tillage for effective weed control, which is the major situation after several years of monocropping under tillage (Bounthong *et al.*, 2005; Tran Quoc *et al.*, 2008). The reduction in production costs is the main reason (28% of answers) given by farmers for expanding their cultivated surface under CA in the south of Sayabouri province (Lestrelin *et al.*, 2012b). Despite no significant differences in grain yields, the differences in costs for land preparation and weed control led to significant differences in economic returns (+10 to 15% higher profits) for maize continuous cultivation under no-till and crop residue management as compared to conventional monocropping under tillage (Tran Quoc *et al.*, 2008; Slaats and Lestrelin, 2009), and explained the rapid and large diffusion of this cropping system.

Similar production costs x increased yields

In mountainous, newly connected to market areas, farmers often practise high input cultivation but without adequate knowledge (Nicetic *et al.*, 2011). Improved crop and input management, and intercropping with legumes have been described to significantly improve maize production and increase profits (Nicetic *et al.*, 2011).

Increased production costs x increased yields

In extensive subsistence and low input-based agriculture, operational costs associated with CA systems are generally higher than under conventional slash and burn systems with such additional observed outlays as the costs of seed of secondary crops, minimum fertilization and/or pesticide use and fencing material. The gains in economic returns rely therefore on the gains in productivity and have been shown to be substantial (Husson *et al.*, 2001), modest (Nicetic *et al.*, 2011), to nil (Affholder *et al.*, 2009) according to the system tested (diversified rotational system versus mulching) and the number of years of experimentation (short to medium term).

In Cambodian rainfed areas long engaged in intensive market-oriented agriculture, Chabierski *et al.* (2011) showed greater economic returns under CA systems than under ConvT systems for maize (+15 to 25% increase) and cassava (+20 to 35% increase) production, due to substantial gains in productivity. However, these increases in productivity were associated with higher investments, which represent the main constraint for a broader diffusion of CA systems (Chabierski *et al.*, 2011).

Lienhard *et al.* (2008) showed for acidic and highly weathered acid savannah soils of northern Laos that grain and forage production could be significantly improved and could lead to significant gains in economic returns, but required higher initial investments (machinery, fertilizers) as compared to traditional tilled and unfertilized production systems.

In addition, fencing costs (material, labour) are in most cases necessary in CA, as opposed to in conventional systems, to ensure an effective protection of relay crops from communal grazing, and are rarely included in economic calculations and comparisons with conventional systems.

Unlike that which is usually described as a major benefit of CA practices (Kassam *et al.*, 2009; Derpsh *et al.*, 2010; Johansen *et al.*, 2012), CA experiences in SEA show that the reduction in production costs is far from being systematic and is a major issue for CA scaling-up in SEA.

Labour productivity and penibility under CA

One of the main household challenges in SEA is certainly how to optimize the use of a limited labour force (Garrity, 1996). The effect of CA on labour requirements and productivity (economic return to labour ratio) is therefore often quite relevant as an economical tool to evaluate and understand farmers' interest and/or disinterest in CA systems.

High labour requirements and low labour productivity are the main reasons for farmers' disinterest in mulching-based CA systems (Affholder *et al.*, 2009; Nicetic *et al.*, 2011). Interestingly, low (27% of respondents) and high (26%) labour charge requirements are both main reasons given by farmers in Laos for extending or not trying CA systems, respectively, depending on their access (or not) to suitable implements (Lestrelin *et al.*, 2012b).

Labour penibility is another key parameter explaining farmer's interest or disinterest in CA systems. Strenuous sowing and spraying were given as the main reasons (with 21% and 10% of respondents, respectively) by Lao farmers not interested in CA systems (Lestrelin *et al.*, 2012b). Contrary to these results, Chabierski *et al.* (2011) reported the reduction of labour penibility for cassava weeding and harvest as a major positive feedback from farmers experimenting with CA systems.

Sharing residues and cover crops biomasses: a difficult but key issue for Conservation Agriculture adoption in South-east Asia

Conservation agriculture requires a critical level of crop residues and cover crops to maintain or enhance soil chemical, physical and biological properties and prevent land degradation (Séguy *et al.*, 2006; Kassam *et al.*, 2009; Derpsh *et al.*, 2010). Sá *et al.* (2001) found that the critical level of dry matter to be returned was 7 Mg ha^{-1} $year^{-1}$ for southern Brazil, but this level is questioned in a context where there is limited access to fertilizers and high pressure on crop residue resources (Affholder *et al.*, 2009).

Livestock plays a crucial role (e.g. food security, saving, cash) in many smallholder farming systems in SEA (Husson *et al.*, 2003a). Several studies have suggested valorizing relay crop production (grains, biomass) into livestock production, as an additional economic incentive for CA adoption: use of forage grass (Husson *et al.*, 2003a; Lienhard *et al.*, 2013) and/or forage legume (Ponsich *et al.*, 2011) biomass for ruminants; use of legume pods (Tuan *et al.*, 2002; Jullien *et al.*, 2008b) or forage grains (Lienhard *et al.*, 2013) for pigs and poultry.

New opportunities for using crop residues for energy production (biofuel, biogas) are also on the rise in SEA (Lal, 2005). As in many other countries (Lal, 2005; Kassam *et al.*, 2009; Derpsh *et al.*, 2010; Johansen *et al.*, 2012), the management and share of crop residues and cover crops between *in situ* recycling, livestock feed and energy supply is a key issue for the diffusion of CA farming systems in SEA.

7.4.3 Conservation Agriculture systems' impact on soil fertility, quality and on the environment

Effect on soil erosion

Soil erosion is deemed as a key reason for CA promotion in SEA sloping areas (Bounthong *et al.*, 2005; Tuan and Doanh, 2008; Mercado *et al.*, 2011). Valentin *et al.* (2008) have shown that mulching significantly reduces runoff and total sediment yield in different catchments in Laos, Thailand and Vietnam. Lestrelin *et al.* (2012b) have shown that soil conservation issues were an important reason for farmers to experiment with CA systems (12% of answers) and/or to expand their cultivated land under CA (9% of answers).

Effect on soil physico-chemical properties

CA has shown to have a positive effect on soil aggregation, which plays a key role in soil organic turnover and soil susceptibility to erosion. Tivet *et al.* (2008) have shown a significant increase (up to 60%) in soil aggregate stability as estimated through the mean weight diameter of aggregates (MWD) for topsoils (0–10 cm) of fields conducted under no-till management (maize monocropping with residue management and 2-year rotation of maize and rice bean) as compared to ConvT and maize monocropping system. Lienhard *et al.* (2013) have found similar results for a 3-year rotation of rice, maize and soybean cultivated under no-till management with cover crops prior to and with main crops, or under ConvT.

Lienhard *et al.* (2013) observed for Laos savannah grassland a significant decrease of topsoil (0–10 cm) C and N content under ConvT as compared to CA management (11% difference after 2 years of cultivation). Despite similar amendments, the sum of exchangeable bases was 1.5-fold higher under CA systems than under ConvT.

Effect on soil biodiversity and biological activity

Several regional studies have shown a significant positive effect of CA systems on soil macrofauna diversity (Husson *et al.*, 2003b; Boyer *et al.*, 2008; Boulakia *et al.*, 2012b), density and biomass (Husson *et al.*, 2003b; Boyer *et al.*, 2008; Tivet *et al.*, 2008). All studies underline notably the positive effect of CA on earthworm populations, with earthworm biomasses higher (up to 80%) under CA as compared to conventional burn and/or tillage system.

Husson *et al.* (2003b) observed similar microbial communities (as estimated by FAME (fatty acid methyl ester) profiles) under a 2-year managed fallow of ruzi grass and a 10-year natural fallow. Boyer *et al.* (2008) observed a significant (+30%) increase in microbial respiration under no-till systems with mulch as compared to bare soils. Lienhard *et al.* (2013) showed a significant decrease (–20%) of soil microbial molecular abundance (as estimated by soil DNA extracts quantification) under ConvT system as compared to CA ones.

7.5 Problems Encountered in Scaling-up Conservation Agriculture in South-east Asia

The factors influencing farmers' decision to adopt (or not) CA practices have been shown to be both highly context specific (e.g. biophysical characteristics, involvement of local elites, extension staff motivation and capacity; e.g. Lestrelin *et al.*, 2012b) and fast changing, e.g. market opportunities, land degradation stage and/or production costs changes (Lestrelin and Castella, 2011).

These two parameters (diversity of situations and highly dynamic environment) are both a chance and a constraint for CA systems design and promotion. One of the best examples for this is the role of rainfed rice in traditional and CA farming systems in SEA: whereas many CA farming systems have been initially based on rainfed rice production in sloping areas (Table 7.2), the recent boom of maize cultivation observed in many SEA countries (Lestrelin and Castella, 2011) has given little chance to these systems to be adopted, at least to date. The ability to develop and sustain on-field adaptative, flexible and effective research and extension support is a major issue for CA up-scaling in SEA.

The absence of universal variables explaining farmers local engagement in CA jeopardizes any attempt to extract general theories on farm-level determinants for CA adoption (Lestrelin *et al.*, 2012b). However, we can point out general constraints learned from this decade of CA experiments in SEA. Some are specific to CA systems (e.g. local unavailability of suitable implements, relay crop and residue management), but most of them are common to all innovations dealing with agricultural intensification in the uplands and with smallholders (e.g. lack of land tenure security, communal land use plan, public resources and support, unadapted credit access).

7.5.1 Local unavailability of suitable implements

The unavailability at local level of suitable equipment for CA implementation, notably for smallholders, is a major constraint already described for other small-scale agricultural contexts (Harrington and Erenstein, 2005; Kassam *et al.*, 2009; Johansen *et al.*, 2012).

Manual sowing in a mulch increases labour force requirements (Affholder *et al.*, 2009; Lestrelin *et al.*, 2012b), penibility (Lestrelin *et al.*, 2012b), induces delays in crop establishment with negative impact on productivity, and increases competition for labour with other farm activities, notably the transplanting of lowland paddy rice (Lienhard *et al.*, 2008). Different no-till planters have been introduced from Brazil and tested/adapted in Laos and Cambodia to respond to a wide range of cultivation system (manual versus mechanized) and farmer's investment capacity (low to medium) (Boulakia *et al.*, 2008; Jullien *et al.*, 2008a). These equipments include (Fig. 7.3): hand jab seeder (Fitarelli) for manual sowing; 1- to 2-line no-till planters (Fitarelli, Knapick) for sowing with two-wheel hand tractors; 2- to 3-line no-till planter (Fitarelli) for sowing with small tractors (<45 hp); and 4- to 7-line no-till planters (Knapick, Vincetudo, Semeato) for sowing with big tractors (>70 hp). These equipments have shown to reduce significantly sowing penibility and improve labour productivity: e.g. labour requirement for maize sowing in a mulch decreased from 15–18 man days ha^{-1} for manual sowing with a bamboo stick (moderate slope conditions) to 6–8 man days ha^{-1} with hand jab, 2 man days ha^{-1} with 2-line no-till planters and less than 1 man day ha^{-1} with 4-line no-till planters (PRONAE, 2009).

Similarly, experiments on spraying equipment (low volume nozzle from Berthoux, France; 20 l-wheel sprayer from Knapick, Brazil (Fig. 7.4d); 200 l-sprayer adapted for two-wheel hand tractor from Campo novo and Rubemaq, Brazil (Fig. 7.4c)) have shown to reduce significantly spraying penibility (from 400–500 l ha^{-1} with traditional backpack sprayer to 150–200 l ha^{-1}) and increase the safety of users (PRONAE, 2009).

The importation process and cost of such equipments, as well as the local need for equipment maintenance and continuous adaptation have highlighted the need for an increased involvement of local (national/regional) manufacturers in the development and deployment of affordable and effective no-till implements.

Fig. 7.3. Examples of no-till planters introduced and evaluated in South-east Asia: (a) hand jab seeder (Fitarelli, Brazil), (b) 2-line seeder (Knapick, Brazil) for two-wheel hand tractors, (c) 2-line seeder (Fitarelli, Brazil) for two-wheel hand tractors and (d) 4-line seeder (Knapick, Brazil) for tractors.

7.5.2 Communal grazing and relay crop protection

Communal grazing after crop harvest is a widespread traditional territory management rule in SEA mountainous areas (Garrity, 1996). Animals are posted far from cultivated areas during the cropping season and brought back after crop harvest, threatening relay crop development and effective residues management.

Relay crops must provide clear economic benefits for smallholders to shift from conventional monocropping to systems with crop association and/or succession. Most successful stories of CA intercropping systems in SEA are associated with edible or commercial beans production and/or forage use for livestock system intensification (Table 7.2). However, successful relay cropping requires a regulation of cattle roaming. Participatory Land Use Planning (PLUP) and Community Agricultural Development Plans (CADP) are important tools to help define new community agreements regarding grazing period and areas, and/or reinforcing local by-laws on cattle roaming (Bourgoin *et al.*, 2011). In Sayabouri province, Laos, the expansion of the maize+rice bean intercropping system has been favoured by community agreements delaying the return of cattle in the fields until rice bean harvest. However, the presence of legume residues also induced a detrimental effect with higher animal concentration observed in these fields as compared to fields without legume residues, leading to soil compaction and lower residue retention (Jullien, Vientiane, 2012, pers. comm.).

Fig. 7.4. Examples of rolling knife and sprayers tested in South-east Asia: (a) and (b) locally produced rolling knifes, (c) 200 l sprayer (Campo novo, Brazil) adapted for two-wheel hand tractors and (d) 20 l wheel sprayer (Knapick, Brazil).

Fencing is therefore often required to supplement local by-laws on cattle roaming and ensure effective management and share of crop residues and cover crops between *in situ* recycling, livestock feed and energy supply.

7.5.3 Un-adapted credit system

Regardless of annual production costs, the practice of CA often requires high initial investments hardly affordable by smallholders in absence of adequate credit support. Credit needs are highly context specific and depend notably on the cultivation system (manual verus mechanized), the productivity of the land (fairly productive versus degraded) and the local prices of commodities:

- *Implements:* if implements for manual cultivation (hand jab seeder, back pack sprayer) are usually affordable by smallholders, implements for mechanized areas (1- to several- line no-till planters, rolling knife, and high volume sprayers) are more expensive, even if locally produced and/or shared in cooperatives.
- *Fertilizers:* soil nutrient deficiency correction is a prerequisite to ensure an effective CA system notably in strongly degraded or naturally poor areas. Investment in organic (green manure) and/or mineral (lime, rock phosphate, micronutrients) amendments can hardly be paid back on a single agricultural campaign (Lienhard *et al.*, 2013).
- *Fencing:* the maintenance of traditional bamboo fences is labour-intensive and has been pointed out as a major constraint

for CA adoption in Laos acid savannah (Lienhard *et al.*, 2008). Fencing with barbed wire and/or living fence is more effective but has a cost that cannot be paid back on a single agricultural campaign.

Smallholders' access to financial capital is a major issue for CA adoption in SEA. With limited guarantees (e.g. land titles) to support their credit demand, Laos farmers have been shown to encounter difficulties in gaining access to bank loans, which are in any case subject to high interest rate and short-term refund period, hence hardly compatible with the time frame required for such investments (Lienhard *et al.*, 2008).

7.5.4 Weed management and herbicide use in Conservation Agriculture systems

Changing from a conventional system to CA changes the nature of weeds and weeding patterns (Kassam *et al.*, 2009; Johansen *et al.*, 2012). The traditional reliance on burning and/or full tillage for initial weed control is incompatible with CA principles of maximum soil covering and minimal mechanical disturbance, respectively. Beyond considering soil disturbance, traditional hoeing of weeds during the crop cycle is hindered under CA by the presence of crop residues, and this leads to increased labour requirements (Jat *et al.*, 2012a, b).

To replace tillage and/or burning for weed control, CA-based projects have promoted slashing (in replacement of hoeing), rolling (in mechanized areas), crop rotations, use of cover crops, adjustment of sowing time and method, use of competitive crop genotypes, arrangement of planting pattern and adjustment of fertilizer strategy and herbicides, which are all part of an integrated weed management strategy (Johansen *et al.*, 2012). Tran Quoc *et al.* (2008) showed that with proper management a 2-year rotation of maize with rice bean (long cycle creeping legume) could replace herbicide use for maize cultivation. But highly specialized local agricultural systems (see below) strongly limit the promotion of diversified and hence effective integrated weed management systems in CA, leading to the misperception that CA systems are herbicide-dependent, whereas they are much more market-dependent.

7.5.5 High specialization of agriculture at local level

If local agriculture in SEA is becoming increasingly integrated to market (ADB, 2011), they are also more and more specialized. Lestrelin and Castella (2011) have shown for Laos that total annual maize production has increased ten-fold between 2000 and 2009, from 117,000 to 1,130,000 t and that the maize crop could represent more than 90% of total rainfed cultivated land in several areas. In Cambodia, Boulakia *et al.* (2010, 2012a) described a high specialization of production systems in the uplands, and the difficulty notably to introduce any crops in rotation with cassava due to cassava high selling prices since 2008 (above US$200 Mg^{-1}).

If higher integration to market is truly a chance for smallholders since it has led to increased monetary income (Lestrelin and Castella, 2011), the high specialization of agriculture is also a strong limitation to develop more ecologically intensive CA systems.

7.5.6 Limited public resources to ensure adequate on-field research, sensitization and technical support

As the proverb says, 'Money is the sinews of war'. SEA countries are not equal regarding public support to research, education and agricultural extension but many of them can be considered as low income countries with limited means to invest in the agricultural development sector.

Despite numerous former worldwide experiences (Séguy *et al.*, 2006; Kassam *et al.*, 2009; Derpsh *et al.*, 2010), CA local adaptation and promotion takes time. Harrington and Erenstein (2005) remind that it is not untypical that CA implement development and adaptation takes at least 10 years of (continuous) research and extension.

If CA economic and environmental benefits can be shown rapidly at field and farm level, the assessment of CA economic and environmental benefits at watershed/regional scale – such scale being crucial to release data for policy makers and private sector – requires higher resources (human, financial) and a longer period of time. Long-term mentoring and technical assistance to farmers (Harrington and Erenstein, 2005) and broad economic and environmental sensitization (Lestrelin *et al.*, 2012b) have been shown to be key factors for CA dissemination. Low governmental salaries and means are also common constraints leading to limited motivation and effective support from agricultural extension agents outside the projects' duration.

7.6 Government Support and Policy Towards Conservation Agriculture

7.6.1 Favourable institutional contexts for sustainable agricultural intensification

Several recent governmental resolutions show the wish in all SEA countries to pursue national economic development through the enhancement of agricultural productivity and commercialization of agriculture but without creating additional burdens on natural resources and agroecosystems.

In Laos, several governmental policies have been promoting CA since 2005: government decree for the promotion of CA as promising agroecological technique (Decree N°554, dated 21/4/2005); circular from the Ministry of Agriculture and Forestry (MAF) to promote CA (circular N°0372/MAF, dated 11/05/2005); and more recently a decree from MAF to integrate CA practices in all rural development programmes managed by the ministry (Decree N°0565/MAF, dated 11/02/2011). In 2009, the National Conservation Agriculture Centre (NCAC) was created within the national agriculture and forestry research institute, and the Maize Development Fund (based on a tax on maize exportation) was created in Sayaboury province to continue CA promotion and technical support to farmers (Panyasiri *et al.*, 2011).

In Cambodia, the Agriculture Strategic Development Plan (ASDP, 2009–2013) prepared by the Ministry of Agriculture, Forestry and Fisheries (MAFF) underlines the need to 'enhance the agricultural productivity, diversification and commercialization in order to reduce poverty and promote the economic growth through high consideration of environmental protection and sustainable natural resource management'.

In Vietnam, the National plan on New Rural Areas Development (2010–2020), approved by the Prime Minister (Decision No. 800/QD-TTg from 4/06/2010), underlines notably the need for: (i) an increased protection of the environment; and (ii) the promotion of appropriate development of agriculture and rural areas according to existing resources (labour, land, forest and marine).

7.6.2 Limited specific and long-term support to research, extension and education on Conservation Agriculture

In all these countries, the shift from pilot projects to national programmes on CA associated with long-term support to research, extension and education on CA has not yet been observed. The first attempts to develop academic curriculum on CA started in 2012 with exchanges between universities from Brazil (Ponta Grossa), France (Institute for Tropical Regions) and SEA (National University of Laos, Kasetsart University in Thailand, Royal Faculty of Cambodia) supported by the French Development Agency (AFD).

7.6.3 CANSEA, a regional initiative to facilitate and increase exchanges of results and experiences on Conservation Agriculture in South-east Asia

The Conservation Agriculture Network in South-East Asia (CANSEA, http://www.cansea.org.vn) is a regional network created in 2009 with the objectives of: increasing and facilitating exchanges of results and experiences between members; proposing mechanisms of cooperation between projects and institutions; and developing common regional projects (Legoupil and Kingkeo, 2011).

The network is based on an agreement signed by eight research and education institutions coming from six South-east Asian countries:

- The General Directorate of Agriculture (GDA) of the Ministry of Agriculture, Forestry and Fisheries (MAFF) in Cambodia;
- The Yunnan Academy of Agricultural Sciences (YAAS) in China (Yunnan province);
- The Indonesian Agency for Agriculture Research and Development (IAARD) in Indonesia;
- The National Agriculture and Forestry Research Institute (NAFRI) in Laos;
- The Kasetsart University in Thailand;
- The Northern Mountainous Agriculture and Forestry Science Institute (NOMAFSI) and the Soils and Fertilizers Research Institute (SFRI) in Vietnam;
- The French Agricultural Research Centre for International Development (CIRAD), which cooperates with all the previous partners in South-east Asia.

Four topics of regional interest have been identified by the members:

- How to better link research (small-scale) and extension (large-scale);
- CA systems for the diversification of rice-based cropping systems in sloping areas;
- CA systems to restore the fertility and productivity of highly weathered and acid soils;
- Curriculum development on CA.

7.7 Efforts/Policies Required for Scaling-up Conservation Agriculture

7.7.1 Convincing donors and policy makers to invest and sustain Conservation Agriculture national initiatives

CA development and scaling-up in SEA requires long and continuous on-field financial and technical support. The sensitization of donors and policy makers to invest and sustain CA national initiatives is a major issue in order not to lose qualified human resources, materials (equipment prototypes, relay crops collection) and knowledge.

7.7.2 Developing and providing suitable machinery at local level

One of the keys for furthering CA among smallholders is the development and deployment of affordable and effective CA implements (no-till planters, rolling knife, and sprayers), notably for two wheel hand tractors that are popular worldwide with small-scale farmers (Johansen *et al.*, 2012).

Local and regional manufacturers have to be identified and associated to build, test and promote implements adapted to both plains and sloping areas, taking stock in addition of former and close experiences (e.g. north China, Bangladesh, India) and informal networks (e.g. 'Two-wheel tractor newsletter' managed by Esdaile, R.J., Australia).

7.7.3 Developing effective and integrated weed and pest management techniques

One of the main public concerns relative to CA practices in SEA is related to herbicide use and misuse by smallholders showing limited knowledge on effective and safe use of chemicals.

Promising results of 'zero glyphosate' cropping systems have been observed in Cambodia by combining mechanical control of biomasses (rolling knife) and spraying of salted solutions (NaCl or KCl associated to vinegar; Séguy, 2010). Such results have to be confirmed and added to the entire toolkit of options available for an effective and integrated weed management. Similarly, there are still limited data regarding the use of natural insect repulsives (neem extracts, wood vinegar) at a large scale (agricultural field versus garden). Such tools have to be developed to widen the current possibilities for integrated and effective pest management.

Institutional mechanisms to promote diversified versus specialized cropping systems have to be developed to propose more ecologically intensive CA systems.

7.7.4 Increasing private sector sensitization and enrolment

There are high expectations regarding the possible role of the private sector in CA development and promotion. The private sector is notably expected to: (i) participate in CA implement development, adaptation, manufacture and marketing; (ii) provide (similarly to conventional agriculture) no-till fee-for-service (e.g. spraying, sowing); (iii) participate in the development of new markets/facilities for secondary crops (e.g. forage seeds market, biogas/biofuel production, livestock increased commercialization); (iv) provide campaign credit to farmers (e.g. contract farming); and (v) provide technical support to farmers in supplement/replacement of national agricultural extension agencies. Increased interactions between public research, private sector and farmers' groups are key challenges for CA scaling-up in SEA.

7.8 Concluding Remarks

Even if still mostly limited to the research sector, the recent CA experiences in SEA have shown that CA could become a viable and accepted alternative to ploughing-based agricultural intensification – and this, even in a context of small-scale farming. Several lessons can be learned from this decade of *in situ* CA experiments:

1. *Agriculture and cropping patterns in SEA are spatially diverse and in constant evolution.* The identification of windows of opportunities for CA, i.e. the key moments for intervention along specific agroecological transition pathways corresponding to successive stages of land use intensification and land degradation, may facilitate the design of appropriate CA technologies and spatially differentiated policies.

2. *Agricultural trajectories often repeat themselves in time and space*, so that lessons can be drawn from past experiences and/or neighbouring countries. The recent CA network for SEA (CANSEA) may therefore play a key role in facilitating the exchanges of results and experiences within the region, hence in CA diffusion.

3. *Increase the participation of the private sector.* There is undoubtedly a need for higher sensitization and enrolment of the private sector to improve the local availability of suitable implements, but also provide credit facilities and/or technical support to farmers group.

4. *Need for long-term active research, training and technical mentoring on CA.* A shift from projects on CA to programmes on CA is required at the national and regional level to capitalize on research results and human resources. Amongst research topics related to the continuous improvement of CA agronomic, economic and environmental performances, the question of enhancing the diversification of farming systems and reducing pesticide use are two important ones.

Acknowledgements

The authors would like to thank all CANSEA members for their contributions and reviewing, and the French Development Agency (AFD) for its financial support to this thematic over the past decade.

Note

[1] The ten referred nation members of the Association of South-east Asian Nations (ASEAN) are: Brunei, Cambodia, Indonesia, Laos, Malaysia, Myanmar, the Philippines, Singapore, Thailand and Vietnam (http://www.aseansec.org); this is extended to 11 nations if taking into consideration the Yunnan province of China, member of the Great Mekong Subregion (GMS) nations (http://www.adb.org/countries/gms/main).

References

ADB (2011) Core Agriculture Support Program Phase II 2011 2015. GMS working group on Agriculture. ADB, Mandaluyong City, the Philippines.

Affholder, F., Jourdain, D., Quang, D.D., Tuong, T.P., Morize, M. and Ricome, A. (2009) Constraints to farmers' adoption of direct-seeding mulch-based cropping systems: A farm scale modeling approach applied to the mountainous slopes of Vietnam. *Agricultural Systems* 103, 51–62.

Boulakia, S., Kou, P., San, S., Leng, V. and Chhit, K. (2008) Five years of adaptative research for upland DMC-based cropping systems creation in Cambodia. In: Chanphengxay, M. *et al.* (eds) *Proceedings of the Regional Workshop on Investing in Sustainable Agriculture: The case of Conservation Agriculture and Direct Seeding Mulch-Based Cropping Systems.* MAF, Vientiane, Laos, pp. 37–49.

Boulakia, S., Vuth, P., Vathana, S., Chabierski, S. and Gilard, O. (2010) Conservation Agriculture in Cambodia: a Triple-Win Option. Paper presented at the conference the Environments of the Poor in the Context of Climate Change and the Green Economy: Making Sustainable Development More Inclusive. New Delhi, India, 24–26 November 2010.

Boulakia, S., Chabierski, S., Kou, P., Sona, S., Kong, R., Leng, V., Sar, V., Chhit, K. and Séguy, L. (2012a) Adaptation of direct seeding mulch-based cropping systems for annual cash crop production in Cambodian rainfed uplands. In: Hauswirth, D. *et al.* (eds) *Conservation Agriculture and Sustainable Upland Livelihoods. Innovations for, with and by Farmers to Adapt to Local and Global Changes – Proceedings of the 3rd International Conference on Conservation Agriculture in Southeast Asia.* CIRAD, Montpellier, France; NOMAFSI, Phu Tho, Viet Nam; University of Queensland, Brisbane, Australia, pp. 92–108.

Boulakia, S., Seguy, L., Tantachasatid, P., Thanisawanyankura, S., Leng, V. and Boyer, J. (2012b) Diversity and structure of soil macrofauna communities under plant cover in a no-till system in Cambodia. In: Hauswirth, D. *et al.* (eds) *Conservation Agriculture and Sustainable Upland Livelihoods. Innovations for, with and by Farmers to Adapt to Local and Global Changes – Proceedings of the 3rd International Conference on Conservation Agriculture in Southeast Asia.* CIRAD, Montpellier, France; NOMAFSI, Phu Tho, Viet Nam; University of Queensland, Brisbane, Australia, pp. 234–235.

Bounthong, B., Tivet, F., Tran Quoc, H., Lienhard, P., Panyasiri, K., Julien, P. and Séguy L. (2005) Direct Seeding Mulch-Based Cropping Systems – A Holistic Research Approach implemented in Northern Laos. Third World congress on Conservation Agriculture, Nairobi, Kenya, 3–7 October 2005.

Bourgoin, J., Castella, J.C., Nanthavong, K., Phatsalin, S., Drouillat, M. and Cottet, L. (2011) Engaging local communities in negotiating their own pathway towards conservation-oriented agricultural practices. In: *Resilient food systems for a changing world. Proceedings of the 5th World Congress on Conservation Agriculture.* WCCA/FSD, Brisbane, Australia, pp. 268–270.

Boyer, J., Tantachasatid, P., Tangtrakanpong, P., Thanisawanyangkura, S. and Séguy, L. (2008) Dynamic and structure of soil macrofauna communities under plant covers used in DMC systems in Sakon Nakhon Province, Thailand. Communication presented at the Regional Workshop on Investing in Sustainable Agriculture: The case of Conservation Agriculture and Direct Seeding Mulch-Based Cropping Systems. Phonsavanh, Laos, 28 October – 1 November 2008.

Castella, J.C. (2012) Agrarian transition and farming system dynamics in the uplands of South-East Asia. In: Hauswirth, D. *et al.* (eds) *Conservation Agriculture and Sustainable Upland Livelihoods. Innovations for, with and by Farmers to Adapt to Local and Global Changes – Proceedings of the 3rd International Conference on Conservation Agriculture in Southeast Asia.* CIRAD, Montpellier, France; NOMAFSI, Phu Tho, Viet Nam; University of Queensland, Brisbane, Australia, pp. 4–21.

Chabierski, S., Tyneth, L., Rada, K., Sona, S., Penot, E. and Boulakia, S. (2011) First Impacts of DMC Adoption among Smallholders in Cambodia. Communication presented at the second International Conservation Agriculture Workshop and Conference in Southeast Asia. Phnom Penh, Cambodia, 4–7 July 2011.

De Schutter, O. (2011) Agroécologie et droit à l'alimentation. Rapport présenté à la 16e session du Conseil des droits de l'homme de l'ONU [A/HRC/16/49]. Geneva, 8 March 2011. Available at: http://www.srfood.org/images/stories/pdf/officialreports/20110308_a-hrc-16-49_agroecology_fr.pdf (accessed 12 October 2012).

Derpsch, R., Friedrich, T., Kassam, A. and Hongwen, L. (2010) Current status of adoption of no-till farming in the world and some of its main benefits. *International Journal of Agricultural and Biological Engineering* 3, 1–25.

FAO (2007) Agriculture and Consumer Protection Department, Food and Agriculture Organization of the United Nations, Rome. Available at: http://www.fao.org/ag/ca (accessed on 15 June 2009).

Garrity, D.P. (1996) Conservation tillage: Southeast Asian perspective. Paper presented at the Conservation Tillage Workshop, Los Baños, the Philippines, 11–12 November 1996.

Giller, K.E., Witter, E., Corbeels, M. and Tittonell, P. (2009) Conservation agriculture and smallholder farming in Africa: The heretics' view. *Field Crops Research* 114, 23–34.

Harrington, L. and Erenstein, O. (2005) Conservation Agriculture and Resource Conserving Technologies. A Global Perspective. In: Abrol, I.P., Gupta, R.K. and Malik, R.K. (eds) *Conservation Agriculture – Status and Prospects*. Centre for advancement of Sustainable Agriculture, New Delhi, India, pp. 1–12.

Hauswirth, D., Boulakia, S., Le Quoc, D. and Dang, V.T. (2011) Designing Sustainable Tea-based Cropping Systems Fitting with Development Goals and Agroecology Concepts for Smallholders in Highlands of Vietnam: Lessons from the ADAM Project. Communication presented at the second International Conservation Agriculture Workshop and Conference in Southeast Asia. Phnom Penh, Cambodia, 4–7 July 2011.

Husson, O., Lienhard, P., Seguy, L., Tuan, H.D. and Doanh, L.Q. (2001) Development of direct sowing and mulching techniques as alternatives to slash-and-burn systems in northern Vietnam. In: *Conservation agriculture, a worldwide challenge. Proceedings of the 1st World Congress on Conservation Agriculture.* Madrid, Spain, pp. 29–33.

Husson, O., Tuan, H.D., Martin, C., Castella, J.C., Lecomte, P., Chabanne, A. and Séguy, L. (2003a) Crop-livestock integration through no-tillage on cover-crop in Vietnam. In: *Producing in Harmony with Nature. Proceedings of the 2nd World Congress on Conservation Agriculture*. Iguassu falls, Parana, Brazil, vol. I, pp. 174–177.

Husson, O., Tuan, H.D., Boyer, J., Chabanne, A., Caesar-Thon That, T.C. and Séguy, L. (2003b) Impacts of direct planting on permanent soil cover (DPPSC) techniques on soil biological activity in Northern Vietnam. In: *Producing in Harmony with Nature. Proceedings of the 2nd World Congress on Conservation Agriculture.* Iguassu falls, Parana, Brazil, vol. II, pp. 460–463.

Husson, O., Séguy, L., Michellon, R. and Boulakia, S. (2006) Restoration of acid soil systems through agroecological management. In: Uphoff, N., Ball, A.S., Fernandes, E., Herren, H., Husson, O., Laing, M., Palm, C., Pretty, J. and Sanchez, P. (eds) *Biological Approach to Sustainable Soil Systems*. CRC Press, Taylor & Francis, Boca Raton, Florida, pp. 343–356.

Jat, R.A., Wani, S.P., Singh, P., Pathak, P., Srinivas, K., Kumar, U., Pavani, E. and Velmurgan, R. (2012a) Effect of conservation agriculture on productivity and economics of different cropping systems under rainfed condition in the semi-arid tropics. In: *Proceedings of 3rd International Agronomy Congress*. New Delhi, India, pp. 888–890.

Jat, R.A., Wani, S.P. and Sahrawat, K.L. (2012b) Conservation agriculture in the semi-arid tropics: prospects and problems. In: Sparks, D.L. (ed.) *Advances in Agronomy* 117, 191–273.

Johansen, C., Haque, M.E., Bell, R.W., Thierfelder, C. and Esdaile, R.J. (2012) Conservation agriculture for small holder rainfed farming: Opportunities and constraints of new mechanized seeding systems. *Field Crops Research* 132, 18–32.

Jullien, F., Tivet, F., Lestrelin, G., Tran Quoc, H., Lienhard, P., Khamhung, A., Rattanatray, B., Panyasiri, K., Chabanne, A., Julien, P. and Séguy, L. (2008a) A farmer group-based approach linking research and development for the promotion of conservation agriculture in the Lao PDR. In: Chanphengxay, M. *et al.* (eds) *Proceedings of the Regional Workshop on Investing in Sustainable Agriculture: The case of Conservation Agriculture and Direct Seeding Mulch-Based Cropping Systems.* MAF, Vientiane, Laos, pp. 196–207.

Jullien, F., Khampa, S., Rattanatray, B., Kenghe, S., Vongvichit, B., Phanlak, V., Philakoun, A. and Tivet, F. (2008b) Improving smallholder's income generation by the integration of DMC by-products into pig raising activities. In: Chanphengxay, M. *et al.* (eds) *Proceedings of the Regional Workshop on Investing in Sustainable Agriculture: The case of Conservation Agriculture and Direct Seeding Mulch-Based Cropping Systems.* MAF, Vientiane, Laos, pp. 95–100.

Kassam, A., Friedrich, T., Shaxson, F. and Pretty, J. (2009) The spread of Conservation Agriculture: justification, sustainability and uptake. *International Journal of Agricultural Sustainability* 7, 292–320.

Lal, R. (1989) Conservation tillage for sustainable agriculture – tropics versus temperate environments. *Advances in Agronomy* 42, 85–197.

Lal, R. (2005) World crop residues production and implications of its use as a biofuel. *Environment International* 31, 575–584.

Legoupil, J.C. and Kingkeo, S. (2011) The Conservation Agriculture Network for Southeast Asia (CANSEA): an Initiative to Develop and Disseminate CA in Southeast Asia. In: *Resilient food systems for a changing world. Proceedings of the 5th World Congress on Conservation Agriculture.* WCCA/FSD, Brisbane, Australia, pp. 286–288.

Lestrelin, G. and Castella, J.C. (2011) Opportunities and challenges for the adoption of conservation agriculture in maize production areas of Laos. In: *Resilient food systems for a changing world. Proceedings of the 5th World Congress on Conservation Agriculture*. WCCA/FSD, Brisbane, Australia, pp. 42–44.

Lestrelin, G., Nanthavong, K., Jobard, E., Keophoxay, A., Lienhard, P., Khambanseuang, C. and Castella, J.C. (2012a) 'To till or not to till?' Opportunities and constraints to the diffusion of Conservation Agriculture in Xieng Khouang Province, Lao PDR. *Outlook on Agriculture* 41, 41–49.

Lestrelin, G., Tran Quoc, H., Jullien, F., Rattanatray, B., Khamxaykhay, C. and Tivet, F. (2012b) Conservation agriculture in Laos: Diffusion and determinants for adoption of direct seeding mulch-based cropping systems in smallholder agriculture. *Renewable Agriculture and Food Systems* 27, 81–92.

Lienhard, P., Tivet, F., Bounkhampone, B., Sosomphou, T., Sayphoummie, S., Phanthavivong, I. and Séguy, L. (2008) Direct seeding mulch-based cropping systems for rice-beef production in the plain of Jars, Xieng Khouang province, Lao PDR: an Example of 'Creation-Validation' Methodological Approach. In: Chanphengxay, M. *et al.* (eds) *Proceedings of the Regional Workshop on Investing in Sustainable Agriculture: The case of Conservation Agriculture and Direct Seeding Mulch-Based Cropping Systems*. MAF, Vientiane, Laos, pp. 210–223.

Lienhard, P., Tivet, F., Chabanne, A., Dequiedt, S., Lelièvre, M., Sayphoummie, S., Leudphanane, B., Chemidlin Prévost-Bouré, N., Séguy, L., Maron, P.A. and Ranjard, L. (2013) No-till and cover crops shift soil microbial abundance and diversity in Laos tropical grasslands. *Agronomy for Sustainable Development* 33(2), 375–384.

MBRLC (2004) *How to Farm your Hilly Land Without Losing your Soil? The sloping agricultural land technology (SALT) farming system*, 1st edn. Asian Rural Life Development Foundation International, Bansalan, the Philippines, How to Series No 1.

Mercado, A., Reyes, M. and Ella, V. (2011) Developing Conservation Agriculture in the Philippines. Communication presented at the second International Conservation Agriculture Workshop and Conference in Southeast Asia. Phnom Penh, Cambodia, 4–7 July 2011.

Nicetic, O., Le Huu, H., Trinh Duy, N., Nguyen Hoang, P., Pham Thi, S., Kirchhof, G. and van de Fliert, E. (2011) Impact of Erosion Prevention Methods on Yield and Economic Benefits of Maize Production in North West Vietnam. Communication presented at the second International Conservation Agriculture Workshop and Conference in Southeast Asia. Phnom Penh, Cambodia, 4–7 July 2011.

Panyasiri, K., Sen, P.T., Baokun, L. and Autfray, P. (2011) Regional conservation agriculture project proposal in degraded annual cropping systems areas in South East Asia. In: *Resilient food systems for a changing world. Proceedings of the 5th World Congress on Conservation Agriculture*. WCCA/FSD, Brisbane, Australia, pp. 324–327.

Ponsich, A., Chabierski, S., Sovann, P., Martin, C. and Rada, K. (2011) Cattle Fattening Opportunities through DMC Adoption in Cambodia. Poster presented at the second International Conservation Agriculture Workshop and Conference in Southeast Asia. Phnom Penh, Cambodia, 4–7 July 2011.

PRONAE (2009) Diversity and functioning of CA equipments. Training leaflet in Lao language. PRONAE document. NAFRI-NCAC, 22 pp.

Roder, W. and Maniphone, S. (1998) Shrubby legumes for fallow improvement in northern Laos: establishment, fallow biomass, weeds, rice yield, and soil properties. *Agroforestry Systems* 39, 291–303.

Sá, J.C.M., Cerri, C.C., Lal, R., Dick, W.A., Venzke Filho, S.P., Piccolo, M.C. and Feigl, B. (2001) Organic matter dynamics and carbon sequestration rates for a tillage chronosequence in a Brazilian Oxisol. *Soil Science Society of America Journal* 65, 1486–1499.

Séguy, L. (2010) Mission d'appui scientifique et technique au projet PADAC pour la création-diffusion de scénarios d'agriculture durable (SCV) au Cambodge. CIRAD, Montpellier, France, 62 pp.

Séguy, L. and Bouzinac, S. (2008) La symphonie inachevée du semis direct dans le Brésil central: Le système dominant dit de 'semi-direct'. CIRAD, Montpellier, France, 214 pp.

Séguy, L., Bouzinac, S. and Husson, O. (2006) Direct-seeded tropical soil systems with permanent soil cover: Learning from Brazilian experience. In: Uphoff, N., Ball, A.S., Fernandes, E., Herren, H., Husson, O., Laing, M., Palm, C., Pretty, J. and Sanchez, P. (eds) *Biological Approach to Sustainable Soil Systems*. CRC Press, Taylor & Francis, Boca Raton, Florida, pp. 323–342.

Slaats, J. and Lestrelin, G. (2009) *Improving cropping systems by introducing Conservation Agriculture: Taking stock of the results and methodology of research-development in southern Sayaboury province, Lao PDR*. PCADR, Vientiane, Laos, 115 pp.

Stibig, H.J., Belward, A.S., Roy, P.S., Rosalina-Wasrin, U., Agrawal, S., Joshi P.K., Hildanus, Beuchle, R., Fritz, S., Mubareka, S. and Giri, C. (2007) A land-cover map for South and Southeast Asia derived from SPOT VEGETATION data. *Journal of Biogeography* 34, 625–637.

Tao, D., Kong, L., Zhu, H., Hu, F., Xu, P., Zhou, J., Deng, X., Li, J. and Deng, W. (2008) Integration of conservation and intensification agricultural technology in Yunnan. Communication presented at the Regional Workshop on Investing in Sustainable Agriculture: The case of Conservation Agriculture and Direct Seeding Mulch-Based Cropping Systems. Phonsavanh, Laos, 28 October – 1 November 2008.

Tivet, F., Tran Quoc, H., Boyer, J., Inthavong, C., Senephansiri, S., Keodouangsy, L., Chounlamountry, T., Khamxaykhay, C., Panyasiri, K. and Séguy, L. (2008) Changes in soil aggregation, soil water holding capacity and soil biological activity under no-till systems and cropping sequence in Lao PDR. In: Chanphengxay, M. *et al.* (eds) *Proceedings of the Regional Workshop on Investing in Sustainable Agriculture: The case of Conservation Agriculture and Direct Seeding Mulch-Based Cropping Systems.* MAF, Vientiane, Laos, pp. 130–144.

Tran Quoc, H., Tivet, F., Senephansiri, S., Keodouangsy, L., Chounlamountry, T., Khamxaykhay, C. and Séguy, L. (2008) Maize yield and profit increase under no-tillage system and crop rotation with leguminous in southern Xayaburi province, Lao PDR. Communication presented at the Regional Workshop on Investing in Sustainable Agriculture: The case of Conservation Agriculture and Direct Seeding Mulch-Based Cropping Systems. Phonsavanh, Laos, 28 October – 1 November 2008.

Tuan, H.D., Husson, O., Chabanne, A., Lienhard, P. and Séguy, L. (2002) Mucuna pruriens followed by maize cultivation. A solution for less degraded and compacted soils. In: PAOPA (ed.) *Scaling-up Innovative Approaches in Agricultural Development*. Agricultural Publishing House, Hanoi, Vietnam, p. 54.

Tuan, H.D. and Doanh, L.Q. (2008) Conservation agriculture on sloping lands in Northern mountainous regions of Vietnam. In: Chanphengxay, M. *et al.* (eds) *Proceedings of the Regional Workshop on Investing in Sustainable Agriculture: The case of Conservation Agriculture and Direct Seeding Mulch-Based Cropping Systems*. MAF, Vientiane, Laos, pp. 27–36.

Valentin, C., Agus, F., Alamban, R., Boosaner, A., Bricquet, J.P., Chaplot, V., De Guzman, T., De Rouw, A., Janeau, J.L., Orange, D., Phachomphonh, K., Phai, D.D., Podwojewski, P., Ribolzi, O., Silvera, N., Subagyono, K., Thiébaux, J.P., Toan, T.D. and Vadari, T. (2008) Runoff and sediment losses from 27 upland catchments in Southeast Asia: Impact of rapid land use changes and conservation practices. *Agriculture, Ecosystems & Environment* 128, 225–238.

Von Uexkull, H.R. and Mutert, E. (1995) Global extent, development and economic impact of acid soils. *Plant and Soil* 171, 1–15.

8 Conservation Agriculture in China

Li Hongwen, He Jin and Gao Huangwen
China Agricultural University, Beijing, China

8.1 Introduction

China is situated in the south-eastern part of the Eurasian continent, and water shortage is one of the main constraints that limit the sustainable development of agriculture. The total amount of water resources in the country ranks sixth in the world and the annual renewable water resources are 2812.4 billion m^3. Agriculture uses 64.4% of the water resource. However, the amount of water resources per capita and per unit cultivated land is far less than the world average. Furthermore, water distribution is spatially uneven. With 42.4% population and 60.2% farmland, Northern China, which mainly includes the north-east ridge tillage area, North China Plain annual double-cropping area, north-west oasis farming area, farming-pastoral area and loess plateau with one crop a year occupies 19.8% of total water resources, and the agriculture mainly relies on rainfall.

China is also one of the countries in the world that has been severely affected by desertification. The area covered by desertification is 370,000 km^2 and direct annual economic loss by desertification amounts to 54 billion yuan. In the main cropping regions, particularly in the dryland areas of Northern China, the soils suffer severe degradation and desertification. Water and wind erosion are the main factors for soil degradation. The threat of water erosion in dryland areas is influenced by the amount and intensity of the rainfall, the type of irrigation, erodibility of the soil, cropping methods and management factors, and erosion control practices. The impact of raindrop or flood irrigation on the soil surface is the most important part of the erosion process. Wind erosion is also caused by sandstorms, which have caused much degradation of farmlands in the past few decades (Zhou, 2004).

Long-term conventional tillage (ConvT), characterized by conventional cultivation using mouldboard plough and rotary hoes, and the removal of crop residues from the fields for use as fodder and household fuel, coupled with the above-mentioned problems, has led to soil, water and nutrient loss and the degradation of the fragile soil resource base, leading to reduced crop productivity. The dryland areas are most affected by soil degradation and loss of productivity. The severe land degradation and serious environmental problems have led the Chinese government to emphasize the need for the implementation of farming practices that contribute to the conservation of soil and water, with tillage as an important component of these practices. An important strategy lies in the use of Conservation Agriculture (CA).

8.2 The Development of Conservation Agriculture in China

8.2.1 Conservation Agriculture development during the first period (1970–1990)

During 1970–1990, Chinese scientists conducted studies on single or selected component CA technologies, and the experiments were successful. However, the CA system based on human power could not be extended due to the lack of suitable no-till seeders. China has two main problems that restricted the use of the available no-till seeders during this period: (i) Chinese farmers have only 0.5 ha crop land per family and the large-size no-till seeders available from overseas were not suitable for such small farmland; and (ii) China needs high yield from unit land due to more people and less land. Therefore CA needs to produce higher yields and the no-till seeders need to work in the fields with a thick layer of crop residues on the soil surface; for example, under direct seeding of wheat in a field with more than 20 t ha^{-1} maize residues no such no-till seeder existed for use.

8.2.2 Conservation Agriculture development during the second period (1991–2000)

In this period, CA development was mostly around the above-stated problems, and the main development was the availability of small, light seeders for use on small-scale farms. The characteristics of the seeders were:

1. The unit weight of the seeders was reduced to 250–400 kg per machine from the common weight of 800 kg per machine of imported overseas no-till seeders; this change matched with the force required (280–600 kg m^{-1}) for Chinese small tractors.
2. The seeder price was reduced to less than 10,000 yuan per machine compared to a cost of more than 80,000 yuan per machine for an imported no-till seeder.
3. Separate vertical placement of fertilizer and seeds, which not only meets the requirement of deep placed fertilizer (under seeds ~5 cm) but also meets the requirement of putting a relatively large amount of fertilizer (600 kg ha^{-1}) required for obtaining higher yields from poor fertility soils.

8.2.3 Conservation Agriculture development during the third period (2001–2009)

Power-driven no-till seeders were developed for adoption of CA in the double-cropping region (Gao *et al.*, 2008).

From 2000, CA expanded to annual double-cropping areas of North China Plain, where winter wheat and summer maize yields are 7 and 9 t ha^{-1}, respectively. The no-till seeders developed for the areas with single cropping and relatively low yields were not suitable for this region. To develop high anti-blockage performance no-till seeders, several mechanisms of power-cutting residue were innovated, for example:

- Strip rotary hoe cutting mechanism, in which the rotary blades cut out stalks and roots to avoid opener blockage, and at the same time loosen the soil for better seedbeds with reduced opener forward force. Each furrow has two rows for seeding and one row for applying fertilizer; only 30% of soil was tilled and 70% remained non-tilled.
- Strip rotary chopping mechanism, in which the rotary blades do not penetrate into the soil and the blades just cut off or beat down the residues, which twist on the chisel opener. The chisel opener digs a deep ditch and the fertilizer is placed at the bottom of it, a double-disc opener re-opens a shallow ditch for the seeds. Seeds and fertilizer are thus placed at two depths.

8.2.4 Conservation Agriculture development since 2009

Following the development of no-till seeders, the area under the CA greatly expanded. By the end of 2011, CA has been extended to 5.67 Mha in China. According to the future plans, CA will be extended to about 22 Mha by the end of 2015. The main targets of CA during this period would be raising the yield and resource-use efficiency through improved technology and equipment.

8.3 Research Results Reported in China

The effects of long-term CA and ConvT were evaluated in the typical cropping areas in China. In these experiments, along with CA, various kinds of resource conservation technologies were used, such as the no/minimum tillage (NT), controlled traffic tillage treatment and permanent raised beds (PRB) (Fig. 8.1).

8.3.1 Description of experimental sites

The annual rainfall ranges from 200 to 600 mm in the loess plateau and one crop is grown annually in this region. Soil in the loess plateau of China is easily erodible and is intensively cropped with dryland winter wheat. Limited crop-available water is one of the major factors constraining agricultural production in the loess plateau, and severe erosion has resulted in degradation of the soils and reduced water-holding capacity of the soils (Zha and Tang, 2003).

In cold and semi-arid north-east China, spring maize is one of the most important grain crops in terms of area and output (Liu *et al.*, 2002). The annual rainfall varies from 400 to 1000 mm and the average cumulative evaporation is ~1800 mm, which is ~4 times higher than the average total rainfall received during the growing season for spring maize. Therefore, the low status of soil moisture in the root-zone usually limits productivity of spring maize in this region. Conserving moisture accumulated in the root-zone during the rainfall season can increase productivity of spring maize in the dry north-east China region.

In annual double-cropping areas of the North China Plain, the annual rainfall is 450–800 mm and the annual cumulative evaporation greatly exceeds the annual rainfall. Since the 1980s, the cropping system in this region has changed from a single crop to a double cropping (winter wheat–summer maize) (Liu, 2004). Therefore, the use of water has greatly increased, and water scarcity has become much more serious at the present time.

In farming–pastoral ecotone of Inner Mongolia, the annual rainfall is 450–500 mm and the annual cumulative evaporation is 1300–1880 mm, which greatly exceeds the annual rainfall. In some parts of the farming-pastoral areas, the annual rainfall is even less than 50 mm (He *et al.*, 2010a). In the last 100 years, large areas of grasslands have been converted into cropland due to increased population and food demand (Zhang *et al.*, 1998). The agriculture–pasture transition region has about 32.8 Mha of land, representing 27.8% of the total land area of Inner Mongolia (LZU, 2005). In this region, the conversion of grassland to cropping combined with insufficient rainfall and wind erosion has resulted in serious soil nutrient depletion and structural deterioration of soils (Liu *et al.*, 2007).

In north-west China areas, water shortage is definitely one of the major constraints to the production of agricultural crops. The average precipitation varies from 40 to 200 mm (Xie *et al.*, 2005) and annual potential evaporation in this region exceeds 1500 mm; the water requirement for one season of spring wheat, the most popular cereal crop in north-west China, is >600 mm.

8.3.2 Effects of Conservation Agriculture on soil water conservation

Infiltration

As indicated in Fig. 8.2, in the Chenghuang village, Linfen City of the loess plateau, soil water infiltration rate under no tillage (NT) and ConvT decreased with time. In the first 3 min of the infiltration test, the differences between the infiltration rates under ConvT and NT plots were negligible, probably due to similar soil physical properties in the

Fig. 8.1. The resource conservation technologies used: (a) controlled traffic and (b) permanent raised beds.

upper layer. However, when the water infiltrated into deeper soil layers, NT plots showed significantly ($P<0.05$) higher infiltration rates than ConvT plots. Consequently, total infiltration under NT was greater, and final (steady state) infiltration rate in NT plots (17.0 mm min^{-1}) was 4 times that in ConvT plots (4.3 mm min^{-1}) (Bai *et al.*, 2008;

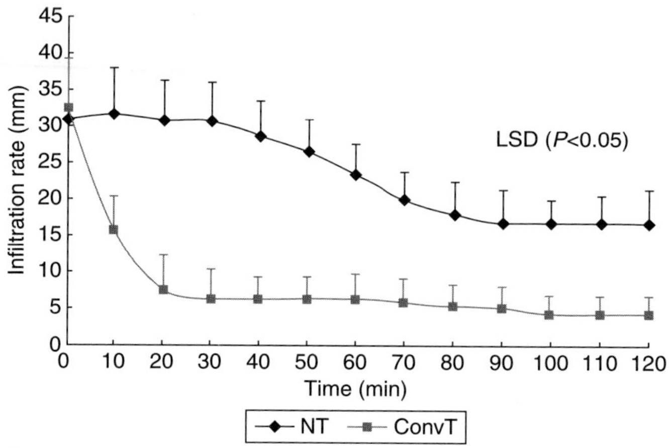

Fig. 8.2. Changes in soil infiltration rate during 120 min under no-till and conventional tillage treatments. NT, no-tillage; ConvT, conventional tillage.

He *et al.*, 2009a; Wang *et al.*, 2009). These results showed that the CA treatment created increased capacity for water storage.

Soil moisture depletion

In Sujiatun (Liaoning province) and Lianxi (Heilongjiang province) in north-east China, the positive effects of PRB and NT on conserving water were obtained compared with ConvT (He *et al.*, 2010b). Soil moisture depletion from any soil layer during a dry cycle is due to the loss of water through evaporation, drainage and lateral seepage. Data on moisture depletion in the soil profile between sowing and harvest in Sujiatun and Lanxi are presented in Table 8.1.

For the top 0–15 cm soil layer, the moisture depletion under ConvT was 6.7% and 4.9% higher than that under PRB and NT in Sujiatun, and this value was 5.6% and 3.1% higher in Lanxi. Under ConvT, the exposure of soil without residue cover and excessive soil disturbance as a result of frequent ploughing were the reasons for the faster rate of moisture depletion in the topsoil layer. Residue cover was effective in reducing the loss of water through evaporation from the soil surface by forming a barrier between the soil surface and the atmosphere, reducing moisture loss (Gupta and Acharya, 1993). In the middle soil layer the trend was reversed; PRB had the highest moisture depletion, particularly at 15–45 cm, significantly ($P<0.05$) increasing moisture depletion by 5.7–8.9% and 6.3–9.1% at 15–30 and 30–45 cm depths compared with ConvT. NT also showed a trend to a greater mean moisture depletion in the middle soil layer, but the difference was not significant. In the deepest soil layer (60–100 cm) the differences in soil moisture depletion among different treatments was small in both Sujiatun and Lanxi. Furthermore, under PRB, capillary continuity is least disturbed due to no-tillage and controlled traffic (Li *et al.*, 2007), and this accelerated the water uptake by the spring maize roots from the surrounding soil.

Soil water content

During 1998–2005 in Linfen of the loess plateau, the controlled traffic treatment had significantly ($P<0.05$) higher mean soil water storage in the 0–150 cm soil layer at wheat sowing than in ConvT (Table 8.2). Compared to ConvT, the mean soil water content in controlled traffic treatments at sowing was over 9.0% higher. Besides the first year of the experiment, water storage in the two controlled traffic treatments was always higher than that in ConvT treatment during the experimental period, with significant difference in the relatively dry years in 1999, 2002 and 2004. The results

Table 8.1. Treatment effects on soil moisture depletion (mm) from different layers during a period of 162 days in Sujiatun and Lanxi (averaged over 2005, 2006 and 2007).

		Soil depth (cm)				
Site	Treatment	0–15	15–30	30–45	45–60	60–100
Sujiatun, Liaoning	PRB	12.1a	12.2b	10.8b	10.8a	26.4a
	NT	12.3a	11.5a	10.2ab	11.1a	25.9a
	ConvT	12.9b	11.2a	9.9a	10.8a	26.0a
Lanxi, Heilongjiang	PRB	12.5a	12.9b	11.9b	11.1a	28.3a
	NT	12.8ab	12.4ab	11.3a	10.8a	28.5a
	ConvT	13.2b	12.2a	11.2a	10.9a	28.0a

PRB, permanent raised beds; NT, no tillage; ConvT, conventional tillage. Means within a column followed by the same letters are not significantly different ($P<0.05$).

Table 8.2. Soil water (mm) in the 0–150 cm soil layer at wheat planting time under three tillage systems during 1998–2005.

Treatments	1998	1999	2000	2001	2002	2003	2004	2005	Mean	S.D.
NTCN	336.3a	332.6a	287.6a	362.8a	336.0a	403.9a	338.8a	355.8ab	344.2a	32.8
STCN	334.3a	337.3a	284.4a	366.2a	347.8a	402.8a	330.1a	358.1a	345.1a	33.9
ConvT	337.3a	251.8b	282.2a	350.8a	283.6b	283.6b	285.3b	342.7b	314.8b	42.3

NTCN, controlled traffic with no tillage and full residue cover; STCN, controlled traffic with shallow tillage and full residue cover; ConvT, random traffic with conventional tillage and partial residue cover. Means within a column in the same year followed by the same letter are not significantly different at $P<0.05$; S.D., standard deviation.

(Table 8.2) also showed that soil water in controlled traffic treatments was less variable (indicated by S.D.) over the years than that in ConvT. This was as a result of higher rainfall storage during the fallow in the controlled traffic treatments, especially in the low rainfall years.

As shown in Gaocheng (Hebei province) in the North China Plain, the mean soil water storage in the surface layer (0–30 cm) in conventional ploughed soils from 1999 to 2009 was 55.8 mm, while in NT soils it was higher, approximately 60.0 mm (Table 8.3). In the dry years of 2001 (annual rainfall: 347 mm), 2004 (annual rainfall: 373 mm), 2006 (annual rainfall: 400 mm) and 2009 (annual rainfall: 389 mm) particularly soil water storages in NT were 49.9, 48.5, 48.1 and 45.9 mm, and in ConvT plots were 40.3, 41.1, 40.2 and 34.3 mm, representing a mean improvement of 19.3% in NT treatment (He *et al.*, 2011).

In the semi-arid agriculture–pasture transition region in Shang Tuhe village (41°06′N, 111°27′E), Wuchuan (Inner Mongolia, China), soil water storage was similar under the NT, ST, RT and ConvT treatments at the beginning (1998) (Table 8.4). However, 3–4 years later the differences among the tillage treatments started to emerge. During the 1998 to 2007 period, mean soil water storage in the 0–30 cm layer was about 10% greater in the NT plots (59 mm) than in the ConvT (54 mm). In the dry years of 2003, 2006 and 2007, soil water storage in NT plots increased on average by 8 mm (19%) (He *et al.*, 2009b).

In Zhangye of Hexi Corridor of northwest China (2005–2007), the mean volumetric water contents in the 0–30 cm soil profile of ConvT, NT and PRB were 0.167, 0.182 and 0.187 $cm^3\ cm^{-3}$, respectively, in 2005 (first year); and the mean soil moisture content in NT and PRB treatments was significantly ($P<0.05$) greater, by about 9.0% and 12.0%, respectively, than in ConvT (Fig. 8.3). In 2006 (second year), the mean volumetric water contents for 0–30 cm soil depth in NT and PRB were 8.7% and 14.2% higher at most growth stages of the wheat crop than in ConvT, which was significant ($P<0.05$).

Table 8.3. Soil water storage (mm) at winter-wheat seeding time of no tillage and conventional tillage at 0–30 cm soil depth from 1999 to 2009.

Treatment	1999	2000	2001	2002	2003	2004	2005	2006	2007	2008	2009
Annual rainfall (mm)	583	596	347	518	614	373	521	400	521	542	389
NT	60.1a	65.0a	49.9a	58.5a	72.5a	48.5a	58.6a	48.1a	77.3a	76.0a	45.9a
ConvT	61.3a	62.2a	40.3b	58.1a	70.6a	41.1b	55.8a	40.2b	74.9a	75.2a	34.3b

NT, no tillage; ConvT, conventional tillage. Values within a column followed by the same letters are not significantly different ($P<0.05$).

Table 8.4. Soil water stored in the surface soil layer (0–30 cm) at time of planting of spring wheat and oat.

Treatment	1998	1999	2000	2001	2002	2003	2004	2005	2006	2007	Mean
NT	41.3a	50.9a	68.8ab	66.8a	63.6a	49.8a	83.5ab	71.7a	55.5a	50.1a	60.2
ST	42.2a	52.2a	71.2a	63.1ab	[illegible]	46.3ab	85.2a	72.8a	51.9a	48.1a	59.2
RT	39.9a	50.6a	69.3ab	59.4bc	61.7a	44.7b	80.1bc	69.4a	52.5a	46.9a	57.5
ConvT	40.7a	50.2a	65.9b	56.2c	56.6b	38.7c	78.1c	63.6b	45.2b	40.6b	53.6

Values are expressed in mm. Values within a column in the same year followed by the same letters are not significantly different ($P<0.05$). NT, no-tillage with straw cover; ST, subsoiling with straw cover; RT, rototilling with straw cover; ConvT, conventional ploughing tillage.

Similar effects were found in 2007, indicating that PRB probably provided more water for wheat growth by maintaining greater soil water content than in ConvT (He *et al.*, 2008).

The higher stored water in the CA systems contributed to a better growth of the winter wheat. At the heading stage of the winter wheat in 2005, the field treated with the CA systems after 9 years had higher water content (Fig. 8.4), while the field treated with the ConvT was cracking because of water shortage.

8.3.3 Effects of Conservation Agriculture on wind erosion

As shown in Table 8.5, in Yanggao of the loess plateau (one crop a year region), the wind-blown sediment transport produced per sample was 44.7% less in the NT plot compared to the ConvT plot. In Lingyuan of the north-east ridge tillage areas, the wind-blown sediment transport per sample of NT was (37.3%) significantly lower than in ConvT. In the North China Plain annual double-cropping areas, the NT land produced 12.1% less dust than NT in Changping; the NT land produced 4.2 g of dust, which was significantly less than the 5.0 g observed in the ConvT plot in Yanqing, while the NT land produced 12.7 g of wind-blown sediment transport per sample in both Zhangbei and Fengning, which indicates 70% deduction ($P<0.05$) as compared to the traditionally tilled land. In the farming–pastoral areas, land under NT produced 34.2% less wind-blown sediment transport per sample than ConvT in Chifeng, while the wind-blown sediment transport per sample of NT was (61.6%) significantly lower than that in ConvT in 2003 in Zhenglanqi. Furthermore, in Hetian of the north-west oasis farming area, the wind-blown sediment transport produced per sample of NT was 7.4 g, which was 92.9% less than in land under ConvT.

8.3.4 Effect of Conservation Agriculture on soil chemical properties

Soil organic matter

In northern China, several studies have demonstrated that long-term CA improved soil organic matter (SOM) as compared to that under ConvT (Table 8.6). In Linfen of Chinese loess plateau, the mean SOM in the 0–30 cm soil layer for NT was 1.5% higher than for ConvT after 16 years of experimentation. In the surface soil layer (0–10 cm), the mean SOM in NT was significantly higher than that in the ConvT treatment. In Beijing, after 8 years, the mean SOM in the 0–10 cm layer of NT was 10.5% higher than that in ConvT. In the 10–20 cm layer, the mean SOM in NT was 13.6% greater than that in the ConvT. In the 20–30 cm soil layer, however, no significant differences were observed among the tillage treatments. In Wuchuan, the mean SOM in the 0–10 cm soil layer was 16.5 g kg^{-1} for NT treatment after 10 years, which was significantly greater than the 13.4 g kg^{-1} observed on ConvT plots. The SOM difference between

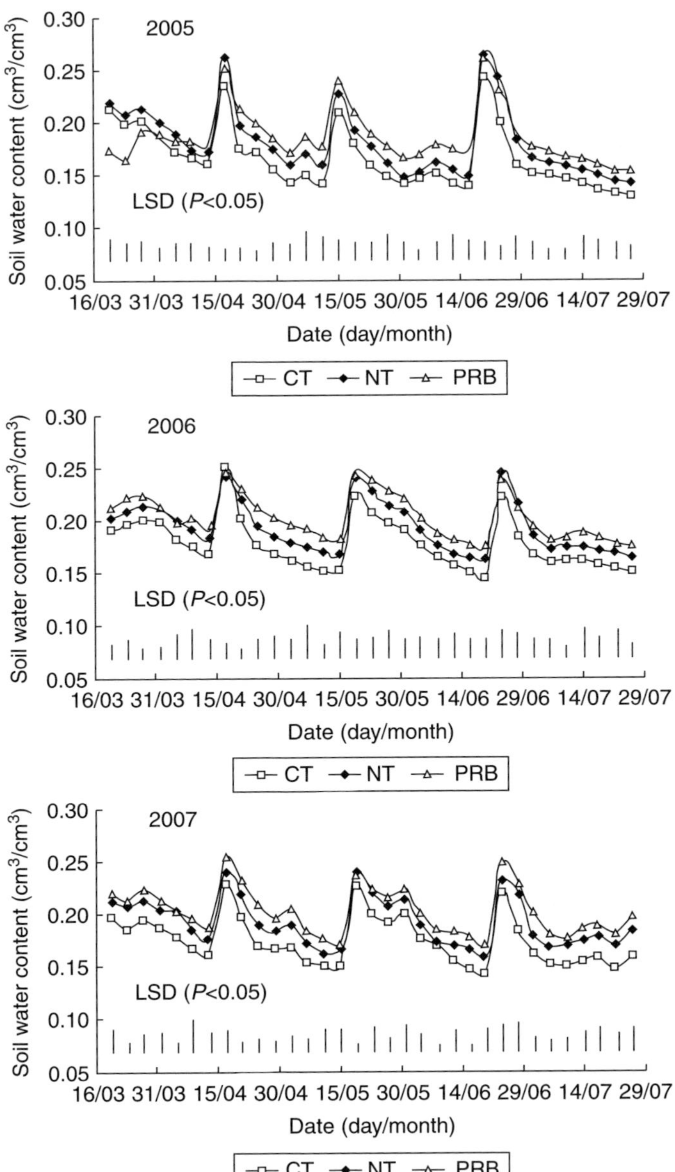

Fig. 8.3. Mean soil volumetric water content in the surface soil layer (0–30 cm) under three treatments during wheat growing period, 2005–2007. ConvT, conventional tillage; NT, no tillage; PRB, permanent raised beds.

(a) (b)

Fig. 8.4. Winter wheat performance: (a) heading stage of winter wheat in Conservation Agriculture treatment; (b) heading stage of winter wheat in conventional tillage treatment.

Table 8.5. Comparison of soil sediment transport under no-till and conventional tillage treatments.

Region	Testing site	Collection period	NT (g)	ConvT (g)	Reduced rate in comparing NT to ConvT (%)
Loess plateau of China	Yanggao, Shanxi	25.3.2004–3.4.2004	8.4	15.1	44.7
North-east ridge tillage areas	Lingyuan, Liaoning	25.3.2004–3.4.2004	10.2	16.3	37.4
North China Plain	Fengning, Hebei	22.3.2002–21.4.2002	12.7	42.5	70.0
	Zhangbei, Hebei	8.4.2002–8.5.2002	12.7	42.5	70.0
	Changping, Beijing	28.3.2005–17.4.2005	16.7	19.0	12.1
	Yanqing, Beijing	16.3.2005–20.3.2005	4.2	5.0	17.0
Farming-pastoral areas	Chifeng, Inner Mongolia	22.4.2003–3.5.2003	4.7	7.1	34.2
	Zhenglanqi, Inner Mongolia	23.3.2003–27.4.2003	11.3	25.0	54.8
	Wuchuan, Inner Mongolia	26.3.2003–6.4.2003	2.9	7.4	61.6
North-west China	Hetian, Xinjiang	16.3.2004–27.4.2004	7.4	105.5	92.9

Table 8.6. Soil organic matter content (g kg^{-1}) in the 0–30 cm soil layers under no-till and conventional tillage treatments in Linfen after 16 years, in Beijing after 8 years and in Wuchuan after 10 years of experimentation.

Sites	Treatment	Soil depth (cm)		
		0–10	10–20	20–30
Linfen, loess plateau	NT	18.2a	11.1b	6.3a
(Wang *et al.*, 2008)	ConvT	15.0b	13.8a	9.2b
Beijing, North China Plain	NT	16.5a	15.9a	13.0a
(Zhang *et al.*, 2009)	ConvT	14.9b	14.0b	12.8a
Wuchuan, farming-pastoral ecotone	NT	16.5a	9.6a	7.0a
(He *et al.*, 2009a)	ConvT	13.4b	7.4b	6.5a

Values within a column in each experimental site followed by the same letters are not significantly different ($P<0.05$).

NT and ConvT declined in the deeper layers, but were still significant at 20 cm depth.

The SOM increase under CA is attributed to the straw input and reduced decomposition of organic matter under NT. The higher SOM in the topsoil layer under CA shows that results from other arid regions can be applied in China. Roldan *et al.* (2005) recorded that CA resulted in SOM increase of up to 33% in the 0–5 cm layer in Mexico. In Germany, Koch and Stockfisch (2006) also reported that under CA SOM increased in the top 0–10 cm soil layer.

Total N

In northern China, soil total N showed the same trend as that observed in SOM in relation to tillage treatments (Table 8.7). Zhang *et al.* (2009) showed that total N in the 0–10 cm soil layer in Beijing was 24.3% higher under NT than under ConvT, and the improvement in total N in the 10–20 cm soil layer was 18.6% under NT after 8 years. No significant differences were apparent in the 20–30 cm layer. In Linfen, Wang *et al.* (2008) indicated that after 16 years, total N in the 0–10 cm layer under NT increased by 51.5% compared to that in ConvT. The differences in total N in the deeper layers (deeper than 10 cm) were not significant. A similar result was reported by He *et al.* (2009a) in Wuchuan after 10 years of NT management, but the differences were significant only in the 0–10 cm soil layer.

Tillage-induced changes in soil organic N are directly related to the changes in soil organic C, and the higher total soil N under NT in our experiments is consistent with the findings of other researchers (e.g. Torbert and Reeves, 1995; Thomas *et al.*, 2007).

Available phosphorus

Under different long-term tillage management treatments, available P in soils under NT was 48.5%, 56.3% and 10.5% higher than under ConvT soils in the 0–10 cm layer in Beijing (Zhang *et al.*, 2009), Linfen (Wang *et al.*, 2008) and Wuchuan (He *et al.*, 2009a), respectively (Table 8.8). In the 10–20 cm

Table 8.7. Soil total N (g kg^{-1}) in 0–30 cm depth in no-tillage and conventional tillage in Linfen (16 years), Beijing (8 years) and Wuchuan (10 years).

		Soil depth (cm)		
Sites	Treatment	0–10	10–20	20–30
Linfen, loess plateau	NT	1.03a	0.67a	0.40a
(Wang *et al.*, 2008)	ConvT	0.68b	0.66a	0.35a
Beijing, North China Plain	NT	1.38a	1.02a	0.62a
(Zhang *et al.*, 2009)	ConvT	1.11b	0.86b	0.58a
Wuchuan, farming-pastoral	NT	0.52a	0.30a	0.25a
ecotone (He *et al.*, 2009a)	ConvT	0.42b	0.24a	0.27a

Values within a column in each experimental site followed by the same letters are not significantly different ($P<0.05$).

Table 8.8. Available P concentrations (mg kg^{-1}) in 0–30 cm soil depth in no-till and conventional tillage treatments in Linfen (16 years), Beijing (8 years) and Wuchuan (10 years).

		Soil depth (cm)		
Sites	Treatment	0–10	10–20	20–30
Linfen, loess plateau	NT	35.0a	10.5a	6.4a
(Wang *et al.*, 2008)	ConvT	22.4b	22.9b	7.6a
Beijing, North China Plain	NT	20.2a	16.9a	11.3a
(Zhang *et al.*, 2009)	ConvT	13.6b	13.9b	11.1a
Wuchuan, farming-pastoral	NT	17.9a	8.3a	5.4a
ecotone (He *et al.*, 2009a)	ConvT	16.2b	10.1b	6.3a

Values within a column in each experimental site followed by the same letters are not significantly different ($P<0.05$).

soil layers, the P content was 54.1% and 17.8% lower under NT than under ConvT in Linfen and Wuchuan, respectively. In the 20–30 cm soil layer the difference was not significant.

Long-term CA management generally leads to a stratification of available P in soils. The accumulation of P in the top layer in NT is attributed to the limited downward movement of particulate-bound P in NT soils and the upward movement of nutrients from deeper layers through nutrient uptake by roots.

8.3.5 Effect of Conservation Agriculture on soil physical properties

Bulk density

Since long-term NT is reported to reduce hard pan, it can decrease soil bulk density. For example, in the loess plateau Wang *et al.* (2008) showed that after 16 years of NT, bulk density to 30 cm depth was 2.2% lower in NT than under ConvT at the Linfen experimental site. In the North China Plain, Zhang *et al.* (2009) demonstrated that the bulk density at 0–30 cm depth in NT was 1.2–2.0% lower than that in ConvT in Beijing after 8 years; He *et al.* (2008, 2009a) also reported a positive effect of NT on bulk density in long-term NT management in the farming pastoral ecotone and north-west oasis farming areas. The lower bulk density in NT can be attributed to the increased soil organic C and biotic activity following long-term NT practice.

Soil pore size distribution

Measurements made in northern China suggest that long-term NT practices positively affect total porosity and aeration porosity. For example, He *et al.* (2009a) demonstrated that after 10 years of study, total porosity in the 0–30 cm depth was 42% on NT plots and 38% on ConvT plots, and the increased porosity was largely due to an increase in macro-porosity and meso-porosity on the NT plots. In the North China Plain, Zhang *et al.* (2009) also showed that after 8 years of implementation, NT registered 10.6% and 8.6% higher aeration porosity than ConvT in the 0–30 cm depth at Daxing and Changping, respectively. This improvement in soil porosity under NT is most probably related to the beneficial effects of SOM caused by NT and residue cover.

Soil water-stable aggregates

Soil aggregation is important as it improves soil structure and reduces soil erosion. Several studies in North China showed that in long-term NT soils, the percentage of water-stable aggregates of the largest size class (>2 mm) was higher than in ConvT soils, and the percentage of water-stable aggregates of the smallest size class (<0.25 mm) was greater in ploughed soil. For example, Li *et al.* (2007) indicated that macro-aggregates constituted 58.6% and 53.5% in 0–10 and 10–20 cm depths, respectively, in the soil under NT compared to 45.1% and 47.4% for ploughed soil after 15 years in the loess plateau. He *et al.* (2009a) found that soils from NT plots contained more macro-aggregates (13–37%) than those under ConvT at 0–30 cm depth. The percentage of micro-aggregates was 25–59% greater in traditional tilled soils. The higher macro-aggregates can be attributed to the greater biological activity and a reduction in breakdown of surface soil aggregates in NT due to residue cover on the soils.

8.3.6 Effect of Conservation Agriculture on crops yields

The improvement in soil characteristics contributes to better crop growth (Fig. 8.5) and greater crop productivity. Many studies have evaluated the effects of CA and ConvT practices on crops yields in northern China.

Winter wheat

Measurements made in northern China suggest that NT is effective in increasing crops yields. In the loess plateau, Chen *et al.* (2008) showed that the mean (1999–2006) yield of NT plots was 9.3% greater than that of traditionally tilled plots. In 3 out of 8 years, NT treatments showed significantly higher yield than in ConvT ($P<0.05$) (Table 8.9). In the North China Plain, He *et al.* (2009b) also demonstrated that winter wheat yield and selected yield components were affected by tillage/residue treatments. The mean yield, grains per spike and thousand kernel weight for NT was 2.2%, 3.0% and 6.3% higher than that in ConvT, respectively.

Summer maize

He *et al.* (2009b) conducted a 2-year (2002–2003) experiment with summer maize in Dingxing of Hebei province, North China Plain, and they found that in the first growing season of 2002 NT treatments produced 5.7% higher yields than in ConvT treatments. In the second growing season (2003), NT again produced the highest yield and significantly ($P<0.05$) enhanced (6.8%) mean maize yield as compared to that in the ConvT treatment. In Beijing, Zhang *et al.* (2009) also showed that average (2004–2007) yield for summer maize in NT was 138 kg ha^{-1} (3.24%) higher than that in ConvT.

Spring wheat and oat

In a 3-year (2005–2007) experiment in Hexi Corridor of north-west China, He *et al.* (2008) showed that spring wheat yield was not significantly affected by different treatments in the first growing season of 2005, but pronounced yield advantage was

Fig. 8.5. Spring wheat growth under Conservation Agriculture (right) and conventional tillage (left) in Gansu, north-west China.

Table 8.9. Winter wheat yield (t ha^{-1}) in no-till and conventional tillage treatments in Linfen of Loess Plateau (1999–2006) (Chen *et al.*, 2008).

Treatments	1999	2000	2001	2002	2003	2004	2005	2006	Mean
NT	3.27a	2.48a	3.08a	3.68a	3.51a	4.01a	2.71a	4.43a	3.40a
ConvT	3.79b	1.46b	2.91a	3.52a	3.64a	4.12a	1.91b	3.50b	3.11b

Means within the same column followed by the same letters are not significantly different at $P<0.05$.

observed in NT plots in 2006. The mean wheat yields in ConvT and NT treatments were 5981 kg ha^{-1} and 6314 kg ha^{-1}, indicating a significant (P<0.05) yield improvement of 5.6% under NT as compared to the ConvT treatment. A similar trend continued in 2007. In Changping of Beijing, Zhang *et al.* (2009) also found tillage treatments significantly affected spring maize yields. Compared with ConvT, NT treatments increased mean (2000–2007) spring maize yields by 276–376 kg ha^{-1} (4.25–5.80%), and the improvement was significant (P<0.05) in 5 of 8 years.

He *et al.* (2009a) showed that the 5-year mean oat yields under NT treatments were 14.6% greater than those in the ConvT plots, and the differences were significant (P<0.05) in 2 out of 5 years.

8.4 Conservation Agriculture Machinery

According to a survey, there are more than 100 factories manufacturing CA machines in China. These machines mainly include manual seeders, no-till seeders driven by animals, 2-wheel tractors and 4-wheel tractors, minimum tillage and weeding machines. Some typical CA machineries are as follows.

8.4.1 Manual seeder

Li Seeder (Fig. 8.6) is a typical manual seeder for seeding in no-till maize and soybean. It can be used in sloping and small farmlands.

Fig. 8.6. (a, b) Li seeder, (c) planting with Li seeder in the field and (d) maize crop planted with Li seeder.

The total weight of Li Seeder is 2.2 kg and one farmer can seed 0.2–0.3 ha day^{-1}.

8.4.2 No-till seeder powered by two-wheel tractor

This seeder (Fig. 8.7) was specially developed by the China Agricultural University and Rogro Machinery Sales Pty Ltd, Australia, for use with a two-wheel tractor on small farmland. It is suitable for use in no-till and residue-covered field for seeding most crops. Variable tine layout is available to seed in different soil and residue conditions, the integral fertilizer box allows accurate fertilizer placement in seed rows and it has integral press wheels. The weight is 150 kg and the working speed is 2.0–4.5 km h^{-1}.

Fig. 8.7. Seeder powered by two-wheel tractor.

Fig. 8.8. BMF-7 no-till wheat seeder.

8.4.3 No-till seeder powered by four-wheel tractor

Under light residue cover conditions (wheat or rice residues), Chinese scientists developed a passive anti-blocking no-till seeder. Under maize residues, the active anti-blocking method was adopted because of its ability to handle maize stubble.

Passive anti-blocking no-till seeder

This no-till seeder normally uses the multi-beam structure to get high trash flow and/or anti-blocking components to handle residues to achieve no-till seeding in wheat or rice residue-covered soils.

2BMF-7 NO-TILL WHEAT SEEDER. the 2BMF-7 no-till wheat seeder (Fig. 8.8) uses the multi-beam structure to achieve anti-blocking. In this design, residue clearance is maximized by mounting three openers on the front, two on the middle and two on the rear bar of the machine. During seeding, the machine uses narrow-point openers and press wheels to place and firm seed and fertilizer at depths of 5 cm and 10 cm, respectively. The machine is set to the 16 cm row spacing, commonly used by local farmers with an operating width of 1.12 m.

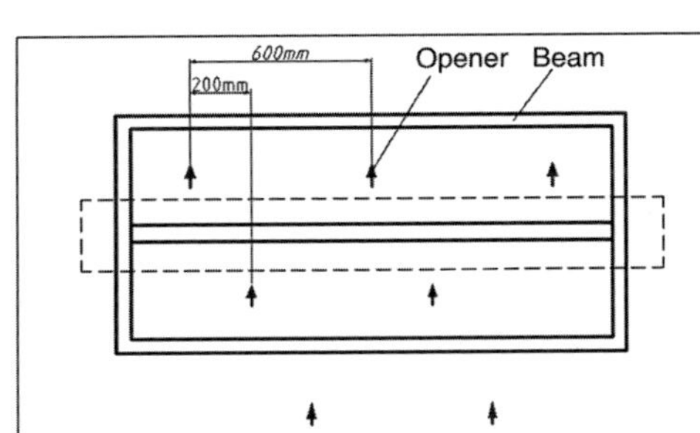

2BMQF-4 NO-TILL CORN SEEDER. The main anti-blocking device for 2BMQF-4 no-till corn seeder (Fig. 8.9) is a disc coulter combining with dual dentate discs. For seeding under NT, the disk coulter cuts the residue and then the following dual dentate discs remove the residue from the seeding row, so the narrow-point opener can easily complete

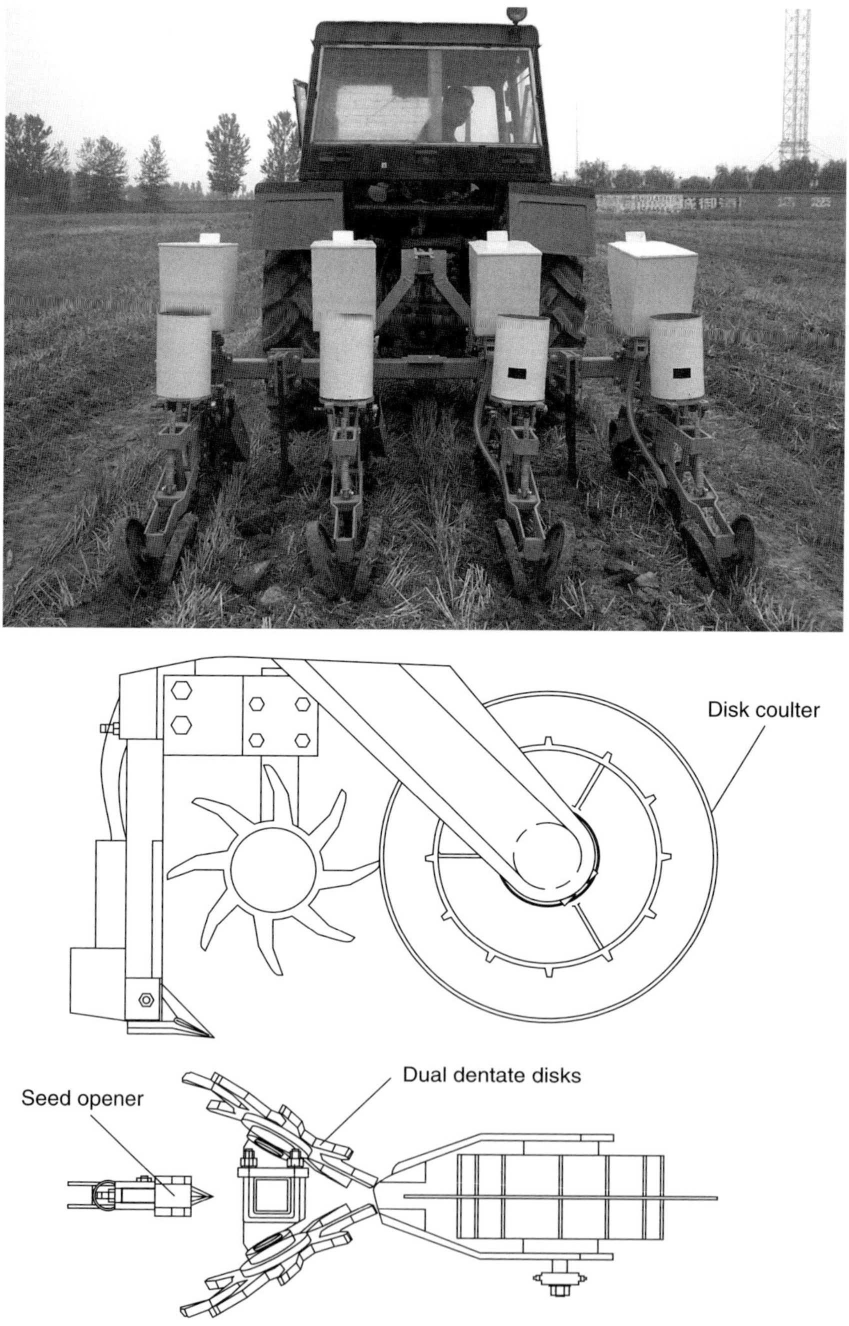

Fig. 8.9. BMQF-4 no-till maize seeder.

NT seeding. Furthermore, the wide row spacing (45–65 cm) in maize also helps the machine to get high trash flow, which is needed for handling residues.

Active anti-blocking no-till seeder

STRIP ROTARY HOE MINI-TILL SEEDER. The strip rotary hoe mini-till seeder uses power-driven rotary blades to not only loosen the seedbed soil but also to cut stalks and crush and break roots so that the openers can pass through easily. A 2BMFS-10 no-till wheat seeder is shown in Fig. 8.10, and the main parameters of the machine are: the rotary tillage width 12 cm, NT width 28 cm, soil disturb rate 30%.

Fig. 8.10. 2BMFS-10 no-till wheat seeder.

In the mini-till seeder, the main functions of the rotary blade are to loosen the soil, cut stalks and crush roots. Normally, in order to reduce blade wearing while loosening soil and crushing roots, the rotary hoe cultivator usually has a low speed (about 200 rpm). On the other hand, if we want to ensure cutting the stalks, the chopper should have a relatively high speed (over 1500 rpm). According to the results from an experiment, 400 rpm could be selected to meet the soil loosening, root crushing requirements with minimum stalk-cutting ability.

Fertilizer application is the other key issue for the mini-till seeder. In order to avoid seed burning, the fertilizer should be immediately placed at the centre bottom of neighbouring furrows after loosening the soil by rotary blade, and then the seeds should be covered by soil (Fig. 8.11). Fertilizer should be put in between two seed rows, and at a greater depth than the seeds, to avoid burning of seeds. According to our experimental results, 5 cm difference between fertilizer and seed rows is helpful for seed germination and root growth.

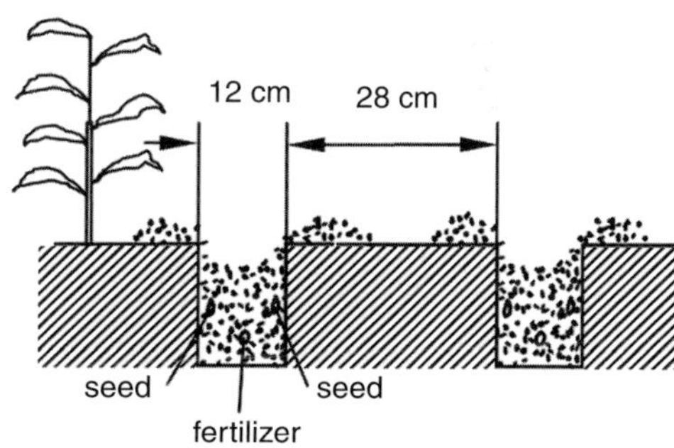

Fig. 8.11. Sketch showing strip rotary hoe mini-seeding.

STRIP CHOP NO-TILL SEEDER. As shown in Fig. 8.12, the power-driven chop blades (2) beside the opener (3) cut off or push away the stalks hanging on the opener and crush the roots. The following tine opener is the disc opener (5), which pushes chopped stalks or grass to the sides, and evenly puts seeds in the soil. The press wheel (6) immediately presses the seeding row. Moving blades do not touch the ground, reducing soil disturbance and power consumption and increased uniformity of seed sowing depth is achieved.

POWERED DISC NO-TILL SEEDER. The general idea for the anti-blocking of the powered disc no-till seeder (Fig. 8.13) is: in the operation, the powered disc, driven by the tractor, cuts the maize root and residues, and opens the furrow, and then the following narrow-point opener can open the furrow further and produce a smooth seedbed for the germination of seeds without getting blocked by residues.

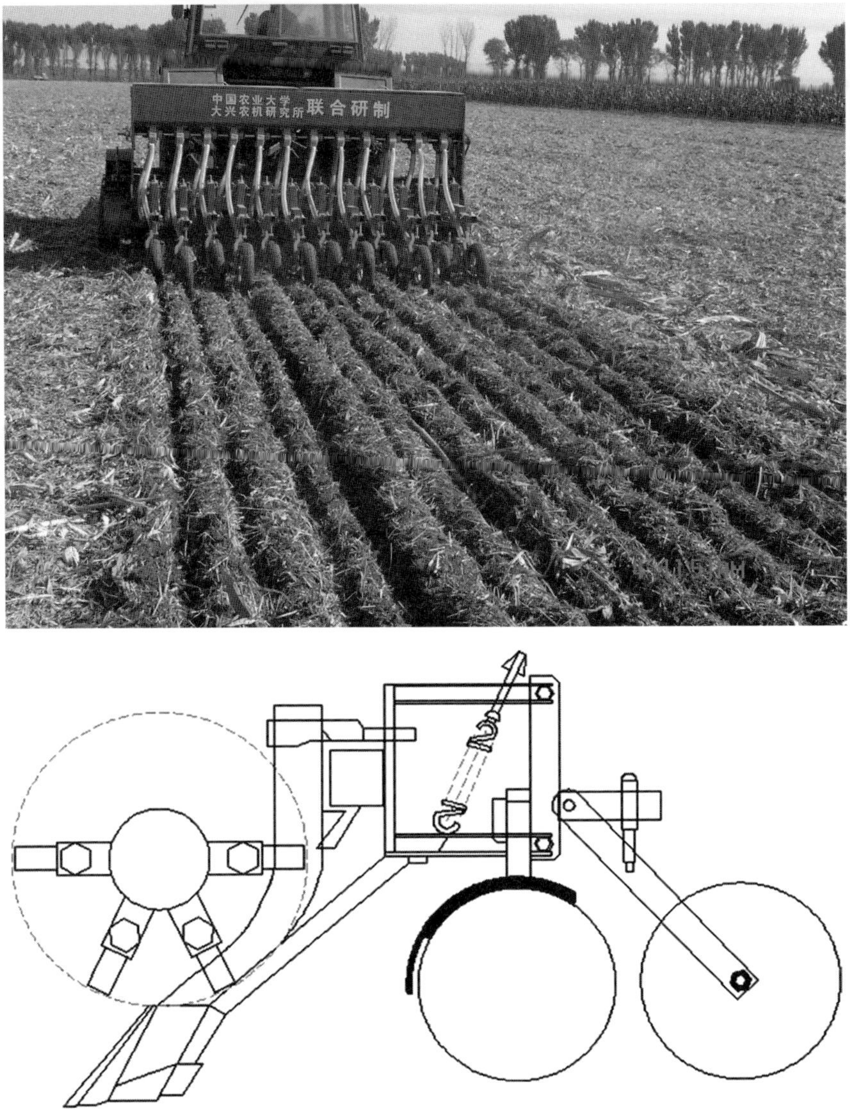

Fig. 8.12. Strip chop no-till seeder.

8.4.4 Minimum tillage and weeding machines

In order to loosen soils and reduce herbicide use, minimum tillage and weeding machines were developed such as sweeper, spring harrow, shallow cultivator etc. (Fig. 8.14). These machines can be used during the fallow season to loosen soils and control weeds.

8.5 Strategies for Conservation Agriculture Development

Conservation Agriculture is an improvement over ConvT, which has lasted several thousand years in China, so the difficulties in application and extension of the CA are evident. Although China has made encouraging achievements in CA, the development

Fig. 8.13. Powered disc no-till seeder.

of CA is still slow, and the following strategy can be helpful in promoting CA.

8.5.1 Support and enabling government policy

Conservation Agriculture is an advanced agricultural technique with significant economic, ecological and social benefits. Its long-term benefits are greater than the short-term benefits; and social benefits are greater than farmer individual benefits. The government, which is in charge of social management and public service, has the responsibility and obligation to promote the application and extension of CA. Governments need to strengthen through support by scientific research, demonstration,

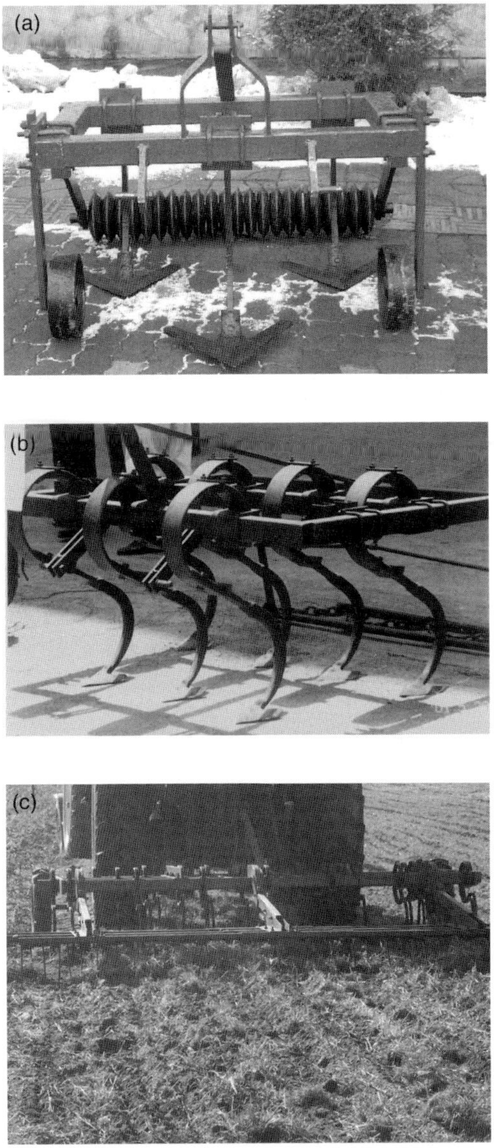

Fig. 8.14. Minimum tillage and weeding machines: (a) sweeper, (b) shallow cultivator and (c) spring harrow.

extension and purchase of agricultural machinery according to current basic needs. Conservation Agriculture techniques and equipment will be gradually developed and improved and new problems will appear in future extension, so governments need to support the research on CA techniques and equipment, and continuously set up special projects to support experts and technicians. Governments need to help the enterprises with the development of suitable equipment and ensure development of relevant agronomic measures; they also need to support the demonstration and extension of the CA, improved farmers' cognition and train technicians and farmers. As a result, technicians will be able to help farmers use various techniques; there is need to set up demonstration areas to encourage farmers by the demonstration effects and facts.

8.5.2 Leading role of demonstration project

Project demonstration areas need to be set up to develop the main technical mode, suitable tested machines and to train technician teams, so as to gain experience for large-scale extension of CA. The demonstration project should be set up in the region where CA is needed, with good basic agricultural mechanization, recognition by local government and the availability of sufficient number of trained technicians. The demonstration project funds, which concentrate on propagation, training and comparative evaluation of components, should be scientific. The funds can be used to train farm machinery contractors and subsidize the machines for farmers. Furthermore, governments need to improve local enthusiasm, drive local fund inputs and encourage local experiment and demonstration.

8.5.3 Interest-driving mechanism

The beneficiaries of CA are the most enthusiastic group to use CA. The participators of CA include government, enterprise, farmer, tractor driver and technician. Governments can mobilize farmers' enthusiasm by propagating positive effects of CA on the agro-environment, yield and income. The driving function of enterprises can be promoted by making the enterprises cognizant of the potential market. The participation and recognition of tractor drivers can be realized from increased benefits by increasing their operation areas and machine use efficiency. Furthermore, the government can prompt the enthusiasm of technicians by funding and proper commendation, so as to promote the enthusiasm of relevant departments and push the forming of a good environment for CA.

8.5.4 Mature techniques

In China, many mature CA technical modes suited to local conditions have been formed. The application of these effective and feasible techniques should be accelerated. In areas with mature conditions, these techniques can be applied by the whole village, township and county. The application effects should be used to stabilize farmer's self-confidence and attract more participants. The relationship between research and extension should be properly handled, which means that we should not only attach importance to scientific research but also not neglect extension. Acceptable manners should be adopted to demonstrate and teach farmers, and accelerate the popularization of techniques. The information dissemination platform with online column, seminars and experience-sharing sessions should be constructed to strengthen technical exchanges, promote information sharing and expand the application effect for outcomes.

8.5.5 Combination of agricultural machinery and agronomy

In China, problems such as weed management, pest control, straw disposition and NT seeding, appear during the implementation of CA. Some problems can be easily solved by agronomic methods and some can be easily solved by machinery, so the agricultural machinery departments need to take full advantages of agronomic experts, and form expert groups involving engineers and agronomists to strengthen instructions to all involved and jointly push the implementation of CA with cooperation from all.

8.6 Conclusions

Following over 20 years of systematic experimentation, demonstration and extension, China has found its own way to develop CA systems. Manufacturing systems for NT equipment for the small- and middle-size farm-holding levels are in place. The Chinese government recognizes the importance of CA, and more and more farmers are accepting it. An increasing number of machinery companies found that CA will bring them new markets. Although the problems exist in developing CA, it is believed that CA will be adopted in larger areas in the near future in China.

References

Bai, Y.H., Chen, F., Li, H.W., Chen, H., He, J., Wang, Q.J., Tullberg, J.N. and Gong, Y.S. (2008) Traffic and tillage effect on wheat production on the Loess Plateau of China: 2. Soil physical properties. *Australian Journal of Soil Research* 46, 652–658.

Chen, H., Bai, Y.H., Wang, Q.J., Chen, F., Li, H.W., Tullberg, J.N., Murray, J.R., Gao, H.W. and Gong, Y.S. (2008) Traffic and tillage effects on wheat production on the Loess Plateau of China: 1. Crop yield and SOM. *Australian Journal of Soil Research* 46, 645–651.

Gao, H.W., Li, H.W. and Li, W.Y. (2008) Development of conservation tillage. *Transactions of the Chinese Society for Agricultural Machinery* 9, 43–48. (In Chinese)

Gupta, R. and Acharya, C.L. (1993) Effect of mulch induced hydrothermal regime on root growth, water use efficiency and quality of strawberry. *Journal of the Indian Society of Soil Science* 41, 17–25.

He, J., Li, H.W., McHugh, A.D., Ma, Z.M., Cao, X.H., Wang, Q.J., Zhang, X.M. and Zhang, X.R. (2008) Spring wheat performance and water use efficiency on permanent raised beds in arid northwest China. *Australian Journal of Soil Research* 46, 659–666.

He, J., Wang, Q.J., Li, H.W., Tullberg, J.N., McHugh, A.D., Bai, Y.H., Zhang, X.M., McLaughlin, N. and Gao, H.W. (2009a) Soil physical properties and infiltration after long-term no-tillage and ploughing on the Chinese Loess Plateau. *New Zealand Journal of Crop and Horticultural Science* 37, 157–166.

He, J., Kuhn, N.J., Zhang, X.M., Zhang, X.R. and Li, H.W. (2009b) Effect of conservation tillages of 10 years of conservation tillage on soil properties and productivity in the farming–pastoral ecotone of Inner Mongolia, China. *Soil Use and Management* 25, 201–209.
He, J., Li, H.W., Wang, Q.J., Gao, H.W., Li, W.Y., Zhang, X.M. and McGiffen, M. (2010a) The adoption of conservation tillage in China. *Annals of New York Academy of Science* 1195, E96–E106.
He, J., Li, H.W., Kuhn, N.J., Wang, Q.J. and Zhang, X.M. (2010b) Effect of ridge tillage, no-tillage, and conventional tillage on soil temperature, water use, and crop performance in cold and semi-arid areas in Northeast China. *Australian Journal of Soil Research* 48, 737–744.
He, J., Li, H.W., Rasaily, R.G., Wang, Q.J., Cai, G.H., Su, Y.B., Qiao, X.D. and Liu, L.J. (2011) Soil properties and crop yields after 11 years of no tillage farming in wheat–maize cropping system in North China Plain. *Soil and Tillage Research* 113, 48–54.
Koch, H.J. and Stockfisch, N. (2006) Loss of soil organic matter upon ploughing under a loess soil after several years of conservation tillage. *Soil and Tillage Research* 86, 73–83.
Li, H.W., Gao, H.W., Wu, H.D., Li, W.Y., Wang, X.Y. and He, J. (2007) Effect of 15 years of conservation tillage on soil structure and productivity of wheat cultivation in northern China. *Australian Journal of Soil Research* 45, 344–350.
Liu, L.J. (2004) Systematic experiments and effect analysis of all year conservation tillage in two crops a year region. PhD dissertation, China Agricultural University, Beijing, China (In Chinese)
Liu, L.Y., Li, X.Y., Shi, P.J., Gao, S.Y., Wang, J.H., Ta, W.Q., Song, Y., Liu, M.X., Wang, Z. and Xiao, B.L. (2007) Wind erodibility of major soils in the farming–pastoral ecotone of China. *Journal of Arid Environments* 68, 611–623.
Liu, X.E., Guo, H.A. and Li, L.C. (2002) The question and developmental countermeasure of breeding for maize in Northeast China. *Journal of Jilin Agricultural Science* 27, 20–23. [In Chinese]
LZU (2005) The Chinese map for agriculture–pasture transition region based on GIS. Lanzou University, Lanzou, China. (In Chinese)
Roldan, A., Salinas-Garcia, J.R., Alguacil, M.M., Diaz, E. and Caravaca, F. (2005) Soil enzyme activities suggest advantages of conservation tillage practices in sorghum cultivation under subtropical conditions. *Geoderma* 129, 178–185.
Thomas, G.A., Dalal, R.C. and Standley, J. (2007) No-till effects on organic matter, pH, cation exchange capacity and nutrient distribution in a Luvisol in the semi-arid subtropics. *Soil and Tillage Research* 94, 295–304.
Torbert, H.A. and Reeves, D.W. (1995) Interactions of traffic and tillage applied to cotton on N movement below the root zone of a subsequent wheat crop. *Soil and Tillage Research* 33, 3–16.
Wang, Q.J., Bai, Y.H., Gao, H.W., He, J., Chen, H., Chesney, R.C., Kuhn, N.J. and Li, H.W. (2008) Soil chemical properties and microbial biomass after 16 years of no-tillage farming on the Loess Plateau, China. *Geoderma* 144, 502–508.
Wang, Q.J., Chen, H., Li, H.W., Li, W.Y., Wang, X.Y., McHugh, A.D., He, J. and Gao, H.W. (2009) Controlled traffic farming with no tillage for improved fallow water storage and crop yield on the Chinese Loess Plateau. *Soil and Tillage Research* 104(1), 192–197.
Xie, Z.K., Wang, Y.J. and Li, F.M. (2005) Effect of plastic mulching on soil water use and spring wheat yield in arid region of northwest China. *Agricultural Water Management* 75, 71–83.
Zha, X. and Tang, K. (2003) Change about soil erosion and soil properties in reclaimed forestland of loess hilly region. *Acta Geographica Sinica* 58, 464–469. (In Chinese)
Zhang, Q., Zhao, X. and Zhao, H.L. (1998) *Sandy Grassland in China*. China Weather Press, Beijing. (In Chinese)
Zhang, X.R., Li, H.W., He, J., Wang, Q.J. and Golabi, M.H. (2009) Influence of conservation tillage practices on soil properties and crop yields for maize and wheat cultivation in Beijing, China. *Australian Journal of Soil Research* 47, 362–371.
Zhou, J.Z. (2004) Experimental study on soil wind erosion and using conservation tillage to reduce wind storm disaster. PhD dissertation, China Agricultural University, Beijing, China. (In Chinese)

9 Conservation Agriculture in Central Asia

Aziz Nurbekov,[1] Akmal Akramkhanov,[2] John Lamers,[3] Amir Kassam,[4] Theodor Friedrich,[4] Raj Gupta,[5] Hafiz Muminjanov,[4] Muratbek Karabayev,[5] Dossymbek Sydyk,[6] Jozef Turok[1] and Malik Bekenov[7]

[1]International Center for Agricultural Research in the Dry Areas (ICARDA) Central Asia and the Caucuses Regional Office, Tashkent, Uzbekistan; [2]Khorezm Rural Advisory Support Service, Khorezm, Uzbekistan; [3]Center for Research Development, Bonn, Germany; [4]Food and Agriculture Organization of the United Nations, Rome, Italy; [5]International Maize and Wheat Improvement Center, New Delhi, India; [6]South-Western Research Institute of Livestock and Crop Production, Chimkent, Kazakhstan; [7]Ministry of Agriculture and Water Management, Bishkek, Kyrgyzstan

9.1 Introduction

The Central Asia region comprises five independent republics: Kazakhstan, Kyrgyzstan, Tajikistan, Turkmenistan and Uzbekistan (Fig. 9.1). The climate in the region is mostly arid and semi-arid and strongly continental, with hot summers and cold winters. Average annual precipitation, which is concentrated in the winter and spring, is about 270 mm, and varies from 600 to 800 mm in the mountainous zone and 80–150 mm in the desert regions. The land area of the five Central Asian countries covers about 393 Mha, which is 98% of the total area (Table 9.1). Kazakhstan with 270 Mha is the largest country, comprising almost 68% of the entire area, followed by Turkmenistan and Uzbekistan, while Kyrgyzstan and Tajikistan are two smaller states, which together constitute 34.25 Mha (9%).

The Central Asia region includes some of the most sparsely populated areas in the world. In 2010, the estimated population was over 62.5 million with Uzbekistan having the highest population of 29.0 million, followed by Kazakhstan with 15.7 million. The remaining three countries have a combined population of 17.8 million. The population is sparsely settled with a highest density of 65 persons km^{-2} in Uzbekistan to the lowest in Kazakhstan of 6 persons km^{-2}. Cropland amounts to 32.6 Mha. Reported average per capita cropland is 0.45 ha (Table 9.1) with the lowest of 0.11 ha and 0.17 ha in Tajikistan and Uzbekistan, respectively, to the highest of 1.45 ha in Kazakhstan.

Agricultural croplands in Central Asia include rainfed and irrigated areas (Kienzler *et al.*, 2012); and consequently, the adoption and adaptation of Conservation Agriculture (CA) practices should be considered according to farming systems in different agroclimatic zones. For example, raised beds[1] which are suitable in the irrigated systems of Central Asia, are less appropriate in rainfed systems. Thus, to the extent possible, we have reviewed hereafter results pertaining to farming systems by agroclimatic zones.

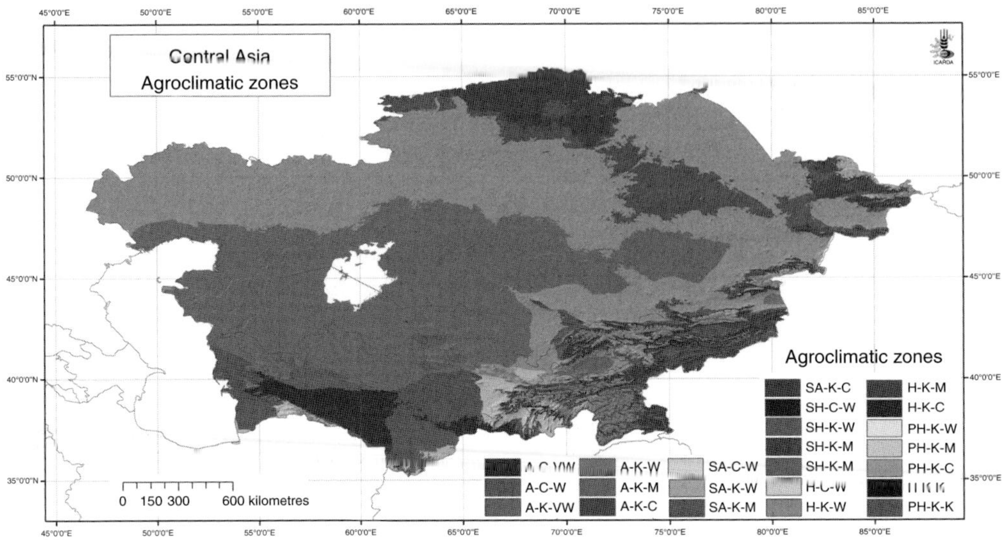

Fig. 9.1. Agroclimatic zones of Central Asia (De Pauw, 2008).

Table 9.1. Land resources and population and agricultural indicators of Central Asia (National statistical books of Kazakhstan, Kyrgyzstan, Tajikistan, Turkmenistan and Uzbekistan, 2010).

Country	Total territory (Mha)	Land area (Mha)	Cropland (Mha)	Cropland (%)	Agric. GDP (%)	Population (million)	Population density (km^{-2})	Rural population (%)	Per capita cropland (ha)
Kazakhstan	272.49	269.7	24.0	8.8	5.3	15.7	6	42.8	1.45
Kyrgyzstan	19.99	19.18	1.4	7.0	25.8	5.2	28	63.4	0.25
Tajikistan	14.255	13.99	0.9	6.3	19.8	7.4	44	71.4	0.11
Turkmenistan	48.81	46.99	1.8	3.7	22.1	5.2	11	72.0	0.27
Uzbekistan	44.74	42.54	4.9	10.9	19.4	29.0	65	63.5	0.17
Total	400.3	392.7	33	8.3	9.9	62.5	15.1	62.6	0.45

9.2 History of Conservation Agriculture in Central Asia

Conservation Agriculture as a term has not been extensively used in Central Asia until the last decade. However, conservation tillage in the rainfed areas and raised bed planting in irrigated areas, formerly researched and applied, could be attributed to be a development towards CA. The development of conservation tillage in rainfed areas of Central Asia was caused by the massive soil erosion occurring at the time when the reclamation of virgin and fallow lands in Kazakhstan became of enormous importance in ensuring food security. In 1954, special surveys by agronomists, soil scientists and land developers examined vast tracts of the land stocks of Kazakhstan. Those expeditions identified more than 21 Mha of arable virgin and fallow lands of which more than 13 Mha are lands of good to medium quality that could be used for agricultural crops, and primarily for planting of cereal crops without significant costs (Dvurechenskiy, 2010).

Hence, the newly introduced system of farming called 'conservation tillage' (CT), unlike conventional tillage (ConvT) farming, radically changed the way of land tillage in the steppes of Kazakhstan, and allowed a reduction in soil degradation on dozens of millions of hectares. After development of new lands, average per annum production of grain in Kazakhstan in the 1961–1965 period increased up to 14.5 Mt (versus 3.9 Mt in the 1949–1953 period), whereas after the introduction of conservation tillage in 1976–1985, per annum production of grain in Kazakhstan increased to 25–27 Mt (Sydyk and Isabekov, 2009a).

During the last decade, the term CA has made its way into research communities of Central Asian countries. For example, since 2006 the Kustanay Research Institute of Agriculture, located in the north-west of Kazakhstan, has gradually moved away from conservation tillage to no-till (NT).[2]

Researchers from the Kustanay Research Institute have achieved complete elimination of mechanical tillage and converted to full CA with all three component practices. At the same time, in the early 20th century the area under irrigation had also been growing rapidly from the 1930s until the 1990s in Central Asia (Fig. 9.2) totalling 8.5 Mha by 1990. This expansion of irrigated farmland combined with poor water management caused a number of environmental problems and devastation of the Aral Sea. After 1990, the growth of irrigated areas slowed significantly in all countries, and during the last 20 years (1990–2010) it increased by only 1 Mha.

The main crop in the irrigated areas in Central Asia during the Soviet era was cotton, covering 80% of the total irrigated area. Some elements of CA such as replacing inversion ploughing in irrigated cotton areas with NT were reported by Kondratjuk (1972). Efforts were made to replace mould-board ploughing with chisel tillage. However, a major factor in assessing the effects was weed control. Ploughing was seen to 'eliminate' weeds in the cotton fields, but chisel tillage on its own did not result in any significant reduction of weeds. A number of studies conducted in different regions and soils reported that weed infestation was substantially higher with NT compared to the ploughed treatments, particularly when lucerne preceded cotton.

Currently, permanent beds and furrow system of soil management practice has received much attention in irrigated areas. Ryzhov *et al.* (1980) reported favourable conditions for the growth of cotton when planted on beds including optimum bulk density, 0.9–3.1°C higher temperatures at the 5 cm depth, and more uniform soil moisture conditions.

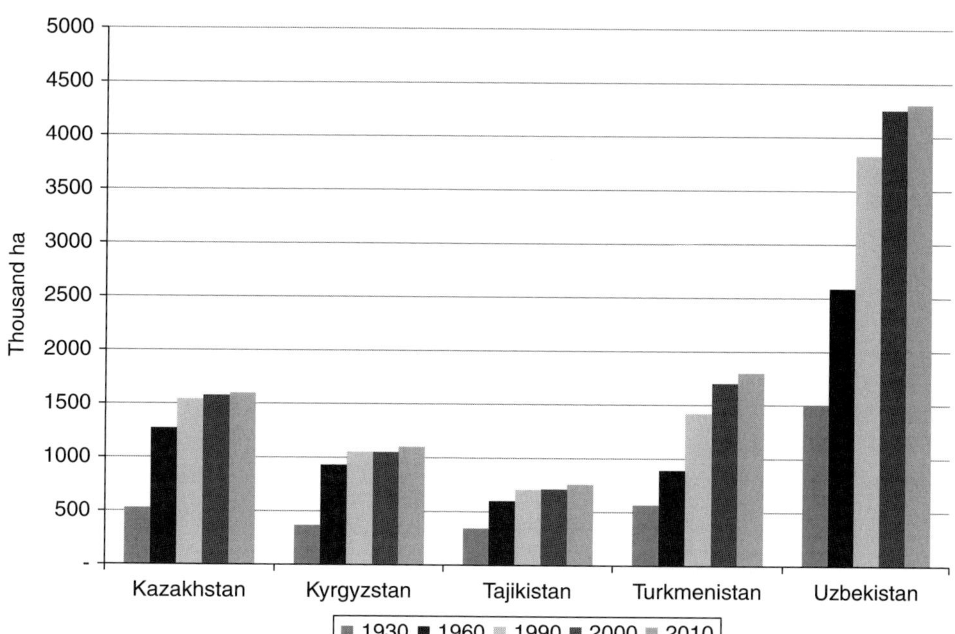

Fig. 9.2. Development of irrigated areas in five countries of Central Asia during 1930–2010.

9.3 Current Status of Conservation Agriculture in the Region

According to the Ministry of Agriculture of Kazakhstan, in 2011, NT and conservation tillage were introduced on an area of 11.7 Mha (Table 9.2), which is 70% of all the area sown to wheat in Kazakhstan (Sydyk *et al.*, 2008). Consequently, in 2011, the country harvested record gross output of grain of 20 Mt corresponding to a yield of 1.7 t ha^{-1} (Sydyk *et al.*, 2008). These results were achieved due to the introduction of CA-related practices, although the area under full CA in Kazakhstan is only 1.7 Mha.

CA is still not widely practised among the farming population in the irrigated areas of the lower half of Central Asia. Current activities are mainly concentrated in research institutes to integrate CA principles and practices into existing production systems.

Over the last 20 years, Uzbekistan has been researching ways of introducing grain crops into existing crop rotation mainly with cotton and lucerne. Earlier, only irrigated cotton or rainfed winter wheat was grown. However, now with well-proven research findings, timely planting of winter wheat in standing cotton has shown promising results. As a result, annual area under planting of winter wheat into standing cotton reached 600,000 ha in Uzbekistan (Qilichev and Khalilov, 2008). In Tajikistan, some elements of CA can be found within several donor-funded projects implemented in the past, but their geographic coverage and number of beneficiaries (mostly farmers) are relatively small. However, there is a claim that direct seeding of winter wheat with minimum soil disturbance after cotton harvest is annually implemented in 25,000–50,000 ha (Hafiz Muminjanov, Ankara, 2012, pers. comm.).

Conservation Agriculture is still not widely practised among the farming population in the irrigated areas of the lower half of Central Asia. However, earlier research on NT bed planting is relevant and useful nowadays as winter wheat has become another strategic crop to provide food security in most countries of the region. The researchers of the South-Western Research Institute for Livestock and Plant Production (Kazakhstan) studied and recommended raised-bed-furrow technology for the cultivation of winter wheat in the central irrigated zone of southern Kazakhstan. Cultivation of winter wheat on raised beds with lowered seeding rate

Table 9.2. Main agroclimatic zones and extent of land area under Conservation Agriculture (Adapted from De Pauw, 2008).

Agroclimatic zones[a]	Total area (ha)	Area planted with elements of CA (ha)	Description of the elements	Source
SA-K-W Semi-arid, cold winter, warm summer	151,387,760	11,700,000	Including 1,700,000 ha of no-tillage, i.e. direct seeding of spring wheat and barley	Ministry of Agriculture of Kazakhstan
A-K-W Arid, cold winter, warm summer	123,027,520	<650,000	Conservation tillage, sowing of winter wheat into standing cotton	Qilichev and Khalilov (2008); H. Muminjanov, Ankara, 2012, pers. comm.
A-C-W Arid, cool winter, warm summer	19,572,560			
SA-C-W Semi-arid, cool winter, warm summer	5,991,600			

[a]For detailed description of the agroclimatic zones, see De Pauw (2008).

between 2.0 and 3.0 million seeds ha^{-1} and application of mineral fertilizers at the rate of $P_{45}N_{90}$ kg ha^{-1} in ridges ensures steady yields of winter wheat with a reduction in production cost.

The demand for food and fodder is expected to continue to grow in countries of Central Asia. A model was proposed by Suleymenov *et al.* (2004, 2006) that grouped the rainfed and irrigated-based zones into three main crop-based production systems: (i) the northern Kazakh steppes; (ii) the warmer foothills of Kyrgyzstan and southern Kazakhstan where a mixture of rainfed and irrigated agriculture is practised; and (iii) Tajikistan, Turkmenistan and Uzbekistan where irrigated bed-and-furrow or basin systems are used (Table 9.3). Wheat, cotton and livestock are the most important commodities in the region. However, with a trend towards diversification, oil crops such as sunflower could also become important (Fig. 9.3). The results of research on adaptive cropping systems and CA conducted since 2003 have been introduced across 230–347 ha in the southern Kazakhstan region.

Several collaborative research and development projects have been implemented to promote CA in Uzbekistan. The projects are demonstrating appropriate management techniques for rehabilitation and improvement of salt-affected and gypsiferous irrigated lands to support food security in the country. Some of the studies and guidelines produced by these projects serve as useful reference materials for other countries in the region. No-till and raised-bed planting practices tested in Karakalpakstan and Tashkent provinces in Uzbekistan proved technically and economically suitable for local conditions and can provide similar or higher crop yields while saving considerable production resources and costs including fuel, seeds and labour. These practices are ready to be disseminated more widely in Uzbekistan.

Some of the striking features of CA experience noted and reported by many farmers in the region include reduction in inputs such as fuel, seed and water and in wear and tear of tractors and machinery. The other benefits include reduced soil erosion due to reduced soil disturbance and soil cover, and enhanced carbon sequestration.

9.4 Research Results Reported in Central Asia

9.4.1 Effect on soil quality under Conservation Agriculture

The impact of CA practices on physical and chemical soil properties has been a subject of many studies in Central Asia. However, most reports are related to different soil properties, CA management and crop rotations. Hence, the numbers of comparable and common research results are too few to make generalizations and further extrapolation to other regions with similar soil and climatic conditions is not yet possible.

Soil physical properties

Egamberdiev (2007) showed that mulching with crop residues improved soil microaggregation in the irrigated areas of Uzbekistan. More recently, the impact of tillage and crop residue management on properties of a silt loam soil under irrigation in Uzbekistan was reported stemming from a rotation of winter wheat and maize for 2 years followed by cotton for another 2 years (Ibragimov *et al.*, 2011). This study compared permanent raised beds (PRB) with limited reshaping, and ConvT (mouldboard ploughing). ConvT cotton and maize were superficially tilled (15–17 cm soil depth) two to five times during each season. Both tillage practices were subjected to either 25% (R25) or 100% (R100) retention of crop residues from the previous crops. After four years, PRB+R100 showed significant differences in soil physical properties and organic carbon compared to ConvT. Irrespective of the amount of crop residues retained, with PRB (R25 and R100) soil bulk density and consolidation in the 0.2–0.3 m depth had increased. Furthermore, the findings showed that in the last year of the study under PRB+R100, the amount of water-stable macro-aggregates were greater compared with those under ConvT and PRB+R25 treatments.

Table 9.3. Salient information about dominant cropping systems in the five Central Asian countries according to agroecological zones (Modified after De Pauw, 2008, Gupta *et al.*, 2009 and Kienzler *et al.*, 2012).

Country/ region	Major production system	Cropping intensity (%)	Growth period (days)	Distinguished features of the agroecology	Production constraints
Kazakhstan (northern parts)	Rainfed spring wheat–fallow systems	60–80, rainfed	210–240	Rainfed cereal systems, steppes, long cold winters	Drought, cold and water stress (precipitation 300–400 mm), soil erosion
Kazakhstan (southern parts)	Extensive cereal–livestock systems Irrigated cotton/ wheat-based systems, rice, rangelands	50–60, rainfed	30–89	Rainfed rangelands with mixed crop–livestock systems, high Mg-soils, saline groundwater	Drought, cold and water stress (precipitation 250–350 mm), 12–14°C, Mg-soil, erosion
Kyrgyzstan (Osh, Chu and Fergana Valley)	Irrigated agriculture on sloped and valley areas	40–60+	60–119	Sloped lands (up to 10%), supplemental irrigation, generally fresh but shallow groundwater table	Drought and heat (precipitation 200–300 mm), saline water use, 16–22°C
Tajikistan (south-west/ north-west)	Irrigated systems (cotton–wheat) Agriculture on sloped land of 5–16%	40–60+	60–150	Pastoral systems/ irrigated agriculture on sloping lands, saline groundwater	Drought (precipitation 250–350 mm), 7–9°C, sloped land, mechanization. Water erosion by irrigation, drainage congestion
Uzbekistan (irrigated)	Irrigated cropping systems, cotton–wheat (mostly furrow irrigation)	>60	60–119	Irrigated crop production, drainage water use, soil salinity, long growing season, double cropping	Drought and heat (precipitation 250–500 mm), 16–20°C, salinity, water erosion
Turkmenistan (irrigated)	Rainfed pastoral/cereal production systems (mostly furrow irrigation)	30–60	30–59	Crop–livestock systems, saline groundwater, overgrazing, soil salinity	Drought and heat (precipitation 200–350 mm), 14–18°C, water scarcity, salinity

Soil organic matter dynamics

Soil organic matter (SOM) adds to structural stability and improves soil moisture-holding capacity (Bot and Benites, 2005). Numerous results from the irrigated areas showed that crop residue retention improves SOM and soil N content (e.g. Egamberdiev, 2007; Nurbekov *et al.*, 2012; Pulatov *et al.*, 2012). Egamberdiev (2007) reported from an operational-scale field trial conducted on 2.85 ha under irrigation in north-west Uzbekistan, which involved comparing four tillage practices and two residue management levels. The treatments also included ConvT practice, PRB, intermediate or semi-permanent beds re-shaped every cropping cycle (IT), and NT with planting on flat soil. In each of these tillage systems, the crop residues were either completely removed at harvest (CR–) or retained on the soil surface (CR+). The findings showed that CA practices increased SOM significantly with corresponding

Fig. 9.3. Crop diversification with no-till maize (a) for livestock feed and sunflower (b) for oil extraction; examples are from Kazakhstan and Uzbekistan, respectively.

improvements in soil structure and greater soil moisture-holding capacities (Egamberdiev, 2007; Pulatov *et al.*, 2012). Yet, the significant increase in SOM content from about an initial 0.57% to about 0.75% (i.e. about 32% of increase) after seven cropping cycles in a cotton–winter wheat rotation, represented a moderate absolute increase. In a study by Ibragimov *et al.* (2011), soil organic C in the 0–0.4 m depth increased annually by as much as 0.70 t ha^{-1} with permanent beds and complete residue retention (PRB+R100), whilst this annual increase was not more than 0.48 t ha^{-1} with ConvT+R100. This is due to the temperature/moisture regime, which in arid regions encourages high soil microbial activity and SOM turnover rates (Sanchez *et al.*, 2004). A literature review by Kienzler (2009) showed no reported changes in SOM over a period of around 30 years for conventional farming practices in the northwest of Uzbekistan. In comparison, a wealth of information on CA practices worldwide shows an increase in SOM (e.g. West and Post, 2002; Sanchez *et al.*, 2004; Govaerts *et al.*, 2006; Corsi *et al.*, 2012), and these results were also confirmed by selected studies in the irrigated areas in Central Asia.

Various research experiments have investigated the impact of different tillage systems on crop productivity. Mohanty *et al.* (2007) reported that regression analyses between crop yield and SOM values for tillage and crop residues in rice–wheat systems revealed that both crops showed a positive yield response to increased levels of SOM. There is general agreement that reduced tillage can increase SOM. The overall findings for the irrigated areas in Central Asia can be summarized as follows: the usual low initial SOM contents can be significantly and rapidly boosted by CA practices, but under the prevailing arid and semi-arid agroclimatic conditions in the region, this increase is proportional to the annual amounts of organic matter added.

Soil salinization

Secondary soil salinization, caused by capillary rise of the groundwater, is a major cause of ongoing cropland degradation in the irrigated areas of Central Asia (Akramkhanov *et al.*, 2012; Tischbein *et al.*, 2012). Overall, research results showed a declining rate of soil salinization increase after crop residue retention during a 4-year study on irrigated cotton–winter wheat rotation in Uzbekistan (Egamberdiev, 2007). The differences in soil salinity at the end between conventional practices (0.52%) and NT (0.39%) were significant. After 4 years, NT system had the lowest soil salinity level (Pulatov *et al.*, 2012).

Bezborodov *et al.* (2010), from 3 years' field research on cotton in Uzbekistan, observed a 20% soil salinity reduction with wheat straw mulch of 1.5 t ha^{-1} compared to the non-mulched treatments.

Reduced soil salinity was also reported by Devkota (2011b) under a combination of raised beds and full crop residue retention of up to 45% in the top 10 cm and by 18% in the top 90 cm soil layer compared to a bare soil common under ConvT systems. When comparing three irrigation modes on PRB (Devkota, 2011b), soil salinity on the top of the beds increased significantly with every-furrow and alternative skip-furrow irrigation compared to permanent skip-furrow irrigation (PSFI). Soil salinity management with PSFI resulted in the least saline area and which in addition appeared on the dry furrows only. Recent simulation findings of the soil-water model Hydrus-1D indicated that although water uptake by cotton or wheat would only marginally benefit from a surface mulch layer, it markedly reduced soil evaporation, capillary rise of groundwater and in turn secondary soil salinization (Forkutsa *et al.*, 2009).

Although research findings in the irrigated areas illustrated that soil salinization with CA practices could not be arrested, the observed reductions in soil salinity increase are none the less relevant in the management of irrigated areas of Central Asia that are subject to land degradation caused by soil salinity. It is known that the application of plant biomass helps to ameliorate salinity and sodicity in the soil (for a review of the subject, see Qadir *et al.*, 2007).

In Turkmenistan, trials with raised beds showed differences in soil salinity between top and bottom of the beds. Electric conductivity of soil extracts (EC_e) on the raised-bed was in the range of 3.42–5.47 mS cm^{-1}. At the bottom of the raised bed, in the furrow, EC_e of 1.49–2.86 mS cm^{-1} was considerably lower than in the raised-bed.

Soil bulk density

Soil bulk density is continuously high on the agenda when NT is discussed. A major concern is the perception that without tillage soil compaction will be high. In experiments, therefore, efforts are constantly made to measure bulk density. Data from rainfed areas of Kazakhstan show that bulk density values in regular sierozems before planting (1.29–1.30 g cm^{-3}) and harvesting (1.32–1.37 g cm^{-3}) under conventional technology were comparable to those under conservation tillage and direct planting of winter wheat, 1.26–1.30 g cm^{-3} and 1.29–1.34 g cm^{-3}, respectively (Sydyk *et al.*, 2009).

Overall, the studies indicated that winter wheat grown in cotton–maize rotation and retaining winter wheat residues positively impacted soil bulk density (ρb, Mg m^{-3}) irrespective of the tillage modes, while in one study it was concluded that cotton cropping without tillage resulted in increased soil compaction (Ibragimov *et al.*, 2011).

9.4.2 Carbon sequestration, climate change adaptation and mitigation

Given the relatively short history of CA in Central Asia, evaluation of the effects on carbon sequestration is limited to few short-term trials. In rainfed areas, the biggest problem is with the fallow period when multiple passes of regular tillage are conducted to control weeds, causing substantial erosion of the topsoil. Under such conditions, results from CKARI (Central Kazakh Agricultural Research Institute) over the 4-year trial period showed that there was on average about 0.2% more SOM in the treatments without tillage during fallow and with direct seeding of the wheat crop with narrow seeder shoes (as opposed to the sweeps used on the conventional equipment; Murat Karabayev, Astana, 2012, pers. comms). This translates into approximately 400 kg C ha^{-1} $year^{-1}$, a level which is consistent with the review of Six *et al.* (2002), who reported an average increase in SOC, under CA practices, at a rate of 325±113 kg C ha^{-1} $year^{-1}$ under a wide range of temperate and tropical conditions.

Climate change poses a major threat to the agricultural production potential of Central Asian countries (IPCC, 2007). Burman and Roy (2011) reported that increased temperature in the future is likely to reduce fertilizer use efficiency. This could lead to increased fertilizer requirement for meeting future food production demands, and may also increase greenhouse gas (GHG) emissions.

Overall, CA systems have a higher adaptability to climate change because of the higher effective rainfall due to higher infiltration and therefore reduced surface runoff and soil erosion as well as greater soil moisture-holding capacity (Saturnino and Landers, 2002; Jat *et al.*, 2012). Thus crop growth under CA systems can continue towards maturity for longer than those under ConvT (Stewart, 2007).

In the northern Kazakhstan region, where much of the annual precipitation is in the form of snow in the winter, CA provides a way of trapping snow evenly on the field which may otherwise move away; and this further permits snow to melt evenly into the soil. In the dry areas of continental Eurasia, one-third or more of the precipitation is not effectively used in tillage-based systems, forcing farmers to leave land fallow to 'conserve' soil moisture, leading to extensive wind erosion of the topsoil from exposed fallow land and to dust emissions and transport over large distances (Brimili, 2008). Under CA, more soil moisture can be conserved than by leaving the land fallow; this allows the introduction of additional crops including legume cover crops into the system (Blackshaw *et al.*, 2007; Gan *et al.*, 2008).

No-till farming also reduces the unnecessarily rapid oxidation of soil organic matter to CO_2 induced by tillage (Reicosky, 2008; Nelson *et al.*, 2009). Together with the addition of mulch as a result of saving crop residues, there is a reversal from net loss to net gain of carbon in the soil, and the commencement of a long-term process of carbon sequestration (West and Post, 2002; CTIC/FAO, 2008; Baig and Gamache, 2009). Expanded across a wide area, CA has the potential to slow/reverse the rate of emissions of CO_2 and other GHG by agriculture (Lal, 2002, 2008). However, there are exceptions to such results but in general there is an increase in soil carbon content under CA systems, as shown from the global meta-analyses by West and Post (2002) and Corsi *et al.* (2012).

With CA, reduced use of tractors and other powered farm equipment results in lower CO_2 emissions. Up to 70% in fuel savings have been reported (FAO, 2008). CA systems can also help reduce the emissions of other relevant GHG such as methane and nitrous oxides, if combined with complementary techniques. Both methane and nitrous oxide emissions result from poorly aerated soils, from severely compacted soils, or from heavy poorly drained soils. CA soil management favours the multiplication of methane-oxidizing bacteria, leading to reduced methane emission (Ceja-Navarro *et al.*, 2010).

The soil is a dominant source of atmospheric N_2O (Houghton *et al.*, 1997). The rate of production and emission of N_2O depends primarily on the availability of a mineral N source. Addition of fertilizer N, therefore, directly results in extra N_2O formation. Nitrogen leaching and nitrogen runoff are minimal under CA systems, and over the longer term CA generally reduces the need for mineral N by 30–50% (Derpsch, 2008; Crabtree, 2010). Thus overall, CA has the potential to lower N_2O emissions as reported by Parkin and Kaspar (2006) and Baig and Gamache (2009).

Although detailed studies are needed to provide additional evidence in the local environment, one can expect similar benefits from reducing GHG emissions when promoting CA practices in Central Asia. Thus, the incentive programme launched by Kazakhstan to promote CA could be seen as initial step to promote carbon sequestration by farmers and reduce GHG emissions. However, for its wider application in the region, government incentive programmes are needed across the region in each country.

9.4.3 Crop yields

Using NT raised-beds, yields of the cotton–wheat system in Uzbekistan did not differ significantly from that under ConvT (Suleymenov *et al.*, 2004, 2006). In the Chu Valley in Kyrgyzstan, yields of irrigated winter wheat varieties 'Manas', 'Intensive' and 'Asyl' were at least equal if not higher under NT raised-bed planting compared to conventional practices from the first year of experimentation (Kienzler *et al.*, 2009b). In addition to this yield response, NT raised-bed planting improved seed germination, and hence the seeding rate could be reduced by 50% (Kienzler *et al.*, 2009b). Winter wheat development was advanced by 2–4 days for each growth stage in raised beds, which advanced ripening by 8–10 days with NT and raised beds (Kienzler *et al.*, 2009b). Irrigated maize on raised beds in southern and south-eastern Kazakhstan was harvested 3–5 days earlier than on conventional flat beds (Ospanbaev and Karabayev, 2009).

Although yields of cotton and winter wheat were equal at the onset of studies in north-western Uzbekistan, the use of NT PRB also showed higher yields than conventional practices in the longer run (Egamberdiev, 2007; Tursunov, 2009). When comparing yields of crops in a cotton–wheat–third crop rotation (Fig. 9.4) on permanent beds and conventional land preparation over three cropping cycles, Devkota (2011b) noted that cotton yield and its yield components were unaffected by both tillage practices immediately after the conversion of conventional to CA practices. But the subsequent rainfed wheat crop yielded 12% more with NT PRB than with ConvT, and the following irrigated maize crop yielded 14% higher grain under permanent beds.

No-till mungbean was successfully grown as a catch crop after winter wheat in the irrigated conditions of Uzbekistan and provided 3.3% yield advantage compared to traditional tillage practice, which was not significantly different. Even without significant differences in crop yields the NT system has an advantage in reduced costs of production (Nurbekov, 2007).

Kurvantaev *et al.* (2004) reported results from raised bed system trials involving planting on preformed raised beds, with and without tillage, in Tashkent region. Tillage treatments were combined with maximum (250:175:125) and minimum (150:125:50) NPK fertilizer rates. Direct drilling on beds with maximum fertilizer rates resulted in higher yields (3.89 t ha^{-1}) compared to that under beds with tillage preparations and maximum fertilizer rates (3.65 t ha^{-1}). Minimum fertilizer rates on direct drilled beds yielded 3.45 t ha^{-1} of cotton, whereas under tilled beds the yield was 3.33 t ha^{-1}. Research conducted in 2002–2004 demonstrated that permanent beds can be implemented without yield penalty in the initial years of switching to NT.

Recent results from Kyrgyzstan show that wheat yields were 29–41% higher on NT raised beds compared to traditional tilled fields. In addition to yield increases, seeding rates can also be reduced by 50%. Similar trials need to be established to adapt the NT technology including seeding depth and weed control in various soils (Pozharskiy and Akimaliev, 2002).

Nurbekov (2007) reported that the rate of nitrogen had no significant effect on the winter wheat yields in either of the two tillage systems, NT and ConvT. The 120 kg ha^{-1} rate was as good as the 140 kg ha^{-1} in traditional tillage with mouldboard ploughs, while NT slightly increased grain yield with higher nitrogen rate. Nurbekov *et al.* (2012) reported that winter wheat yield was higher in the treatment involving NT compared to other treatments.

Sydyk *et al.* (2009) studied the possibility of direct seeding and the ways of reducing tillage in cultivating winter wheat. They showed that it was possible to produce winter wheat in rainfed areas of southern Kazakhstan through NT direct seeding with mandatory application of mineral fertilizers and herbicides. Several varieties proved to be most suitable for direct seeding in rainfed areas of southern Kazakhstan (Sydyk and Isabekov, 2009b).

Research on planting of winter wheat before harvest of cotton crop has been carried out in Yavan and Gozimalik districts of Tajikistan. The results of the field observation on wheat growth and development suggest that the proposed technology has some advantages in making savings in seed quantity and increased yield (Sanginov and Khamikov, 2003).

Overall, reported yield responses to CA practices for the region vary, apparently depending on the crop, and land preparation is necessary to ensure a smooth conversion from conventional to CA practices (e.g. Devkota, 2011b). It has been demonstrated that proper field preparation, including

Fig. 9.4. (a) No-till winter wheat in rainfed areas of Kazakhstan and (b) mung bean grown as a catch crop with retention of surface residues in irrigated areas of Uzbekistan.

levelling to advance the implementation of CA practices after years of ConvT use, was conducive in bypassing the often observed yield reduction when changing from conventional to conservation practices. In many studies it was not mentioned whether or not such preparations had been taken into account (Devkota, 2011b).

9.4.4 Runoff, infiltration, soil water content and soil conservation

Soil moisture in the rainfed areas is highly influenced by snow cover during winter periods. At the same time, despite minor slopes of the fields in the northern Kazakhstan the size of the fields is large, leading to melting snowwater accumulation and runoff that causes considerable soil erosion. Residues retained on the soil surface can protect against erosion. The difference in erosion between the tilled fallow plots and the untilled plots was extremely marked: erosion in plots without fallow tillage was only 12% of that measured on the tilled plots (Murat Karabayev, Astana, 2012, pers. comms).

The effects of NT on soil moisture accumulation in the rainfed areas of northern Kazakhstan were also studied in the research farm of the North-West Research Institute (Zarechnoe) (Sydyk *et al.*, 2008). Moisture accumulation in 1 m layer was 160 mm in traditional fallow and 239 mm in treatment with 'coulisse' fallow, the field that is planted with a cover crop instead of leaving it bare. Treatment with tillage using flat sweeps that leaves standing stubble accumulated only 73 mm of soil moisture reserves, while standing stubble without any tillage accumulated 127 mm. Accordingly, under minimum winter precipitation, coulisse fallow and high stubble allowed maximum accumulation of snow cover and uniform snow-melt: on coulisse fallow, the height of the snow reached a highest value of 40 cm, on stubble 33 cm, on traditional fallow 8 cm and fallow tillage with flat sweeps only 3 cm.

9.4.5 Insect pest and disease dynamics

Although weed and pest management have always been given high attention in the Soviet period, little research has been directed so far on these aspects for CA systems in Central Asia. Some authors and farmers reported recurring problems with weed infestations, particularly with NT practices, but a closer look recurrently showed that herbicides have either not been used sufficiently or inappropriate herbicides were used. The preliminary results of testing different herbicides and application rates for rainfed maize (Kienzler *et al.*, 2009b) indicate an efficient weed management with the broad spectrum herbicide Stomp (pendimethaline) at 5 l ha^{-1} before germination and the systemic herbicide Dialen (2,4-D and dicamba) at 1 l ha^{-1} during the vegetative period. This double treatment increased application efficiency to around 83%. This research certainly deserves more attention given the reoccurring shortages in availability of suitable herbicides at the local markets as well as the high costs that discourage farmers to use appropriate herbicides (Murat Karabayev, Astana, 2012, pers. comms).

9.4.6 Nutrient use efficiency

Preliminary findings in Uzbekistan (Egamberdiev, 2007) on the dynamics of soil nitrogen content suggest that crop residue retention must be complemented with nitrogen (N) fertilizer application. This seems particularly true at the onset of conversion from conventional to CA practices to counterbalance any N immobilization caused by residue retention (Hickmann, 2006; Sommer *et al.*, 2007). Yet, few studies in the irrigated areas have addressed this aspect. Devkota (2011b) compared over the course of three cropping cycles, the impact of ConvT versus PRB. In each cycle two crop residue management treatments were included, namely complete retention of residues and complete removal of residues. In addition, he compared the effects of three N application rates for the different crops, which obviously differed according to crops: for example, no application or N-0 was compared to low-N, which in the case of cotton amounted to 125 kg N ha^{-1} but in maize and wheat to 100 kg N ha^{-1}. The high-N treatments for cotton were 250 kg N ha^{-1}, but for maize and wheat were 200 kg N ha^{-1}. The findings illustrated that N application significantly increased crop yields at each cycle with both tillage practices, but resulted in higher nitrogen use efficiency with NT permanent beds compared to ConvT practices for cotton (42%), wheat (12%) and maize (82%). Furthermore, when using the wet and dry irrigation (WAD)-mode, N losses occurred irrespective of the crop retention level. After three cropping cycles, which involved two times rice and one time wheat, cumulative N losses amounted to more than 350 kg N ha^{-1} even when all residues were retained, owing to both leaching and denitrification. However, these N losses for a major part occurred during the two seasons of flooded rice cultivation.

Overall, an appropriate N management seems to be of paramount importance in the irrigated areas of Central Asia for both conventional and CA practices since high N_2O emissions occur with conventional cotton, wheat and rice cultivation, which in all crops peaked when mineral N applications were immediately followed by irrigation (Scheer *et al.*, 2008).

As tillage is reduced or avoided altogether, there is less mineralization of N, often one of the apparent major 'benefits' of intensive tillage. Especially in soils that have low levels of available N, this may result in moderate to severe N deficiency in crops grown without tillage, particularly where considerable levels of crop residues remain on the surface. This N shortage is generally overcome with the application of approximately 20–30 kg ha^{-1} of N fertilizer for a few years, until SOM levels increase and a new level of SOM turnover and N mineralization is established (Murat Karabayev, Astana, 2012, pers. comms).

While NO_3-N levels were adequate in the tilled treatments, P levels were very low in all treatments. Phosphorus fertilizer (60 kg ha^{-1} P_2O_5) was applied to the conventionally tilled fallow and the reduced tillage fallow in the autumn of the fallow season, but even so P levels were low in these treatments. In the other three treatments with untilled fallows, P fertilizer was applied at a rate of 20 kg ha^{-1} P_2O_5 at seeding, after the samples had been taken for the analyses. Previous research had shown that 60 kg ha^{-1} P_2O_5 in the fallow year was sufficient for the whole of the 4-year rotation, and so this common practice and recommendation continued. However, after Perestroika (Independence), the unfavourable economic conditions resulted in little fertilizer use, even on the Central Experimental Base of the research institutes, and this in turn resulted in low soil P levels. This fact underscores one of the big problems affecting research in Kazakhstan. The country has a rich history of very good and meticulous agricultural research. However, the focus of this research, and the recommendations that emanated from it, were oriented towards a very different philosophy of agricultural production and economic circumstances. Consequently, it will require considerable effort in revisiting the past research results and reinterpret them, or repeat much of the research to develop new recommendations (Murat Karabayev, Astana, 2012, pers. comms).

9.4.7 Crop rotation

To diversify the rainfed steppe zones of northern Kazakhstan, recommendations were developed for the four- to five-field crop rotations with different sequences of agricultural crops, which avoids the traditional fallow year in an otherwise wheat monocropping system:

1. Canola for fodder–wheat–pea–wheat.
2. Sudan grass–wheat–chickpea–wheat.
3. Pea–wheat–oilseed rape–wheat.
4. Fallow (coulisse)–wheat–maize for grain–wheat.
5. Fallow (coulisse)–buckwheat–rape for fodder–wheat.
6. Fallow (coulisse)–oilseed rape–wheat–sunflower–wheat.
7. Fallow (coulisse)–wheat–barley–wheat–oats.

In these crop rotation systems monocropping of wheat is interrupted by leguminous and oilseed crops. It is difficult to overestimate the importance of crops such as oilseed rape, safflower, linen, sunflower and soybean for the northern region. The demand for such products for both the domestic and export markets is favourable. Besides, the grain growers benefit from them not only economically but also from the viewpoint of disease and weed management. Their use significantly increases the productivity of each hectare of arable land due to production of diverse products from raw crops to spring wheat. Most of those crops are good predecessors for the main crop spring wheat.

For example, research on the introduction of safflower showed good potential in diversifying crops to fit into the spring-wheat production system. The best preceding crops for safflower were winter and spring crops, leguminous, raw and industrial crops, as well as annual and perennial grasses. In direct planting, the cultivation of safflower after winter wheat, as the second crop after fallow, is recommended.

9.4.8 Economic returns

Although the findings from the research on CA for the irrigated areas in Central Asia are encouraging, there are few reports on the economic aspects of CA. This is particularly important since the economic profitability of CA practices varies over time and, like any production system, can be site-specific, which necessitates site-specific analyses (Knowler *et al.*, 2001). Furthermore, CA practices regularly require long-term investments (e.g. in direct seeders and planters), but it is unrealistic to expect that capital investments for increasing the efficiency of natural resources use will alone be sufficient for convincing farmers to switch from tillage-based agriculture to CA (Knowler *et al.*, 2001).

Although no significant effects of reduced soil disturbance on cotton or wheat yields had been observed for instance in Uzbekistan, the initial yield loss that allegedly occurs when introducing CA was also not observed, while savings in operational costs were achieved immediately (Egamberdiev, 2007; Tursunov, 2009). The values were highest under CA with crop residue retention, which amounted to UZS1,288,000 (ca. US$1075) per ha accumulated over 3 years. While using the results of three consecutive growing seasons, a dominance analysis showed that CA had much higher potential than the conventional practices owing to higher total variable costs and lower gross margins (GM). Cumulative gross margin analysis showed higher GM in all CA practices tested as compared to ConvT. Dominance analysis further revealed the advantage of the CA practices over ConvT because of the lower total variable cost and higher GM (Tursunov, 2009). Thus, adopting CA practices on the irrigated soils of Central Asia can improve the sustainability in agricultural production, provide benefits to farmers and reduce the threat of food insecurity. The financial analyses from Devkota (2011a) over four seasons of a rice-winter wheat rotation subject to a change from flood-irrigated to water-saving irrigation methods, showed a reduction in overall production, GM and benefit:cost ratio, which were highest with ConvT combined with continuous flood irrigation and lowest with permanent-bed and zero-till plantings while retaining all crop residues. Devkota (2011a) concluded that as long as irrigation water cost was not charged to farmers, it is unlikely that an economically-driven change in attitude will occur.

Introduction of CA technologies has been shown to lower production cost, raise profitability of winter wheat production and accordingly facilitate sustainable development of agriculture in different forms of agricultural entities in the south of Kazakhstan (Sydyk *et al.*, 2009). It has been found that in south Kazakhstan direct planting of winter wheat provides cost reduction by 28–44%. According to Fileccia (2009), considering both the cost savings and the yield gains, the economic efficiency of wheat production with NT technology resulted in an average improved net profit per hectare of over 50% in Kazakhstan. Meanwhile, the application of mineral fertilizers at the rate of $P_{30}N_{50}$ and the treatment of crops by herbicides facilitate the growth of conventional net income by 85.2% and reduces the production cost (Sydyk *et al.*, 2009).

In calculating energy efficiency in cultivating wheat in rainfed areas, Sydyk *et al.* (2008) determined the cost of aggregate energy directly associated with fulfilment of field operations described in 'technology maps' based on energy equivalents. It was found that about 7–22% energy expense was for soil treatment and planting; 3–4% for application of mineral fertilizers, 4–5% for application of herbicides, 40–45% for harvesting and transportation of crop and 34–36% for postharvest treatment of grain (Sydyk *et al.*, 2008).

It was determined that application of mineral fertilizers and herbicides, in South Kazakhstan province, did not require significant energy expenditure, but gave highest payback of energy resources, where energy efficiency ratios were 1.31 and 1.52 (Sydyk *et al.*, 2008).

It was revealed that under an irrigated farming system of southern Kazakhstan, raised-bed NT direct planting of winter wheat is a promising technology of CA. Significant reduction in cost for production of grain can increase the amount of conventional net income almost 1.4 times (Sydyk and Isabekov, 2009a).

9.5 Challenges Encountered in Scaling Conservation Agriculture in Central Asia

Several challenges that hinder the spread of CA in Central Asia can be recognized. They are elaborated in the following sections.

9.5.1 Government policies and institutional support

Preceding sections indicate that the governments in Central Asia do not have clear-cut policies on which kind of agricultural paradigm they wish to support to meet their future needs for food security, ecosystem services, climate change adaptability and mitigation as well as to respond to higher costs of energy and production inputs, and environmental degradation. The current status is to continue with tillage-based agriculture as much as possible. Only Kazakhstan took a policy decision to promote and support NT farming for rainfed production through subsidy on equipment (see Box 9.1). However, its policies towards CA and CA-based ecosystem management have some way to go. While there are research institutions or some researchers in some institutions who have been active in CA-related research, by and large research institutions do not explicitly implement a comprehensive CA-based research programme.

The Central Asian and Caucasus Association of Agricultural Research Institutes (CACAARI) in its statement on regional research priorities recognizes the need for capacity development in research and extension in the area of CA; but it is one topic amongst several, reflecting perhaps that CA is an option amongst several other technologies rather than an approach that involves a paradigm change in the way farming is carried out, and the mainstreaming of CA research to generate new knowledge on the

Box 9.1. Support for Conservation Agriculture in Kazakhstan – Subsidy and research

In the Republic of Kazakhstan, the state policy is oriented to the expansion of sowing areas under Conservation Agriculture. Moreover, in agricultural research, the priority area of study is resource and water-saving technology (Conservation Agriculture) of cultivation of agricultural crops in all regions of the country.

In compliance with the Resolutions of the Government of the Republic of Kazakhstan No.221 dated 4 March 2011 and No.938 dated 22 August 2011, the Ministry of Agriculture identified a flexible strategy of subsidizing farmers.

The amount of subsidies in case of using Conservation Agriculture is significantly higher (3–4 times) versus conventional technology. Government subsidies for adopting CA practices also have accelerated adoption. For example in 2011, the Government subsidies for adopting no-till practices were slightly over US$6 ha^{-1} Kazakhstan (Kazakhstan Farmers Union, 2011, Kienzler *et al.*, 2012).

Regrettably, in irrigated farmlands in southern Kazakhstan, CA technologies are being introduced slowly. It is believed that the main reasons are lack of planting machines and a lack of knowledge by the farmers of no-till technologies.

Respectively, agricultural researchers for the last years often (two or three times a year) are organizing Farmers' Days, training workshops and scientific-practical conferences with the invitation of foreign scientists.

different aspects of CA management as well as the benefits that are possible from CA at the farm, community and landscape level. Research is one amongst several institutional responsibilities that need to be aligned towards generating new knowledge regarding CA so that the full potential of CA can be harnessed with locally formulated practices to suit the diversity of ecological and socio-economic contexts. Others involve extension, input suppliers including machinery and equipment, and output value chain and market access. In addition, several other institutions exist to address issues related to agriculture such as irrigation and water resource management, natural resource management and land degradation, livestock, climate change adaptability and mitigation. In general, public institutions are expected to operate within the policy environment of governments, and similarly private institutions have to align themselves to government strategies. Given the almost complete lack of official policy on CA in the Central Asia region, public and private institutions can by and large decide independently on what kind of agriculture to promote and support, resulting in confusion and wastage of human and financial resources.

An enabling government policy and institutional environment is needed to promote the mainstreaming of CA. This in practice requires that all the stakeholders must become engaged in the management of production and of the natural resource base in a sustainable manner. However, it is also necessary for the government to create an enabling environment to promote farmers' interest in undertaking sustainable soil and production management as well as the maintenance of ecosystem services. For this, farmers must be assisted to empower themselves by forming associations so that farmers can work together in testing CA practices and sharing experiences and results as well as in articulating their needs for equipment, information, advice and incentives. Also, there should be effective integrated development planning and policies backed up by relevant research and advisory/extension systems, and the mobilization of private sector stakeholders, for both rainfed and irrigated systems (Kassam et al., 2012, unpublished results).

Providing policy and institutional support to farmers for CA adoption is an important necessary step in establishing ecological sustainability of production systems. When CA can be adopted over large areas such as watersheds and provinces, landscape-level benefits can be harnessed through appropriate schemes. Such schemes could be for carbon offset trading, e.g. in Alberta, Canada, or for water-related services in the Paraná Basin III, Brazil, or for erosion control, e.g. in olive groves in Andalucía, Spain (Kassam et al., 2013).

9.5.2 Changing the tillage mindset

One of the biggest challenges to the widespread adoption of CA in northern Kazakhstan is that of changing the tillage mindset. This has been the case in all other countries where CA has spread, and we believe Kazakhstan will be no different. However, there is clear evidence that the system works under the conditions of the region; and there are some hard-working enlightened individuals who see that the principles of CA are not only functional, but important to halt the marked, albeit slow, soil and land degradation in the region. There are currently some 13.5 Mha of CA in Canada (Friedrich et al., 2012), much of it under conditions similar to that in northern Kazakhstan and elsewhere in Central Asia. Thus farmers in the region can benefit from both the positive experiences and lessons learned by their Canadian counterparts (Karabayev et al., 2012).

9.5.3 Skills required to operate Conservation Agriculture equipment

In addition to the change in mindset, all the skills that are required under ConvT management are also required for management of CA systems. The major differences are the operations of NT seed drills and herbicide sprayers.

Operation of NT seed drills requires the knowledge of the variety of the openers and coulters and their effects on the groove shape and seed placement. Groove shape and seed placement play important roles in seed germination under moist soil conditions. To master such skills, the operator must have deeper knowledge of different soil types, whereas in a ConvT system field preparations for sowing are uniform in terms of the use of machinery. Depending on the surface residue levels, the operator should be able to select the appropriate coulter types and make necessary adjustments to seed the NT crop.

Traditionally, herbicide application in Central Asia is done largely with air-blast sprayers, therefore there is limited knowledge of other types of sprayers such as rotary plate, boom and ultralow volume that produce different sizes of droplets. In CA, boom sprayers are widely used, which are fitted with different types of nozzles to target leaves. Exploitation of boom sprayers requires good understanding of nozzle types, angles produced by nozzles to ensure good coverage, pressure, preparation of solutions amongst other factors. There is also a need for improving legislation and developing the national capacity on pesticide application equipment registration, inspection and operator licensing.

9.5.4 Availability and accessibility to suitable implements

Numerous experiments with locally made and imported seeders have been conducted and seeders have been tested for the common raised-bed systems as well as flat seeding. In irrigated cotton–wheat systems, the replacement of mouldboard ploughing with conservation tillage reduced cotton yield, but not of wheat (Suleymenov *et al.*, 2004). Hence, a modified system was suggested: the use of the mouldboard plough for cotton and the use of conservation tillage for wheat. Prior to introducing CA practices, seeding equipment was adapted in Uzbekistan (Egamberdiev, 2007; Tursunov, 2009). As a first step, seedbed preparation and planting/seeding was tested in north-western Uzbekistan (Tursunov, 2009). The modifications in an imported Indian NT seeder included the introduction of a seeding-depth regulator, appropriate soil openers for planting into the hard and mulched soil, the seeding blade that now is suitable for various crops, and an adoption of the row distance regulator. The modified seeder became suitable for planting cotton and wheat on permanent beds (Tursunov, 2009).

In a 5-year study, Ospanbaev and Karabayev (2009) concluded that the use of a raised-bed seeder advanced the possibilities of crop planting by up to 30 days compared to conventional systems, which is a substantial encouragement for the spread of CA practices. In another joint farmer-researcher managed trial in Uzbekistan, implements for the bed-and-furrow system (BFS) typical for local cotton production and NT technologies were compared (Pulatov *et al.*, 2001). The research focused on the performance of NT and BFS planters and the effects of sowing with NT drill, BFS planter and ConvT on crop yield, irrigation and income. Findings from NT and BFS planting showed that savings in time and labour as well as the user-friendly machine construction and the simple technology appealed to farmers and researchers. The use of implements suitable for CA practices increased yields through an earlier establishment of the crops and decreased crop establishment costs through a reduction in tillage costs, which was underlined by the participating farmers (Pulatov *et al.*, 2001).

Evidence worldwide shows that a widespread adoption of CA practices is unlikely if the suitable equipment is not readily available at acceptable costs (e.g. Knowler *et al.*, 2001; Friedrich and Kassam, 2009). Although national policies in Central Asian countries prioritize agriculture, the necessity to increase the accessibility and affordability of locally made CA implements suitable for seeding in untilled and mulched soils and in the presence of stubbles and/or a cover crop is still underestimated. Moreover, practices such as land levelling and NT raised bed planting can provide employment opportunities to jobless rural youths and employment in small-scale manufacturing and transport-related sectors as shown in other countries (Gupta and Sayre, 2008).

The 'Matyushkov' seeding shoes still cause considerable soil movement and longer (front to back), narrower shoes which cause less lateral soil velocity would be an improvement. Recently, chisel points from India have been imported into Kazakhstan, and a new modification to the standard seeding shoes has been made by Dr Dvurechenskiy. Manufacture of these was tried at both the Agromash factory in Astana, Kazakhstan, and a factory in Omsk, Siberia, with the latter giving better results because of the hardness of the steel used and therefore the extended life of the shoes (Karabayev *et al.*, 2012).

9.5.5 Residue supply and management

In virtually the entire Central Asian region, crop stubbles, essential for CA, are either burned due to a lack of suitable, powerful tractors for ploughing or, more commonly, residues are removed and fed to livestock. Some of the Central Asia nations such as Uzbekistan still have tillage regulations at present that limit the possibility for farmers to leave crop residues on the field. Studies thus far have therefore compared mainly the cases of 100% crop residue retention or no retention (e.g. Egamberdiev, 2007; Kienzler *et al.*, 2009a; Tursunov, 2009; Devkota, 2011a, b; Ibragimov *et al.*, 2011). Research on intermediate levels, rates and residue management practices have been usually beyond the scope of these initial studies. Only Devkota (2011b) concluded from her findings in a cotton–wheat–maize rotation that the retention of all crop residues after each cropping cycle is unnecessary to improve soil quality. The mulch layer from a retention of 8–10 t ha^{-1} wheat straw obstructed seeding, irrigation and fertilizer management in a cotton–winter wheat rotation. The retention of 14 t ha^{-1} standing residues on permanent beds for rice reduced soil temperature and resulted in a delayed germination and reduced yields (Devkota, 2011a). Previous studies outside Central Asia (FAO, 2000) indicated that the retention of 4 t ha^{-1} crop residues was sufficient for CA practices. However, given the scarcity of findings, additional research is needed to clarify this component. Research should in this case concentrate on identifying suitable and manageable levels of partial residue retention and residue management so as to achieve the expected agronomic benefits and consider the alternative demands from farmers.

After independence in 1991 from the Soviet Union, Uzbekistan and Turkmenistan maintained the notion of strategic crops under a state order system, while in the rest of the Central Asian states, the order was abolished or replaced by other crops. Turkmenistan and Uzbekistan still regulate and own the majority of the economic and land resources, while Kyrgyzstan, Kazakhstan and Tajikistan have introduced a certain level of land privatization. Nevertheless, after almost two decades since the changes were introduced, the knowledge, design and equipment available in the different countries in many aspects still 'mimic' the former Soviet agricultural system. The most common crop rotations in Central Asia such as cotton–wheat, wheat–fallow or wheat–rice rotations (Gupta *et al.*, 2009) thus leave little scope for diversifying the system especially under the current agricultural legislation prevalent in some Central Asian nations, thereby failing to harness the benefits of crop rotations, which is an important component of CA practices. Also, in the absence of private land tenure, farmers for instance in Uzbekistan and Turkmenistan refrain from CA practices for a longer timespan, although after only few years typical environmental benefits of CA emerge such as an increase in SOM (Egamberdiev, 2007; Funakawa *et al.*, 2007; Sommer and De Pauw, 2010).

9.5.6 Weed management

Weed infestation is not only common in CA, but rather CA causes a change in the dynamics of weed growth that are already present in traditional production systems. Effects of crop rotation on weeds have been intensively studied in cotton-growing areas during the period 1975–1984 (Tursunkhodjaev and Bolkunov, 1981; Ismailov, 2004) with different combinations of cotton–lucerne–wheat rotations. The principles still hold true that crop rotation helps to suppress weeds and appropriate strategy is needed for CA as well.

Weed control is one of the principal reasons for soil tillage, and when tillage is reduced or avoided, weed control is one of the major management challenges that must be tackled. As CA became more readily possible with the advent of herbicides one can expect that in most instances, in the first years of CA, the use of chemicals for weed control may increase. However, the principal herbicide used for weed control in the growing crop or prior to crop establishment is glyphosate – a herbicide for the total control of weeds. Glyphosate is relatively benign environmentally: it has very low mammalian and invertebrate toxicity; it is tightly bound to clay particles in the soil and so is not leached; and is broken down by soil microbes, generally within about 3 months. As soil erosion is drastically reduced under CA, the chance of glyphosate getting into waterways from CA fields is very low, and even then it is so tightly bound to the clay particles that it is not released into the water. However, one concern is the widespread use of glyphosate in CA systems and the appearance of glyphosate-resistant weeds: populations of 11 weeds resistant to glyphosate have been reported worldwide (International Survey of Herbicide Resistant Weeds, 2006).

The effect of different types of herbicides on productivity of NT winter wheat was studied by Nurbekov (2007) in Karakalpakstan, Uzbekistan. The overall weed infestation observed in conventionally tilled wheat with application of Puma Super in spring was essentially equal to that found in NT wheat with Dafosat applied in the autumn followed by spring-applied Puma Super. Some recommendations on herbicide applications to control specific weeds have already been developed for Kazakhstan. In northern regions of Kazakhstan, during the early growth of wild oats (usually when soils warm up to 10–12°), it is recommended that glyphosate (which has uniform impact) should be applied before planting of cereals. Herbicides could be applied at minimum dose – up to 1.0 l ha^{-1}. Favourable environment, such as mass sprouting of wild oats, cool weather, sufficient soil moisture, provides highly efficient suppression of this weed.

Meanwhile, application of these herbicides in minimum dose costs 2.0–2.7 times less than the use of counter-wild oats herbicides, and is 1.6 times cheaper compared to crop management activities aimed at control (Sydyk *et al.*, 2008). Moreover, since this method does not require large number of machines, practically any farmer can afford it.

When the herbicide Target was applied at the rate of 1 l ha^{-1} in NT directly seeded winter wheat, high yield of 4.0–4.4 t ha^{-1} was achieved in the rainfed areas in high rainfall years, whereas in medium rainfall years the yield was in the range 2.6–3.2 t ha^{-1}. In 2006–2008, the application of herbicide Aroma (50% emulsifiable concentrate) at 1.5 and 2.0 l ha^{-1} with direct seeding, demonstrated greater efficiency. Treatment at the rate of 1.5 l ha^{-1} reduced the number of weeds down to 24.8 plants m^{-2} from the initial number of 124 plants m^{-2}, while under the higher rate these numbers were 21.6 plants m^{-2} and 128.2 plants m^{-2}, respectively. Reduction of fresh biomass of weeds compared to control fluctuated within 74.0–74.6%, depending on the rates of herbicide treatments.

In safflower, preplanting treatment by the herbicides Dual Gold 960 emulsifiable concentrate at a rate of 1.5 l ha^{-1} allowed reduction of number of weeds by 93% and biomass by 96%, providing high yield of oilseeds (1.45 t ha^{-1}) (Sydyk and Isabekov, 2009b). It was found that the treatment of safflower crop by herbicide Dual Gold 960 emulsifiable concentrate (1.5 l ha^{-1}) resulted in increase of thousand kernel weight (TKW) by 4.7 g or 14.5%. Thus, under direct planting at the rate of 1.0 l ha^{-1}, productivity of safflower increased by 0.46 t ha^{-1}; in raising the herbicide rate to 1.5 l ha^{-1}, productivity of safflower comprised 1.18 t ha^{-1} and productivity increased by 0.6 t ha^{-1}. The highest yield of safflower, 1.35 t ha^{-1}, was obtained in treatment with herbicide Pivot (10%) at a rate of 0.8 l ha^{-1} at the sprouting stage, a yield increase of 0.77 t ha^{-1} (Sydyk *et al.*, 2009). Application of Pivot (10%) at 0.5 l ha^{-1} resulted in the average yield of 1.08 t ha^{-1} and the yield increased by 0.5 t ha^{-1}.

The other two principles of CA, NT and maintenance of soil cover, also contribute to suppressing weeds in CA systems that promote integrated weed management. Not tilling the soil promotes the rotting of the weed seed bank in the soil over time, and avoids the burying of weed seeds into the soil, which can protect them. Similarly, mulch cover can suppress weeds and also helps to kill weed seeds with humic acids that are released from the decomposing organic residues. Little work has been done in the region on integrated weed management, and should be encouraged in the future.

In should be noted that the quality control and certification of chemicals, including herbicides, is still not fully in place. Thus, very often low-quality and hazardous herbicides are used by the farmers. On the other hand, the prices for herbicides are high and not all farmers, especially small-scale farmers, can afford their application.

9.6 Prospects for Conservation Agriculture in Central Asia

Conservation Agriculture is one of the most promising agricultural land use options that have been developed in our time. Conservation Agriculture is more an approach to sustainable agroecosystem management than a production technology because it offers a way to produce more with less while at the same time preserves and enhances many of the ecological functions a natural soil has to offer in a natural ecosystem. Conservation Agriculture also offers economic benefits to farmers who apply it. Generally, an immediate cost reduction due to reduced cultivation and machinery operations can be felt right after the introduction of CA. There are a number of challenges that CA faces throughout the largely agricultural region of Central Asia including the lack of: crop and farming system diversification on small-size farming areas; knowledge about CA systems among extension and technical staff; knowledge about CA at decision-making levels; farmers' ability to decide on diversified crop rotations; and implements needed for use in CA. Nevertheless, farmers in the region of Central Asia are now becoming increasingly aware of CA as a new, promising farming paradigm. Awareness comes in the form of accepting NT as a viable system in growing crops as opposed to the earlier total rejection of agriculture without tillage. Particularly for irrigated areas, large programmes by different institutions need to be carried out to adapt CA to local conditions and to generate research results to advise farmers accordingly. For example, in Uzbekistan and Kazakhstan, the governments provide research grants to institutions, and they have approved a number of applications from different research groups for addressing water and crop issues in CA systems.

Only Kazakhstan has managed to implement supportive policies for CA, and as a result the area under CA-based practices increased from 0 ha in 2000 to 1.6 Mha in 2011 with continued expansion according to a recent assessment conducted by CIMMYT (FAO, 2012). Usually manufacturers, importers and dealers are proactive with the objective of increasing the demand for CA implements. Yet, the present political systems in Central Asia indicate that the public rather than the private sector is now being called upon to initiate and lead such efforts.

Agriculture in the region is diverse, and has a great potential to revitalize the withered economies of the Central Asian countries via improved productivity (efficiency) and higher total output through CA-based agriculture development. After independence in 1991, the production of fodder crops such as maize and lucerne sharply decreased along with reduction in area under rice and vegetables (melon). Conservation Agriculture will have to shoulder the largest burden of making sustainable intensification of production systems a reality for food, fodder and fibre crops and livestock in Central Asian countries.

The demand for food and fodder production will continue to grow in Central Asian countries. Wheat, cotton and livestock are the most important agricultural commodities in the region, and with a trend to diversification, oil crops such as rapeseed, sunflower, safflower and soybeans could likewise become important commodities, similar to the Canadian model.

The evidence from Central Asian countries shows that CA practices are suitable for the existing major cropping systems. However, most of the results come from collaborative projects largely initiated and funded by international organizations. Conservation Agriculture is not a single or uniform technology that can be immediately applied anywhere in a standard manner. Rather, it represents a set of principles that encourage the formulation of locally adapted practices, approaches and methods, which need to be tested, evaluated and then adopted or implemented under various biophysical and socio-economic conditions. Further research is necessary, for example, to study in detail the effects of various CA crop rotations and mulch cover on weed management, nutrient, pest and water management, on residue levels, sowing depth, dates, density, and on fertilizer and irrigation rates; and impact assessment on livelihoods and environmental conditions including the potential of integrating trees and livestock into CA farming systems, particularly with small-scale farmers. To make results applicable on a wider scale, state programmes should become more active in conducting research, training and extension.

Considerable knowledge has been generated about CA practices in the Central Asian region, first in rainfed areas and, more recently, in irrigated areas. In fact, the potential of CA for sustainable agricultural development has been demonstrated in the region. Building the technical and scientific capacity of national partners will be essential for moving to widespread CA adoption and uptake. Researchers, extension workers and farmers will continue exchanging experience and knowledge about the new CA methods. Consequently, for the foreseeable future, facilitating national development strategies for up-scaling of CA, conducting training courses with national partners for capacity development, promoting farmer associations and facilitating stakeholder engagement through national and regional platforms in supporting CA adoption and uptake should remain a high priority in the efforts undertaken by FAO, ICARDA, CIMMYT and other international organizations such as IFAD, ADB, EU and national donors, to promote CA in the region.

Notes

[1] We define permanent raised beds as raised beds that were prepared and used during a previous season but subsequently used also for growing the next crop on the same beds. Therefore, we differentiate between raised beds that are not permanent (fresh beds prepared every season) and those that are permanent.

[2] No-till consisting of direct drilling as the only mechanical operation disturbing the soil surface. All other operations usually employed under 'conservation tillage' in the rainfed areas of Kazakhstan, such as sweep tillage, discing and harrowing, are thus not included.

References

Akramkhanov, A., Kuziev, R., Sommer, R., Martius, C., Forkutsa, O. and Massucati, L. (2012) Soils and soil ecology in Khorezm. In: Martius, C., Rudenko, I., Lamers, J.P.A. and Vlek, P.L.G. (eds) *Cotton, Water, Salts and Soums – Economic and Ecological Restructuring in Khorezm, Uzbekistan*. Springer, New York, pp. 37–58.

Baig, M.N. and Gamache, P.M. (2009) *The economic, agronomic and environmental impact of no-till on the Canadian Prairies*. Alberta Reduced Tillage Linkages, Canada.

Bezborodov, G.A., Shadmanov, D.K., Mirhashimov, R.T., Yuldashev, T., Qureshi, A.S., Noble, A.D. and Qadir, M. (2010) Mulching and water quality effects on soil salinity and sodicity dynamics and cotton productivity in Central Asia. *Agriculture, Ecosystems, Environment* 138, 95–102.

Blackshaw, R.E., Harker, K.N., O'Donovan, J.T., Beckie, H.J. and Smith, E.G. (2007) Ongoing development of integrated weed management systems on the Canadian Prairies. *Weed Science* 56, 146–150.

Bot, A. and Benites, J. (2005) The importance of soil organic matter: the key to drought resistant soil, sustained food and production. FAO Soils Bulletin 80. Available at: http://www.fao.org/docrep/009/a0100e/a0100e00.htm (accessed 7 January 2013).

Brimili, W. (2008) A case of extreme particulate matter concentrations over Central Europe by dust emitted over the Southern Ukraine. *Atmospheric Chemistry and Physics* 9, 997–1016.

Burman U. and Roy, M.M. (2011) Climate change mitigation and conservation agriculture. Available at: http://jaibbihar.com/conference/Uday%20Burman%20and%-20M.M.%20Roy.pdf (accessed 17 January 2013).

Ceja-Navarro, J.A., Rivera-Orduna, F.N., Patino-Zuniga, L., Vila-Sanjurjo, A., Crossa, J., Govaerts, B. and Dendooven, L. (2010) Phylogenetic and multivariate analyses to determine the effects of different tillage and residue management practices on soil bacterial communities. *Applied and Environmental Microbiology* 76, 3685–3691.

Corsi, S., Friedrich, T., Kassam, A., Pisante, M. and Sa, J. de M. (2012) *Soil Organic Carbon Accumulation and Greenhouse Gas Emission Reduction from Conservation Agriculture: A Literature Review. Integrated Crop Management Vol. 16*. Plant Production and Protection Division, FAO, Rome.

Crabtree, B. (2010) Search for Sustainability with No-Till Bill in Dryland Agriculture. Crabtree Agricultural Consulting, Australia. Available at: http://www.no-till.com.au (accessed 17 January 2013).

CTIC/FAO (2008) *Mitigating climate change: conservation agriculture stores soil carbon. Recommendations of the Conservation Agriculture Carbon Offset Consultation*. CTIC, West Lafayette.

De Pauw, E. (2008) ICARDA regional GIS datasets for Central Asia: Explanatory notes. GIS Unit Technical Bulletin, International Centre for Agricultural Research in the Dry Areas (ICARDA).

Derpsch R. (2008) No-tillage and Conservation Agriculture: A Progress Report. In: Goddard, T., Zoebisch, M., Gan, Y., Ellis, W., Watson, A. and Sombatpanit, S. (eds) *No-Till Farming Systems*.World Association of Soil and Water Conservation, Special Publication No. 3., WASWAC, Bangkok, pp. 7–39.

Devkota, K. (2011a) Resource utilization and sustainability of conservation based rice-wheat cropping systems in Central Asia. PhD dissertation, ZEF/Rheinische Friedrich-Wilhelms-Universität Bonn, Germany.

Devkota, M. (2011b) Nitrogen management in irrigated cotton-based systems under conservation agriculture on salt-affected lands of Uzbekistan. PhD dissertation, ZEF/Rheinische Friedrich-Wilhelms-Universität Bonn, Germany.

Dvurechenskiy, V.I. (2010) *Vozdelivaniye Zernovykh Kultur na Osnove Novoy Vlagosberegayuschey Tekhnologii i Soveremennoy Tekhniki*. Izdatelskiy Dom (in Russian).

Egamberdiev, O.J. (2007) Dynamics of irrigated alluvial meadow soil properties under the influence of resource saving and soil protective technologies in the Khorezm region. PhD dissertation, National University of Uzbekistan, Tashkent (in Uzbek).

FAO (2000) *Crops and drops*. Land and Water Division of the Food and Agriculture Organization, FAO, Rome, Italy.

FAO (2008) Integrated crop management. In: *Proceedings of the International Technical Workshop on Investing in Sustainable Crop Intensification: The Case for Improving Soil Health*, Vol. 6. FAO, Rome, p. 134.

FAO (2012) CA adoption worldwide. FAO AQUASTAT conservation agriculture website. Available at: http://www.fao.org/ag/ca/6c.html (accessed 7 November 2011).

Fileccia, T. (2009) Conservation agriculture and food security in Kazakhstan. Working Paper, FAO Investment Centre Division, Rome.

Forkutsa, I., Sommer, R., Shirokova, Y.I., Lamers, J.P.A., Kienzler, K., Tischbein, B., Martius, C. and Vlek, P.L.G. (2009) Modelling irrigated cotton with shallow groundwater in the Aral Sea Basin of Uzbekistan: II. Soil salinity dynamics. *Irrigation Science* 27, 319–330.

Friedrich, T. and Kassam, A.H. (2009) Adoption of Conservation Agriculture Technologies: Constraints and Opportunities. In: *Proceedings of the IV World Congress on Conservation Agriculture*, ICAR, New Delhi, India, 4–7 February, 2009.

Friedrich, T., Derpsch, R. and Kassam, A. (2012) Overview of the global spread of Conservation Agriculture. Field Actions Science Reports, Special Issue 6, 1–7. Available at: http://factsreports.revues.org/1941 (accessed 7 January 2013).

Funakawa, S., Yanai, J., Takata, Y., Karbozova-Saljinikov, E., Akshalov, K. and Kosaki, T. (2007) Dynamics of water and soil organic matter under grain farming in Northern Kazakhstan – Toward sustainable land use both from the agronomic and environmental viewpoints. In: Lal, R., Suleymenov, M., Stewart, B.A., Hansen, D.O. and Doraiswamy, P. (eds) *Climate Change and Terrestrial Carbon Sequestration in Central Asia*. Taylor & Francis, London, pp. 279–332.

Gan, Y., Harker, K.N., McConkey, B. and Suleymenov, M. (2008) Moving towards no-till practices in Northern Eurasia. In: Goddard, T., Zoebisch, M., Gan, Y., Ellis, W., Watson, A. and Sombatpanit, S. (eds) *No-Till Farming Systems*. Special Publication No. 3. World Association of Soil and Water Conservation, Bangkok, pp. 179–195.

Govaerts, B., Sayre, K.D., Ceballos-Ramirez, J.M., Luna-Guido, M.L., Limon-Ortega, A., Deckers, J. and Dendooven, L. (2006) Conventionally tilled and permanent raised beds with different crop residue management: Effects on soil C and N dynamics. *Plant Soil* 280, 143–155.

Gupta, R. and Sayre, K. (2008) Conservation agriculture in South Asia – Some lessons learnt. *Professional Alliance for Conservation Agriculture, Newsletter* 3, New Delhi, India, pp. 1–3.

Gupta, R., Kienzler, K., Martius, C., Mirzabaev, A., Oweis, T., De Pauw, E., Qadir, M., Shideed, K., Sommer, R., Thomas, R., Sayre, K.D., Carli, C., Saparov, A., Bekenov, M., Sanginov, S., Nepesov, M. and Ikramov, R. (2009) *Research prospectus: A vision for sustainable land management research in Central Asia*. ICARDA Central Asia and Caucasus Program. Sustainable Agriculture in Central Asia and the Caucasus Series 1, CGIAR-PFU, Tashkent, Uzbekistan.

Hickmann, S. (2006) Conservation agriculture in northern Kazakhstan and Mongolia. FAO, Agricultural and Food Engineering Working Document 4.

Houghton, J.T., Meira Filho, L.G., Lim, K., Trennton, I., Mamaty, I., Bonduki, Y., Griggs, D.J. and Callander, B.A. (1997) *Revised 1996 IPCC Guidelines for National Greenhouse Gas Inventories, 1–3*. Intergovernmental Panel on Climate Change. WMO/UNEP, Cambridge University Press, Cambridge, UK.

Ibragimov, N., Evett, S., Essenbekov, Y., Khasanova, F., Karabaev, I., Mirzaev, L. and Lamers, J.P.A. (2011) Permanent beds versus conventional tillage in irrigated Central Asia. *Agronomy Journal* 103, 1002–1011.

International Survey of Herbicide Resistant Weeds (2006) Glyphosate resistant weeds. Available at http://www.weedscience.org/glphosate.gif (accessed 5 May 2012).

IPCC (2007) 4th Assessment Report of the Inter-Governmental Panel on Climate Change. Available at: http://www.ipcc.ch/publications_and_data/ar4/wg2/en/ch18s18-4-3.html (accessed 5 January 2013).

Ismailov, U.E. (2004) *Nauchnie Osnovi Povisheniya Plodorodiya Pochvy*. Bilim, Nukus, Uzbekistan (in Russian).

Jat, R.A., Wani, S.P. and Sahrawat, K.L. (2012) Conservation agriculture in the semi-arid tropics: prospects and problems. In: Sparks, D.L. (ed.) *Advances in Agronomy* 117, 191–273.

Karabayev, M., Wall, P., Sayre, K. and Morgounov, A. (2012) Conservation agriculture adoption in Kazakhstan: History, Status and Outlooks. CIMMYT report.

Kassam, A., Basch, G., Friedrich, T., Shaxson, F., Goddard, T., Amado, T., Crabtree, B., Hongwen, L., Mello, I., Pisante, M. and Mkomwa, S. (2013) Sustainable soil management is more than what and how crops are grown. In: Lal, R. and Stewart, B.A. (eds) *Advances in Soil Sciences: Principles of Sustainable Soil Management in Agroecosystems*. Taylor & Francis, London.

Kazakhstan Farmers Union (2011) Stop the tractor! I till no more! Available at http://sfk.kz/index.php?id=9&kid=10 (accessed 7 September 2011) (in Russian).

Kienzler, K. (2009) Improving the nitrogen use efficiency and crop quality in the Khorezm region, Uzbekistan. PhD thesis, University of Bonn, Bonn, Germany.

Kienzler, K., Saparov, A., Bekenov, M., Kholov, B., Nepesov, M. and Ikramov, R. (2009a) *Final report – Part I. Sustainable Land Management Research Project 2007-2009*. ICARDA Central Asia and Caucasus Program, Tashkent, Uzbekistan.

Kienzler, K., Saparov, A., Bekenov, M., Kholov, B., Nepesov, M. and Ikramov, R. (2009b) *Final report – Part II. Sustainable Land Management Research Project 2007-2009*. ICARDA Central Asia and Caucasus Program, Tashkent, Uzbekistan.

Kienzler, K.M., Lamers, J.P.A., McDonald, A., Mirzabaev, A., Ibragimov, N., Egamberdiev, O., Ruzibaev, E. and Akramkhanov, A. (2012) Conservation agriculture in Central Asia – What do we know and where do we go from here? *Field Crops Research* 132, 95–105.

Knowler, D., Bradshaw, B. and Gordon, D. (2001) *The Economics of Conservation Agriculture*. Land and Water Division, FAO, Rome, Italy.

Kondratjuk, V.P. (1972) *Soil Tillage for Cotton Planting in Central Asia*. Cotton Research Institute, Ministry of Agriculture USSR, FAN Tashkent (in Russian).

Kurvantaev, R., Kuziboev, O. and Solieva, N. (2004) The effect of conservation tillage technology on cotton yield. In: *Proceedings of the Conference on New Technologies to Increase Soil Fertility*. Soil Research Institute, Tashkent (in Russian).

Lal, R. (2002) Carbon sequestration in dryland ecosystems of West Asia and North Africa. *Land Degradation and Development* 13, 45–59.

Lal, R. (2008) Carbon sequestration in dryland ecosystems. *Environmental Management* 33, 528–544.

Mohanty, M., Painuli, D.K., Misra, A.K. and Ghosh, P.K. (2007) Soil quality effects of tillage and residue under rice-wheat cropping on a Vertisol in India. *Soil and Tillage* 92(1–2), 243–250.

Nelson, R.G., Hellwinckel, C.M., Brandt, C.C., West, T.O., Ugarte De La, T. and Marland, G. (2009) Energy uses and carbon dioxide emissions from cropland production in the United States, 1990–2004. *Journal of Environmental Quality* 38, 418–425.

Nurbekov, A. (2007) *Final report: Sustainable agricultural practices in the drought affected region of Karakalpakstan (Phase II)*. FAO/TCP/3102 (A). ICARDA-CAC.

Nurbekov, A., Suleymenov, M., Friedrich, T., Taher, F., Ikramov, R. and Nurjanov, N. (2012) Effect of tillage methods on productivity of winter wheat in the Aral Sea Basin of Uzbekistan. *Journal of Arid Land Studies* 22, 255–258.

Ospanbaev, J. and Karabayev, M.K. (2009) Outlook for not-till technologies of crop growing in South and Southeast Kazakhstan. In: Suleymenov, M., Kaskarbayev, J.A., Skoblikov, V.F. and Dashkevich, S.M. (eds) *No-till With Soil Cover and Crop Rotation: A Basis for Policy Support to Conservation Agriculture for Sustainable Production Intensification*. Astana-Shortandy, Kazakhstan, pp. 195–199.

Parkin, T.B. and Kaspar, T.C. (2006) Nitrous oxide emissions from corn-soybean systems in the mid-west. *Journal of Environmental Quality* 35, 1496–1506.

Pozharskiy, V. and Akimaliev, M. (2002) Bed-planting winter wheat in Chu valley of the Kyrgyz Republic. In: *Proceedings of International Workshop on Conservation Agriculture for Sustainable Wheat Production in Rotation with Cotton in Limited Water Resource Areas*. 14–18 October 2002, Tashkent, Uzbekistan, pp. 123–126.

Pulatov, A., Choudhary, A. and Akramkhanov, A. (2001) Status of conservation tillage practices in Uzbekistan. In: Gill, M.A. (ed.) *Proceedings of International Workshop on Conservation Agriculture for Food Security and Environment Protection in Rice-Wheat Cropping Systems*. Lahore, Pakistan.

Pulatov, A., Egamberdiev, O., Karimov, A., Tursunov, M., Kienzler, S., Sayre, K., Tursunov, L., Lamers, J.P.A. and Martius, C. (2012) Introducing conservation agriculture on irrigated meadow alluvial soils (Arenosols) in Khorezm, Uzbekistan. In: Martius, C., Rudenko, I., Lamers, J.P.A. and Vlek, P.L.G. (eds) *Cotton, Water, Salts and Soums – Economic and Ecological Restructuring in Khorezm, Uzbekistan*. Springer, New York, pp. 195–217.

Qadir, M., Oster, J.D., Schubert, S., Noble, A.D. and Sahrawat, K.L. (2007) Phytoremediation of sodic and saline-sodic soils. *Advances in Agronomy* 96, 197–247.

Qilichev, A.H. and Khalilov, N. (2008) Go'za qator oralariga ekilgan kuzgi bugdoy hosidorligi va don sifati. *Journal AGRO-ILM* 2, 163–168 (in Uzbek).

Reicosky, D.C. (2008) Carbon sequestration and environmental benefits from no-till systems. In: Goddard, T., Zoebisch, M., Gan, Y., Ellis, W., Watson, A. and Sombatpanit, S. (eds) *No-Till Farming Systems*. Special Publication No. 3. World Association of Soil and Water Conservation, Bangkok, pp. 43–58.

Ryzhov, S.N., Kondratjuk, V.P. and Pogosov, Y.A. (1980) *Cotton Growing on Beds and Ridges*. Cotton Research Institute, Ministry of Agriculture USSR, FAN Tashkent (in Russian).

Sanchez, J.E., Harwood, R.R., Wilson, T.C., Kizilkaya, K., Smeenk, J., Parker, E., Paul, E.A., Knezek, B.D. and Robertson, G.P. (2004) Managing soil carbon and nitrogen for productivity and environmental quality. *Agronomy Journal* 96, 769–775.

Sanginov, S. and Khakimov, A. (2003) Planting winter wheat to growing cotton. In: *Proceedings of the First Central Asian Wheat Conference*. Almaty, Kazakhstan.

Saturnino, H.M. and Landers, J.N. (2002) *The environment and zero tillage*. APDC/FAO, p. 144

Scheer, C., Wassmann, R., Kienzler, K., Ibragimov, N., Lamers, J.P.A. and Martius, C. (2008) Methane and nitrous oxide fluxes in annual and perennial land-use systems of the irrigated areas in the Aral Sea Basin. *Global Change Biology* 14, 1–15.

Six, J., Feller, C., Denef, K., Ogle, F.M., de Moraes Sa, J.C. and Albrecht, A. (2002) Soil organic matter, biota and aggregation in temperate and tropical soils – effects of no tillage. *Agronomie* 22, 755–775.

Sommer, R. and De Pauw, E. (2010) Organic carbon in soils of Central Asia – status quo and potentials for sequestration. *Plant Soil* 338, 273–288.

Sommer, R., Wall, P.C. and Govaerts, B. (2007) Model-based assessment of maize cropping under conventional and conservation agriculture in highland Mexico. *Soil and Tillage Research* 94, 83–100.

Stewart, B.A. (2007) Water conservation and water use efficiency in drylands. In: Stewart, B., Fares Asfary, A., Belloum, A., Steiner, K. and Friedrich, T. (eds) *Proceedings of the International Workshop on Conservation Agriculture for Sustainable Land Management to Improve the Livelihood of People in Dry Areas*. ACSAD and GTZ, Damascus, Syria, pp. 57–66.

Suleymenov, M.K., Akhmetov, K.A., Kaskarbayev, J.A., Khasanova, F., Kireyev, A., Martynova, L.I. and Pala, M. (2004) Developments in tillage and cropping systems in Central Asia. In: Ryan, J., Vlek, P.L.G. and Paroda, R. (eds) *Agriculture in Central Asia: Research for Development*. ICARDA, Aleppo, Syria, pp. 188–211.

Suleymenov, M.K., Pala, M., Paroda, R., Akshalov, K., Martynova, L.I. and Medeubaev, R. (2006) New technologies for Central Asia. *Caravan* 23, 19–22.

Sydyk, D.A. and Isabekov, B.B. (2009a) Ongtustik Qazaqstan olysining talimi jerlerinde topyraqty endemey maqsaryny tikeley egu kezinde ylgal qorynyng jinaqtay ereksheligi. *Journal of Jarshi* 7, 31–36 (in Kazakh).

Sydyk, D.A. and Isabekov, B.B. (2009b) Effectivnost gerbitsidov pri minimalizatsii obrabotki pochvy. *Jounal Vestnik Selskokhoziyaystvennykh Nauk* 8, 22–23 (in Russian).

Sydyk, D.A., Jamalbekov, M.N., Karabaleva, A.D., Medeubaev, R.M., Sydykov, M.A. and Isabekov, B.B. (2009) *Resursosberegauyshyaya Tekhnologiya Vozdelyvanaiya Selskokhozyastevennykh Kultur na Yuge Kazakhstana*. Jebe-Disayin, Shymkent, Kazakhstan (In Russian).

Sydyk, D.S., Jarasov, Sh., Sydykov, M.A., Isabekov, B. and Babakhodjaev, A.T. (2008) *Rekomendatsii po Resursosberegauyshey Technologii Vozdelyvaniya Zernovykh Kolosovykh Kultur v Usloviyakh Bogarnogo i Oroshaemogo Zemledeliya Yujnogo Kazakhstana*. Jebe-Disayin, Shymkent, Kazakhstan (in Russian).

Tischbein, B., Awan, U.K., Abdullaev, I., Bobojonov, I., Conrad, C., Forkutsa, I., Ibrakhimov, M. and Poluasheva, G. (2012) Water management in Khorezm: current situation and options for improvement (hydrological perspective). In: Martius, C., Rudenko, I., Lamers, J.P.A. and Vlek, P.L.G. (eds) *Cotton, Water, Salts and Soums – Economic and Ecological Restructuring in Khorezm, Uzbekistan*. Springer, New York, pp. 69–92.

Tursunkhodjaev, Z.S. and Bolkunov, A.S. (1981) *Nauchnie Osnovy Khlopkovykh Sevooborotov*. Mekhnat, Tashkent.

Tursunov, M. (2009) Potential of conservation agriculture for irrigated cotton and winter wheat production in Khorezm, Aral Sea Basin. PhD dissertation, ZEF/Rheinische Friedrich-Wilhelms-Universität Bonn, Germany.

West, T.O. and Post, W.M. (2002) Soil organic carbon sequestration rates by tillage and crop rotation: a global data analysis. *Soil Science Society of American Journal* 66, 1930–1946.

10 Conservation Agriculture in West Asia

Nasri Haddad,[1] Colin Piggin,[2] Atef Haddad[3] and Yaseen Khalil[3]
[1]*West Asia regional Program, International Center for Agricultural Research in the Dry Areas, Amman, Jordan;* [2]*Australian Centre for International Agricultural Research (ACIAR), Canberra, Australia;* [3]*Diversification and Sustainable Intensification of Production System Research Program, International Center for Agricultural Research in the Dry Areas, Aleppo, Syria*

10.1 Introduction

The West Asia (WA) region covers a range of different agroecosystems, but has rather common problems: the region suffers from scarcity of water, recurrent droughts, degradation of natural resources, poverty, especially in rural areas, and difficulties in marketing the surplus local products. WA holds valuable genetic diversity of food and feed species of global importance, but these are seriously threatened by rapid loss. Different agroecosystems are found in each country of the region, with a predominance of rainfed agriculture and livestock systems. The irrigated areas in Syria and Iraq have begun to suffer from water scarcity due to depletion of groundwater, along with increased salinity.

The dryland cropping systems in WA are characterized by frequent soil tillage, grazing, burning or harvesting of crop stubbles and late sowing of winter cereals and legumes to allow for weed control by cultivation. Yields of field crops are generally low, with low rainfall water use efficiency (WUE; Piggin *et al.*, 2011).

Soil degradation is a significant threat to rural livelihoods and food security in the dry areas of WA. Extractive farming systems, inappropriate soil and crop management and/or the impact of drought and/or erosion exacerbate soil degradation. Tractor use has spread in the region rapidly, and deep ploughing using disc or mouldboard ploughs is a common practice. However, deep tillage is not sustainable in the dry areas of the region.

In an effort to understand the causes of soil degradation, researchers and farmers have begun to question the need for tillage prior to seeding; and this has resulted in the initiation and development of soil conservation research. The early focus was on conservation tillage (CT) practices, with minimum soil disturbance to reduce soil erosion and conserve soil moisture. Such systems include direct drilling, direct seeding, minimum tillage, stubble seeding, zero or no-tillage etc.

Conservation Agriculture (CA) is a holistic concept for sustainable management of agricultural land. It helps achieve a high degree of environmental and economic sustainability of farming. Additionally, by reducing labour requirements and by increasing the farm income, CA improves farm family income in the longer term (Friedrich and Kienzle, 2008).

Conservation Agriculture is one approach suited for sustainable agriculture. Conservation Agriculture is based on: (i) minimizing soil disturbance by mechanical tillage and thus facilitates seeding directly into untilled soil; (ii) maintaining year-round organic matter cover over the soil, by retained residues from the previous crop; and (iii) diversifying crop rotations, adapted to local environmental conditions. Conservation Agriculture functions best when all the three key features are adequately combined together at the farm (Friedrich and Kienzle, 2008; Kassam and Friedrich, 2010; Avci, 2011; Mrabet, 2011; Piggin and Devlin, 2012).

This chapter reviews the research on CA in WA countries, which started in an integrated form during the last three decades in selected countries including Turkey and Iran, and by international institutions such as the International Center for Agricultural Research in the Dry Areas (ICARDA) in Syria and the Arab Center for the Study of Arid Zones and Dry Lands (ACSAD). Other countries such as Jordan, Lebanon and Iraq took to CA fairly recently. We highlight major successes and constraints for scaling out CA, and suggest measures to overcome the constraints.

10.2 Conservation Agriculture Research Results in West Asia

The major rainfed field crops grown in WA countries are cereals, wheat and barley, usually grown in rotation with food legumes such as lentil and chickpeas, and to a lesser extent summer vegetables such as melons and tomatoes. Fallow is very common, especially where rainfall is low and drought prevails. Therefore, most of the CA research work related to these crops. Farming systems in WA countries are characterized by smallholder farmers, with holdings of 1–3 ha. Rainfed farmers are in general poor, risk adverse and with limited access to inputs and technological innovations. Government and private sector investment in dryland farming is very low with support focused more on irrigated agriculture where returns on the investment are relatively high and assured.

Early work on CA in WA was initiated in Turkey with evaluating minimum tillage and zero tillage (ZT) and comparing the tillage tools with crop residue management to conserve soil moisture and reduce severe wind erosion (Avci, 2011).

Overall, CA systems have a higher adaptability to uncertain climate because of higher effective rainfall from higher infiltration rates, and reduced runoff, flooding and soil erosion, as well as improved storage of water in the soil profile (Jat *et al.*, 2012). The advantage of CA over conventional agriculture (ConvA) with tillage in terms of greater soil moisture-holding capacity and therefore duration of plant-available soil moisture is illustrated by Derpsch *et al.* (1991), who showed that soil moisture conditions in the rooting zone throughout the growing seasons under CA are better than under both minimum and conventional tillage (ConvT). Thus crops under CA systems are more frequently able to grow to maturity than those under ConvT. In addition, the period in which available nutrients can be taken up by plants is extended, increasing the nutrient use efficiency.

Research on CA in WA is fairly recent as the earlier work was focused on testing tillage options, and not on a holistic approach of CA. The earliest work was conducted in Turkey, which compared primary tillage with the mouldboard plough or disc harrow (Avci, 2011). In the last 3–4 decades, research work has been carried out in some countries in WA including Turkey and Iran, and by ICARDA in northern Syria as well as in Jordan, Iraq and Lebanon.

10.2.1 Conservation Agriculture and soil quality

Many studies worldwide indicated that CA reduces soil erosion. Research has shown that CA increases the stability of soil aggregates and thus resistance to erosion (Jat *et al.*, 2012). The most effective method for erosion control is the elimination of cultivation and the maintenance of surface residues. Intensive tillage decreases the amount of crop residues returned to the soil and accelerates the decomposition of organic matter. No-tillage (NT; direct seeding) on the other hand improves the quantity of soil organic matter (SOM), which leads to greater aggregation and improved soil fertility in the long term. Crop yields under CA are usually greater than under ConvT, especially on well-drained soils.

Research carried out by GIZ and the American University of Beirut (AUB), and by selected farmers in Beqaa in Lebanon showed that accumulation of organic material in CA fields was up to 5%, compared to 3% in ConvA (Jouni *et al.*, 2012).

10.2.2 Water conservation

Improved WUE is an important objective in the dry areas of WA since rainfall is generally low and variable. Conserving soil moisture and reducing water loss are critical to improving productivity of crops in these areas. Moisture availability is usually higher under CA because improved soil structure increases water infiltration and storage and reduces water loss by evaporation (Jat *et al.*, 2012; Piggin and Devlin, 2012).

Most researchers attribute higher yields under CA to better soil moisture conditions for an extended period of time. Improved soil moisture under CA is due mainly to increased infiltration and reduced evaporation as a result of residue cover on the soil surface. Results of the research conducted in WA countries confirm these findings. For example, research conducted by ICARDA in northern Syria, in areas with 300 mm of rainfall per cropping season, showed that ploughing can enhance evaporation loss of one-third of the rainwater received. The most productive farmers can harvest perhaps 2–3 t ha^{-1} of wheat from rainfed land when CA is practised, compared to 1 t ha^{-1} under conventional practices (Piggin and Devlin, 2012).

In Lebanon, CA was practised as a full package where winter crops were planted, harvested and field maize was then sown in summer directly in the soil, without any land preparation – this helped to conserve soil moisture by up to 4% of water content (Bashour and Bachour, 2008). In addition, seeds were planted 20 days earlier. The summer planting of CA maize with drip irrigation saved up to 60% of the fuel consumption for water pumping (Bashour and Jouni, 2009). Crop residues played an important role in preventing water evaporation and keeping the soil moist throughout the summer season. Research at the Agricultural Research and Education Center (AREC) of the AUB showed maize yield of 7 t ha^{-1} under CA and drip irrigation compared to 5 t ha^{-1} under ConvA (Haddad *et al.*, 2010).

10.2.3 Water use efficiency

Water use efficiency is the ratio of crop grain yield to effective rainfall, and is an indicator of the effectiveness of conversion of moisture into grain in the cropping system (Mrabet, 2011).

In Iran, under CA direct seeding (ZT) through a soil cover of crop residues, results showed improved WUE of wheat in wheat–chickpea (Hemmat and Eskandari, 2004a), wheat–fallow (Hemmat and Eskandari, 2004b) and continuous wheat (Hemmat and Eskandari, 2006) rotations compared to ConvT (mouldboard ploughing followed by discing). The WUE was higher under direct seeding than under ConvT in the three rotations tested; the WUE values were 4.9, 5.4 and 4.4 kg grain ha^{-1} mm^{-1} under direct seeding compared to 3.8, 4.4 and 3.0 kg grain ha^{-1} mm^{-1} under ConvT for the three rotations, respectively.

More comprehensive research was conducted at ICARDA in north Syria comparing different tillage systems in three-course rotations commonly used by farmers in this area, which is typical of the mild lowlands of WA. Pala *et al.* (2000) compared tillage systems: conventional deep disc ploughing and chisel cultivation (20 cm deep), CT by ducks-foot cultivator (20 cm blade opening at 10–12 cm deep) and direct seeding. Tillage effects on the crop performance over 12 years (with 234–504 mm annual rainfall during 1986–1997) were similar in both rotations: durum wheat–lentil–melon and bread wheat–chickpea–melon. WUE of each crop within both rotations followed the crop yield trends (Table 10.1), with no significant differences in the mean and range of WUE levels of the different tillage systems. Conservation tillage for all crops and direct seeding for legumes in particular performed well. Direct seeding of wheat had a slightly lower WUE because of the wider row spacing compared to conventional drilling. However, the use of a new direct seeder with 15–20 cm

Table 10.1. Water use efficiency (kg ha^{-1} mm^{-1}) under four tillage systems of a three-course rotation with durum and bread wheat at Tel Hadya, Syria (1986–1997) (Modified from Pala *et al.*, 2000).

	Tillage systems			
Three-course crop rotations	Zero-till	Disc plough	Chisel	Duck-foot
Durum wheat–lentil–melon rotation				
Durum wheat	6.4 (3.6–9.2)[a]	7.3 (4.2–9.9)	7.2 (4.1–10.6)	7.1 (3.8–9.4)
Lentil	3.8 (1.9–5.2)	3.2 (1.6–4.7)	3.5 (0.9–4.8)	3.5 (1.4–4.7)
Melon	3.7 (0.0–12.5)	8.6 (3.5–19.5)	9.2 (2.6–20.2	9.9 (2.4–19.5)
Bread wheat–chickpea–melon rotation				
Bread wheat	7.7 (4.3–10.2)	8.9 (4.5–12.8)	8.9 (5.5–12.9)	9.1 (5.2–12.9)
Chickpea	3.0 (1.2–4.5)	2.8 (1.1–4.3)	3.1 (0.9–4.4)	3.3 (1.2–4.5)
Melon	3.7 (0.0–13.9)	9.9 (2.0–22.3)	10.9 (2.4–20.1)	11.3 (2.8–25.0)

[a]Values in parentheses show the range of WUEs across the years.

row spacing showed WUE and crop yields comparable to that under ConvT.

In a 2-year FAO-supported CA project (2010–2011 in Jordan), WUE under direct seeding was higher for wheat and vetch (*Vicia sativa*) in all the five locations examined (Shakhatreh, 2011).

10.2.4 Soil physical, chemical and biological properties

In a 2-year experiment in Iran, Shirani *et al.* (2002) evaluated the effects of ConvT (mouldboard ploughing) and reduced tillage (disc harrowing) at three farmyard manure rates on the soil physical properties and maize (*Zea mays* L.) yield. Tillage systems had no significant effect on soil physical properties and maize biomass yield. Conventional tillage increased root penetration compared to under disc harrowing.

Mohammadi (2011) studied the effects of tillage and fertilization on wheat production in Iran and found that direct seeding increased microbial biomass carbon compared to ConvT. Conventional tillage decreased soil organic matter (SOM) and degraded soil structure, perhaps due to reduced soil microbial activity. The reduced tillage system produced the highest grain yield.

Soil organic matter was increased by CT particularly by direct seeding from 0.8 to 0.95% in a three-course wheat–lentil–summer crop rotation, and from 1.3 to 2.8% in a two-course wheat–lentil cropping system (Ryan *et al.*, 2003). Long-term tillage trials undertaken by ICARDA showed that after 8 years of direct seeding, soil carbon levels were 2.8 t ha^{-1} higher compared to that under ConvT.

10.2.5 Crop yields

There has been much research comparing grain yields and economic returns under various conservation and ConvT systems. Average yields for direct seeding are comparable or higher than under ConvT. In cases where direct seeding recorded fewer yields than ConvT, the contributing factors were poor weed control, low fertilizer N efficiency from broadcast application, and poor stand establishment due to inadequate seeder performance under the presence of a thick layer of residues on the soil surface. However, such problems can and have been overcome with proper operation of direct-seeding seed drills, and giving proper attention to residue management and weed control.

Comparing different tillage systems under the north-western temperate continental climate of Iran, Hemmat and Eskandari (2004a) found that wheat grain yields were 35% higher under reduced compared to ConvT. Grain yields under direct drilling in standing stubble were similar to those in reduced tillage (chisel plough) and superior to yields under ConvT (mouldboard plough): 1717, 1796 and

1301 kg ha^{-1}, respectively. The study concluded that winter wheat grain yields were significantly influenced by tillage practice. Reduced tillage or direct seeding in a winter wheat–fallow rotation gave higher yields.

In another 3-year experiment with a winter wheat–chickpea rotation in Iran, the highest yield of wheat was obtained with minimum tillage, whereas the yield of chickpea was greatest under direct seeding without or with residues. Reduced tillage gave 14 and 27% higher wheat and chickpea yields, respectively, compared to under ConvT. Average wheat grain yield with NT and minimum tillage was significantly greater (27–57%) than under reduced or ConvT (Hemmat and Eskandari, 2004b).

Similar results to those reported in Iran were obtained in a 10-year experiment conducted in semi-arid central Jordan (Khattari *et al.*, 2011). In the three-course rotation (wheat–lentil–watermelon), the highest grain and straw yields of wheat were obtained with chisel ploughing, followed by sweep plough early in the season before the rains. Both ploughing operations complement each other – the chisel loosens and pulverizes the soil surface and cuts off cracks, and the sweep removes weeds. Such conditions provide better rain infiltration and storage and thus the maintenance of soil moisture for longer duration during plant growth. It was concluded that sweep-plough tillage after mouldboard or chisel ploughing was adequate for seedbed preparation and controlling weeds in wheat fields. It improved infiltration and storage of soil moisture, while sweep tillage alone was adequate for land preparation for lentil and vetch. Addition of N fertilizer (20 kg ha^{-1}) was important for wheat, but not necessary for lentil and vetch, where residual N after wheat seemed adequate for their initial N need. The effect of wheat residue on the yield of different crops was not pronounced, perhaps due to the low quantity of N added.

In an FAO-supported project focused on CA in Jordan during 2010 and 2011, direct seeding gave consistently higher grain yields than conventional practice (Shakhatreh, 2011). In 2010, in trials at five locations in the north where annual rainfall was around 400 mm, the average grain yield was 1650 kg ha^{-1} under direct seeding compared to 1300 kg ha^{-1} under conventional practice traditionally used by farmers. Over five locations in the central region of Jordan where annual rainfall was around 350 mm, ZT gave grain yield of 1361 kg ha^{-1} compared to 1127 kg ha^{-1} for farmers' practice. In south Jordan, where rainfall was relatively low (<300 mm), results were not consistent amongst the five locations; and the average grain yield over the five locations indicated that the yield under CA exceeded those under ConvA by 20% (982 and 786kg ha^{-1}, respectively). There were similar trends in yield during the 2010/11 growing season.

In trials at ICARDA in northern Syria, Pala *et al.* (2000) reported rather small tillage effects on crop performance in durum wheat–lentil–melon and bread wheat–chickpea–melon rotations over 12 years (rainfall 234–504 mm during 1986–1997). Mean (and most annual) yields of durum and bread wheat (data not shown) differed little between the two deep tillage treatments (disc plough and chisel plough at 20 cm depth) and one shallow tillage (ducks-foot at 10–12 cm) systems. Yields under direct seeding tended to be less than under other tillage systems, and the yields declined with time, probably due to later sowing (about 2 weeks delay) of the crops under wider row spacing (30 rather than 18 cm spacing) and weed build up (Table 10.2).

Yields of lentils in rotation with cereals had significant differences across all four tillage systems, with deep discing showing the poorest results. Yields in minimum (ducks-foot) tillage or direct seeding were highest, reportedly because of the 30 cm row spacing used for direct seeding, which suited the legumes.

A conclusion from this early agronomy work, done at ICARDA before 2005, was that the conventional deep tillage (mouldboard) should be replaced by shallow tillage for cereals following legumes as little residues are left on the soil surface. However, deep tillage should be continued to be used for legumes following cereals to incorporate cereal residues and allow better sowing of the legumes.

In another trial at ICARDA during the 2005/06 growing season (Pala *et al.*, 2008), undertaken under a project of the Australian

Table 10.2. Effect of tillage systems on crop yields (t/ha) in durum wheat–lentil–watermelon rotation in Tel Hadya, Syria (1986–1997; Pala *et al.*, 2000).

	Tillage systems					Standard
Crops	Zero-till	Disc plough	Chisel	Ducks-foot	Mean	error of mean
Durum wheat	2.08 (0.98–3.88)[a]	2.41 (1.24–4.25)	2.40 (1.25–4.33)	2.37 (1.31–4.09)	2.31	0.041
Lentil	0.09 (0.34–1.53)	0.08 (0.15–1.33)	0.85 (0.21–1.47)	0.85 (0.29–1.24)	0.85	0.018
Watermelon	1.04 (0.00–4.73)	2.39 (0.91–7.13)	2.59 (0.75–7.23)	2.70 (0.64–6.88)	2.18	0.138

[a]Values in parentheses show the range of crop yields across the years.

Centre for International Agricultural Research (ACIAR) and funded by the Australian Agency for International Development (AusAID), yields in large unreplicated plots (5 ha) under direct seeding were 78% higher for chickpea and 44% higher for wheat than under ConvT. This was most likely a consequence of better infiltration of rainwater, and conservation and utilization of moisture in the direct seeding treatment (Table 10.3).

The earlier trials at ICARDA (2005/06) compared direct seeding and ConvT, with no pre-sowing cultivation under direct seeding and pre-sowing cultivations (mouldboard ploughing followed by two tine cultivations) with conventional cultivation. All other management (seed drill, sowing time, seeding rate, depth, row spacing and crop management) was identical for both treatments. Better yields from the direct seeding treatment provided encouragement for the direct seeding research to continue and further evaluate and adapt the technology, encourage the local availability of direct-seeding machinery and promote uptake of the technology by farmers. Benefits of direct seeding obtained in other countries, such as on time, fuel and machinery savings, better soil structure, more available soil moisture, higher organic matter for better soil quality and higher yield potential in dry years with timely planting, needed further local research. However, the experiences gained indicate that direct seeding was indeed suitable and highly relevant to the WA region.

Tillage and management systems that retain crop residues on the soil surface, such as direct seeding and minimum tillage, might play a significant role in water conservation, reducing soil loss, increasing yield and sustaining long-term production. Avci (2005) reported results on the comparison of NT and ConvT treatments in different cropping systems in Ankara, Turkey (Table 10.4).

Table 10.3. Wheat and chickpea grain yield (t ha^{-1}) under zero and conventional tillage in large plots (5 ha) at the ICARDA Station, Tel Hadya, Aleppo, Syria (2005/06; 290 mm seasonal rainfall).

Tillage practices	Wheat	Chickpea
Conventional tillage	1.12	0.76
Zero-till direct sowing	1.78	1.35
Percentage yield increase	44	78

Results indicated that wheat grain yields were equal or better under direct seeding compared to under ConvT in the first two cropping seasons; however, in the 2002/03 season, the wheat yields under ConvT were greater than under direct seeding in all cropping systems. It was suggested that favourable rainfall at planting in the 2002/03 season eliminated the advantage of water conservation by direct seeding compared to that obtained in other seasons, although this could not explain as to why the yields were much lower under direct seeding than under ConvT.

At the AUB, Yau (2009) evaluated the performance of safflower under direct seeding with various N fertilization and crop rotation treatments in a 2-year rainfed field experiment in the semi-arid northern Bekaa valley in Lebanon. Mean seed yield under direct seeding was similar to that under ConvT. N fertilization did not increase seed yield. The 1-year rotation study did not show a significant effect of previous crops on seed yield, although there were more weeds, less crop growth and shorter plant height during

Table 10.4. Effects of no-tillage and conventional tillage treatments on wheat yields (t ha^{-1}) under different crop rotations, 2001–2003 cropping seasons, Ankara, Turkey (Avci, 2005).

	2000/01 (rainfall: 216 mm)		2001/02 (rainfall: 403 mm)		2002/03 (rainfall: 375 mm)	
Cropping systems	No-tillage	Conventional tillage	No-tillage	Conventional tillage	No-tillage	Conventional tillage
Wheat–fallow	1.9	1.9	2.9	2.6	2.7	3.6
Wheat–chickpea	1.5	1.4	2.7	2.3	2.2	3.2
Wheat–wheat	2.1	1.7	1.9	2.0	2.0	3.0

the early growth stage following safflower than following barley or chickpea. The study showed that safflower was suitable to grow under direct seeding and that there was no need to apply N to safflower when grown following fertilized cereal crops.

In simulating the effect of direct seeding and crop residue retention on water relations and yield of wheat under rainfed semi-arid Mediterranean conditions, Sommer *et al.* (2012) found that the beneficial effect of early planting on grain and straw yields of wheat could be clearly identified by crop model simulation and verified by field observations. A positive effect of direct seeding on crop yield was visible in the simulation in 25 of 30 years; however, given the high year-to-year variability, this was not statistically significant. With similar yields, direct seeding was still attractive as it reduced the cost of agronomic management compared to ConvT. This provides an incentive for the adoption by farmers, as long as there is no yield penalty or a disproportionate increase in costs related to weed control by herbicides. The two tested levels of residue management showed no impact on yield, possibly because both treatments only differed in the amount of standing residue (stubble), while loose residue left on the soil surface after harvest was the same. About 55% of the seasonal precipitation was lost by unproductive soil evaporation, while direct seeding and residue retention had only a minor mitigating effect.

The positive effects of CA often lead to higher yields in the long run, but can increase yields in the first season of implementation, depending on the crop, especially when starting the rotation with vetch, an annual leguminous crop (Jouni *et al.*, 2012).

10.2.6 Economic returns

In an FAO-supported project in Jordan (Shakhatreh, 2011), the costs and returns under ZT and ConvT were evaluated using partial budget analysis. The production costs were lower for ZT than ConvT due to savings in fuel and labour as a result of eliminating ploughing. There were also savings from reduced seed rate. Overall, the income of farmers increased on average by US$19 ha^{-1}.

Pala *et al.* (2000) reported that deep discing or mouldboard ploughing required two and three times more fuel per hectare than chisel ploughing and ducks-foot cultivation, respectively. Results support a preference for minimum tillage over deep tillage on the grounds of both energy-use efficiency and increased net revenue. Implements for minimum tillage, unlike those for direct seeding, are readily available to farmers in the WA region.

Based on the results of research trials and experience with promoting farmer adoption of direct seeding over 6 years (2005/06 to 2010/11) in Iraq and Syria under the ACIAR-funded project, Piggin and Devlin (2012) concluded that benefits of not ploughing included savings in expenditure on fuel, labour and seed, and increased returns from higher yields as a result of improved water use.

Cost savings and increased earnings were estimated from trials, demonstrations and farmer adoption experiences, and used to calculate profitability from adopting a conservation cropping package of direct seeding, early sowing and reduced seeding rates. In Syria, farmer profitability improved by US$220 ha^{-1}, and in Iraq reduced ploughing and CA improved profitability by US$355 ha^{-1}.

Apart from economic benefits, CA produces environmental benefits such as better soil health, reduced greenhouse gas emissions (GHG) and better water use, and cleaner air and surface waters; and indirect benefits such as social equity and rural development (Piggin and Devlin, 2012).

Conservation Agriculture is not only applied in field crops. It is widely distributed in orchard systems, especially olives – where leguminous crops, especially local vetch, can be used as a cover crop, in addition to eliminating ploughing activities. A saving of US$2000 ha^{-1} was achieved over a 3-year study of olives, while also reducing synthetic fertilizer use due to biological fixation from vetch (Jouni and Adada, 2010).

10.3 Problems Encountered with Conservation Agriculture in West Asia

10.3.1 Weed infestation

In conventional cultivation cropping systems, tillage is used to control weeds present prior to seeding. With NT, pre-sowing weeds need to be controlled using alternative means. The key to the development of NT practice was the development in the 1960s–1970s of non-selective, non-residual herbicides such as paraquat, diquat and especially glyphosate used as pre-sowing herbicides. The research at ICARDA has shown that managing weeds is not difficult as observed in CA trials in Syria and Iraq, especially as few weeds emerge in the dry Mediterranean summer, and a pre-sowing glyphosate application is only occasionally needed.

10.3.2 Plant diseases

Leaving stubble and other residues from previous crops on the soil surface raises the risk of increased incidence of pests and diseases. Therefore, crop rotations are necessary in CA; however, there has been little research in WA on disease effects in CA. In one recent study at ICARDA in Syria, Ahmed *et al.* (2012) evaluated differences between ConvT and ZT systems on: (i) the incidence and impact of nematode and fungal diseases in chickpea and lentil in a long-term wheat–chickpea–barley–lentil rotation; and (ii) the reaction of different lentil genotypes to fusarium wilt and downy mildew.

In two seasons (2008/09 and 2010/11), no significant differences were observed between tillage practices, crops and planting dates and their interactions for mean presence of cyst nematode disease. The mean of cyst nematode disease incidence ranged from 7.3% on early planted lentil under ConvT to 14.5% in late planted chickpea under ZT. Tillage practices significantly (P<0.05) affected *Ascochyta* blight incidence in chickpea but did not affect its severity. The incidence ranged from 4 to 22.5% under early planted chickpea in both tillage practices. Moreover, the mean severity range was rated 3.2 and 5.5 in the early planted ConvT and direct seeding, respectively. In the comparative evaluation of lentil genotypes in 2008/09, 2009/10 and 2010/11, the combined analysis showed significant growth/yield differences (P<0.05) amongst genotypes, but there were no tillage × disease interactions with fusarium wilt or downy mildew reactions. All the genotypes showed less than 10% fusarium wilt mortality indicating high levels of resistance. In this study, soil-borne and foliar diseases were of low importance under conventional or CT treatments in effective crop rotation. Moreover, cool-season legume genotypes with good disease resistance and high yields under ConvT also showed good performance and adaptation under ZT, with no indication that special varieties were needed for use under ZT.

10.4 Constraints and Prospects

A traditional and common perception amongst farmers in WA is that cropping requires soil cultivation. Many smallholder farmers resist change that is poorly understood and contradicts what they believe.

Friedrich and Kienzle (2008) considered that accessibility of the affordable and good quality equipment suited for the local needs of farmers, producers and entrepreneurs with intention in favour of the CA practices, is an important element for the CA adoption. This is regardless of whether farmers operate with animal traction or use tractors and seeders. In many cases where CA is newly introduced, the establishment of a commercial supply chain for such equipment does not happen spontaneously and requires special attention – including technical assistance from national and, if possible, international development partners.

The two factors necessary for the successful adoption of CA under dryland farming are appropriate technologies and favourable policy environment. Locally made low-cost seeders are needed and also require local markets for providing repair and technical services to the farmers (Piggin and Devlin, 2012).

Belloum (2008), in a review on CA in the Arab region, summarizes the constraints to the adoption of CA in the region as follows:

> CA is a new system and requires time for its integration and adoption. CA is a system that will need a relatively long time to show benefits to farmers, and the environmental impacts of CA are not well appreciated by farmers, because securing food in the short term has a higher priority than soil conservation or regeneration that can be only achieved in the medium and long term.

Other constraints include: (i) the lack of knowledge among the farmers about CA and how to develop and promote it; (ii) the lack of knowledge about herbicides and their use to replace tillage for weed control; and (iii) how an enabling policy environment can advance CA beyond the status of a promising idea and make it the paradigm for a new agriculture that is profitable, socially equitable, ecologically sustainable and resilient in the face of climate change.

Experience has found, however, that CA sells itself when supported by a logical approach to develop technology, overcome constraints and promote awareness and adoption in a participatory development and extension programme with relevant local institutions, private industry and farmers (Piggin and Devlin, 2012).

10.5 Government Support and Policy Towards Scaling-up Conservation Agriculture

The transition from conventional cultivation to CA demands a combination of technological and institutional innovations. Rajeswari *et al.* (2005), in their analysis of the policy and institutional requirements of such a transition, highlighted the need for policy analyses to understand how conservation technologies integrate with other technologies, and policy instruments and institutional arrangements that promote/repress CA. They emphasized that the role of new institutional arrangements is more evident in CA systems compared to ConvA. Transition to CA is possible only *if* and *when* the agricultural knowledge community, including all its stakeholders in agriculture and allied sectors acknowledge, adapt, enable and adopt these institutions, processes or ways of working.

From recent experience in the ACIAR–AusAID project in WA, the adoption of CA is encouraged where technology is verified and sensibly adapted to suit local conditions, stakeholder awareness is enhanced, local constraints are overcome and users are encouraged to test and then adopt the technology. Although no specific policy changes or actions have been undertaken in the involved WA countries (Iraq and Syria) to promote direct seeding, there has been considerable adoption of CA and local production of direct seeders. With supportive policies, including the spread of information, the easy availability of loans or subsidies for the purchase of direct seeders and financial support to manufacturers, the adoption of CA would be faster.

10.6 Successful Adoption: a Story from Iraq and Syria

ICARDA has undertaken efforts to develop and promote CA in northern Iraq and Syria in a long-term (2005–2015) project to improve rainfed crop production, its profitability and sustainability. This is supported by ACIAR and AusAID, with implementation by the

Iraq Ministry of Agriculture and University of Mosul in collaboration with the Universities of Adelaide and Western Australia and the Department of Agriculture and Food in Western Australia (Piggin, 2009; Jalili *et al.*, 2011; Piggin *et al.*, 2011).

The project undertook many research trials, several long-term, for evaluating various aspects of agronomy and crop management under conventional cultivation and direct seeding, to verify and adapt the technology under local conditions. Some major findings for the main regional crops (bread wheat, durum wheat, barley, lentil and chickpea) were that grain and biomass yields under direct seeding were similar to or better than under conventional cultivation, and yields were better with early (October–November) planting with low (50–100 kg ha^{-1}) seed rates. Varietal evaluation showed some significant grain yield differences due to tillage, crop and variety effects, but there were no significant interactions between tillage, crop and variety – with similar patterns of variety performance within crops under ZT and ConvT. If a variety performs well under conventional management in a particular agricultural environment, it will perform well under the CA. Importantly, this suggested that the current varieties/lines grown by farmers and developed in breeding programmes can be used, and that new varieties are not needed before direct seeding can be adopted (Piggin *et al.*, 2011).

The lack of effective, affordable direct-seeding seeders was identified as a major constraint to direct-seeding adoption. Direct-seeding technology and requirements were discussed with local village seeder-manufacturers and farmers in Iraq and Syria and local fabrication commenced during 2007–2009. Small 2.3-m wide three-point-linkage (3PL) and wider 4-m trailed and 3PL direct-seeding seeders were fabricated with tines having narrow points, wide spacing, spring release and seed/fertilizer delivery. In Iraq, farmers also developed direct-seeding modification kits for local 3.6-m John Shearer-type seeders. The performance of the imported and local seeders was compared in 2008/09 at ICARDA; all seeders worked well and yields of wheat, barley, lentil and chickpea were similar for Indian, German and three local Syrian direct-seeding seeders. Prices were about US$1250 for direct seeders' modification in Iraq and US$2500 for 2.3-m and US$6000 for 4-m seeders in Syria, although the prices have increased since then due to unrest and difficulties in accessing steel in the region. During 2008–2012, Syrian machinery suppliers continued to manufacture and improve the affordable local direct-seeding seeders; a total of about 90 have been made, with 15 for ICARDA, 35 for farmers, 25 exported to Iraq and 14 exported to Morocco, Tunisia and Algeria. Iraqi counterparts also further developed their kits to modify existing seeders for direct seeding with locally manufactured knife-points and press wheels, together with locally available or imported tines. Some 21 local seeders have now been converted to direct seeders in northern Iraq and, with imported seeders, the total number of direct-seeding seeders in this region now exceeds 50. This local manufacture and availability of direct-seeding seeders has been an important factor in driving adoption, as imported seeders are too expensive and large for small farmers in the region (Jalili *et al.*, 2011; Piggin *et al.*, 2011).

Project experience shows that affordable and effective seeders for use in CA can be built locally, thus overcoming a significant and commonly reported constraint to the adoption of CT. Aside from lower costs, an advantage of locally produced machinery is that it is easy to repair using locally available spare parts and mechanical skills (Piggin and Devlin, 2012).

The project worked with farmers in their fields, and followed a flexible approach in introducing and implementing CA, by demonstrating options and letting farmers decide. The introduction of CA requires radical change in the perceptions by farmers, policy makers and planners, but it does not necessarily mean immediate and fundamental changes in the farmer practices. One issue addressed in a flexible manner was residue retention, as crop residues are an important feed source for small ruminants in this region. The project recommendation was that farmers should not burn stubble and retain as much as possible in the field, recognizing that plant roots and crowns and manure from grazing livestock contribute to the SOM pool.

Local farmers in Iraq and Syria were supported while trying adoption of direct seeding in a participatory approach with project partners, research-extension institutions, private industry and non-government organizations. Direct seeding technology was explained and local seeders made available to the interested farmers for testing without charge; farmers were supplied the tractors and inputs. Comparisons were made on-farm or with neighbours. Farmers generally found that yields were better with direct seeding than under conventional cultivation, even where there was little stubble. In a survey of the 43 Syrian farmers using ZT in 2008/09, 100% responded that they:

- Saved ploughing costs (US\$30–40 ha^{-1}), time, seed and soil moisture and obtained good early germination;
- Obtained a higher yield from direct seeding than from their own or neighbours' conventionally cultivated fields;
- Were keen to continue direct seeding, given direct drills were accessible;
- Were interested to buy/modify a drill.

In Syria in 2006/07, three farmers began testing of direct seeding on a combined area of 15 ha. The following year, the number of farmers and the area doubled – with six farmers working on 30 ha. In 2008/09, there were 43 Syrian farmers using 2075 ha of ZT. In 2009/10, use of ZT had increased to 119 Syrian farmers and 4918 ha. This roughly tripled in 2010/11, to about 380 farmers on 15,000 ha. Surveys conducted in Syria during 2011/12 in collaboration with the Syrian Directorate of Extension showed that 537 farmers sowed 20,574 ha of ZT in 652 fields (Fig. 10.1). However, due to security difficulties, access to all regions was not possible, and total adoption is thought to exceed 30,000 ha, which is double the area of the previous year. Some 70% of this area was spontaneous adoption by farmers using their own, rented or borrowed seeders.

In neighbouring Iraq, the spread of direct seeding has been only slightly less remarkable, with the number of farmer practitioners expanding six-fold, from 12 in 2006/07 to about 70 in 2011/12, and the area expanding by 150 times from 52 ha in 2006/07 to about 7800 ha in 2011/12 (Fig. 10.1). Some 80% of this area was spontaneous adoption.

In Lebanon, while working both with researchers and farmers, the aim was to move forward from direct seeding into CA as a package, i.e. preserve crop residues as a main CA factor. This was achieved at Akkar and

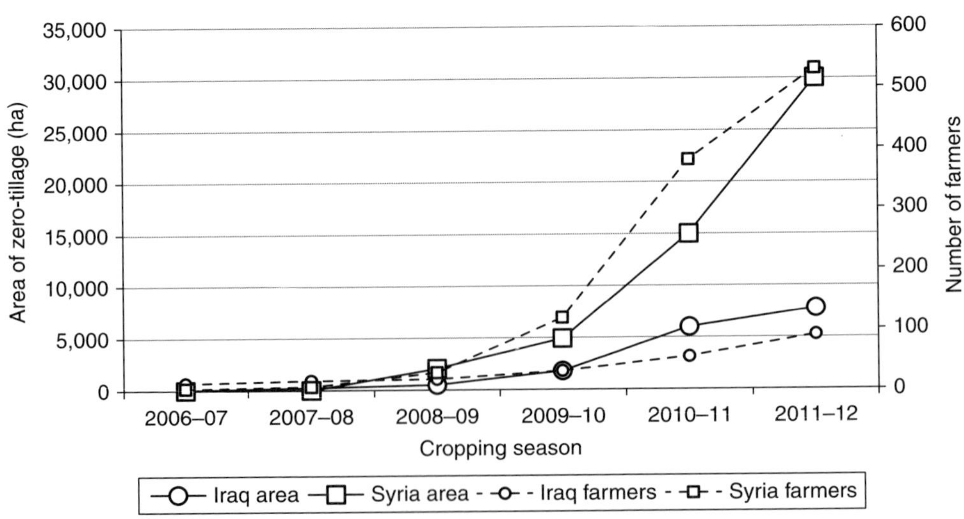

Fig. 10.1. Area and numbers of farmers that adopted zero-till in northern Iraq and Syria between 2006 and 2012 under the ACIAR Iraq project.

AREC due to successive crop rotations with direct seeding and controlling grazing activities (Jouni and Hansmann, 2010).

Adoption of CA was reported in Lebanon (Jouni *et al.*, 2012), where the number of farmers who adopted the system increased from four in the 2007/08 growing season to 150 farmers within four growing seasons (Table 10.5). The area increased from 4 to 1500 ha for the same period. Furthermore, the saving due to adoption of the system was US$450,000 for the indicated area.

ACSAD as a specialized Arab organization within the framework of the League of Arab States, with the support of GTZ, has started a regional project to spread CA practices in Arab countries. This project has already gained popularity among Syrian farmers, which can be attributed to good awareness of the developing programme and the selection of realistic objectives. The number of farmers that joined the project in Syria increased from two in 2006 to 27 farmers in five Syrian provinces in 2007 (Belloum, 2008).

An effective way to promote CA is to demonstrate it through experiences of successful adopters. Policy makers, extension personnel and farmers can benefit from the lessons learned from neighbouring countries with similar agricultural environment and economy. Two such countries that have learned from each other about CA in a structured way are Iraq and Syria. Here the project lent seeders to the collaborating farmers and other partners to raise awareness and experience of ZT. In order to upscale the technology, countries need to formulate plans and policies that encourage the local production, distribution and maintenance of seeders (Piggin and Devlin, 2012).

Table 10.5. Numbers of farmers, area planted and the saving made by applying Conservation Agriculture during four growing seasons in Lebanon.

Growing season	Area (ha)	Number of farmers adopted CA	Saving (US$)
2007/08	4	4	–
2008/09	560	60	168,000
2009/10	1100	100	330,000
2010/11	1500	150	450,000

10.7 Concluding Remarks

The introduction, wide adoption and sustainability of CA in areas where it is not yet practised is a serious challenge despite the positive results and good uptake by farmers and technical staff in selected parts of the world. Experience and lessons learned from other countries whose farmers have successfully adopted the system should be considered when shifting to a new system. Some important issues to be considered are as follows.

- Imperfect adoption. Many adopters do not fully apply the three essential practices of CA: not tilling the soil, retaining soil cover and diverse rotations. At first, results may be encouraging, even in imperfect systems, but incomplete adoption is not sustainable and creates problems such as soil compaction or increased weed infestation in the medium and long term.
- Adoption of CA by small farmers was not spontaneous, but was a result of intense collaborative efforts by researchers, extension personnel, the private sector and participating farmers.
- CA is not just a technique. The success of the system is attributable to direct drilling as a component of an integrated development approach including education and training, marketing, social mobilization and diversification.
- Direct seeding, which is the major element in CA, is not merely a technical package. Direct-seeding success stories are created by farmer-driven adaptation of the system to their specific circumstances and requirements in their respective farm environments.
- Good farm management skills (technical, organizational and financial) are crucial for successful ZT farming.
- Private–public collaboration can enhance the adoption and increase resources available to support large and small farmers.

In its work with partners in the Middle East and Central Asia, ICARDA found that introducing CA begins best with a research and development programme over 3–5 years.

The main tasks of such a programme are to: (i) verify and adapt the system to local conditions through agronomic research; (ii) develop local capacity to fabricate and maintain effective and affordable ZT seeders; and (iii) facilitate extensive participatory collaboration among researchers, extension personnel, NGOs, private industry and farmers to raise awareness of the system, share lessons learned and guide adaption of CA to local conditions.

A strong lesson learned is the need to maintain flexibility. Like other revolutionary concepts, CA is also prone to an overlay of dogma that ill-serves spread of the technology into developing countries. Researchers and extension officers should keep an open mind to avoid steering the local evolution of CA away from its optimal form. Individual countries and agricultural regions do not need a list of prescriptions or a recipe, to which they must adhere without fail, on exactly how much to plough or not plough or on what percentage of stubble to leave in the field. Rather, they need to appreciate the concept and understand the options and their benefits and costs in the overall system, so they are better prepared to select options and adapt the system to their specific situation.

Steps in a logical sequence of activities to develop and promote direct drilling, developed and used in Iraq and Syria, which might provide some approaches for consideration in other areas, are given below.

Organize participatory research and demonstrations with local partners. Once farmers and extension personnel understand CA in principle, they need to see it in practice. This requires participatory research and demonstration ensuring the collaboration of scientists, extension officers, economists, policy makers and farmers. Farmers should have access to ZT seeders, without cost or payment, to enable them to test, evaluate and gain experience with the technology on their own farms. Experience in Iraq and Syria to date is that once farmers use the technology themselves, with its many benefits – such as similar or increased yields, lowered costs, time-efficiency and increased sustainability of resources – few revert to conventional cultivation. This leads many to seek their own machinery and this expands the market for local ZT seeder production (Piggin and Devlin, 2012).

References

Ahmed, S., Piggin C., Haddad, A., Kumar, S., Khalil, Y. and Bejiga, G. (2012) Nematode and fungal diseases of food legumes under conservation cropping systems in northern Syria. *Soil and Tillage Research* 121, 68–73.

Avci, M. (2005) Zero and minimum tillage as alternatives to conventional cultivation in dryland fallow/wheat and annual cropping systems in central Anatolia. In: Pala, M., Beukes, D.J., Dimes, J.P. and Myers, R.J.K. (eds) *Proceedings of a Workshop on Management of Improved Water Use Efficiency in the Dry Areas of Africa and West Asia.* ICARDA, Aleppo, Syria and ICRISAT, Patancheru, India, pp. 89–100.

Avci, M. (2011) Conservation tillage in Turkish dryland research. *Agronomy for Sustainable Development* 31, 299–307.

Bashour, I. and Bachour, R. (2008) Field experiments on conventional and conservation agriculture in Lebanon. Poster presentation - FAO Meeting on CA. Rome/Italy.

Bashour, I. and Jouni, K. (2009) Comparing conventional and conservation agriculture studies in Lebanon. Presentation – Faculty of Agricultural and Food Sciences, American University of Beirut, September 2009, Beirut, Lebanon.

Belloum, A. (2008) Conservation agriculture in the Arab region between concept and application. In: Stewart, B., Fares Asfary, A., Belloum, A., Steiner, K. and Friedrich T. (eds) *Proceedings of the International Workshop on Conservation Agriculture for Sustainable Land Management to Improve the Livelihood of People in Dry Areas.* ACSAD and GTZ, Damascus, Syria, pp. 11–24.

Derpsch, R., Roth, C.H., Sidiras, N. and Kopke, U. (1991) *Controle da erosa o no Paraná, Brasil: Sistemas de cobertura do solo, plantio direto e prepare conservacionista do solo.* GTZ, Eschborn, Germany.

Friedrich, T. and Kienzle, J. (2008) Conservation agriculture: impact on farmers' livelihoods, labour, mechanization and equipment. In: Stewart, B., Fares Asfary, A., Belloum, A., Steiner, K. and Friedrich, T. (eds)

Proceedings of the International Workshop on Conservation Agriculture for Sustainable Land Management to Improve the Livelihood of People in Dry Areas. ACSAD and GTZ, Damascus, Syria, pp. 25–36.

Haddad, N., Hansmann, B., Bashour, I. and Jouni, K. (2010) Conservation agriculture in Lebanon, field corn under drip irrigation. In: *Presentation Seminar on Sustainable Agriculture.* Chamber of Commerce and Industry and Agriculture, Zahleh, March 2010, Lebanon.

Hemmat, A. and Eskandari, I. (2004a) Conservation tillage practices for winter wheat–fallow farming in the temperate continental climate of north-western Iran. *Field Crops Research* 89, 123–133.

Hemmat, A. and Eskandari, I. (2004b) Tillage system effects upon productivity of a dryland winter wheat-chickpea rotation in the northwest region of Iran. *Soil and Tillage Research* 78, 69–81.

Hemmat, A. and Eskandari, I. (2006) Dryland winter wheat response to conservation tillage in a continuous cropping system in north-western Iran. *Soil and Tillage Research* 86, 99–109.

Jalili, S., Fathi, G., Al Rijabo, A. Piggin, C. and Desbiolles, J. (2011) Farmer innovation: seeder fabrication and uptake of zero tillage in Iraq. In: Proceedings of the 5th World Congress on Conservation Agriculture, Brisbane, Australia, 26–29 September 2011, pp. 64–65. Available at: http://aciar.gov.au/files/node/13993/farmer_innovation_seeder_fabrication_and_uptake__23210.pdf or http://aciar.gov.au/WCCApapers (accessed 20 December 2012).

Jat, R.A., Wani, S.P. and Sahrawat, K.L. (2012) Conservation agriculture in the semi-arid tropics: prospects and problems. In: Sparks, D.L. (ed.) *Advances in Agronomy* 117, 191–273.

Jouni, K. and Adada, F. (2010) Conservation agriculture in olive orchards in Lebanon. In: Poster Presentation, 4th Mediterranean Meeting on Conservation Agriculture. Setif, Algeria. Options Méditerranéennes, A no. 96, IV RencontresMéditerranéennes du Semis Direct (May 2010).

Jouni, K. and Hansmann, B. (2010) Conservation agriculture in Lebanon, field observations: *GIZ Annual Report.* Beirut, Lebanon.

Jouni, K., Hansmann, B. and Bashour, I. (2012) Conservation agriculture in Lebanon. In: *Proceedings of Workshop on Conservation Agriculture: A System to Promote Sustainable Production Intensification (SPI)*, Beirut, Lebanon, 38 pp.

Kassam, A. and Friedrich, T. (2010) Conservation agriculture: concepts, worldwide experience, and lessons for success of CA-based systems in the semi-arid Mediterranean environments. *Options Méditerranéennes A* 96, 11–51.

Khattari, S., Snober, B., Battikhi, A., Pala, M. and Katkhuda, N. (2011) Wheat, lentil, and vetch yield in three crop rotations under different tillage, crop residue management and nitrogen fertilization in Mushagar (semi-arid) in central Jordan. *Jordan Journal of Agricultural Science* 7, 644–655.

Mohammadi, K. (2011) Soil microbial activity and biomass as influenced by tillage and fertilization in wheat production. *American-Eurasian Journal of Agricultural and Environmental Sciences* 10, 330–337.

Mrabet, R. (2011) No-tillage agriculture in West Asia and North Africa. In: Tow, P.G., Cooper, I.M., Partridge, I. and Birch, C.J. (eds) *Rainfed Farming Systems.* Springer, Dordrecht, the Netherlands, pp. 1015–1042.

Pala, M., Harris, H.C., Ryan, J., Makboul, R. and Dozom, S. (2000) Tillage systems and stubble management in a Mediterranean-type environment in relation to crop yield and soil moisture. *Experimental Agriculture* 36, 223–242.

Pala, M., Haddad, A. and Piggin, C. (2008) Challenges and opportunities for conservation cropping: ICARDA experience in dry areas. In: Stewart, B., Fares Asfary, A., Belloum, A., Steiner, K. and Friedrich, T. (eds) *Proceedings of the International Workshop on Conservation Agriculture for Sustainable Land Management to Improve the Livelihood of People in Dry Areas*, GTZ-ACSAD-AAAID-FAO-UNEP-ROWA, Damascus, Syria, pp. 165–181.

Piggin, C. (2009) Improving agriculture with zero tillage cropping systems in Iraq. *Issues* 89, 11–13.

Piggin, C. and Devlin, M. (2012) Conservation agriculture: opportunities for intensifying farming and environmental conservation. *Research to Action:* 2. ICARDA, Aleppo, Syria.

Piggin, C., Haddad, A. and Khalil, Y. (2011) Development and promotion of zero tillage in Iraq and Syria. In: Proceedings of the 5th World Congress on Conservation Agriculture, Australian Centre for International Agricultural Research, Brisbane, Australia, *pp.* 304–305. Available at: http://aciar.gov.au/WCCApapers (accessed 10 January 2013).

Rajeswari, S., Rasheed, V., Andrew, J.H. and Sunita, S. (2005) Policy and institutional requirements for transition to conservation agriculture: an innovation systems perspective. In: Abrol, I.P., Gupta, R.K. and Malik, R.K. (eds) *Conservation Agriculture – Status and Prospects.* Centre for Advancement of Sustainable Agriculture, New Delhi, India, pp. 224–232.

Ryan, J., Masri, Z., Diekman, J. and Pala, M. (2003) Organic matter and nutrient distribution following conservation tillage, straw management and compost application under dryland Middle Eastern conditions. In: *American Society of Agronomy Annual Meeting*, 2–6 November 2003, Denver, Colorado.

Shakhatreh, Y. (2011) Sustainable agriculture production through promotion of conservation agriculture in Jordan. *TCO/Jordan/3201*. NCARE, Amman, Jordan, 39 pp.

Shirani, H., Hajabbasi, M.A., Afyuni, M. and Hemmat, A. (2002) Effects of farmyard manure and tillage systems on soil physical properties and corn yield in central Iran. *Soil and Tillage Research* 68, 101–108.

Sommer, R., Piggin, C., Haddad, A., Hajdibo, A., Hayek, P. and Khalil, Y. (2012) Simulating the effects of zero tillage and crop residue retention on water relations and yield of wheat under rainfed semiarid Mediterranean conditions. *Field Crops Research* 132, 18–32.

Yau, S.K. (2009) Compatibility of safflower in no-till systems. *Australian Grain* 19(1), May/June.

11 Conservation Agriculture in Eastern and Southern Africa

Patrick C. Wall,[1] Christian Thierfelder,[2] Amos Ngwira,[3] Bram Govaerts,[4] Isaiah Nyagumbo[2] and Frédéric Baudron[5]

[1]*Independent International Consultant, Bahías de Huatulco, Oaxaca, México;* [2]*International Maize and Wheat Improvement Center (CIMMYT), Harare, Zimbabwe;* [3]*Department of Agricultural Research Services, Chitedze Research Station, Lilongwe, Malawi;* [4]*International Maize and Wheat Improvement Center (CIMMYT), México DF, México;* [5]*International Maize and Wheat Improvement Center (CIMMYT), Addis Ababa, Ethiopia*

11.1 Introduction

It is generally understood and well documented that conventional farming practices with frequent ploughing gradually degrade the physical structure of tropical soils (Brady and Weil, 2007) leading to increased soil erosion and decreased chemical quality of tropical soils. These processes are the same, but possibly somewhat slower, in subtropical areas. There have been numerous efforts, many ongoing, to develop functional Conservation Agriculture (CA) systems in eastern and southern Africa (E&S Africa) to overcome the negative effects of tillage-induced degradation, led by multiple institutions including public, commercial, religious, research and development organizations. In general, all efforts have endeavoured to implement systems based on the three principles of CA: minimal soil disturbance, soil cover with living plants or crop residues, and crop rotation (FAO; http://www.fao.org/ag/ca). However, often only one or two of these principles have been applied and the techniques, focus and methodologies employed have been as diverse as the organizations supporting the efforts. Recently some papers have suggested that CA is only applicable to small pockets of farmers in E&S Africa, and that the systems are constrained by numerous challenges that are considered insurmountable (Giller *et al.*, 2009; Andersson and Giller, 2012). Those developing CA systems in the quest for sustainable agricultural production for farmers in Africa, smallholders and commercial farmers alike, have often acknowledged these same problems (e.g. Wall, 2007), and have dedicated considerable efforts to overcoming them, considering, therefore, that the challenges are surmountable. Given the confusion that these two different positions have provoked among many of those labouring to advance African agriculture, this chapter examines research results from E&S Africa in an effort to develop a clearer picture of the future of CA systems in the region.

11.1.1 Conservation Agriculture – what it is and what it is not

Conservation Agriculture is not a technology but rather a way of conducting agriculture. The term 'conservation agriculture' has been coined to describe the principal differences between the CA system and other (tilled) systems – not to describe all of the components of a functional CA system. In CA the degradative components are removed from conventionally tilled (ConvT) agricultural systems – tillage that damages soil structure and breaks down soil organic matter (SOM), insufficient return of organic matter to the soil and lack of protection of the soil surface, and monoculture are replaced with minimum soil disturbance, crop residue retention and crop rotation. All of the other components of productive agricultural systems, such as adequate nutrition, optimum seeding dates and plant populations, adequate weed control etc., are as much a part of productive CA systems as they are of tilled systems, and should be fine-tuned within the context of the new CA system based on sound agronomic decision-taking. Furthermore, CA is not a low-external-input system (Wall, 2009), as suggested by Gowing and Palmer (2008) – it is a highly productive system and functions poorly with poor management, just as, or even worse than, ConvT systems. Just removing the degradative components from unproductive tilled situations is unlikely to result in productive systems. The CA components are embedded in an overall system such as that developed by the World Agroforestry Center (ICRAF) in collaboration with several national programmes combining CA with agroforestry options, notably with *Faidherbia albida*, that reaps the benefits of both the CA for the intercrops and the trees (Garrity *et al.*, 2010). The Golden Valley Research Trust (GART) and the Conservation Farming Unit (CFU) of the Zambian National Farmers Union (ZNFU) have a well-developed demonstration of *F. albida* intercropped in a CA system on the GART farm north of Lusaka (pictured in Garrity, 2011).

The principles of CA appear to have wide applicability, functioning in different continents, latitudes, soil types and with many different crops and cropping systems. However, the way in which the principles are applied depends on farmer circumstances and neighbouring farmers may use very different techniques to practise sustainable CA systems (Wall, 2007). Many of the experiences with CA in E&S Africa have used 'best bet' CA packages based on systems developed elsewhere, and then have compared these with current ConvT systems, without a phase of participatory adaptation. The confusion over the applicability of CA is also compounded by the use of different terms (Mazvimavi and Twomlow, 2009), especially Conservation Farming (CF) and Conservation Tillage (CT). Conservation Farming is a term used in southern Africa to describe a particular form of CA with small basins (covering 8–15% of the field surface) dug in the same place each year, and inputs and seed concentrated in these basins, while CT, as strictly defined, refers to any system that maintains at least 30% soil cover with residues after seeding (Soil Science Glossary Terms Committee, 2008). As such, CT may include complete disturbance of the surface soil and therefore, depending on the level of soil disturbance, CT systems may not classify as CA systems. Unfortunately many authors use the terms CA and CT interchangeably, complicating the interpretation of results.

11.1.2 The need for a change in farming systems in eastern and southern Africa

Human beings seldom change unless there is an important reason to do so. What are the problems with current farming systems that suggest the need to embark on the difficult task of knowledge development and system change among millions of smallholder farmers in E&S Africa? Farmers in the region commonly complain of declining yields and rising costs of production although the causes of these are not always obvious or apparent to many farmers.

Excessive nutrient mining over most of Africa (Stoorvogel *et al.*, 1993) is acute, and adequate plant nutrition is often cited as the most limiting factor to crop production in E&S Africa, while at the same time fertilizer use is very low (about 20 kg ha^{-1} of nutrients in 2009/10 calculated from FAOSTAT http://faostat3.fao.org/home/index.html). Fertilizer use by smallholders is not just a function of availability and affordability, but also of both production and market risk (Morris *et al.*, 2007). Smallholder farmers, in particular, are averse to risk given their precarious financial situation and their poor access to credit – if fertilizer application to a crop is perceived as risky, it will not be applied (Rockström *et al.*, 2002). One of the major causes of risk in E&S Africa is the risk of moisture stress, which is often more a function of inefficient use of rainfall than of insufficient or poorly distributed rainfall *per se*. Across the semi-arid tropics of sub-Saharan Africa, between 70 and 85% of rainfall is lost to surface runoff, deep drainage and evaporation rather than being used by crops for productive transpiration (Rockström *et al.*, 2002) while in Zimbabwe 30% of rainfall may be lost to runoff alone (Elwell and Stocking, 1988). As a result of global warming and climate change, increased variability of seasonal distribution of rainfall is expected throughout the region coupled with a reduction in rainfall in much of the region (Lobell *et al.*, 2008), factors that will aggravate the inefficiencies in rainfall use noted above.

E&S Africa tops the list of regions affected by land degradation, a long-term decline in ecosystem function measured in terms of net primary productivity (Bai *et al.*, 2008) closely linked to rural household food insecurity and poverty (Malley *et al.*, 2006). Elwell's comments on the causes of land degradation in Zimbabwe undoubtedly apply to the rest of the region: 'High rates of erosion and land degradation are a result of inappropriate tillage and cropping systems which have resulted in SOC (soil organic carbon) and soil structural reduction' (Elwell, 1989). Although tillage with a hand hoe accounts for 80% of the cultivated area in E&S Africa (Sims *et al.*, 2012), this still results in soil structural breakdown and the formation of hard pans (Douglas *et al.*, 1999 quoted by Bot and Benites, 2001) and severe hard pans are common in manually cultivated farms in Malawi and Mozambique (e.g. Materechera and Mloza-Banda, 1997).

Soil erosion and the loss of SOM are intrinsically linked to soil chemical and biological quality. Continent-wide, 5 Mg ha^{-1} of Africa's soils are lost to lakes and oceans each year (Angima *et al.*, 2003). The surface soil is the most fertile and the loss of SOC from the top few centimetres of soil has a disproportionately large effect on soil infiltrability and nutrient supply (Mills and Fey, 2003). Estimated annual on-farm losses of SOC through sheet erosion in Zimbabwe were over 500 kg ha^{-1} together with approximately 50 kg ha^{-1} nitrogen and 8 kg ha^{-1} phosphorus (Elwell and Stocking, 1988).

Conventional management practices of smallholder farmers lead to organic matter decline and loss of fertility of the land (Zingore *et al.*, 2005). A study of South African soils cultivated for 0–85 years showed decreased SOC content, regardless of the duration of cultivated cropping, and cultivated soils had 10–75% less SOM than uncultivated areas (Du Preez *et al.*, 2011).

Agricultural systems are complex, multi-component systems adjusted and adapted to local conditions and farmer circumstances. Therefore the transfer of an agricultural system from one place to another is unlikely to be successful, but rather systems need to be tailored to local conditions (Wall, 2007). In the case of CA, tailoring technological changes to local conditions and farmer circumstances while following the principles of CA requires well-developed farmer participatory adaptive research, taking into account farmer preferences and management.

Normally farmers seek other benefits rather than yield *per se* – even subsistence farmers seek to sell excess produce for economic gain to help them access food supplies that they do not produce, and so achieve food security. Neither CA alone, nor any other agricultural technology, will solve all of the problems of smallholder farmers in E&S Africa, an enabling environment of adequate input and produce markets (the U-impact pathway of Dixon *et al.*, 2007), policies, research and information support are all also required (see section 11.5).

Smallholder farmers are generally not well linked to knowledge systems, and often have little access to new information and knowledge outside the community. While access to information is changing rapidly with the spread of cell phone and other information technologies, knowledge development (the application of information to understand and apply new ideas) needs more direct and constant contact. With respect to agricultural systems, farmer-participatory research followed and accompanied by farmer-to-farmer information exchange has proved to be an effective means of building knowledge in smallholder farming communities: 'farmers should be inspired to experiment, test, learn and think for themselves' (Bolliger *et al.*, 2005).

11.2 History of Conservation Agriculture in Eastern and Southern Africa

Most of the cropped area of E&S Africa follows a maize mixed or an agro-pastoral millet/sorghum agricultural system (Dixon *et al.*, 2001) and so much of the history of CA in the region is linked to maize-based systems.

Some of the earliest experiences in E&S Africa with CA were in the highlands of Kenya in the mid-1970s, at about the same time that farmers were starting to work with CA in southern Brazil. In an effort to conserve rainwater and to reduce production costs several large farmers began with zero tillage (ZT; Apina *et al.*, 2007) and many continue to practise CA today.

At about this time (1976) the Small Grains Institute of the Agricultural Research Council of South Africa started research on CA, with trials conducted over the maize-growing areas of South Africa (Berry *et al.*, 2001). Farmer-managed demonstrations were also started in the early 1970s by the extension branch of the Ministry of Agriculture of Rhodesia (now Zimbabwe) but with little success, leading to the conclusion that CA (no-till) systems were not adapted to local conditions (Oldrieve, undated).

In the 1982/83 season Oldrieve began experiments with CA on the farm he managed in north-eastern Zimbabwe and also developed systems and conducted outreach programmes for smallholder farmers in Zimbabwe (Oldrieve, 1993). The following season trials to evaluate CA were initiated at the Agricultural Research Trust (ART) farm near Harare, Zimbabwe with trials designed to evaluate CA systems (MacRobert *et al.*, 1995). These trials provided important information for commercial farmers and, despite the earlier reticence, adoption of CA began, driven by rising fuel and mechanization costs (Nyagumbo, 2008) and, prior to land reform starting in 2001, approximately 20% of commercial farmers in central Zimbabwe were applying the principles of CA (Oldrieve, n.d.).

In 1988 the German Development Corporation (GTZ) initiated a research project in Zimbabwe on CA (ConTill/Agritex), a project that provided much important information to underpin future CA efforts. However, again results were not universally positive and CA was not incorporated into the agenda of the Ministry of Agriculture. In 1995 the World Bank asked Oldrieve from Zimbabwe to attend a workshop in Zambia to plan a new CA initiative. The workshop was attended by the Zambian Minister of Agriculture, demonstrating marked political commitment, and the outcome of the workshop was the establishment of the Conservation Farming Unit of the Zambian National Farmers Union.

In 1995/96 the Planting without Ploughing project focused on smallholder farmers in KwaZulu-Natal, South Africa, installed large numbers of CA demonstration plots. The extension branch of the Ministry of Agriculture of Mozambique (DNEA) started installing demonstrations of CA together with the national research institute (INIA) in the 1996/97 season in the Manica and Nampula provinces, in a project with Sasakawa Global 2000 (SG2000) and Monsanto (Nhancale *et al.*, 2006).

It was at about this time that various important donors began to show increased interest in CA, especially the German Development Corporation (GTZ), the Regional Land Management Unit of the Swedish International Development Agency (RELMA), the World Bank and the Food and Agriculture Organization of the United Nations (FAO). This interest resulted in the organization of an international workshop on 'Conservation Tillage (sic) for Sustainable Agriculture' held in Harare, Zimbabwe, 22–27 June 1998. This workshop had many outcomes, including the establishment of the African Conservation Tillage Network (ACT) – originally financed by GTZ – and the increased interest of donors and national programmes in pursuing CA projects. Following the workshop major new projects were initiated in eastern Africa, in Ethiopia, Kenya, Uganda and Tanzania.

11.3 Current Status of Conservation Agriculture in Eastern and Southern Africa

Worldwide, the adoption of CA systems by smallholder farmers has lagged well behind the adoption on large, mechanized farms: only 0.3% of the area under no-till (NT) worldwide is on smallholder farms (Derpsch *et al.*, 2010). This is not unprecedented as smallholders are less able to invest in new equipment, are more risk averse than large farmers, generally have fewer links to new information systems and, importantly, manage more complex farming systems, generally mixed crop–livestock systems (Wall, 2007). In the Americas and Australia the CA movement was largely driven by farmers (Ekboir, 2002), but smallholders generally do not have the resources or linkages that enable them to take hold of the reins of development.

Recent estimates of the number of CA practitioners and the area under CA in E&S Africa are shown in Table 11.1. Much of the area in South Africa, and some of the areas in Kenya and Sudan, are likely to be on large mechanized farms, but the remainder is almost entirely on smallholder farms. It appears that well over 500,000 farmers in E&S Africa are currently using CA on at least part of their farm. The importance of this number is that the concepts of CA have reached at least half a million minds, and there are likely more farmers in E&S Africa managing CA systems today than there are CA farmers in the USA, the country with the largest area of CA in the world.

Table 11.1. Recent estimates of the use of Conservation Agriculture practices on farms in eastern and southern Africa.

Country	Year of estimate	Area ('000 ha)	No. of farmers ('000)
Sudan	2009[a]	10	
Kenya	2009[a]	15	
Tanzania	2009[a]	6	
Malawi	2012[b]	14	84
Mozambique	2009[a]	9	
Zambia	2009[c]		150
Zambia	2009[a]	40	
Zimbabwe	2009[a]	7.5	
Zimbabwe	2012[d]	141	371
South Africa	2008[e]	368	
Total	Most recent	603	

Source: [a]Derpsch *et al.*, 2010; [b]J. Chisui, Lilongwe, Malawi, 2012, pers. comm.; [c]Aagard, 2009; [d]L.S. Marongwe, Zimbabwe, 2012, pers. comm.; [e]Derpsch and Friedrich, 2009.

11.4 Research Results from Eastern and Southern Africa

Seldom have all three principles of CA been part of the systems applied and reported in the literature from the region. In analysing research results we have included all reports where soil disturbance has been kept to a minimum and some residues have been left on the soil surface. Often common farmer practices have been compared to CA treatments using different fertilizer levels, causing problems in interpretation of results (Baudron *et al.*, 2007; Thierfelder and Wall, 2012). While this may be valid for farmer demonstration plots we prefer to remove the fertilizer variable and apply the same fertilizer to all systems, unless one system does in fact require more fertilizer than the others. For the purposes of this chapter we have only considered results where fertilizer levels have been the same across the different systems.

11.4.1 Yield and economic benefits

There is a relative wealth of information available on the effects of CA systems on crop yields in E&S Africa, understandably especially with respect to maize. A comparison of various CA systems compared to ConvT maize systems on farmers' fields in E&S Africa shows marked yield benefits to CA in the majority of cases (Table 11.2). Results from eight different countries generally show that yields under CA are equal to or higher than ConvT. Of the six results (out of 40 shown in Table 11.2) where CA resulted in maize yields 10% or more below the yield of the ConvT treatment, two were not fertilized (Enfors *et al.*, 2011), one was probably not fertilized but was after a green manure

Table 11.2. Effects of Conservation Agriculture practices on maize yields in eastern and southern Africa. Results from trials on farmers' fields, mostly farmer managed. Plots were fertilized and surface residues were retained unless otherwise indicated.

Country	Place	Soil type	Seasons of data	Fertilized[a]	Surface residues	Weed control[b]	Average rainfall[c]	Mean yield[d]	CA treatment[e]	% Yield increase[f]	Reference	Notes
South Africa	Thukela Basin, KZN	Sandy loam	1		Prob none	H+M	710	1900	Rip line	140	Kosgei *et al.*, 2007	
Tanzania	Arusha and Arumeru	Various	4		No	Manual	1100	2800	Rip line	54	Rockström *et al.*, 2009	46 sites
Zimbabwe	Zimuto	Ferralic arenosol	6			H+M	620	1250	DS	54	Thierfelder and Wall, 2012	
Zambia	Malende, Monze District	Sandy loam	3			H+M	758	6400	DS	51	Thierfelder *et al.*, 2013	Second 3 years
Kenya	Masai farm, W.Kenya	Ferralic arenosol	1	Prob not	No	Manual	1800	1850	NT	47	Boye and Albrecht, 2005	
Malawi	Ntonda section	Fluvisol	3			H+M	800	4100	DS	42	Ngwira *et al.*, 2012a	
Malawi	Balaka market	Fluvisol, 80% sand	3			H+M	800	4200	DS	40	Ngwira *et al.*, 2012a	
Tanzania	Arusha and Arumeru	Various	4		No	Manual	1100	2600	Basins	39	Rockström *et al.*, 2009	46 sites
Zambia	Malende, Monze District	Sandy loam	3			H+M	758	6400	Rip line	38	Thierfelder *et al.*, 2013	Second 3 years
Zimbabwe	11 Districts		1	Manure	Prob none	Manual			Basins	36	Twomlow *et al.*, 2009	
Zimbabwe	Zimuto	Ferralic arenosol	6			H+M	620	1250	Rip line	29	Thierfelder and Wall, 2012	
Malawi	Lemu	Luvisol – sandy loam	5			H+M	935	5200	DS	26	Ngwira *et al.*, 2012c	
Tanzania	Arusha and Arumeru	Various	4	No	No	Manual	1100	1750	Rip line	25	Rockström *et al.*, 2009	46 sites

Malawi	Ntonda section	Fluvisol	3			H+M	800	3800	DS	23	Ngwira *et al.*, 2012a	
Malawi	Balaka market	Fluvisol, 80% sand	3			H+M	800	3850	DS	22	Ngwira *et al.*, 2012a	
Malawi	Zidyana	Luvisol – sandy clay loam	5			H+M	1375	5200	DS	19	Ngwira *et al.*, 2012c	
Tanzania	NE – Makanya catchment	Ferralsols – infertile	3	Manure		M	562	600	Rip line	18	Enfors *et al.*, 2011	Three bad seasons; crop failure. Both also had cover crops
Zambia	Kayowozi, Chipata District	Acrisol	4			H+M	950	3200	Dibble stick	15	Thierfelder *et al.*, 2013	
Tanzania	NE – Makanya catchment	Ferralsols – infertile	2	Manure		M	562	2100	Rip line	14	Enfors *et al.*, 2011	Two good seasons; also had cover crops
Ethiopia	708 farms in highlands		5		30% rec.	H+M		4700	NT	12	Ito *et al.*, 2007	708 sites across highlands; 708 paired plots
Zimbabwe	Hereford	Chromic luvisol	5			H+M	850	5150	DS	11	Thierfelder and Wall, 2012	
Kenya	Leuro farm, W Kenya	Nito-humic Ferralsol	1	Prob not	No	M	1800	2100	NT	10	Boye and Albrecht, 2005	
Zimbabwe	11 Districts		1	No	Prob none	M			Basins	10	Twomlow *et al.*, 2009	

Continued

Table 11.2. Continued.

Country	Place	Soil type	Seasons of data	Fertilized[a]	Surface residues	Weed control[b]	Average rainfall[c]	Mean yield[d]	CA treatment[e]	% Yield increase[f]	Reference	Notes
Zimbabwe	11 Districts		1	Man+28N	Prob none	M			Basins	9	Twomlow *et al.*, 2009	
Uganda	753 demon-strations		1			H+M		4800	Hoe	9	Findlay *et al.*, 2001	
Kenya	Meru South, Central Kenya	Humic nitisols	3			H+M	1500	3235	Hoe holes	7	Guto *et al.*, 2012	Medium fertility soils
Ethiopia	Tigray, 3 sites	Various	5		No	M	500	1500	Rip+Ss	6	Rockström *et al.*, 2009	25 sites
Zambia	Malende, Monze District	Sandy loam	3			H+M	758	3650	DS	3	Thierfelder *et al.*, 2013	First 3 years
Kenya	Meru South, Central Kenya	Humic nitisols	3			H+M	1500	4000	Hoe holes	2	Guto *et al.*, 2012	Good fertility soils
Kenya	Masai farm, W Kenya	Ferralic arenosol	1	Prob not	Only	M	1800	2800	NT	2	Boye and Albrecht, 2005	
Ethiopia	Tigray, 3 sites	Various	5	No	No	M	500	1250	Rip+Ss	1	Rockström *et al.*, 2009	19sites
Kenya	Meru South, Central Kenya	Humic nitisols	3		No	H+M	1500	4000	Hoe holes	1	Guto *et al.*, 2012	Good fertility soils
Zimbabwe	Hereford	Chromic luvisol	5			H+M		5150	Rip line	−1	Thierfelder and Wall, 2012	
Kenya	Meru South, Central Kenya	Humic nitisols	3		No	H+M	1500	3235	Hoe holes	−5	Guto *et al.*, 2012	Medium fertility soils
Zambia	Malende, Monze District	Sandy loam	3			H+M	758	3650	Rip line	−5	Thierfelder *et al.*, 2013	First 3 years
Kenya	Leuro farm, W Kenya	Nito-humic Ferralsol	1	Prob not	gmcc	M	1800	7250	NT	−20	Guto *et al.*, 2012	

Tanzania	NE – Makanya catchment	Ferralsols – infertile	2	No	No	M	562	2100	Rip line	–20	Enfors *et al.*, 2011	Two good seasons
Zimbabwe	11 Districts		1	28N TD	Prob none	M			Basins	–22	Twomlow *et al.*, 2009	
Tanzania	NE – Makanya catchment	Ferralsols – infertile	3	No	No	M	562	400	Rip line	–24	Enfors *et al.*, 2011	Three bad seasons; crop failure
Kenya	Meru South, Central Kenya	Humic nitisols	3			H+M	1500	2160	Hoe holes	–26	Guto *et al.*, 2012	Poor fertility soils
Kenya	Meru South, Central Kenya	Humic nitisols	3		No	H+M	1500	2160	Hoe holes	–33	Guto *et al.*, 2012	Poor fertility soils

[a]Man, manure; [b]H, herbicide; M, manual weeding; [c]approximate average annual rainfall (mm); [d]approximate mean yield of trial kg ha^{-1}; [e]DS, mechanical direct seeder; NT, no-tillage; rip line, seeded in rip line; Ss, sub-soiled; [f]Effect of CA treatment on yield (%) compared to ConvT with same fertilizer.

Table 11.3. Effects of Conservation Agriculture practices on maize yields in eastern and southern Africa. Results from researcher-managed trials on research stations. Plots were fertilized and surface residues were retained unless otherwise indicated.

Country	Place[a]	Soil type	Seasons of data	Fertilized[b]	Surface residues	Weed control[c]	Average rainfall[d]	Mean yield[e]	CA treatment[f]	% Yield increase[g]	Reference	Notes
Zimbabwe	Henderson RS	Arenosol	2			H+M	880	4300	Rip line	41	Thierfelder *et al.*, 2012a	
	Matopos RS	Arenosol	2	Man+50N		M	590	900	Rip line	40	Mupangwa *et al.*, 2007	
	Henderson RS	Arenosol	2			H+M	880	4300	Basins	39	Thierfelder *et al.*, 2012a	
	Makoholi RS	Ferralic	2			?	620	1600	Mulch-rip	37	Moyo, 1998	
	Henderson RS	Arenosol	2			H+M	880	4300	DS	32	Thierfelder *et al.*, 2012a	
	Matopos RS	Arenosol	2	Man+50N		M	590	900	Basins	30	Mupangwa *et al.*, 2007	
	Henderson RS	Arenosol	6			H+M	880	2600	Basins	25	Thierfelder pers. comm.	Harare, December 2012
	Henderson RS	Arenosol	6			H+M	880	2600	Rip line	22	Thierfelder and Wall, 2012	
	Henderson RS	Arenosol	6			H+M	880	2600	DS	11	Thierfelder and Wall, 2012	
	Makoholi RS	Ferralic	3			?	620	1600	Mulch-rip	10	Munodawafa and Zhou, 2008	Marked seasonal differences seasons
	Matopos RS	Cambisol	2	Man+50N		M	590	1400	Rip line	8	Mupangwa *et al.*, 2012	1st year maize in each of 2 seasons
	Domboshawa FTC	Sand	8		?	?	750	3600	Mulch-rip	6	Munyati, 1997	CA better in 5 of 8 years
	Matopos RS	Cambisol	1	Man+50N		M	590	2100	Basins	4	Mupangwa *et al.*, 2012	Mean of monoculture and rotations
	Henderson RS	Arenosol	2			H+M	880	3800	Basins	4	Thierfelder and Wall, 2009	Marked seasonal differences

	Matopos RS	Cambisol	2	Man+50N	M	590	2600	Basins	3	Mupangwa *et al.*, 2007	
	Henderson RS	Arenosol	2		H+M	880	3800	DS	1	Thierfelder and Wall, 2009	Marked seasonal differences
	Henderson RS	Arenosol	2		H+M	880	3800	Rip+Ss	0	Thierfelder and Wall, 2009	
	Makoholi RS	Ferralic	6		Prob M	475	2800	Mulch-rip	–1	Chuma and Hagmann, 1995	
	Matopos RS	Cambisol	2	Man+50N	M	590	2600	Rip line	–5	Mupangwa *et al.*, 2007	
	Matopos RS	Cambisol	1	Man+50N	M	590	2100	Rip line	–11	Mupangwa *et al.*, 2012	Mean of monoculture and rotations
	Matopos RS	Cambisol	2	Man+50N	M	590	1400	Basins	–13	Mupangwa *et al.*, 2012	1st year maize in each of 2 seasons
	Domboshawa FTC	Granite	1		M	438	650	Mulch-rip	–71	Vogel *et al.*, 1994	Severe drought
Zambia	Monze FTC	Lixisol	2		H+M	660	5000	Basins	26	Thierfelder and Wall, 2009	
	Monze FTC	Lixisol	4		H+M	750	4000	DS	25	Thierfelder *et al.*, 2012a	
	Monze FTC	Lixisol	2		H+M	660	5000	DS	18	Thierfelder and Wall, 2009	
	GART Farm	Humic	2		M	850	4900	Hoe holes	0	Gill *et al.*, 1992	
Mozambique	Susendenga RS	Haplic	5		H+M	1085	2500	DS	–2	Thierfelder *et al.*, 2012b	
Malawi	Chitedze RS	Chromic	4		H+M	960	5250	Hoe holes	2	Thierfelder *et al.*, 2012b	
	Bunda College	Oxic	3		M	857	4600	Hoe holes	–7	Materechera and Mloza-Banda, 1997	Wide spacing: 0.91 m × 0.91 m

[a]RS, Research Station; FTC, Farmer Training Centre; [b]Man, manure; [c]H, herbicide; M, manual weeding?, not known; [d]approximate average annual rainfall (mm); [e]approximate mean yield of trial in kg ha^{-1}; [f]DS, mechanical direct seeder; NT, no-tillage; Rip line, seeded in rip line; Ss, sub-soiled; [g]effect of CA treatment on yield (%) compared to ConvT with same fertilizer.

cover-crop (gmcc) (Boye and Albrecht, 2005), one received only a modest nitrogen top dressing and no other nutrients (Twomlow *et al.*, 2009) and two came from the poor soils that Guto *et al.* (2012) studied in Kenya. It appears that adequate soil fertility levels are important for the successful functioning of CA systems – we hypothesize that not only are the benefits of CA restricted when crop nutrition is limiting, but also that increased biomass production is important to achieve the potential of CA systems, both through the production of sufficient residues for ground cover, as well as increased return of organic matter to the soil to improve soil physical, chemical and biological fertility.

The 23 reports of yield increases (>10% above ConvT) in farmers' fields (Table 11.2) come from a diverse set of conditions, including a range of soil types and annual rainfall (from about 500 to 1500 mm). Most were fertilized (19 of 23) and had residues retained as mulch (15 of 23) and 14 included chemical weed control, often complemented by manual weeding.

Overall the effects of CA on yield in trials from research stations (Table 11.3) have been positive and only in five of the cases shown were yields under CA > 10% less than under ConvT. In general, yield benefits of CA systems from on-farm trials (Table 11.2) were, however, clearer as compared to the ones from research stations (Table 11.3).

There is not as much information about the management of other crops under CA in E&S Africa as there is for maize. However, results (Table 11.4) suggest that cotton, cowpea, sorghum, wheat and even teff, can yield just as well under CA as they can with tillage. Achieving acceptable plant stands was a problem on some of the reported cowpea and sorghum work, highlighting the need for adequate equipment and understanding of the management of the CA system before embarking on comparative studies.

Increased yield is not necessarily a prerequisite for CA adoption: cost and labour reductions may be just as important and in the final analysis it is the balance of benefits as perceived by the farmer that will define adoption. Economic analysis is probably not only important but also the best analysis for comparisons of systems where several components vary. Economic analysis is able to integrate results where systems have different seeding dates, fertilizer levels, use different equipment etc.

The three main components of comparisons of the economics of ConvT and CA systems in E&S Africa are the cost of tillage, the cost of weed control and crop yield. The first two of these are also the factors that have the biggest impact on labour demands. Where herbicides have been used in Malawi, labour costs were lower in CA systems than the normal farmer production practices (Table 11.5) by between 28 (Ngwira *et al.*, 2012a) and 63% (Ito *et al.*, 2007). These labour savings were partially offset by increased input costs, but because of increased yields net returns per hectare were increased by US\$130–370 (Ito *et al.*, 2007; Ngwira *et al.*, 2012a, c) resulting in a mean 60% increase in net benefits, while returns to labour (US\$ day^{-1}) were increased by 100% in CA systems with continuous maize and by 92% in CA maize systems intercropped with legumes (mean data from Ngwira *et al.*, 2012a, c).

Where herbicides are not used, labour savings from land preparation may be offset by labour requirements for weeding (Jat *et al.*, 2012a, b). In Zimbabwe, labour for weeding a mulch-rip treatment was similar to that used in the tilled plots at two sandy soil sites (Vogel, 1994), but 10–25% more labour was used in two 'best bet' CA treatments on sandy soils previously abandoned due to degradation (Siziba, 2007). However, even in the latter case net benefits were higher in CA systems once tillage costs were included.

In northern Tanzania work with gmcc intercropped with maize (Mariki, 2004) without fertilizer and using glyphosate herbicide only in the first three seasons, weed pressure declined with time due principally to the competition of the gmcc, probably largely the increased ground cover. In the first season of this work, 11% more labour (178 person days ha^{-1}) was used in the no-till plots with gmcc (*Mucuna pruriens* and *Lablab purpureus*) than in the tilled plots (160 person days ha^{-1}) but four seasons later

Table 11.4. Effects of Conservation Agriculture practices on yields of crops other than maize in eastern and southern Africa. Surface residues were retained unless indicated.

Crop	Country	Place[a]	Soil type	Seasons of data	Fertilized[b]	Surface residues	Weed control[c]	Average rainfall[d]	Mean yield[e]	CA Treatment[f]	% yield increase[g]	Type of trial[h]	Reference
Cotton	Zimbabwe	North-east	Sandy loam	3[i]	No		M	550	700	Hoe holes	–1	OF-FM	Baudron *et al.*, 2012a
		Save Valley	Cambisol	6	High		M	Irrigated	3070	Hoe holes	–14	OS	Gwenzi *et al.*, 2008
Cowpea	Zimbabwe	Matopos RS	Cambisol	2	Man+50N		M	590	800	Basins	7	OS	Mupangwa *et al.*, 2012
										Rip line	6	OS	
				1	No		M	580	300	Rip line	–34	OS	Mashingaidze *et al.*, 2012
										Basins	–45	OS	
Sorghum	Zimbabwe	North-east	Sandy loam	3[j]	No		M	550	1070	Hoe holes	–1	OF-FM	Baudron *et al.*, 2012a
		Matopos RS	Cambisol	1	Manure		M	580	3500	Basins	–37	OS	Mashingaidze *et al.*, 2012
										Rip line	–12		
					Man+50N		M	590	1900	Basins	–21	OS	Mupangwa *et al.*, 2012
										Rip line	9		
Soybean	Zimbabwe	Harare	Sandy clay	4	Yes	+ and –[k]	Yes	Suppl. irrig.	3800	DS	1	OS	MacRobert *et al.*, 1995
Teff	Ethiopia	Different sites	Various	4	Yes	30%[l]	H+M		1180	?	11	PP-FM	Ito *et al.*, 2007
		Tigray	Various	1	?		H	500–800	800	PB	–35	OF-FM	Nyssen *et al.*, 2010
Wheat	Ethiopia	Different sites	Various	4	Yes	30%?	H+M		2380	Rip line	9	PP-FM	Ito *et al.*, 2007
	Ethiopia	Tigray	Various	2	?		H	500–800	1250	PB	90	OF-FM	Nyssen *et al.*, 2010
	Zimbabwe	Harare	Sandy clay	5	Yes	+ and –	H+M	Irrigated	6300	DS	2	OS	MacRobert *et al.*, 1995

Continued

Table 11.4. Continued.

Crop	Country	Place[a]	Soil type	Seasons of data	Fertilized[b]	Surface residues	Weed control[c]	Average rainfall[d]	Mean yield[e]	CA Treatment[f]	% yield increase[g]	Type of trial[h]	Reference
	South Africa	SW Cape	Sand	6	Yes	Burned	H	411		DS	0	OS	Agenbag and Maree, 1991
	Zimbabwe	Save Valley	Cambisol	6	High		M	Irrigated	4400	Hoe holes	–4	OS	Gwenzi *et al.*, 2008
	South Africa	Near Bethlehem	Plinthosol	9	Yes	+ and –	Mech+H	650	2000	Conv. drill	–6	OS	Du Preez *et al.* 2001

[a]RS, Research Station; [b]Man, manure; [c]H, herbicide; M, manual weeding; Mech, mechanical weeding; [d]approximate average annual rainfall (mm); [e]approximate mean yield of trial in kg ha^{-1}; [f]DS, mechanical direct seeder; NT, no-tillage; Rip line, seeded in rip line; Ss = sub-soiled; Conv. drill, conventional seed drill; ?, not known; [g]effect of CA treatment on yield (%) compared to ConvT with same fertilizer; [h]OF-FM, on-farm, farmer managed; OS, on-station; PP-FM, paired plots, farmer managed; [i]18, 28 and 23 farms in the 3 years; [j]FTC = Farmer Training Centre; on 18, 28 and 23 farms in the 3 years, respectively; [k]mean of burned and unburned plots; [l]farmers were recommended to leave at least 30% of the residues, but residue cover was not evaluated.

Table 11.5. Comparison of labour use (person days ha^{-1}) in Conservation Agriculture and conventional tillage maize systems in eastern and southern Africa. All plots fertilized unless otherwise indicated.

Country	Site(s)	Fertilizer	Weed control	CA treatment[a]	Land prep., planting and fertilizer application ConvT	Land prep., planting and fertilizer application CA	Weed control ConvT	Weed control CA	Total ConvT	Total CA	Reference	Notes
Malawi	Various		Glyphosate, alachlor + atrazine	Hoe holes	39	14	19	0	68	25	Ito *et al.*, 2007	Assume $2/day wage
	Balaka + Ntonda		Glyphosate, alachlor + atrazine	Hoe holes					65	44	Ngwira *et al.*, 2012a	
	Lemu + Zidyana		Glyphosate	Hoe holes					62	49	Ngwira *et al.*, 2012c	Legume intercrop
			Glyphosate, alachlor + atrazine						62	40		
Tanzania	Karatu +Hanang	No	Glyphosate (3 of 6 seasons)	DS					161	114	Mariki, 2004; Steiner, 2013 unpubl.	GMCC intercrop
Zimbabwe	ART, Harare		Herbicides	DS					22	13	MacRobert *et al.*, 1995	
	Shamva		Glyphosate	DS′	8	2	29	32	61	62	Siziba, 2007	
				Rip line		5		27		55		
Zambia	Various	Variable	Manual	Basins	97	113	58	81	176	211	Haggblade and Tembo, 2003	
				Rip line	14	26	27	35	48	77		
	Magoye			Dibble			12	57			Muliokela *et al.*, 2001	
				Basins			12	46				
Zimbabwe	Various			Basins	29	46	26	52	77	116	Mazvimavi and Twomlow, 2009	
	Zimuto			DS	8	2	19	31	38	47	Siziba, 2007	
				Rip line		5		25		41		
	Domboshawa			Mulch-rip			55	61			Vogel, 1994	
	Makoholi						29	24				

labour use was 45% lower in the CA plots than in the tilled plots (90 person days ha^{-1} and 162 person days ha^{-1}, respectively).

More efficient machinery use has been one of the drivers of CA adoption on mechanized farms in the Americas. There are little published data on machinery use in CA in E&S Africa, but on the ART farm near Harare, machinery costs for land preparation in CA as compared to ConvT were reduced by 97% (Steiner, 2002) and over the whole crop season by 66% (MacRobert *et al.*, 1995).

Labour is an important component of the CF system in Zambia, Zimbabwe and Malawi. Farmers are advised to dig basins during the dry season to reduce labour requirements and ensure that the basins are ready when the first planting rains arrive. However, digging basins involves considerable labour (28–34 person days ha^{-1}) in the first year (Mazvimavi and Twomlow, 2009; Umar *et al.*, 2012) although as the basins are dug in exactly the same place each year the difficulty and time for digging is reduced in subsequent years (Baudron *et al.*, 2007). In areas where the basins have not given large returns this results in disadoption (Baudron *et al.*, 2007) and leads to complaints about the labour requirements (Haggblade and Tembo, 2003), and even referring to the basin system as 'dig and die' (Andersson and Giller, 2012). Although digging of basins involves about 2.5 times more labour than building ridges for planting (34 versus 13 person days ha^{-1}), because of the increased maize yield the returns to labour ($ day^{-1} worked) in Zambia were five times higher in the basin system than with ConvT (Umar *et al.*, 2012).

11.4.2 The importance of mulch and crop rotation

There is little consistent information of the importance of mulch and crop rotation in CA systems in E&S Africa. In central Kenya residues generally had positive effects on yield of minimum tillage (CA) plots, although this effect was small (0.6 t ha^{-1} extra grain yield), seldom significant and only marginally economic on good and medium soils and uneconomic on poor soils (Guto *et al.*, 2012). In the drier conditions of south-western Zimbabwe (Mupangwa *et al.*, 2012) the effects of mulch were variable, but there was a significant yield response to mulch levels up to 4 t ha^{-1} in rip-line seeded maize, sorghum and cowpea. A survey of farmer trials with planting basins in Zimbabwe (Mazvimavi and Twomlow, 2009) showed major positive effects of residues on crop yield both under CA (74%; 1.7 t ha^{-1} increase over yields without residues) and ConvT (92%; 1.4 t ha^{-1} yield increase). Given that the issue of residues is such a central one to the adaptation and adoption of CA systems it appears that the quantification of the benefits of the residues under different conditions needs to be addressed more by research.

Crop rotation is one of the three pillars of CA, but is often the last to be incorporated into the system by farmers, often because of a lack of adequate markets for alternative crops (Thierfelder and Wall, 2010a). Although one of the main reasons for crop rotation in CA systems is to avoid problems of pests and diseases harboured on the residues (Baudron *et al.*, 2012b), there may also be marked yield benefits associated with crop rotation under CA conditions (Fig. 11.1). Only maize grain yield in the maize phase of the rotation is shown in Fig. 11.1, and a full economic analysis is necessary to ascertain the profitability of the rotations. Legumes are often preferred for rotations because of the benefits of biological nitrogen fixation, but non-legume crops may also benefit the following maize crop as evidenced by the 10% yield increase in yield of maize in a maize–cotton rotation in Monze FTC (Fig. 11.1).

In areas where there is intense land pressure, farmers may prefer to intercrop rather than rotate their crops. Some results from farm trials in Zimbabwe (Thierfelder *et al.*, 2012a) and Malawi (Ngwira *et al.*, 2012a) show intermediate yield benefits (less than the effect of rotation) in maize due to intercropping, but more information on the effects of intercrops is needed, including information on their effects on diseases and pests.

In some parts gmcc are an integral part of productive CA systems. The feasibility of gmcc is closely linked to water availability. If there is sufficient moisture for an economic

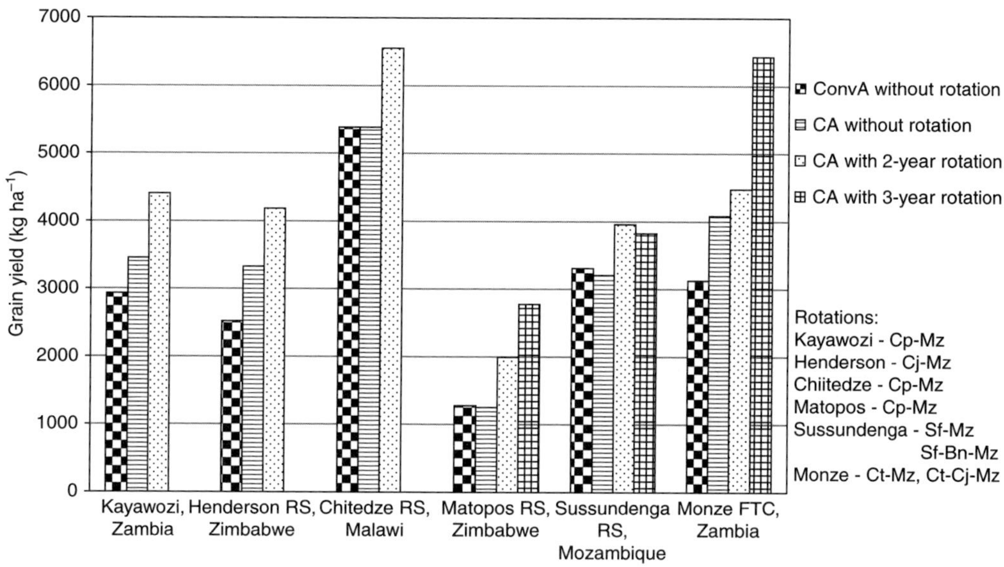

Fig. 11.1. The effect of Conservation Agriculture and crop rotation on maize yield on farmers' fields (Kayawozi) and five research stations in southern Africa (Adapted from Thierfelder *et al.*, 2012a, 2013 and Mupangwa *et al.*, 2012). Mz = Maize; Cp = Cowpea; Cj = *Crotalaria juncea* (sunnhemp); Sf = sunflower; Bn = common bean; Ct = cotton.

crop it is unlikely that farmers will sow a gmcc unless it can be demonstrated that profits are greater with the gmcc than with only economic crops. In northern Tanzania, which has a bimodal rainfall pattern, in a comparison of ConvT monoculture maize, a NT maize–gmcc rotation and a no-till maize–gmcc intercrop (where the legume gmcc continued to grow during the unreliable short rains), all without fertilizer application, maize yields with the intercrop (1.7 t ha^{-1}) were 350% higher than with the ConvT monocrop (0.5 t ha^{-1}) (Mariki, 2004). A large part of the yield response was undoubtedly due to N fixation by the gmcc. Rotation only increased yields by another 11% over the maize yields with the intercrop, unlikely therefore to be of interest to farmers.

11.4.3 Effect of Conservation Agriculture on soil quality (physical, chemical and biological)

A meta-analysis of the effects of CA on SOC in the developing world, including Africa (Govaerts *et al.*, 2009), found little evidence of increases in SOC under CA. However, in the published reports we have analysed from E&S Africa CA has increased SOC compared to ConvT in practically all cases (Chuma and Hagmann, 1995; Haynes *et al.*, 2003; Boye and Albrecht, 2005; Chivenge *et al.*, 2007; Nyamadzawo *et al.*, 2007; Oicha *et al.*, 2010; Du Preez *et al.*, 2011; Mchuru *et al.*, 2011; Guto *et al.*, 2012; Ngwira *et al.*, 2012b, c; Thierfelder and Wall, 2012; Thierfelder *et al.*, 2012a, b, 2013) although differences are relatively seldom statistically significant. Only two instances were found in the literature where CA treatments had lower SOC than the cultivated soils: the 0–10 cm horizon of the Zidyana soils in Nkhotakhota Province of Malawi after 6 years of CA (Ngwira *et al.*, 2012c) (SOC was higher under CA in these soils in the 10–30 cm horizon) and the 0–5 cm horizon of the sandy soils at the Masai farm site in western Kenya after 5 years of NT (Boye and Albrecht, 2005). However, recent studies comparing SOC under CA and ConvT in southern Africa have shown little difference in SOC despite several years under CA conditions

(Thierfelder, unpublished data). Obviously further work is required to better understand SOC dynamics after a change to CA.

Importantly, many of the published reports have compared CA and ConvT effects on SOC over the whole tillage horizon (top 20 cm for animal traction tillage and top 30 cm for mechanical traction tillage) or more, thus permitting a valid comparison as ConvT systems may distribute SOC over this horizon, whereas in CA, surface accumulation of SOC is common. Most reports that have considered the whole tillage horizon have found significant increases in total SOC under CA compared to ConvT once the CA system had been practised for several years. In Zimbabwe there were significant increases in SOC after 10 years of mulch-ripping (15% on clay soil, 62% on a sandy soil) (Chivenge *et al.*, 2007), while in CA systems established with an animal traction direct seeder there were significant increases in SOC in the 0–30 cm horizon after only 3–4 years of CA at three diverse sites in Zimbabwe: 17% higher on a sandy soil at Henderson Research Station, 13% higher on farmers' fields on a clay loam at Hereford Farm near Shamva and a huge 93% increase on the extremely sandy soils (>90% sand) around Chikato Village near Masvingo (Thierfelder and Wall, 2012). In Malawi, a survey of the fields of 48 farmers managing CA systems showed higher SOC levels in the 0–20 cm horizon of CA plots compared to ConvT plots, but only after 4 (44% higher SOC) or 5 (74% higher SOC) years of CA were the differences statistically significant (Ngwira *et al.*, 2012b).

As expected from the generally higher levels of SOC under CA systems, the proportion of water-stable aggregates is also usually higher in CA systems in E&S Africa (Chuma, 1993; Boye and Albrecht, 2005; Thierfelder and Wall, 2010b), although there were no differences in aggregate stability between CA and ConvT systems in farmers' fields in Malawi (Ngwira *et al.*, 2012c) and Ethiopia after 1 year of permanent beds (Oicha *et al.*, 2010). Differences in aggregate stability of sandy soils may not be apparent using the aggressive wet sieving technique (e.g. Boye and Albrecht, 2005) and may require less aggressive tests, such as the dispersion test, to show differences in aggregation (Thierfelder and Wall, 2012).

11.4.4 Conservation Agriculture and soil water balance

One of the major benefits of CA is the effects on water balance, through effects on infiltration, evaporation, soil water-holding capacity, soil compaction and crop rooting characteristics. Changes in these components will also affect drainage, and possibly nutrient losses, as well as runoff and soil erosion. There are now considerable data on many of these processes under conditions of E&S Africa.

Thirty-nine sets of data from Kenya (3), Malawi (6), Zambia (12), Zimbabwe (17) and South Africa (1) show an overall increase in infiltration rate of 67% under CA conditions, although data obviously include different numbers of sites, years of results etc. Of all these results the only three that do not show a positive effect of CA on infiltration rates are the results of Guto *et al.* (2012) from Meru South in central Kenya, obtained with ring infiltrometers that were pushed into the soil. We suggest that this method is not appropriate for comparing infiltration rates between treatments that have different effects on the structure of the soil surface, as infiltration is affected greatly by conditions in the surface layer (0–1 cm) and if this layer is disturbed by inserting infiltration rings results will be questionable. Comparisons of the effects of CA and tillage on infiltration rates can best be achieved with (small, portable) rainfall simulators (e.g. Thierfelder *et al.*, 2005) or with the simple 'time to pond' procedure (Govaerts *et al.*, 2006) without soil surface disturbance. With these methods CA systems have universally shown greater water infiltration rates than tilled systems in southern Africa. Residues on the soil surface protect the soil from the impact of raindrops and reduce soil sealing and crust formation. In the absence of crop residues, untilled (CA) soils, especially degraded soils, can have lower infiltration rates and be less productive than tilled soils (Wall, 1999; Govaerts *et al.*, 2005) and so it is not surprising that where little mulch is kept infiltration will be reduced and tillage to break the soil crust can give beneficial effects (Baudron *et al.*, 2012a).

Soil water-holding capacity is a function of soil texture, porosity and SOC content. As SOC content increases the soil will hold more water (Hudson, 1994). Water that cannot be held by the soil should drain if there is no dense layer that impedes water flow. Although drainage may also take nutrients from the root zone it is a positive process in that it allows aeration of the root zone. As water infiltration is increased by CA, drainage is likely to increase during periods of excess rainfall above the soil water-holding capacity (Moyo and Hagmann, 1994; Nyagumbo, 2002; Munodawafa and Zhou, 2008; Thierfelder and Wall, 2009). However, where drainage is restricted, CA may lead to excess moisture in the profile, waterlogging and anoxia (Rusinamhodzi *et al.*, 2011). Adaptation of CA practices for these conditions, including the possibility of permanent raised beds, is required.

Little work has been done to directly measure water evaporation from the soil surface under conditions of E&S Africa. However, between 60 and 75% of precipitation in semi-arid South Africa may be lost to unproductive evaporation (Bennie and Hensley, 2001), while across semi-arid sub-Saharan Africa losses to evaporation of 30–50% of precipitation have been calculated (Rockström *et al.*, 2002). Obviously any reduction in evaporation could play a major role in increasing rainfall-use-efficiency (RUE), but we have found no reports of consistent benefits of CA in reducing evaporation (see Bennie and Hensley, 2001). The fact that soil is generally moist under a cover of crop residues suggests that evaporation is reduced but this needs confirmation and quantification in order to prioritize actions to enhance RUE.

Increased water infiltration results in reduced runoff. Studies from Ethiopia (Gebreegziabher *et al.*, 2009; Nyssen *et al.*, 2010), sandy soils in Kenya (Boye and Albrecht, 2005), Zimbabwe (Vogel, 1992; Moyo and Hagmann, 1994; Vogel *et al.*, 1994; Moyo, 1998; Nyagumbo, 1998, 2002; Munodawafa and Zhou, 2008; Thierfelder and Wall, 2009) and South Africa (Kosgei *et al.*, 2007; Mallet quoted by Bennie and Hensley, 2001) show an average 51% reduction in runoff with CA (range 14–95%). However, runoff was higher with NT treatments on a heavy soil in central Kenya (Boye and Albrecht, 2005) (although runoff plots were very small (1 m^{-2}) and only gmcc residues were left on the surface) and on sandy soils with less than a 2% slope in a semi-arid environment in South Africa (Bennie *et al.*, 1994 quoted by Bennie and Hensley, 2001) where over 4 years 12.1% of the rainfall was lost to runoff from the NT plots with residue retention, while only 7.1% was lost from the ploughed plots.

Runoff water leads to erosion, with the amount of erosion depending on the erodability of the soil and the velocity of water running across it (erosivity). Data on the effect of CA treatments on erosion are shown in Table 11.6. On average 12.6 Mg ha^{-1} soil was lost from tilled treatments compared to 2.9 Mg ha^{-1} from CA treatments – a 77% reduction in erosion with CA. Not only is runoff reduced by CA, but residues on the soil surface reduce the velocity of water flow, allowing suspended solids to be deposited. On a very sandy site in Zimbabwe erosion was reduced by 99% by CA, and the distribution of textural classes in the sediment was changed. Silt+clay concentration in the sediment from the CA plots (47%) was higher than from the tilled plots (30%) because the larger sand particles had more time to settle. This is of course relative and a lot less clay and silt was lost overall from the CA plots (0.09 Mg ha^{-1}) than from the tilled plots (10.29 Mg ha^{-1}) (Moyo, 1998).

11.4.5 Conservation Agriculture and soil biological activity

Because of the lack of disturbance by tillage coupled with a more constant food supply from the residues, soil biological activity may be increased by CA. Earthworms are one of the indicators of enhanced soil biological activity and play an important role in incorporating organic matter and in increasing porosity and root channels. Data from South Africa (Haynes *et al.*, 2003), Zambia (Thierfelder and Wall, 2010b), Malawi

Table 11.6. Soil loss (Mg ha^{-1}) from Conservation Agriculture and conventional tillage systems in eastern and southern Africa. Surface residues were kept unless otherwise indicated.

Country	Place	Soil type	Treatment description	Years of data	Average rainfall[a]	Soil loss Mg ha^{-1}		Reference
						ConvT	CA	
Ethiopia	Near Mekelle, Tigray	Vertisol	CA = permanent beds[b]	1	466	19.5	4.7	Gebreegziabher *et al.*, 2009
	Tigray, Gum Selasa	Vertisol	CA = Derdero + (permanent beds)	2	500–800	14.6	3.0	Nyssen *et al.*, 2010
	Tigray, May Zegzeg	Vertisol		1	500–800	10.4	8.2	
Kenya	Leuro farm	Nito-humic Ferralsol	Continuous maize	1	1800	2.1	1.9	Boye and Albrecht, 2005
			Improved fallow (gmcc)			0.3	0.2	
	Masai farm	Ferralic Arenosol	Continuous maize	1	1800	2.0	2.9	
			Improved fallow (gmcc)			0.7	0.9	
Zimbabwe	Makoholi RS	Ferralic Arenosol	CA = no-till tied ridges	3	547	2.6	0.7	Vogel, 1992
	Domboshawa RS	Sand		3	858	4.1	1.7	
	MakoholiRS	Ferralic Arenosol		1	483	34.3	0.2	Moyo, 1998
				3	547	33.0	2.0	Munodawafa and Zhou, 2008
	Domboshawa RS	Sand		8	858	41.0	8.0	Munyati, 1997
	Henderson RS	Arenosol	CA= Animal traction direct seeder	2	880	7.2	4.5	Thierfelder and Wall, 2009
			CA = Rip line seeded with intercrop				4.1	
Mean						12.8	3.1	

[a]Approximate average annual rainfall (mm); [b]not clear whether residues were retained.

(Ngwira *et al.*, 2012c) and Tanzania (Mariki, 2004: Steiner, 2013, unpublished data) all show significant increases in earthworm populations under conditions of CA with residue retention. On average across all sites there was a five-fold increase, with 22 earthworms m^{-2} (range 2–48) under ConvT and 113 earthworms m^{-2} (range 40–400) under CA. In the Zambian data the eight-fold increase in earthworm populations was after only 3 years in a CA system.

Generally CA systems increase soil microbial biomass. In results from KwaZulu-Natal, South Africa, soil microbial biomass (C) was

40% higher than ploughed fields after 3 years of CA (Chaplot *et al.*, 2012) and 50% higher after 25 years. In Malawi, there were significant reductions in soil microbial biomass (C and N) on farmers' fields after 2–3 years of CA, but then significantly higher soil microbial biomass after 4 and 5 years of CA (Ngwira *et al.*, 2012b). Farmers often talk about a transition period of 2–3 years after initiating CA, after which the system starts to improve, which can probably be attributed to the gradual increase in soil biological activity.

11.4.6 Conservation Agriculture and weeds, pests and diseases

There is little published information on the effects of CA on weeds, pests and diseases. One of the principal reasons for tillage is to control weeds, and so a move to CA systems obviously leads to weed problems. Earlier we highlighted this aspect in the discussion of herbicide and labour use. Some results from South America suggest that weed pressure declines with time as long as weed control is good (Skóra Neto, 1993) and Baudron *et al.* (2007) report that GART in Zambia holds to a 50% reduction in weed pressure after 5 years of CA practice with good weed control. There are also reports that residue cover helps control weeds but the results shown in Fig. 11.2 suggest that the levels of residue cover for any reasonable weed control would be excessively high. Mulch may also affect the weed spectrum, and in Tanzania the populations of problematic weeds such as *Digitaria* spp. and *Cyperus* sp. were markedly reduced by CA with gmcc (Steiner, 2013, unpublished).

In the 1986/87 season in South Africa, just as adoption of CA was beginning to take hold in KwaZulu-Natal, a diplodia maize cob rot epidemic (caused by *Stenocarpella maydis*) was worse on CA fields and caused a setback to the adoption of CA (Fowler, 1999). This can be expected from necrotrophic diseases (that survive on dead tissue – the residues) and stresses the importance of crop rotation in CA systems to guard against disease. The maize varieties at that time were not resistant to the diplodia, but since then maize varieties resistant to the disease, and also to grey leaf spot (GLS), another residue-borne disease, have become available.

There are few consistent reports of increased insect damage in CA systems in E&S Africa. Termites are often cited as a problem

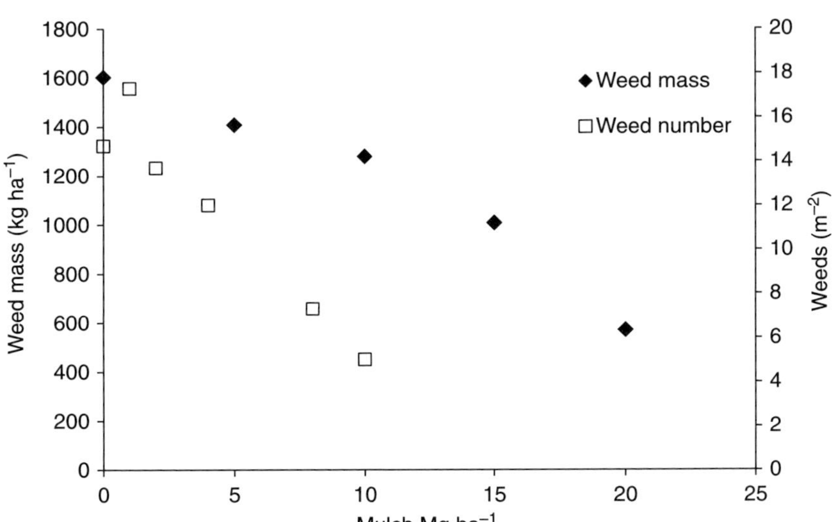

Fig. 11.2. Effect of mulch quantity on weed numbers at Matopos Research Station, Zimbabwe (adapted from Twomlow *et al.*, 2009) and weed mass at the GART Research Farm in Zambia (adapted from Gill *et al.*, 1992).

because they consume the residues but at the same time they open up channels that aid in water infiltration. Reports of damage to the maize crop are very variable, probably because termite species differ greatly from place to place. Although there are reports of increased damage with CA, our observations in Malawi show clearly more damage to the maize crop in conventionally ridged plots without residues.

White grubs (Scarabidae larvae) are a sporadic pest that can do considerable damage by killing young plants. There are reports that damage due to white grubs may be worse in CA than in conventional systems, e.g. the report in Giller *et al.* (2009) based on field observations in a few fields in Mozambique. We have seen similar situations in Malawi, but it appears that crop rotation with legumes overcomes the problem (Christian Thierfelder, pers. comm., March 2013).

11.5 Problems Encountered in Scaling-out Conservation Agriculture in Eastern and Southern Africa

Numerous problems have been encountered in scaling-out even simple, single component technological innovations (e.g. new maize varieties) in E&S Africa and therefore it is not surprising that the adoption of a complex system change such as CA has been slow. While the principal technological challenges of CA involve weed control, adequate equipment for seeding, residue and nutrient management, these are surmountable problems that have technical solutions. However, the practical limitations to widespread adoption, described by Wall (2007), are principally economic, organizational, social and legal:

- Mind-set;
- Knowledge of the CA system;
- Residue retention, and competition for scarce residues;
- Physical and financial access to inputs;
- Availability of adapted equipment;
- Capacity building among farmers, researchers and extension agents;
- Development of innovation systems around CA;
- Land tenure;
- Support to farmers for environmental services.

Overcoming these problems will require concerted efforts in resolving the weak links in the value chains surrounding maize and accompanying crops in E&S Africa. A major step in this direction is the development of local innovation systems involving multiple agents representing all major stakeholders in the value chains (including agents involved in both input and output markets – the U-impact pathway), and especially farmers (Ekboir, 2002; Wall *et al.*, 2002). Moving away from the common linear model of agricultural technology development and dissemination to a farmer participatory model within a local innovation system will help overcome many of the limitations noted above. However, innovation systems do not normally develop spontaneously, they need to be catalysed (Wall *et al.*, 2002), and in the absence of other catalysts the capacity of extension agents and researchers to catalyse, build and nurture local innovation systems will be extremely important. As CA becomes more widespread there should be an important change in the focus of adaptive CA research embedded in these innovation systems, moving away from comparisons between ConvT and CA systems, and rather concentrating on the identification and resolution of problems within the developing CA systems, while strategic research to understand the processes underpinning the successful application of CA systems is needed to aid in the development of adapted systems. At the same time the focus of extension agents should change from being conduits of information to farmers, to rather becoming catalysts of local innovation systems and facilitators of farmer-to-farmer information exchange – farmers believe information they get from other farmers far more readily than they do from those that do not rely on farming for their livelihoods.

Maintaining crop residues on the soil surface is an issue in smallholder systems worldwide, as smallholders generally manage complex mixed crop/livestock systems. The need for understanding and demonstrating the importance of crop residues for soil cover in E&S Africa was noted above. However, in other environments, especially on degraded soils with a tendency to form a surface crust, direct seeding of crops into untilled soil without surface mulch leads to reduced yields compared to tilled systems (Wall, 1999; Govaerts *et al.*, 2005; Enfors *et al.*, 2011; Baudron *et al.*, 2012a); soil crusts need to be broken by tillage or their development reduced by the surface protection offered by mulch cover. However, leaving mulch on the soil surface implies direct competition for residues with different household enterprises, especially livestock. Residues of cereal crops are at best a low quality maintenance feed, meaning that draught animals are weak after the dry season and tillage with animal traction is a slow and lengthy process. Options for system intensification under CA include possibilities of producing higher quality feed on part of the land (as allowed by the increased staple crop productivity in CA systems), thus allowing some of the low-quality cereal crop residues to be retained for mulch (Thierfelder and Wall, 2011), although smallholder farmers generally do not like to grow fodder crops (K. Steiner, 2013, pers. comm.). However, as with many other smallholder systems worldwide, communal grazing is the norm in E&S Africa (e.g. Mtambanengwe and Mapfumo, 2005 in Zimbabwe), meaning that an individual farmer cannot protect the residues on his or her own fields without considerable cost of both capital and goodwill (Erenstein, 2002). Residue retention becomes a social issue and needs to be dealt with through the innovation system at the community level. This will not be easy, not least of all because in many parts it is the richer and more powerful members of the community who own more cattle and benefit most from communal grazing rights. However, there are some examples from Tanzania and Zimbabwe of community action that has resulted in restrictions to communal grazing in E&S Africa (Wall, 2007) – these need to be multiplied.

Equipment for manual CA systems is not a major issue as the ubiquitous hoe is a functional tool for seeding in CA systems, as is the pointed (or dibble) stick (Ngwira *et al.*, 2012c). However, improved versions of these basic tools such as the Chaka hoe for more effective digging of planting basins (Conservation Farming Unit, 2007) facilitate more efficient crop establishment under CA. Punch planters made in Brazil, Tanzania and China have as yet not been widely accepted in E&S Africa, although tests continue. Animal traction equipment, largely from Brazil, has been introduced into many countries and functions well, but price restricts their greater use. Government and international support is needed to stimulate local production in the region and to adapt the equipment to local 'materials, conditions, economic circumstances and skill levels' (Sims *et al.*, 2012) and to ensure local capacity for maintenance, repair and the supply of spare parts. Four-wheel tractors are used on the relatively few large commercial farms in the region, and imported equipment has generally been used for CA experiences. Some service provision to smallholders has also occurred. However, more recently interest has grown, especially in eastern Africa, for two-wheel tractors and prototype equipment for CA using two-wheel tractors is being manufactured in both Kenya and Tanzania.

Once adapted equipment is available, markets for the equipment and other inputs needed for productive agricultural systems, together with functional credit markets, are required. Markets in many smallholder farming areas are weak due to large seasonal variation in demand and supply; stabilizing demand through more productive, lower risk (more stable) production systems will help make input supply and produce purchase enterprises more attractive to small entrepreneurs, but systematic support to fair and competitive market development and maintenance will be required. Although the public sector should take a lead and monitor this process, non-governmental and aid programmes have an opportunity to play a leading role in stimulating development through support to market development for inputs, outputs and credit.

11.6 Prospects for the Widespread Adoption of Conservation Agriculture in Eastern and Southern Africa

The combined effects of advancing soil degradation, climate change and rising prices of inputs, will increase the pressure on farmers, researchers and development agencies in E&S Africa to develop and practise more efficient and sustainable farming systems. To think that this change will be easy would be illusionary, but nevertheless it is inevitable. Although agricultural technology, based on the principles of CA, can address many of the challenges, without changes in infrastructure and markets, both for inputs and produce, widespread adoption of these technologies is unlikely. The Zambian example of political support and the involvement of all of the major stakeholders (policy makers, donors, input suppliers, trainers) in an innovation system to develop and support CA (Baudron *et al.*, 2007) is an example of a functional, albeit imperfect, U-impact pathway. Increased efforts are needed to develop local innovation systems around CA, focusing on the efforts of innovative farmers and the bottlenecks in the value chains surrounding the principal farm enterprises. We have shown in this chapter that CA systems do function acceptably in many situations in E&S Africa, although there is still considerable scope for improvement and local adaptation, necessarily with farmer involvement. However, more effort is now needed on the other components of the value chains to ensure widespread adoption.

11.7 Concluding Remarks – Is the Glass Half-Full or Half-Empty?

There is a wealth of scientific data showing that tillage results in soil and land degradation, reduced SOM and soil structural breakdown, leading to decreased soil biological activity and water infiltration, as well as increased water runoff and soil erosion. Although much of the testing of CA systems in E&S Africa has not followed a process of local adaptation and system development prior to their comparison with ConvT systems that have been adapted by and with farmers over decades, most of the reports show yield benefits to the CA systems, as well as important benefits in increased SOM and water infiltration, and reduced runoff and erosion. Given that rainfall is both the motor to agricultural production in good seasons and the brake in poor seasons, increasing water availability in poor seasons reduces risk – one of the key limitations to input use in smallholder production systems in E&S Africa.

However, there are also reports of failures of the applied CA systems. It is often difficult to understand just how these CA systems were established, and therefore what led to the failure of the system – in other reports the reason is clear, including crop stand establishment, lack of residues and/or lack of plant nutrients. However, it is the reaction to these failures that appears to divide the constituency dedicated to improving farm household livelihoods in E&S Africa. There are some (e.g. Giller *et al.*, 2009; Baudron *et al.*, 2012a) who suggest that CA systems (not just the one that was tested) do not work in that environment or for that group of farmers – the glass is half empty – while others, including ourselves, hold to the premise that this offers an opportunity to find out how better to manage systems based on the principles of CA under those particular conditions – the glass is half full. This is probably a healthy debate, but the alternative of continuing with soil tillage (as suggested by Giller *et al.*, 2009 and Baudron *et al.*, 2012a) is very unhealthy, especially for the farm families of the future who will depend on the ever degrading soils (unless this is stopped) for their livelihoods. Current scientific evidence suggests that CA systems offer the best option we have today for the sustainable production of field crops.

Acknowledgements

P.C. Wall developed this work as an Independent International Consultant under contract with the International Maize and Wheat Improvement Center (CIMMYT), and the research was conducted under the CGIAR Research Program on MAIZE.

References

Aagard, P. (2009) Maize production and conservation farming in Zambia - an update, Lusaka, Zambia. Available at: http://conservationagriculture.org/uploads/pdf/MAIZE-PRODUCTION-AND-CF-IN-ZAMBIA.pdf (accessed 7 January 2013).

Agenbag, G.A. and Maree, P.C.J. (1991) Effect of tillage on some soil properties, plant development and yield of spring wheat (*Triticum aestivum* L.) in stony soil. *Soil and Tillage Research*, 21(1–2), 97–112.

Andersson, J.A. and Giller, K.E. (2012) On heretics and God's blanket salesmen: contested claims for Conservation Agriculture and the politics of its promotion in African smallholder farming. In: Sumberg, J. and Thompson, J. (eds) *Contested Agronomy: Agricultural Research in a Changing World*. Earthscan, London, pp. 1–21.

Angima, S., Stott, D.E., O'Neill, M.K., Ong, C.K. and Weesies, G.A. (2003) Soil erosion prediction using RUSLE for central Kenyan highland conditions. *Agriculture, Ecosystems and Environment* 97(1–3), 295–308.

Apina, T., Wamai, P. and Mwangi, P.K. (2007) Laikipia District. In: Kaumbutho, P. and Kienzle, J. (eds) *Conservation Agriculture as Practised in Kenya: two case studies*. African Conservation Tillage Network, Centre de Coopération Internationale de Recherche Agronomique pour le Développement, Food and Agriculture Organization of the United Nations, pp. 2–56.

Bai, Z.G., Dent, D.L., Olsson, L. and Schaepman, M.E. (2008) *Global Assessment of Land Degradation and Improvement 1. Identification by remote sensing. Report 2008/01*, Wageningen, the Netherlands, 78 pp.

Baudron, F., Mwanza, H.M., Triomphe, B. and Bwalya, M. (2007) *Conservation agriculture in Zambia: a case study of southern province, Nairobi*. African Conservation Tillage Network, Centre de Coopération Internationale de Recherche Agronomique pour le Développement, Food and Agriculture Organization of the United Nations, 57 pp.

Baudron, F., Tittonell, P., Corbeels, M., Letourmy, P. and Giller, K.E. (2012a) Comparative performance of conservation agriculture and current smallholder farming practices in semi-arid Zimbabwe. *Field Crops Research* 132,117–128.

Baudron, F., Andersson, J.A., Corbeels, M. and Giller, K.E. (2012b) Failing to yield? Ploughs, conservation agriculture and the problem of agricultural intensification: an example from the Zambezi Valley, Zimbabwe. *Journal of Development Studies* 38(3), 393–412.

Bennie, A.T.P. and Hensley, M. (2001) Maximizing precipitation utilization in dryland agriculture in south Africa – a review. *Journal of Hydrology* 241(1–2), 124–139.

Berry, W.A.J., Birch, E.B., van Rensburg, J.J., Fowler, R.M. and Findlay, J.B.R. (2001) A case study of conservation and no-tillage technology transfer – KwaZulu Natal, South Africa. In: Garcia-Torres, L., Benites, J. and Martínez-Vilela, A. (eds) *Conservation Agriculture, A Worldwide Challenge. First World Congress on Conservation Agriculture*. Vol. II: *Offered Contributions*. XUL, Cordoba, Spain, pp. 743–748.

Bolliger, A., Damgaard Hansen, K. and Fowler, R. (2005) Constraints limiting smallholder adoption of conservation agriculture: some observations based on three south African smallholder-orientated programmes. In: *Proceedings of the III World Congress on Conservation Agriculture, Nairobi, Kenya*. On CD.

Bot, A. and Benites, J. (2001) *Conservation Agriculture: case studies in Latin America and Africa*. Food and Agriculture Organization of the United Nations, Rome.

Boye, A. and Albrecht, A. (2005) Soil and water conservation by crop rotation with leguminous shrubs – a case study on runoff and soil loss under natural rainfall in western Kenya. In: *Proceedings of the III World Congress on Conservation Agriculture, Nairobi, Kenya*. On CD.

Brady, N.C. and Weil, R.R. (2007) *The Nature and Properties of Soils*, 13th edn. Prentice Hall, New Jersey.

Chaplot, V., Mchunu, C.N., Manson, A., Lorentz, S. and Jewitt, G. (2012) Water erosion-induced CO_2 emissions from tilled and no-tilled soils and sediments. *Agriculture, Ecosystems and Environment* 159, 62–69.

Chivenge, P.P., Murwira, H.K., Giller, K.E., Mapfumo, P. and Six, J. (2007) Long-term impact of reduced tillage and residue management on soil carbon stabilization: Implications for conservation agriculture on contrasting soils. *Soil and Tillage Research* 94(2), 328–337.

Chuma, E. (1993) Effects of tillage on erosion-related properties of a sandy soil in semi-arid Zimbabwe. In: Kronen, M. (ed.) *Proceedings of the Fourth Annual Scientific Conference, SADC Land and Water Management Research Programme*. SACCAR, Gabarone, Botswana, pp. 331–349.

Chuma, E. and Hagmann, J. (1995) Summary of results from on-station and on-farm testing and development of conservation tillage systems in the semi-arid Masvingo, Zimbabwe. In: Twomlow, S. (ed.) *Proceedings of a Technical Workshop on Soil and Water Conservation for Smallholder Farmers in Semi-Arid Zimbabwe*. Belmont Press, Masvingo, Zimbabwe, pp. 41–60.

Conservation Farming Unit (2007) *Conservation Farming and Conservation Agriculture Handbook for Hoe Farmers in Agro-Ecological Regions I and IIa - Flat Culture*. ZNFU, Conservation Farming Unit, Lusaka, Zambia, 60 pp.

Derpsch, R. and Friedrich, T. (2009) Global overview of conservation agriculture adoption. In: *4th World Congress on Conservation Agriculture: Innovations for Improving Efficiency, Equity and Environment*. ICAR, New Delhi, India. Available at: http://www.fao.org/ag/ca (accessed 15 October, 2011).

Derpsch, R., Friedrich, T., Kassam, A. and Li, H. (2010) Current status of adoption of no-till farming in the world and some of its main benefits. *International Journal of Agricultural and Biological Engineering* 3(1), 1–26.

Dixon, J., Hellin, J., Erenstein, O. and Kosina, P. (2007) U-impact pathway for diagnosis and impact assessment of crop improvement. *Journal of Agricultural Science* 145,195–206.

Dixon, J., Gulliver, A. and Gibbon, D. (2001) Farming systems and Poverty. In: Hall, M. (ed.) *Improving Farmers' Livelihoods in a Changing World*. FAO and World Bank, Rome and New York, p. 412.

Du Preez, C.C., Steyn, J.T. and Kotze, E. (2001) Long-term effects of wheat residue management on some fertility indicators of a semi-arid Plinthosol. *Soil and Tillage Research* 63(1–2), 25–33.

Du Preez, C.C., Van Huyssteen, C.W. and Mnkeni, P.N.S. (2011) Land use and soil organic matter in south Africa 2: A review on the influence of arable crop production. *South African Journal of Science* 107(5/6), 2–9.

Ekboir, J. (2002) Developing no-till packages for small-scale farmers. In: Ekboir, J. (ed.) *World Wheat Overview and Outlook*. CIMMYT, Mexico City, Mexico, pp. 1–38.

Elwell, H.A. (1989) *Soil Structure under Conservation Tillage*. Commercial Grain Producers Handbook, Harare, Zimbabwe, pp. 33–39.

Elwell, H.A. and Stocking, M.A. (1988) Loss of soil nutrients by sheet erosion is a major hidden farming cost. *Zimbabwe Science News* 22(7/8), 79–82.

Enfors, E., Barron, J., Makurira, H., Rockström, J. and Tumbo, S. (2011) Yield and soil system changes from conservation tillage in dryland farming: a case study from north eastern Tanzania. *Agricultural Water Management* 98(11),1687–1695.

Erenstein, O. (2002) Crop residue mulching in tropical and semi-tropical countries: an evaluation of residue availability and other technological implications. *Soil and Tillage Research* 67(2), 115–133.

Findlay, J.B.R., Modestus, K.W., Lawrence-Brown, D. and Miheso, V. (2001) The introduction of conservation tillage practices to small-scale farmers in Kenya, Tanzania and Uganda soils. In: Garcia-Torres, L., Benites, J. and Martínez-Vilela, A. (eds) *Conservation Agriculture, A Worldwide Challenge. First World Congress on Conservation Agriculture*. Vol. II: *Offered Contributions*. XUL, Cordoba, Spain, pp. 73–76.

Fowler, R. (1999) Conservation tillage research and development in south Africa. In: Kaumbutho, P.G. and Simalenga, T.E. (eds) *Conservation Tillage with Anilmal Traction. A Resource Book of the Animal Traction Network for Eastern and Southern Africa (ATNESCA)*. Harare, Zimbabwe, pp. 51–60.

Garrity, D.P. (2011) Making Conservation Agriculture ever green. In: *Resilient Food Systems for a Changing World: Proceedings of the 5th World Congress on Conservation Agriculture. 25–29 September 2011, Brisbane, Australia*. Delhi, India, pp. 14–15.

Garrity, D.P., Akinnifesi, F., Ajayi, O., Weldesemayat, S., Mowo, J., Kalinganire, A., Larwanou, M. and Bayala, J. (2010) Evergreen Agriculture: a robust approach to sustainable food security in Africa. *Food Security* 2, 197–214.

Gebreegziabher, T., Nyssen, J., Govaerts, B., Getnet, F., Behailu, M., Haile, M. and Deckers, J. (2009) Contour furrows for in situ soil and water conservation, Tigray, Northern Ethiopia. *Soil and Tillage Research* 103(2), 257–264.

Gill, K.S., Arshad, M.A., Chivundu, B.K., Phiri, B. and Gumbo, M. (1992) Influence of residue mulch, tillage and cultural practiceson weed mass and crop yield from three field experiments. *Soil and Tillage Research* 24, 211–223.

Giller, K.E.,Witter, E., Corbeels, M. and Tittonell, P. (2009) Conservation agriculture and smallholder farming in Africa: The heretics' view. *Field Crops Research* 114(1), 23–34.

Govaerts, B., Sayre, K. and Deckers, J. (2005) Stable high yields with zero tillage and permanent bed planting? *Field Crops Research* 94(1), 33–42.

Govaerts, B., Sayre, K. and Deckers, J. (2006) A minimum data set for soil quality assessment of wheat and maize cropping in the highlands of Mexico. *Soil and Tillage Research* 87(2), 163–174.

Govaerts, B., Verhulst, N., Castellanos-Navarrete, A., Sayre, K.D., Dixon, J. and Dendooven, L. (2009) Conservation Agriculture and soil carbon sequestration: between myth and farmer reality. *Critical Reviews in Plant Sciences* 28(3), 97–122.

Gowing, J.W. and Palmer, M. (2008) Sustainable agricultural development in sub-Saharan Africa: the case for a paradigm shift in land husbandry. *Soil Use and Management* 24(1), 92–99.

Guto, S.N., Pypers, P., Vanlauwe, B., de Ridder, N. and Giller, K.E. (2012) Socio-ecological niches for minimum tillage and crop-residue retention in continuous maize cropping systems in smallholder farms of central kenya. *Agronomy Journal* 104(1), 188–198.

Gwenzi, W., Gotosa, J., Chakanetsa, S. and Mutema, Z. (2008) Effects of tillage systems on soil organic carbon dynamics, structural stability and crop yields in irrigated wheat (*Triticum aestivum* L.)–cotton (*Gossypium hirsutum* L.) rotation in semi-arid Zimbabwe. *Nutrient Cycling in Agroecosystems* 83(3), 211–221.

Haggblade, S. and Tembo, G. (2003) *Conservation Farming in Zambia. EPTD Discussion Paper No. 108.* International Food Policy Research Institute, Washington DC, 128 pp.

Haynes, R.J., Dominy, C.S. and Graham, M.H. (2003) Effect of agricultural land use on soil organic matter status and the composition of earthworm communities in KwaZulu-Natal, South Africa. *Agriculture, Ecosystems and Environment* 95(2–3), 453–464.

Hudson, B. (1994) Soil organic matter and available water capacity. *Journal of Soil and Water Conservation* 49(2), 189–194.

Ito, M., Matsumoto, T. and Quiñones, M. (2007) Conservation tillage in Sub-Saharan Africa: The experience of Sasakawa Global 2000. *Crop Protection* 26, 417–423.

Jat, R.A., Wani, S.P., Singh, P., Pathak, P., Srinivas, K., Kumar, U., Pavani, E. and Velmurgan, R. (2012a) Effect of conservation agriculture on productivity and economics of different cropping systems under rainfed condition in the semi-arid tropics. In: *Proceedings of 3rd International Agronomy Congress*. New Delhi, India. pp. 888–890.

Jat, R.A., Wani, S.P. and Sahrawat, K.L. (2012b) Conservation agriculture in the semi-arid tropics: prospects and problems. In: Sparks D.L. (ed.) *Advances in Agronomy* 117, 191–273.

Kosgei, J.R., Jewitt, G.P.W., Kongo, V.M. and Lorentz, S.A. (2007) The influence of tillage on field scale water fluxes and maize yields in semi-arid environments: A case study of Potshini catchment, South Africa. *Physics and Chemistry of the Earth, Parts A/B/C* 32(15–18), 1117–1126.

Lobell, D.B., Burke, M.B., Tebaldi, C., Mastrandrea, M.D., Falcon, W.P. and Naylor, R.L. (2008) Prioritizing climate change adaptation needs for food security in 2030. *Science* 319, 607–610.

MacRobert, J.F., Winkfield, R.A. and Pilbrough, S. (1995) Conservation tillage on the Agricultural Research Trust Farm. In: Vowles, M. (ed.) *Conservation Tillage: A handbook for commercial farmers in Zimbabwe.* LaserPrint, Harare, Zimbabwe, pp. 101–108.

Malley, Z.J.U., Semoka, J.M.R., Kamasho, J.A. and Kabungo, C.V. (2006) Participatory assessment of soil degradation in the uplands of southwestern Tanzania: Implications for sustainable agriculture and rural livelihoods. *International Journal of Sustainable Development and World Ecology* 13(3), 183–197.

Mariki, W.L. (2004) *The Impact of Conservation Tillage and Cover Crops on Soil Fertility and Crop Production in Karatu and Hanang Districts of Northern Tanzania. TFSC/GTZ Technical Report 1999–2003.* Arusha, Tanzania, 38 pp.

Mashingaidze, N., Madakadze, C., Twomlow, S., Nyamangara, J. and Hove, L. (2012) Crop yield and weed growth under conservation agriculture in semi-arid Zimbabwe. *Soil and Tillage Research* 124, 102–110.

Materechera, S.A. and Mloza-Banda, H.R. (1997) Soil penetration resistance, root growth and yield of maize as influenced by tillage system on ridges in Malawi. *Soil and Tillage Research* 41(1–2), 13–24.

Mazvimavi, K. and Twomlow, S. (2009) Socioeconomic and institutional factors influencing the adoption of conservation farming by vulnerable households in Zimbabwe. *Agricultural Systems* 101, 20–29.

Mchuru, C.N., Lorentz, S., Jewitt, J., Manson, A. and Chaplot, V. (2011) No-till impact on soil and soil organic carbon erosion under crop residue scarcity in Africa. *Soil Science Society of America Journal* 75(4), 1503–1512.

Mills, A.J. and Fey, M.V. (2003) Declining soil quality in South Africa: effects of land use on soil organic matter and surface crusting. *South African Journal of Science* 99(9–10), 429–436.

Morris, M., Morris, M., Kelly, V.A., Kopicki, R.J. and Byerlee, D. (2007) *Fertilizer Use in African Agriculture: Lessons Learned and Good Practice Guidelines*. The World Bank., Washington DC, 146 pp.

Moyo, A. (1998) The effect of soil erosion on soil productivity as influenced by tillage, with special reference to clay and organic matter losses. In: *Proceedings of the 9th ISCO Conference, Bonn, 26–30 August, 1996.*

Moyo, A. and Hagmann, J. (1994) Growth-effective rainfall in maize production under different tillage systems in semi-arid conditions and granitic sands of southern Zimbabwe. In: Jensen, B.E. *et al.* (eds) *Soil Tillage for Crop Production and Protection of the Environment. Proceedings of the 13th International*

Conference of the International Soil Tillage Research Organization (ISTRO). Aalborg, Denmark, pp. 475–480.

Mtambanengwe, F. and Mapfumo, P. (2005) Organic matter management as an underlying cause for soil fertility gradients on smallholder farms in Zimbabwe. *Nutrient Cycling in Agroecosystems* 73(2–3), 227–243.

Muliokela, S.W., Hoogmoed, W.B., Stevens, P. and Dibbits, H. (2001) Constraints and possibilities for conservation farming in Zambia. In: Garcia-Torres, L., Benites, J. and Martínez-Vilela, A. (eds) *Conservation Agriculture, A Worldwide Challenge. First World Congress on Conservation Agriculture*. Vol. II: *Offered Contributions*. XUL, Cordoba, Spain, pp. 61–65.

Munodawafa, A. and Zhou, N. (2008) Improving water utilization in maize production through conservation tillage systems in semi-arid Zimbabwe. *Physics and Chemistry of the Earth, Parts A/B/C* 33(8–13), 757–761.

Munyati, M. (1997) Conservation tillage for sustainable crop production: Results and experiences from on-station and on farm research in Natural Region II (1988-1996). *Zimbabwe Science News* 31(2), 27–33.

Mupangwa, W., Twomlow, S., Walker, S. and Hove, L. (2007) Effect of minimum tillage and mulching on maize (*Zea mays* L.) yield and water content of clayey and sandy soils. *Physics and Chemistry of the Earth* 32(15–18), 1127–1134.

Mupangwa, W., Twomlow, S. and Walker, S. (2012) Reduced tillage, mulching and rotational effects on maize (*Zea mays* L.), cowpea (*Vigna unguiculata* (Walp) L.) and sorghum (*Sorghum bicolor* L. (Moench)) yields under semi-arid conditions. *Field Crops Research* 132, 139–148.

Ngwira, A., Sleutel, S. and Neve, S. (2012a) Soil carbon dynamics as influenced by tillage and crop residue management in loamy sand and sandy loam soils under smallholder farmers' conditions in Malawi. *Nutrient Cycling in Agroecosystems* 92(3), 315–328.

Ngwira, A.R., Aune, J.B. and Mkwinda, S. (2012b) On-farm evaluation of yield and economic benefit of short term maize legume intercropping systems under conservation agriculture in Malawi. *Field Crops Research* 132, 149–157.

Ngwira, A.R., Thierfelder, C. and Lambert, D.M. (2012c) Conservation agriculture systems for Malawian smallholder farmers: long-term effects on crop productivity, profitability and soil quality. *Renewable Agriculture and Food Systems* 1–14.

Nhancale, I.T., Zandamela, C.B. and Massinga, R. (2006) Agricultura de conservaçâo em Moçambique: Um breve historial. In: Zandamela, C.B., Wall, P.C. and Nhancale, I.T. (eds) *Mapeando o caminho para o futuroda agricultura da conservaçâo em Moçambique. Memorias do seminario*. Agrodec, Maputo, pp. 7–11.

Nyagumbo, I. (1998) Experiences with conservation tillage practices in southern and eastern Africa: a regional perspective. In: Benites, J. *et al.* (eds) *Conservation Tillage For Sustainable Agriculture: International Workshop 22–27 June 1998*. GTZ, Eschborn, pp. 73–86.

Nyagumbo, I. (1999) Conservation tillage for sustainable crop production systems: experiences from on-station and on-farm research in Zimbabwe (1988-1997). In: Kaumbutho, P.G. and Simalenga, T.E. (eds) *Conservation Tillage with Animal Traction. A resource book of the Animal Traction Network for Eastern and Southern Africa (ATNESCA)*. Harare, Zimbabwe, pp. 108–115.

Nyagumbo, I. (2002) Effects of three tillage systems on seasonal water budgets and drainage of two Zimbabwean soils under maize. DPhil thesis, Department of Soil Science and Agricultural Engineering, University of Zimbabwe, Harare, 270 pp.

Nyagumbo, I. (2008) A review of experiences and developments towards conservation agriculture and related systems in Zimbabwe. In: Goddard, T. *et al.* (eds) *No-Till Farming Systems*. World Association of Soil and Water Conservation (WASWC), Bangkok, pp. 345–372.

Nyamadzawo, G., Chikowo, R., Nyamugafata, P., Nyamangara, J. and Giller, K.E. (2007) Soil organic carbon dynamics of improved fallow-maize rotation systems under conventional and no-tillage in Central Zimbabwe. *Nutrient Cycling in Agroecosystems* 81(1), 85–93.

Nyssen, J., Govaerts, B., Araya, T., Cornelis, W.M., Bauer, H., Haile, M., Sayre, K. and Deckers, J. (2010) The use of the marasha ard plough for conservation agriculture in Northern Ethiopia. *Agronomy for Sustainable Development* 31(2), 287–297.

Oicha, T., Cornelis, W.M., Verplancke, H., Nyssen, J., Govaerts, B., Behailu, M., Haile, M. and Deckers, J. (2010) Short-term effects of conservation agriculture on Vertisols under tef (*Eragrostis tef* (Zucc.) Trotter) in the northern Ethiopian highlands. *Soil and Tillage Research* 106(2), 294–302.

Oldrieve, B. (n.d.) *Conservation Agriculture and Smallholder Farming in Africa: A practical farmer's view*. 10 pp.

Oldrieve, B. (1993) *Conservation Farming for communal, small scale, resettlement and co-operative farmers of Zimbabwe: A Farm Management Handbook*. Prestige Business Services (Pvt) Ltd, Harare.
Rockström, J., Barron, J. and Fox, P. (2002) Rainwater management for increased productivity among smallholder farmers in drought prone environments. *Physics and Chemistry of the Earth* 27, 949–959.
Rockström, J., Kaumbutho, P., Mwalley, J., Nzabi, A.W., Temesgen, M., Mawenya, L., Barron, J., Mutua, J. and Damgaard-Larsen, S. (2009) Conservation farming strategies in East and Southern Africa: yields and rain water productivity from on-farm action research. *Soil and Tillage Research* 103, 23–32.
Rusinamhodzi, L., Corbeels, M., Wijk, M.T., Rufino, M.C., Nyamangara, J. and Giller, K.E. (2011) A meta-analysis of long-term effects of conservation agriculture on maize grain yield under rain-fed conditions. *Agronomy for Sustainable Development* 31(4), 657–673.
Sims, B.G., Thierfelder, C., Kienzle, J., Friedrich, T. and Kassam, A. (2012) Development of the conservation agriculture equipment industry in sub-Saharan Africa. *Applied Engineering in Agriculture* 28(6), 1–11.
Siziba, S. (2007) Assessing the adoption of conservation agriculture in Zimbabwe's smallholder sector. PhD thesis, Institute of Agricultural Economics and Social Sciences in the Tropics and Sub-tropics, Faculty of Agriculture, University of Hohenheim, 175 pp.
Skóra Neto, F. (1993) Controle de plantas daninhas através de coberturas verdes consorciados com milho. *Pesquisa Agropecuaria Brasileira* 28(10), 1165–1171.
Soil Science Glossary Terms Committee (2008) *Glossay of Soil Science Terms*. Soil Science Society of America, Madison, Wisconsin, 92 pp.
Steiner, K. (2002) *The economics of conservation tillage*. African Conservation Tillage Network, Information Series No. 2, GTZ/ACT, Harare, 4 pp.
Stoorvogel, J.J., Smaling, E.M.A. and Jansen, B.H. (1993) Calculating soil nutrient balances in Africa at different scales. 1. Supra-national scale. *Fertilizer Research* 35(3), 227–235.
Thierfelder, C. and Wall, P.C. (2009) Effects of conservation agriculture techniques on infiltration and soil water content in Zambia and Zimbabwe. *Soil and Tillage Research* 105, 217–227.
Thierfelder, C. and Wall, P.C. (2010a) Investigating Conservation Agriculture (CA) Systems in Zambia and Zimbabwe to mitigate future effects of climate change. *Journal of Crop Improvement* 24(2), 113–121.
Thierfelder, C. and Wall, P.C. (2010b) Rotation in Conservation Agriculture systems of Zambia: effects on soil quality and water relations. *Experimental Agriculture* 46(3), 309–325.
Thierfelder, C. and Wall, P.C. (2011) Reducing the risk of crop failure for smallholder farmers in Africa through the adoption of Conservation Agriculture. In: Bationo, A. *et al.* (eds) *Innovations as Key to the Green Revolution in Africa*. Springer, the Netherlands, pp. 1269–1277.
Thierfelder, C. and Wall, P.C. (2012) Effects of conservation agriculture on soil quality and productivity in contrasting agro-ecological environments of Zimbabwe. *Soil Use and Management* 28, 209–220.
Thierfelder, C., Amezquita, E. and Stahr, K. (2005) Effects of intensifying organic manuring and tillage practices on penetration resistance and infiltration rate. *Soil and Tillage Research* 82, 211–226.
Thierfelder, C., Cheesman, S. and Rusinamhodzi, L. (2012a) A comparative analysis of conservation agriculture systems: Benefits and challenges of rotations and intercropping in Zimbabwe FCR. *Field Crops Research* 137, 237–250.
Thierfelder, C., Cheesman, S. and Rusinamhodzi, L. (2012b) Benefits and challenges of crop rotations in maize-based conservation agriculture (CA) cropping systems of southern Africa. *International Journal of Agricultural Sustainability* (July), 1–17.
Thierfelder, C., Mwila, M. and Rusinamhodzi, L. (2013) Conservation agriculture in eastern and southern provinces of Zambia: Long-term effects on soil quality and maize productivity. *Soil and Tillage Research* 126, 246–258.
Twomlow, S., Hove, L., Mupangwa, W., Masikati, P. and Mashingaidze, N. (2009) Precision Conservation Agriculture for vulnerable farmers in low-potential zones. In: Humphreys, L. (ed.) *Proceedings of the Workshop on Increasing the Productivity and Sustainability of Rainfed Cropping Systems of Poor, Smallholder Farmers, Tamale, Ghana, 22–25 September 2008*. Colombo, Sri Lanka, pp. 37–54.
Umar, B.B., Aune, J.B., Johnsen, F.H. and Lungu, I.O. (2012) Are smallholder Zambian farmers economists? A dual-analysis of farmers' expenditure in Conservation and Conventional Agriculture Systems. *Journal of Sustainable Agriculture* 36(8), 908–929.
Vogel, H. (1992) Effects of conservation tillage on sheet erosion from sandy soils at two experimental sites in Zimbabwe. *Applied Geography* 12, 229–242.
Vogel, H. (1994) Weeds in single-crop conservation farming in Zimbabwe. *Soil and Tillage Research* 31, 169–185.

Vogel, H., Nyagumbo, I. and Olsen, K. (1994) Effect of tied-ridging and mulch ripping on water conservation in maize production on sandveld soils. *Der Tropenlandwirt* 95, 33–44.

Wall, P.C. (1999) Experiences with crop residue cover and direct seeding in the Bolivian highlands. *Mountain Research and Development* 19(4), 313–317.

Wall, P.C. (2007) Tailoring Conservation Agriculture to the needs of small farmers in developing countries: an analysis of issues. *Journal of Crop Improvement* 19, 137–155.

Wall, P.C. (2009) Strategies to overcome the competition for crop residues in southern Africa: some light at the end of the tunnel. In: *Innovations for Improving Efficiency, Equity and Environment. 4th World Congress on Conservation Agriculture. Lead Papers.* New Dehli, pp. 65–70.

Wall, P.C., Ekboir, J.M. and Hobbs, P.R. (2002) Institutional aspects of Conservation Agriculture. Paper presented at the *International Workshop on Conservation Agriculture for Sustainable Wheat Production in Rotation with Cotton in Limited Water Resource Areas, Tashkent, Uzbekistan*, 13–18 October 2002.

Zingore, S., Manyame, C., Nyamugafata, P. and Giller, K.E. (2005) Long-term changes in organic matter of woodland soils cleared for arable cropping in Zimbabwe. *European Journal of Soil Science* 56, 727–736.

12 Conservation Agriculture in North Africa

Hakim Boulal,[1] Mohammed El Mourid,[2] Habib Ketata[2] and Ali Nefzaoui[2]
[1]*International Center for Agricultural Research in the Dry Areas (ICARDA), Rabat, Morocco;* [2]*International Center for Agricultural Research in the Dry Areas (ICARDA), Tunis, Tunisia*

12.1 Introduction

North Africa's Mediterranean climate is characterized with hot and dry summers and wet winters. Rains generally fall during the period of October–April, but both rainfall amount and distribution vary highly from year to year. Agriculture productivity and sustainability, therefore, are highly dependent on weather and soil conditions in the region. A large proportion of total land is located in the arid and semi-arid zones. Cereals, primarily wheat and barley, represent the main cultivated crops, dominantly grown under rainfed conditions. In the past 20 years, drought has become more frequent, resulting in increased yield variability and declining cereal production in the region. Agricultural scientists in the region have increased efforts to counter these production-inhibiting factors through research in the different countries and in collaboration with international institutions involved in agricultural research and development. Encouraging results have been achieved in the 1970s and 1980s, including development of high-yielding cultivars and improved agricultural technologies, e.g. crop rotations and management of soil fertility, diseases, pests and weeds.

Despite the good performance of these technologies both on research stations and in farmers' fields, adoption is still hindered by the lack of resources among smallholder farmers who represent the dominant category of farmers in North Africa. Conservation Agriculture (CA), referred to under different labelling (e.g. no-tillage or NT, zero tillage or ZT, direct seeding) is an agricultural technology that combines minimum or no soil disturbance, direct seed-drilling, cover crops or residue retention, and crop diversification through crop rotations (Kassam *et al.*, 2010; Nefzaoui *et al.*, 2011). Conservation Agriculture should appeal to smallholder farmers because of lower cost of crop production. It also reduces the advent of high soil erosion and low soil organic matter (SOM) stemming from excessive soil cultivation (Roose, 1991; Ibrahimi *et al.*, 2005; Jat *et al.*, 2012). Conservation Agriculture has emerged as an alternative system to circumvent the disadvantages of conventional tillage (ConvT) (Kassam and Friedrich, 2011; Nefzaoui *et al.*, 2011).

Although CA in different forms has been tested since the 1960s in various regions, its introduction and assessment in North Africa is more recent. Early CA work in North Africa started in the mid-1980s in Morocco, through research and training programmes of young Moroccan scientists funded by USAID (Ryan *et al.*, 2007). Moroccan scientists assessed NT versus ConvT on-station in semi-arid areas (Bouzza, 1990; Kacemi, 1992; Mrabet, 1997). This was followed by the local development of a seed-drill prototype at INRA-Morocco, which encouraged the evaluation of NT on a larger scale at the farm level (El Gharras et al., 2009a). INRA-Morocco scientists teamed up with partners of a French NGO (FERT) to disseminate NT in a wheat-based system of central Morocco (Vadon et al., 2006), involving around 100 farmers and covering an area of 900–1200 ha (El Gharras et al., 2009b). Moroccan scientists also participated in an ICARDA (International Center for Agricultural Research in the Dry Areas) regional programme on integrated natural resource management to assess NT in wheat-based systems of semi-arid areas in Morocco. AAAID (Arab Authority for Agricultural Investment and Development) funded another CA programme involving the testing in Morocco of a Brazilian made seed drill (El Gharras et al., 2009b). Although 80 farmers participated in this programme with more than 1200 ha planted in one season, difficulties related to the costs of the drill and weed control put an end to this programme.

In Tunisia, CA work started in the 1990s with the assessment of 'direct sowing' in collaboration with CIRAD (International Center of Agricultural Research for Development) with funding by AFD (French Agency for Development) (Baccourri, 2008; Raunet et al., 2003). This CA programme implemented in a wheat-based system evolved through two phases: the first (1999–2005) phase involved the participation of two national institutions, namely the Academic Agricultural Education School at Kef (ESAK) and the Technical Cereal Center (CTC), which conducted on-farm experiments in both sub-humid and semi-arid regions of the country. A Tunisian private company imported Brazilian NT seed drills that were made available at moderately high prices to well-informed, large-farm owners. APAD, a local NGO, joined the two initiating institutions (ESAK and CTC – now transformed into a larger institution 'INGC') to further promote CA among farmers. The total area managed under NT reached 6000 ha in 2008 and about 12,000 ha in 2011 (Baccouri, 2008; Nefzaoui et al., 2011).

In Algeria, CA work started in the mid-2000s in wheat-based cropping systems following the Tunisian example through the importation of 16 NT seed drills (Kheyar et al., 2007; Nefzaoui et al., 2011; Zaghouane and Makhlouf, 2011). By 2011, the cultivated area under CA reached 5559 ha with durum wheat and chickpea as main crops (Zaghouane and Makhlouf, 2011).

Although CA is not widely adopted by farmers yet, the results achieved so far point to its potential success in the North African region. In fact, NT circumvents late sowing often imposed in ConvT system by heavy rains at sowing time, and reduces crop production costs, and therefore can be successfully disseminated among farmers, especially if promoting policies are introduced by decision makers in the region.

The current chapter addresses the status of CA in North Africa, and provides a review of preliminary results and the prospects of CA in the region.

12.2 Research Results Reported in North Africa

12.2.1 Soil quality

Due to the high erosion rates in North African soils, improving or maintaining soil quality in the region is a high priority for scientists and agriculturalists. Research conducted in North Africa in general showed that NT improves soil quality relative to ConvT (Ben Moussa-Mechraoui et al., 2010; Moussadek et al., 2011a). The retention of crop residues as permanent soil cover under NT protects soil and improves SOM near the soil surface (Bessam and Mrabet, 2003; Abdellaoui et al., 2010; Angar et al., 2010; Bouzrara et al., 2010; Moussadek et al., 2011a; Jat et al., 2012). In wheat-based systems of North Africa, it is difficult to maintain a full soil cover following crop harvest, as straw is generally removed upon harvest and stubbles are grazed through the summer, when other feed resources are scarce. However, Ibno-Namr and Mrabet (2004) observed that 50% residue cover is sufficient to increase significantly soil organic carbon (SOC) content near the soil surface as compared to ConvT. In fact, a 4-year NT with 50% residue retention and soil incorporation resulted in a SOC value of 1.8% versus 1.4% under ConvT (Ibno- Namr and Mrabet, 2004).

The increase in SOM in the soil surface improves aggregate stability (Saber and Mrabet, 2002; Bouajila and Gallali, 2008; Moussadek *et al.*, 2011a) and decreases soil compaction (Abdellaoui *et al.*, 2010). Results from a Calcixeroll soil in semi-arid Morocco showed that aggregate stability increases with depth as residue cover increases under NT (Lahlou and Mrabet, 2001). Jemai *et al.* (2012) reported significant improvement of SOC and structural soil properties as a result of 3-year and 7-year NT in a semi-arid region of Tunisia.

However, maintaining NT over a long period increases the risk of soil compaction in certain soil types. In particular, soil compaction becomes a major constraint in clay soils as plots are grazed in moist conditions. In clay soils of Morocco, Kacemi *et al.* (1992) reported that NT tends to increase soil compaction in the upper 40 cm as compared to tilling using disc- and mouldboard plough. However, leaving residues on the soil surface tends to improve soil structural properties near the soil surface and decreases bulk densities. Compared to ConvT, significant decreases of bulk density under NT were reported in clay loam soils by Abdellaoui *et al.* (2010) and Jemai *et al.* (2012) at 0–8 cm and 0–10 cm depth, respectively.

No-till-induced improvement of soil chemical properties in the upper soil layer were reported by Ibno-Namr and Mrabet (2004). Significant changes in soil pH were also observed following the introduction of NT. A slight acidification near the soil surface was observed in wheat-based systems after 11 years of NT management as compared to ConvT (Mrabet *et al.*, 2001). However, an increase in the residue cover under NT did not affect soil pH, with similar pH values recorded for residue cover rates of 0%, 50% and 100% in 4-year testing of NT in a semi-arid region of Morocco (Ibno-Namr and Mrabet, 2004).

In addition to physical and chemical soil properties, soil biological activity is an important component of soil quality. In fact, changes in physical and chemical properties often affect soil biological activity. Errouissi *et al.* (2011) showed that NT increased soil invertebrate fauna drastically relative to ConvT (Fig. 12.1).

12.2.2 Carbon sequestration

In general, SOC in cultivated lands of North African countries is low (Brahim *et al.*, 2011). On an average basis, less than 1 g SOC 100 g^{-1} was observed in cultivated land of Tunisia

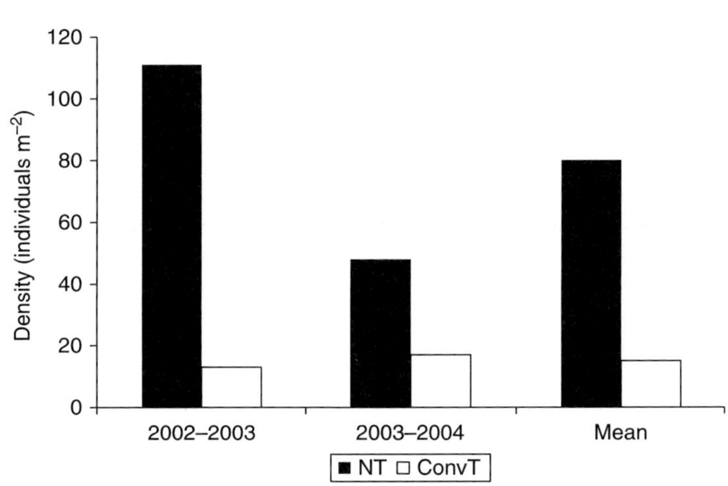

Fig. 12.1. Effect of tillage management (NT: no-tillage; ConvT: conventional tillage) on density of soil invertebrates in northern Tunisia. For each cropping season, data are the average of two sites (Mahassen and Krib) (Adapted from Errouissi *et al.*, 2011).

(Bouajila and Gallali, 2008; Gallali *et al.*, 2010). Several results confirm that switching from ConvT to NT increases SOC (Mrabet *et al.*, 2001; Annabi *et al.*, 2011; Moussadek, 2012). Following 11 years of NT, Mrabet *et al.* (2001) observed 13.6% increase of SOC in the 0–20 cm soil layer, corresponding to a sequestration of 0.5 t C ha^{-1} $year^{-1}$. When the soil surface is permanently covered with crop residues, Brahim *et al.* (2009) found that NT can store up to 0.9 t C ha^{-1} $year^{-1}$. No-till caused a stratification of SOC (Fig. 12.2), with highest SOC concentrations found in the upper soil surface (Bessam and Mrabet, 2001; Ibno-Namr and Mrabet, 2004; Brahim *et al.*, 2009; Bouzrara *et al.*, 2010). Conservation Agriculture research in Morocco, Algeria and Tunisia therefore confirms an increased SOC storage under NT versus ConvT throughout North Africa; however, the magnitude of SOC sequestration is affected by site, rotation, soil type and timespan of NT application. In sub-humid regions of Tunisia, Ben Moussa-Mechraoui *et al.* (2010) observed a slight but non-significant increase in SOC after 4 years of NT management as compared to ConvT. In the same region, the average SOC storage in the upper 10 cm layer after 7 years of NT in a wheat–maize rotation was found to be 18 t C ha^{-1} as opposed to 12 t C ha^{-1} in ConvT (Brahim *et al.*, 2009). In the semi-arid areas of Morocco, Bessam and Mrabet (2003) reported an increase in SOC storage in the 0–2.5 cm layer of 5.62 t ha^{-1} and 7.21 t ha^{-1} after 4 years and 7 years of NT, respectively. According to Moroccan research results, it seems that the highest C sequestration is obtained on heavy clay soil (Mrabet, 2010). Comparing SOC storage in three soil types of central Morocco (Vertisol, Cambisol and Luvisol), Moussadek (2012) reported that 5 years of NT resulted in a SOC increase of 10% in Vertisol, 8% in Cambisol and 2% in Luvisol, corresponding to SOC storage in the upper 30 cm soil layer of 31.89 t ha^{-1} in Vertisol, 30.76 t ha^{-1} in Cambisol and 25.41 t ha^{-1} in Luvisol.

Cropping sequence plays an important role in SOC sequestration under NT (Mrabet *et al.*, 2001). In semi-arid areas, higher SOC storage was observed in wheat–forage (vetch/oat mixture)–fallow rotation due to greater production of residues as compared to wheat–fallow rotation (Mrabet *et al.*, 2001). According to Mrabet *et al.* (2003), the increase in SOC content is proportional to the intensity of residue cover. On a long-term basis, retaining a high residue cover therefore increases SOC.

12.2.3 Crop yield

While rainfall amount and distribution are the main determining factors of wheat productivity in rainfed wheat-based systems of North Africa, yield level can be improved through proper crop management and the judicious use of suitable inputs. The introduction of NT into North African cropping systems showed that yields can be improved without tillage. The NT system outperforms ConvT over longer periods of adoption, with rates varying with climate conditions. In semi-arid and sub-humid regions, grain yield of wheat has gradually improved following the introduction of the NT system (Angar *et al.*, 2010). Results of 5-year testing in a sub-humid region of Algeria showed that grain yield of durum wheat was significantly higher in the NT versus the ConvT starting the third year of testing (Abdellaoui *et al.*, 2010). Other authors reported equal or higher grain yield of wheat in NT versus ConvT in semi-arid areas of North Africa (Angar *et al.*, 2010; Chennafi *et al.*, 2011; Mrabet, 2011). Table 12.1 shows short- and long-term effects of NT on grain yield in North Africa. In most cases, better yield in the NT system is observed after a longer period of adoption. In contrast, the first year following the introduction of the NT system, yields are generally reduced in comparison with ConvT.

Comparing various tillage treatments over four seasons in a semi-arid region of Morocco, Mrabet (2000) observed that grain yields of wheat under NT were equal to those obtained using a chisel plough or deep tillage, and superior to yields obtained under rotovating, conventional off-set discing, stubble mulching or sub-surface traditional tillage. Significant NT beneficial effect on grain yield of durum wheat was observed in

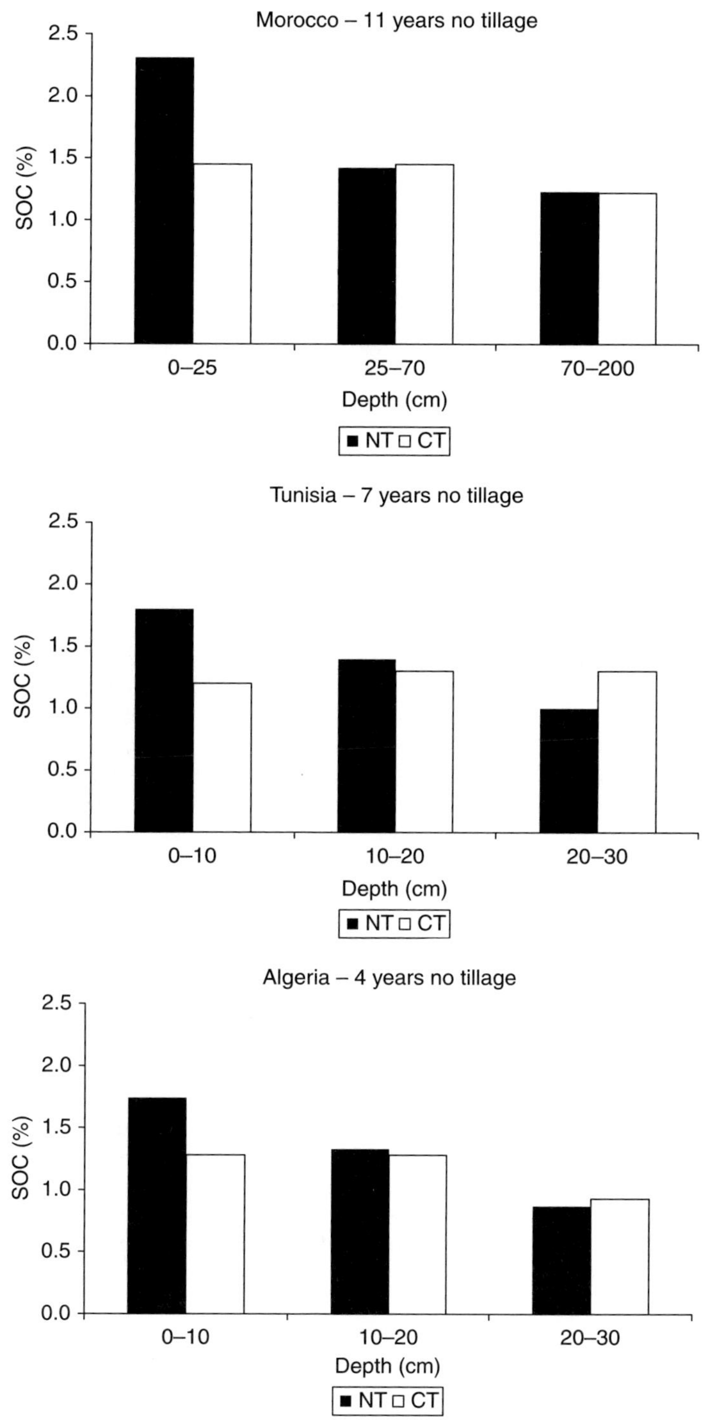

Fig. 12.2. Soil organic carbon (SOC) content at different soil depths under no tillage (NT) and conventional tillage (CT) in wheat-based systems of Morocco, Tunisia and Algeria (Adapted from Saber and Mrabet (2002); Jemai *et al.* (2012) and Bouzrara *et al.* (2010)).

Table 12.1. Grain yield of annual crops under no-tillage (NT) and conventional tillage (ConvT) in North African countries.

Country	Climate class	Years of NT	Crop	NT (t ha^{-1})	ConvT (t ha^{-1})	References
Algeria	Sub-humid	1	Durum wheat	2.84	3.85	Abdellaoui *et al.* (2006)
	Sub-humid	1	Durum wheat	1.01	1.35	Abdellaoui *et al.* (2010)
	Sub-humid	3	Durum wheat	3.85	2.89	Abdellaoui *et al.* (2010)
	Semi-arid		Durum wheat	2.00	1.80	Benniou (2012)
	Semi-arid	2	Barley	1.47	1.12	Bouguendouz (2010)
	Semi-arid	1	Durum wheat	2.24	2.32	Chennafi *et al.* (2011)
	Semi-arid	2	Durum wheat	3.02	2.54	Chennafi *et al.* (2011)
Tunisia	Sub-humid	2	Durum wheat	1.90	2.20	Angar *et al.* (2010)
	Sub-humid	5	Durum wheat	4.75	4.00	Angar *et al.* (2010)
	Sub-humid	9	Durum wheat	4.60	3.00	Angar *et al.* (2010)
	Semi-arid	3	Durum wheat	0.98	0.64	Angar *et al.* (2010)
	Semi-arid	6	Durum wheat	2.80	2.60	Angar *et al.* (2010)
	Semi-arid	10	Durum wheat	3.50	3.30	Angar *et al.* (2010)
Morocco	Semi-arid	3	Bread wheat	1.40	0.40	Benaouda *et al.* (2006)
	Semi-arid	4	Bread wheat	1.13	0.99	Mrabet (2000)
	Semi-arid	8	Bread wheat	3.21	2.85	Mrabet (2011)
	Semi-arid	2	Lentil	1.70	1.50	Moussadek (2012)

a semi-arid region of Algeria with a grain yield of 2.0 t ha^{-1} for NT versus 1.8 t ha^{-1} for ConvT (Benniou, 2012). Furthermore, the grain yield advantage is significantly higher as residue retention is greater. In the Algerian site of Setif, Chennafi *et al.* (2011) reported that grain yield of durum wheat under NT increased from 1.85 t ha^{-1} under bare soil conditions to 2.91 t ha^{-1} under 60% residue retention.

In dry seasons, El Gharras *et al.* (2009b) observed better yield performance for NT compared to ConvT. On-farm results from central Morocco where seasonal rainfall was 200 mm showed that wheat grain yield was four times greater for NT relative to ConvT (Benaouda *et al.*, 2006). In a drier season in the same region, a grain yield of 1 t ha^{-1} was obtained under NT versus total crop failure under ConvT (El Gharras *et al.*, 2009b). In a semi-arid area with less than 300 mm annual rainfall, the contribution of stored water to final grain yield of wheat after fallow is equivalent to the contribution of seasonal rainfall (Bouzza, 1990). Keeping residues on the soil surface without tillage decreases soil evaporation and increases water storage for better crop growth and development (Bouzza, 1990). The adoption of a wheat–fallow rotation instead of continuous wheat in the semi-arid areas increases grain yield of wheat under NT as compared to ConvT (Mrabet, 2011). However, the effectiveness of residue retention on yield depends on climatic conditions. In the semi-arid conditions of Morocco, total retention or partial removal of residues increases wheat grain yield compared to ConvT (Mrabet, 2011). During a dry cropping season in central Morocco, farmers testing NT were able to achieve an average wheat grain yield of 1.3 t ha^{-1} where grain yield under ConvT was negligible (El Brahli and Benazzouz, 2004).

In more humid seasons, as in two consecutive cropping seasons in central Morocco with an average rainfall of 700 and 500 mm, NT had no significant effect on wheat grain yield relative to ConvT (Moussadek, 2012). High levels of weed infestation in the humid seasons depressed grain yield and prevented the favourable effect of NT in those seasons.

12.2.4 Soil and water conservation

Soil erosion is a widespread phenomenon in North Africa where soils generally are prone to degradation (Lahlou, 1996; Ibrahimi *et al.*,

2005; Bai and Dent, 2007). In cultivated land, the most sustainable practice to limit erosion is to decrease the intensity of tillage and to limit grazing. Several research results indicate that NT reduces the negative effects of water erosion on soil. Results from North African countries show that NT with residue retention on the soil surface improves soil and water conservation (Bouzza, 1990; Moussadek *et al.*, 2011a). Experiences in wheat-based systems of Morocco showed that no tillage with 50% soil coverage with crop residues reduces water runoff and soil loss by over 50% as compared to ConvT (Moussadek *et al.*, 2011a). Measurement of soil losses in a sloping cultivated land of Morocco showed that for a rainfall event of 48 mm during 1 month, soil losses were 10 t ha^{-1} under NT and 16 t ha^{-1} under ConvT (Duchemin *et al.*, 2008). Such beneficial effect of NT is attributed to the improvement in water infiltration under this system (Angar *et al.*, 2010), likely due to improved porosity of the topsoil (Lahlou *et al.*, 2005; Jemai *et al.*, 2012). Crop residues also contribute to soil protection through reduced raindrop impact. However, in certain conditions, water infiltration may decrease under NT due to increased soil compaction, particularly in clay soils (Moussadek, 2012).

Improvement in water infiltration under NT system contributes to increased water retention (Bouzza, 1990). In a calcixeroll soil of Morocco, Mrabet (1997) showed that wilting point was reached more quickly under mouldboard ploughing as compared to NT (8 days versus 32 days). Suitable crop rotation can improve soil moisture in the semi-arid areas. For example, a clean fallow can store 84 mm versus 30 mm only in a weedy fallow (Bouzza, 1990).

Further research is needed to better assess soil erosion under NT versus ConvT in semi-arid regions, taking into consideration the intensity of stubble grazing and the ploughing direction in ConvT of sloping lands.

12.2.5 Climate change mitigation and adaptation

While North African countries emit low levels of CO_2 (between 1.5 and 3.5 t CO_2 per inhabitant $year^{-1}$), they are particularly vulnerable to climate change (Agoumi, 2003). Conservation Agriculture has the potential to significantly reduce gas emission by reducing the machine passes in tillage operations and by increasing carbon sequestration. Little information is available on the effect of NT on CO_2 emission in North Africa. A recent study conducted in Morocco showed that NT decreases CO_2 emissions compared to ConvT particularly within 24 h of the tillage operation (Moussadek *et al.*, 2011b). On a longer-term basis, higher carbon sequestration through NT will decrease CO_2 emission. Research conducted in North Africa shows that CA can have a beneficial effect on climate change by increasing SOC storage (Bessam and Mrabet, 2003; Moussadek, 2012). Nefzaoui *et al.* (2011) indicated that CA contributes both to climate change mitigation through reduced greenhouse gas emission (GHG) and enhanced C sequestration and to adaptation through soil water retention and infiltration and increased water use efficiency.

12.2.6 Environmental benefits

Nitrate pollution related to excessive use of fertilizers was observed in certain areas of North Africa (El Ayni *et al.*, 2012). Herbicide transport in runoff can be large under excessive tillage and unprotected soil structures. Water transport of chemicals to rivers and oceans is reduced by the improvement of infiltration associated with residue retention on soil surface, and the reduced water erosion under NT. In Morocco where dams represent important water reservoirs, decreasing runoff intensity through CA helps to reduce the risk of high sediment rate in these reservoirs, and to allow more water for irrigation.

12.2.7 Weed management

Weed management under NT includes agronomic, chemical and mechanical interventions. Residues on the soil surface form a barrier for soil– herbicide contact. In the first years of NT adoption, herbicides have to be applied more frequently than in ConvT (Jat *et al.*, 2012).

In wheat-based systems, the rate of weed infestation in NT plots can reach 3 to 5 times higher than those of ConvT plots. In Algeria, Abdellaoui *et al.* (2006) recorded a population of 979 plants m^{-2} of broadleaf weeds in NT plots against 245 plants m^{-2} in ConvT plots. Early rains in the season increase weed infestation before sowing, thus requiring a pre-sowing herbicide. A post-emergence herbicide application is also recommended to control the emerging weeds after sowing. In food legume crops, the use of a non-selective pre-planting herbicide combined with hand weeding after crop emergence can significantly decrease the weed population (El Brahli and Mrabet, 2001).

12.2.8 Insect-pest and disease dynamics

Introduction of CA changes the dynamics of pests and diseases (Duveiller *et al.*, 2007). Residue cover under NT changes soil microorganisms and pathogens in the surface soil layer. The development of root fungal diseases is common under NT, particularly in poorly drained soil (Acevedo *et al.*, 2009). Research on diseases and insect pests under NT conditions showed an increase of root diseases, including take-all (*Gaeumannomyces graminis* var. *tritici*) in wheat crops, particularly in sub-humid areas (Abdellaoui *et al.*, 2006). However, there is a general lack of research on diseases and insect-pests under NT versus ConvT in North Africa. In fact, most insects and diseases are common to both tillage systems. In semi-arid and sub-humid areas of North Africa, hessian fly (*Mayetiola destructor* (Say)) is the predominant insect in wheat and barley crops (Lhaloui, 2001); the major foliar diseases are septoria leaf blotch (*Septoria tritici*) and leaf rust (*Puccinia recondita*) in wheat (Nsarellah *et al.*, 2000; Jlibene, 2009).

12.2.9 Nutrient use efficiency

Crop response to fertilizers under NT may be quite different from that under ConvT. In particular, the retention of residues on the soil surface affects fertilizer management, including fertilizer placement, and time and rate of application. In fact, most of nutrient availability differences between ConvT and NT occur in the seed placement zone. The non-disturbance of soil under NT leads to higher concentrations of immobile nutrients such as phosphorus (P) and potassium (K) in the upper soil layers. Higher total nitrogen (N), extractable P and exchangeable K were observed in the upper 0–7 cm layer under NT relative to ConvT in soil maintained under NT for 11 consecutive years in Morocco (Mrabet *et al.*, 2001). A similar trend was observed in the upper 0–20 cm following 4 years of NT in northern Tunisia (Ben Moussa-Mechraoui *et al.*, 2010). Ibno-Namr and Mrabet (2004) observed that a period of 4 years of continuous NT is sufficient to increase total N in the upper 0–2.5 cm layer. However, the nutrient differences between NT and ConvT decrease with depth, and become zero or change in favour of ConvT at deeper layers (Mrabet *et al.*, 2001). Moreover, total N under NT tends to be higher near the soil surface due to increased retention of crop residues (Ibno-Namr and Mrabet, 2004). Rotation management under NT also affects nutrient availability. In the semi-arid areas, higher level of total N was observed for NT both in continuous wheat and in wheat–forage–fallow rotation (Mrabet *et al.*, 2001). In favourable regions, the introduction of leguminous species in crop rotations can drastically reduce the need for fertilizer N. In the NT system, the N fertilizer must be placed under the crop residues to make it more available to the growing plants.

12.2.10 Input use efficiency

Farmers generally wish to achieve the maximum effect of production inputs with the minimum cost possible. Reducing tractor passes under NT decreases energy input per area and per unit yield. In fact, energy use efficiency assessed in terms of grain or biomass yield per unit of energy is improved under NT (Jat *et al.*, 2012). Using direct drills of good quality allows to reduce seed rate, without affecting grain yield, as shown in experiments with bread wheat in semi-arid Morocco (El Gharras *et al.*, 2008).

12.2.11 Biodiversity

Conservation Agriculture contributes to diversity of soil fauna and flora. In particular, crop rotation increases the biodiversity of the cropping system and reduces the buildup of diseases and insect pests. In Algeria, Benniou (2012) reported a large change of weed species 3 years after the introduction of NT with a dominance of broadleaf weeds and grasses. In Morocco, El Brahli and Mrabet (2001) observed that NT increases both annual (i.e. brome) as well as perennial weed species.

12.2.12 Economics

Economic benefits of the NT system have been reported in on-farm trials in North Africa (Ben-Salem *et al.*, 2006; Vadon *et al.*, 2006; Nefzaoui *et al.*, 2011). Early work on conservation tillage in central Morocco showed that the introduction of NT system saves time and energy by reducing machine passes for land preparation. Furthermore, improved seedling establishment enables reducing seed rate and decreases seed cost. In a semi-arid region of Morocco with less than 200 mm rainfall, the best grain yield of bread wheat was obtained with the lower seed rate of 60 kg ha^{-1} under NT (El Gharras *et al.*, 2008), although farmers in this region still use seed rate of 180 kg ha^{-1}. Although fungicide and herbicide costs generally are higher for NT compared to ConvT (Ben-Salem *et al.*, 2006), total production costs are reduced under NT. In Morocco, production cost is reduced by about US$60 ha^{-1} through NT as compared to ConvT (Benaouda *et al.*, 2006). Using a Brazil-made NT seed drill, farmers from central Morocco involved in a research-cum-development programme funded by AAAID, saved about 40 l fuel ha^{-1} (El Gharras *et al.*, 2009b). In Tunisia, economic studies undertaken over the period 2001–2004 showed that NT decreases the cost of land preparation before sowing by 26% for durum wheat, 24% for bread wheat and 2% for barley as compared ConvT (Ben-Salem *et al.*, 2006). Marginal economic gains were 183% for bread wheat, 31% for barley and 9% for durum wheat (Ben-Salem *et al.*, 2006). Similar results were reported for barley in Algeria where the introduction of NT decreased the production cost by 28% relative to ConvT (Bouguendouz, 2010).

12.3 Difficulties in the Adoption of Conservation Agriculture in North Africa

12.3.1 Residue management

In North Africa, residues represent a substantial source of revenue for farmers taking into account the fact that most farmers are also livestock producers. While typical CA recommends a minimum of 30% residue retention, crop residues are much valued as animal feed in integrated crop–livestock systems. After the wheat crop is harvested, straw is baled and the remaining stubble is generally grazed throughout the dry season (Fig. 12.3).

In rainfed wheat-based systems, overgrazing and low biomass production reduces the amount of crop residues left on the soil. In the dry season, the price of straw peaks, thus putting a very high pressure on available stubbles that are soon grazed and disappear, leaving a bare soil. Magnan *et al.* (2012) reported that stubble value may reach one-fourth the value of cereal production in a normal rainfall year, and three-quarters that value in a drought year. In such a situation, the NT system loses the beneficial effects of a soil cover with crop residues listed above. Such a competition for crop residues from the livestock sector poses a major challenge to CA development and adoption in North Africa. Nefzaoui *et al.* (2011) proposed a compromise involving partial stubble grazing.

12.3.2 Availability of suitable implements

The non-availability of low-cost direct seed drills is a major constraint to the development and adoption of CA in North Africa. Attempts have been made to solve

Fig. 12.3. Stubble grazing, an old and persisting practice in cereal-based farming systems, poses a great challenge to Conservation Agriculture adoption in North Africa.

the problem through local manufacturing. In Morocco, INRA scientists worked in the early 2000s with a local manufacturer to produce a prototype of a local direct seed-drill with a width of 2.4 m. A total of 24 units were produced and used to plant on-farm demonstrations and motivate farmers to adopt CA (El Gharras and Idrissi, 2006). However, these seed-drills lacked precision and require further adjustment. In Tunisia and Algeria, researchers working with extension institutions and farmers, used well-designed direct seed-drills from Brazil that were acquired by government institutions and also supplied to a limited number of farmers who could afford the relatively high price of these seed-drills. Efforts are now made to produce low-cost and effective direct seed-drills (Fig. 12.4) through collaboration among the three countries and with international research and development organizations.

12.3.3 Mindset of farmers

For centuries farmers have tilled the soil to prepare a clean seedbed and to get rid of weeds. More recently, North African countries encouraged agricultural research and the dissemination of new research-derived technologies. In the 1980s, a large tillage operation was launched by the Moroccan government and disseminated over the country. The decrease and irregularity of rainfall in the past few decades and the adoption of deep tillage increased the risk of erosion and water loss via evaporation. However, switching from deep tillage to minimum or NT has met little enthusiasm among farmers who have a history-long habit of tilling their lands before sowing. To overcome the traditional mindset of farmers, a large operation for the promotion of CA has been initiated among farmers, primarily cereal growers, in Morocco, Tunisia and Algeria. Constraints

Fig. 12.4. Locally made low-cost direct seed drills should help disseminate Conservation Agriculture technology in North Africa (the displayed direct seed drill has been designed and manufactured in Morocco).

including the availability of low-cost direct seed-drills and livestock competition for crop residues delayed the adoption of CA in the region. While it is not easy to quickly change the mindset of farmers, overcoming the other constraints of CA in North Africa seems to be possible in the near future given the increasing area under CA in the three North African countries.

12.3.4 Skill requirements

Since the early 2000s, scientists from North Africa participated in international training programmes on CA organized by CIMMYT, GTZ, ACSAD and ICARDA. Overall, very few scientists are specialized in CA. However, the multidisciplinary approach adopted in CA projects conducted in the region is helping to overcome this constraint. Capacity building focused on farmer training in CA remains an important factor for the promotion and adoption of CA in North Africa. This includes weed management, acquiring and using suitable low-cost seed-drills, and intelligent residue management that takes into account the specificities of North African farmers and their environments.

12.3.5 Yield reduction

No-till effectiveness varies with soil type, climate and age of the system. In the first year, grain yield of wheat is often lower under NT relative to ConvT (Abdellaoui *et al.*, 2010). These results were observed in a sub-humid region of Algeria where durum wheat yield under NT was 25% less than in ConvT (Abdellaoui *et al.*, 2010) and in Tunisia where the yield under NT was 15% lower compared to ConvT (Angar *et al.*, 2010). Such low yield performance of NT during the first years of NT is generally attributed to the high weed density (Abdellaoui *et al.*, 2010). In semi-arid areas, yield reduction in the first years of NT has been reported in a few cases depending on soil type and previous crop. For example, in the JemaaShaim of Morocco, sweep tillage produced higher yield than NT. However, in clay loam soils, grain yield of wheat was higher in NT than in sweep tillage for eight consecutive cropping seasons (Mrabet, 2000). Mekhlouf *et al.* (2011) reported that in the first years, grain yield of bread wheat under NT was lower relative to ConvT if the previous crops were wheat or oat; however, the yield improved when lentil was the previous crop.

12.3.6 Weed infestation

Weed infestation is one of the major constraints to NT adoption. High weed infestation associated with the NT technology requires an increased herbicide application, adding to the production cost incurred by farmers. However, such expenses will gradually decrease over the NT years with the declining weed seed bank in the soil (Abdellaoui *et al.*, 2010). Furthermore, crop rotation can contribute to control weeds effectively in NT systems and reduce the use of expensive herbicides.

12.3.7 Diseases and insect-pests

Continuous cropping and residue retention on the soil surface offer opportunities for wider spread and greater severity of diseases and insect-pests in NT relative to ConvT. However, except for the tendency of increased frequency of root diseases in NT versus ConvT (Abdellaoui *et al.*, 2006), there is little evidence of a major constraint to CA adoption emanating from the difference between the two systems with respect to diseases and insect-pests.

12.3.8 Lack of suitable policies

No clear government policies on CA are in place in the three North African countries, which has delayed the adoption of this technology by farmers. For example, the critical issue of residue retention versus grazing can be resolved if policies are established to promote and make available low-cost or subsidized alternative feed resources such as feed blocks. Similarly, firm and clear decisions on arresting soil degradation and erosion can enhance CA adoption in erosion-prone regions.

12.4 Government Support and Policy Towards Conservation Agriculture

12.4.1 Research support

In North Africa, research and technical institutions led the first efforts to initiate CA research-development programmes. Since the mid-1980s, national research institutions started soil and water conservation programmes. However, the major research support to CA initiatives comes from international funds or via bilateral collaboration. More input should be deployed at the national level through regional programmes to support CA. Relevant research related to pests and weed management, breeding, residue management, rotation, soil and water conservation is indeed required to enhance dissemination and adoption of CA in the region. Collaboration between the three North African countries within a regional programme will help the development of complementary and harmonized research and training projects.

12.4.2 Incentives for mechanization

Even though NT proved highly rewarding at the experimental level, the lack of incentives from governments of the three countries led farmers to continue practising ConvT. Although regular seed-drill machines are subsidized by Moroccan and Tunisian governments, the imported NT drills are still too costly for the smallholder farmers who account for 80% of all farmers in the three countries. Incentives to local manufacturers, including subsidies, will help the development of local low-cost NT seed-drills and boost the adoption of CA in North Africa.

12.4.3 Promotional campaigns

Promotional campaigns in support of CA have been initiated in the early 2000s in the three North African countries. These were primarily led by a limited number of technical persons, including researchers, university professors and students, some extension staff, well-informed farmers and a few private enterprises. Associations, led by elite farmers, work in collaboration with the other stakeholders to disseminate CA practices and advantages. These include the associations APAD in Tunisia, AGENDA in Morocco and Trait-d'Union in Algeria. A few field days are organized by these limited stakeholders to train and demonstrate CA practices to motivated and willing farmers. However, the contribution of governments to these efforts remains very modest, which reflects the low pace at which CA adoption progresses among farmers. Such adoption remains overwhelmingly dependent on external funding and the good will of a limited number of stakeholders.

12.5 Efforts/Policies Required for Scaling-up Conservation Agriculture

12.5.1 Active research

Current CA research in North Africa is primarily conducted by a small number of institutions, involving a limited number of researchers (agronomists, breeders, mechanical engineers, pathologists, weed control scientists, entomologists and students). Research is primarily conducted in the wheat-based systems of both semi-arid and more humid regions. In addition to on-station research experiments, on-farm validation of research results is undertaken in larger demonstration plots, with participation of willing and generally well-trained farmers and a number of extension staff and other stakeholders. Most research programmes focus on the evaluation of NT as a package compared to ConvT. Residue management, rotations, soil quality, runoff and infiltration, and yields under NT make up the core of the current research programmes in North Africa. Assessment of a new NT seed-drill prototype is ongoing in Morocco. However, the research thrust on CA remains dependent on the goodwill of a limited number of stakeholders, and on external support both in terms of technical knowhow and research funding, which explains the ill coverage of important topics including the socio-economics and the livestock dimension, both of which greatly influence the adoption rate of CA in the region.

12.5.2 Identifying suitable cover crops and augmenting residue supply

In the semi-arid areas of North Africa, livestock raising is intimately associated with cereal crop production. While crop residues form a solid pillar of the CA technology, they are considered a necessary feed for livestock during the lean period when natural feed resources are very scarce. Both the straw derived from cereal crop harvest and the stubble remaining in the field are targeted by livestock keepers. It is therefore imperative to find additional feed resources during the dry summer season if some plant material is to be kept on the ground to protect the soil and improve its properties. Some success has been reported in Tunisia by introducing sorghum and oat as forage crops under NT systems (Ben-Hammouda *et al.*, 2009). In Morocco, forage crops such as barley and vetch/oat mixture introduced in 3-year rotations (wheat–forage–fallow) have led to at par or better performance as compared to the wheat–fallow rotation (Mrabet and Bouzza, 1997; Mrabet *et al.*, 2001). Taking advantage of late season rains, to grow summer forages (such as sorghum) on one hand and introducing forage crops within the wheat-based system, on the other, are two complementary options that diversify the feed resources for the integrated cereal–livestock system of North Africa. These and other options, such as partial stubble grazing referred to in Nefzaoui *et al.* (2011), can contribute to alleviating the constraints to CA adoption in North Africa.

12.5.3 Developing and providing suitable machinery

The development of low-cost direct seeders well adapted to the specific conditions of Maghreb countries has been investigated in Morocco. In Algeria and Tunisia, early NT research was based on the use of imported Brazilian direct seed-drills that proved effective, but judged too expensive. The effectiveness of direct seed-drills depends largely on the performance of the coulter, the opener and the press-wheel (El Gharras *et al.*, 2009a). Bahri (1992) recommends a hoe-type direct seed-drill in hard and dry soil conditions. From past results, the direct seed-drill prototype developed in Morocco has a 2–3 m effective working width with 20 cm furrow spacing and includes a coulter to cut residues plus a press wheel to cover and compact soil around seeds (El Gharras *et al.*, 2009a). According to El Gharras and Idrissi (2006), the number of prototypes manufactured in Morocco is not economically sustainable and the cost effectiveness is not yet achieved. Further efforts are still needed involving both researchers as well as local manufacturers to develop suitable low-cost seed-drills adapted to the North African farming conditions. Service providers can play a vital role for smallholder farmers whose small-sized farms prohibit the purchase of personal direct seeders.

12.5.4 Developing effective and integrated weed management techniques

Weed management remains a crucial component of CA. To minimize the use of herbicides, it is important to adopt an integrated method. El Brahli and Mrabet (2001) reported that preplanting herbicide application may not be required if weeds had not emerged. Varieties with rapid early growth can effectively shade out and suppress germinating weeds (Serraj and Siddique, 2012) and should be selected for CA. El Brahli and Mrabet (2001) reported that delaying planting food legume crops (e.g. faba bean, chickpea and lentils) and using a pre-planting herbicide decreases weed infestation. However, weeds still remain a major challenge in NT systems, and require more research and testing to overcome such a challenge and boost CA adoption among North African farmers, especially those with limited resources. Subsidizing herbicides may still be needed, especially for smallholder farmers convinced of CA effectiveness for food security and benefits to natural resources.

12.5.5 Developing effective and integrated insect-pest and disease management techniques

The adoption of an integrated pest management technique helps reduce CO_2 emission (Lal, 2004) and decreases both the use of pesticides and the cost of fuel to run spraying equipment. The introduction of resistant varieties decreases the negative effects of diseases and insects on wheat grain yield (Yahyaoui *et al.*, 2000). As an integral component of CA, crop rotation brings in crop diversity and reduces crop vulnerability to diseases and insects by breaking pest build-up.

12.5.6 Technology dissemination through training, field days and media

Innovative participatory approaches based on on-farm demonstrations and field days, and taking into consideration the specific characteristics of North Africa, proved effective in promoting the dissemination and adoption of CA in this region. Given the experience accumulated in the past decades, both the feedback as well as the original ideas and initiatives of local farmers can help develop specific CA options that are easily embraced by North African farmers. Collaboration and exchange of opinions and experiences among the various stakeholders is essential for successful application and adaptation of CA principles. Extension agents, farmers and researchers need to collaborate for implementing a CA package adapted to local conditions. Taking into account that the direct seed-drill is a critical component of CA, the development of farmer field schools and field days centred on seed-drill machinery is essential for successful training of young farmers on the use of NT seed-drills. Previous experience on NT systems shows that pilot demonstration plots can be effectively used as training sites for neighbouring farmers as well as young extension agents.

12.6 Concluding Remarks

Experiments conducted over the past 25 years in North African countries tend to confirm that CA is a suitable and viable system to improve soil and water resources and to contribute to their sustainable use. Conservation Agriculture has been shown to improve North African soils, to arrest soil erosion and to reduce the cost of crop production. Improvement in grain yield in wheat-based systems is being observed a few years following initiation of CA adoption. However, some challenges still need attention and perseverance. Weed control may impose a prohibitive cost on smallholder farmers, and needs to be addressed using integrated low-cost technologies, including early planting, competitive weed-depressing cultivars and suitable crop rotations. The high cost of successfully imported direct seed-drills make them unavailable to the majority of North African farmers and requires sustained efforts to locally develop effective, low-cost NT seed-drills. The competition between livestock and CA for crop residues calls for a concerted action of both crop producers and livestock owners (often being the same persons) to reach a balanced solution that serves the needs of both CA and livestock. These challenges call for more efforts and collaboration among all stakeholders and, most importantly, the decision makers who need to develop and enact CA-promoting policies.

References

Abdellaoui, Z., Fettih, S. and Zaghouane, O. (2006) Etude comparative de l'effet du semis direct et du labour conventionnel sur le comportement d'une culture de blé dur. *Options Méditerranéennes, Serie A* 69, 115–120.

Abdellaoui, Z., Teskrat, H., Belhadj, A. and Zaghouane, O. (2010) Etude comparative de l'effet du travail conventionnel, semis direct et travail minimum sur le comportement d'une culture de blé dur dans la zone subhumide.*Options Méditerranéennes, Serie A* 96, 71–87.

Acevedo, E., Martínez, E. and Silva, P. (2009) Constraints to zero tillage in Mediterranean environments. In: *Proceedings of 4th World Congress on Conservation Agriculture on Innovations for Improving Efficiency, Equity and Environment*, New Delhi, India, pp. 195–206.

Agoumi, A. (2003) *Vulnerability of north African countries to climatic changes: adaptation and implementation strategies for climate change* (ed. International Institute for Sustainable Development). IISD, Winipeg, Manitoba, Canada, p. 11.

Angar, H., Ben Hadj Salah, H. and Ben-Hammouda, M. (2010) Semis direct et semis conventionnel en Tunisie: les résultats agronomiques de 10 ans decomparaison. *Options Mediterranéennes, Serie A* 96, 53–59.

Annabi, M., Bahri, H., Chibani, R., Angar, H., Bahri, B. and Ben Hadj Salah, H. (2011) Soil carbon storage under no-tillage practice in Northern Tunisia. In: *Proceedings of 5th World Congress on Conservation Agriculture*, Brisbane, Australia, Poster presentation.

Baccouri, S. (2008) Conservation agriculture in Tunisia. Conservation Ag. Carbon Offset Consultation, West Lafayette, USA, October 2008. FAO-CTIC.

Bahri, A. (1992) Evaluation of opener and press wheel combinations on a no-till grain drill when seeding wheat. MSc thesis, University of Nebraska, Lincoln, Nebraska.

Bai, Z.G and Dent, D.L. (2007) Land degradation and improvement in Tunisia. 1. Identification by remote sensing. *Report 2007/08*, ISRIC – World Soil Information, Wageningen, the Netherlands, p. 45.

Benaouda, H., El Gharras, O., Vadon, B. and Farouq, E. (2006) Le semis direct: Un système en extension à la coopérative agricole Khemisset Chaouia. *Options Mediterranéennes, Serie A* 69, 173–175.

Ben-Hammouda, M., M'Hedhbi, K., Cheikh M'hamed, H. and Ghouili, H. (2009) Direct drilling is behind agronomy of opportunity in Tunisia. In: *Proceedings of 4th World Congress on Conservation Agriculture on Innovations for Improving Efficiency, Equity and Environment*, New Delhi, India, pp. 110–116.

Ben Moussa-Mechraoui, S., Errouissi, F., Ben-Hammouda, M. and Nouira, S. (2010) Comparative effects of conventional and no-tillage management on some soil properties under Mediterranean semi-arid conditions in northwestern Tunisia. *Soil and Tillage Research* 106, 247–253.

Benniou, R. (2012) Agriculture conservation role of moisture and soil organic matter semi-arid. *Journal of Materials and Environment Science* 3(1), 91–98.

Ben-Salem, H., Zaibet, L. and Ben-Hammouda, M. (2006) Perspectives de l'adoption du semis direct en Tunisie, une approche économique. *Options Mediterranéennes, Serie A* 69, 69–75.

Bessam, F. and Mrabet, R. (2001) Time influence of no tillage on organic matter and its quality of a vertic Calcixeroll in a semiarid area of Morocco. In: Garcia-Torres *et al.* (eds) *Proceedings of 1st World Congress on Conservation Agriculture*. Madrid, Spain, pp. 281–286.

Bessam, F. and Mrabet, R. (2003) Long-term changes in soil organic matter under conventional and no-tillage systems in semiarid Morocco. *Land Use and Management* 19, 139–143.

Bouajila, A. and Gallali, T. (2008) Soil organic carbon fractions and aggregate stability in carbonated and no carbonated soils in Tunisia. *Journal of Agronomy* 7(2), 127–137.

Bouguendouz, A. (2010) Effet de trois itinéraires techniques sur l'élaboration du rendement de l'orge (*Hordeum vulgare* L.) sous les conditions semi-arides des hauts plateaux Algériens. *Options Méditerranéennes, Serie A* 96, 221–226.

Bouzrara, S., OuldFerroukh, M.E.H. and Bouguendouz, A. (2010) Influence du semis direct et des techniques culturales simplifiées sur les propriétés d'un sol de la ferme piloteSersour (Sétif). *Options Méditerranéennes, Serie A* 96, 123–129.

Bouzza, A. (1990) Water conservation in wheat rotations under several management and tillage systems in semi-arid areas. PhD dissertation, University of Nebraska, Nebraska, 125 pp.

Brahim, N., Gallali, T. and Bernoux, M. (2009) Effect of agronomic practices on the soil carbon storage potential in Northern Tunisia. *Asian Journal of Agricultural Research* 3(3), 55–66.

Brahim, N., Gallali, T. and Bernoux, M. (2011) Carbon stocks by soils and departments in Tunisia. *Journal of Applied Sciences* 11(1), 46–55.

Chennafi, H., Hannachi, A., Touahria, O., Fellahi, Z.E.A., Makhlouf, M. and Bouzerzour, H. (2011) Tillage and residue management effect on durum wheat (*Triticum turgidum* (L.)Thell. spp. *turgidum*conv. *durum* (Desf.) MacKey) growth and yield under semi arid climate. *Advances in Environmental Biology* 5(10), 3231–3240.

Duchemin, M., Benmansour, M., Nouira, A. and Gallichand, J. (2008) Mesure et modélisation de l'érosion hydrique des sols agricoles au Maroc et au Quebec. In: Roose, E., Albergel, J., De Noni, G., Laouina, A. and Sabir, M. (eds) *Efficacité et gestión de l'eau et de la fertilité des sols en milieu semi-aride*. AUF, EAC et IRD editeurs, Paris, p. 425.

Duveiller, E., Singh, R.P. and Nicol, J.M. (2007) The challenges of maintaining wheat productivity: pests, diseases, and potential epidemics. *Euphytica* 157, 417–430.

El Ayni, F., Cherif, E., Manoli, S., Assimacopoulos, D., Jrad, A. and Trabelsi-Ayadi, M. (2012) Impact of agriculture on a Tunisian coastal aquifer and possible approaches for a better water management. *Sixteenth International Water Technology Conference*, IWTC 16 2012, Istanbul, Turkey, pp. 1–14.

El Brahli, A. and Benazzouz, S. (2004) *Rapport d'activité annuel*. Centre Régional de la recherche agronomique de Settat, Morocco.

El Brahli, A. and Mrabet, R. (2001) La jachère chimique. Pour relancer la céréaliculture non irriguée en milieu semi-arid Marocain. *Acte de la journée nationale sur le désherbage*. INRA, Settat, Morocco, pp. 133–145.

El Gharras, O. and Idrissi, M. (2006) Contraintes technologiques au développement du semis direct au Maroc. *Options Mediterranéennes, Serie A* 69, 120–124.

El Gharras, O., El Brahli, A., Benaouda, H. and El Gharous, M. (2008) Le semis direct. Rapport interne de campagne 2007-2008, INRA-AAAID, INRA Settat, Morocco.

El Gharras, O., El Brahli, A. and El Mourid, M. (2009a) No-till system applied to northern Africa rain-fed agriculture: case of Morocco. In: *Proceedings of 4th World Congress on Conservation Agriculture on Innovations for Improving Efficiency, Equity and Environment*. New Delhi, India, pp. 41–50.

El Gharras, O., El Brahli, A., El Aissaoui, A. and El Hantaoui, N. (2009b) Le semis direct pour une agricultura pluviale de conservation. In: *Procedings of Symposium International Agriculture durable en région Méditerranéenne (AGDUMED)*, Rabat, Morocco, pp. 249–256.

Errouissi, F., Ben Moussa-Machraoui, S., Ben Hammouda, M. and Nouira, S. (2011) Soil invertebrates in durum wheat (*Triticum durum* L.) cropping system under Mediterranean semi arid conditions: a comparison between conventional and no-tillage management. *Soil and Tillage Research* 112(2), 122–132.

Gallali, T., Brahim, N. and Bernoux, M. (2010) Soil organic carbon density and storage in Tunisia. In: *Proceedings of 19th World Congress of Soil Science on Soil Solutions for a changing World.* Brisbane, Australia, pp. 24–27.

Ibno-Namr, K. and Mrabet, R. (2004) Influence of agricultural management on chemical quality of a clay soil of semi-arid Morocco. *Journal of African Earth Sciences* 39, 485–489.

Ibrahimi, S., Damnati, B., Radakovitch, O., Hassouni, K. and Simon, B. (2005) Using conversion models to estimate soil erosion and deposition rates, from the ^{137}CS measurements in cultivated soils (north Morocco). *Revista de la Sociedad Geologica de Espana* 18(3–4), 217–224.

Jat, R.A., Wani, S.P. and Sahrawat, K.L. (2012) Conservation agriculture in the semi-arid tropics: prospects and problems. In: Sparks D.L. (ed.) *Advances in Agronomy* 117, 191–273.

Jemai, I., Ben Aissa, N., Ben Guirat, S., Ben-Hammouda, M. and Gallali, T. (2012) On-farm assessment of tillage impact on the vertical distribution of soil organic carbon and structural soil properties in a semi-arid region in Tunisia. *Journal of Environmental Management* 113, 488–494.

Jlibene, M. (2009) *Amélioration génétique du blé tender au Maroc à l'aube du 21ème siècle.* INRA Morocco edt, 80 pp.

Kacemi, M. (1992) Water conservation, crop rotations, and tillage systems in semiarid Morocco. PhD dissertation, Colorado State University, Fort Collins, Colorado, 200 pp.

Kacemi, M., Hilali, H. and Monroe, G. (1992) Effect of different tillage methods on bulk density, penetrability and aggregate size distribution on a clay soil. *Hommes, Terre et Eaux* 22(86), 96–102.

Kassam, A. and Friedrich, T. (2011) Conservation agriculture: global perspectives and developments. In: *Proceedings of Regional Conservation Agriculture Symposium,* Johannesburg, South Africa, pp. 1–33.

Kassam, A., Friedrich, T. and Derpsch R. (2010) Conservation agriculture in the 21st Century: a paradigm of sustainable agriculture. In: *Proceedings of the European Congress on Conservation Agriculture – Towards Agro-Environmental Climate and Energetic Sustainability,* Madrid, Spain, pp. 19–68.

Kheyar, M.O., Amara, M. and Harrad, F. (2007) La Mécanisation de la céréaliculture Algérienne: Constat et perspectives. *Annales de l'Institut National Agronomique - El-Harrach* 28(1–2), 95–102.

Lahlou, A. (1996) Environmental and socio-economic impacts of erosion and sedimentation in north Africa. In: *Proceedings of the Exeter Symposium on Erosion and Sediment Yield: Global and Regional Perspectives.* IAHS Publ. 236, pp. 491–500.

Lahlou, S. and Mrabet, R. (2001) Tillage influence on aggregate stability of a Calcixeroll soil in semiarid Morocco. In: Garcia-Torres, L., Benites, J. and Martinez-Vilela, A. (eds) *Proceedings of 1st World Congress on Conservation Agriculture*, Madrid, Spain, pp. 249–254.

Lahlou, S., Ouadia, M., Malam Issa, O., Le Bissonnais, Y. and Mrabet, R. (2005) Modification de la porosité du sol sous les techniques culturales de conservation en zone semi-aride Marocaine. *Étude et Gestion des Sols* 12(1), 69–76.

Lal, R. (2004) Carbon emission from farm operations. *Environment International* 30, 981–990.

Lhaloui, S. (2001) *System Wide Program on Integrated Pest Management Morocco IPM Pilot Site.* Project report 2000-2001, INRA Morocco-ICARDA IPM project, 55 pp.

Magnan, N., Larson, D.M. and Taylor, J.E. (2012) Stuck on stubble? The non-market value of agricultural by-products for diversified farmers in Morocco. *American Journal of Agricultural Economics* 1–15; doi: 10-1093/ajae/aas057 (http://ajae.oxfordjournals.org).

Mekhlouf, A., Makhlouf, M., Achiri, A., Ait Ouali, A. and Kourougli, S. (2011) Etude comparative de l'effet des systèmes de travail du sol et des précédents culturaux sur le sol et le comportement du blé tendre (*Triticum aestivum* L.) en conditions semi-arides. *Agriculture* 2, 52–65.

Moussadek, R. (2012) Impacts de l'agriculture de conservation sur les propriétés et la productivité des vertisols du Maroc Central. PhD Thesis, University of Ghent, Belgium, 206 pp.

Moussadek, R., Mrabet, R., Zante, P., Lamachère, J.M., Pépin, Y., Le Bissonnais, Y., Ye L., Verdoodt, A. and Van Ranst, E. (2011a) Effets du travail du sol et de la gestion des résidus sur les propriétés du sol et sur l'érosion hydrique d'un Vertisol Méditerranéen. *Canadian Journal of Soil Science* 91, 627–635.

Moussadek, R., Mrabet, R., Dahan, R., Douaik, A., Verdoodt, A., Van Ranst, E. and Corbeels, M. (2011b) Effect of tillage practices on the soil carbon dioxide flux during fall and spring seasons in a Mediterranean Vertisol. *Journal of Soil Science and Environmental Management* 2(11), 362–369.

Mrabet, R. (1997) Crop residue management and tillage systems for water conservation in a semiarid area of Morocco. PhD Thesis, Colorado State University, Fort Collins, Colorado, 220 pp.

Mrabet, R. (2000) Differential response of wheat to tillage management systems in a semi-arid areas of Morocco. *Field Crops Research* 66, 165–174.

Mrabet, R. (2010) Climate change and carbon sequestration in the Mediterranean basin: Contributions of no-tillage systems. *Options Méditerranéennes, Serie A* 96, 2010 – IV *Rencontres Méditerranéennes du Semis Direct*, pp. 165–184.

Mrabet, R. (2011) Effects of residue management and cropping systems on wheat yield stability in a semiarid Mediterranean clay soil. *American Journal of Plant Sciences* 2, 202–216.

Mrabet, R. and Bouzza, A. (1997) *Conservation de l'eau sous différentes rotations céréalières et système de gestion des résidus de récolte en semi-aride*. Rapport annuel INRA, Settat, Morocco, pp. 42–46.

Mrabet, R., Ibno-Namr, K., Bessam, F. and Saber, N. (2001) Soil chemical quality changes and implications for fertilizer management after 11 years of no-tillage wheat production systems in semi-arid Morocco. *Land Degradation and Development* 12, 505–517.

Mrabet, R., El Brahli, A., Anibat, I. and Bessam, F. (2003) No-tillage technology: research review of impacts on soil quality and wheat production in semiarid Morocco. *Options Méditerranéennes* 60, 133–138.

Nefzaoui, A., Ketata, H. and El Mourid, M. (2011) Agricultural Technological and Institutional Innovations for Enhanced Adaptation to Environmental Change in North Africa. In: Young, S.S. and Silvern, S.E. (eds) *International Perspectives on Global Environmental Change*. ISBN 978-953-307-815-1, pp. 57–84.

Nsarellah, N., Nachit, M. and Lhaloui, S. (2000) Breeding durum wheat for biotic stresses in the Mediterranean region. *Options Méditerranéennes, Série A, Séminaires Méditerranéens* 40, 341–347.

Raunet, M., Richard, J.F. and Rojat, D. (2003) *Premiers résultats d'introduction du semis direct sous couvert et lutte anti-érosive en Tunisie*. Bulletin du Reseau Erosion, IRD Ed., France.

Roose, E. (1991) Conservation des sols en zones Méditerranéennes: Synthèse et proposition d'une nouvelle stratégie de lutte antiérosive: la CGES. *Cah. Orstom, Sér. Pédol.*, XXVI(2), 145–181.

Ryan, J., El Mourid, M., Shroyer, J.P. and El Gharous, M. (2007) The dryland agriculture applied research project in Morocco: a perspective 12 years after completion. *Journal of Natural Resources and Life Sciences Education* 36, 120–128.

Saber, N. and Mrabet, R. (2002) Impact of no tillage and crop sequence on selected soil quality attributes of a verticcalcixeroll soil in Morocco. *Agronomie* 22, 451–459.

Serraj, R. and Siddique, K.H.M. (2012) Conservation agriculture in dry areas. *Field Crops Research* 132, 1–6.

Vadon, B., Lamouchi, L., Elmay, S., Maghfour, A., Mahnane, S., Benaouda, H. and El Gharras, O. (2006) Organisations paysannes: Un levier pour développer l'agriculture de conservation au Maghreb. *Options Mediterranéennes, Serie A* 69, 87–99.

Yahyaoui, A., Hakim, S., Al-Naimi, M. and Nachit, M.M. (2000) Multiple disease resistance in durum wheat (*Triticum turgidum* L. var. durum). *Options Méditerranéennes, Série A* 40, 387–392.

Zaghouane, O. and Makhlouf, M. (2011) Promoting no-till practices to stabilize durum wheat yields and sustain agricultural production in semi-arid regions of Algeria. In: *Proceedings of 5th World Congress on Conservation Agriculture (WCCA)*. Brisbane, Australia, Full text of Poster, 4 pp.

13 Conservation Agriculture in West and Central Africa

Patrice Djamen Nana,[1] Patrick Dugué,[2] Saidi Mkomwa,[3] Jules Benoît Da Sansan,[1] Guillaume Essecofy,[4] Harouna Bougoum,[5] Ibrahima Zerbo,[5] Serge Ganou,[6] Nadine Andrieu[2] and Jean-Marie Douzet[7]

[1]ACT, Ouagadougou, Burkina Faso; [2]CIRAD, UMR Innovation, Montpellier, France; [3]ACT, Nairobi, Kenya; [4]CIHEAM/IAM Montpellier, France; [5]Université Polytechnique de Bobo Dioulasso/Institut de Développement Rural (UPB/IDR), Burkina Faso; [6]Université de Ouagadougou, Burkina Faso; [7]CIRAD, UR SCA, Ouagadougou, Burkina Faso

13.1 Introduction

The 2012 edition of the report on the state of food insecurity in the world revealed that 26.8% of the 870 million undernourished people in the world live in sub-Saharan Africa, where hunger rates have been rising by 2% per year since 2007 (FAO *et al.*, 2012). Like in other African sub-regions, yield gaps in West and Central African smallholder farming are among the largest in the world (Tittonell and Giller, 2012). Poor agricultural productivity in this sub-region is attributed to several causes including among others the inherent low soil fertility (Pieri, 1989), the lack of access to agricultural inputs, increased labour demands caused by lack of mechanization, the small size and increasing fragmentation of farms and the lack of capital to invest in building productive soils in harsh environments (Tittonell and Giller, 2012).

Frequent poor yields and market failures of agricultural products often result in social unrest such as dramatic hunger riots recorded in 2008 in several countries. The general tendency is dependency on rainfed agriculture, which is the reason why in 2011 the poor rainfall pattern resulted in severe crop failure and accentuation of food insecurity, leading to malnutrition and starvation especially in the Sahel region. It is estimated that a five-fold increase in agricultural production in West and Central (WCA) Africa by 2050 will be necessary to meet the demand for food (Collomb, 1999). WCA farmers have the challenge and opportunity to produce enough food not only for their own consumption, but also to feed the growing urban population.

The governments of WCA countries, sometimes under the pressure of civil society organizations, are increasingly recognizing that the massive importations of cereal and the use of international aid to address the deficit of the national food production cannot be an efficient and sustainable solution. Many policy makers despite disappointing past experiences still encourage the 'green revolution' models based on tillage mechanization and use of chemical inputs to increase agricultural productivity (Dugué *et al.*, 2012a). This solution has shown its shortcomings as it is not accessible to many smallholders who are rather poor with very weak connection to markets of agricultural inputs and outputs. Moreover, the conventional green revolution approach can lead to the degradation of the environment (Abba *et al.*, 2006), decrease in biodiversity and accentuation of climate change, which is already an issue in WCA (World Bank, 2012) where the rainfall pattern is erratic with a high spatial variability (Akponikpè *et al.*, 2011). According to experts, a decline in annual rainfall has been observed in WCA since the end of the 1960s, with a decrease of 20 to 40% noted in the periods 1931–1960 and 1968–1990, resulting in negative consequences for crop and livestock production (IPCC, 2007). This situation raises serious concerns as WCA agriculture is mostly rainfed. Hence, there is a need to find adequate means that can allow the sustainable increase of agricultural productivity under a changing climate and limited access to external farm inputs. Improving the resilience of WCA agriculture to climate change is necessary more so since agriculture is a key economic sector of countries of this sub-region. It is well-thought-out that changes in existing farming practices from rainfed and green revolution models to a more sustainable intensification of agricultural production model (FAO, 2011) will lead to better food availability, while preserving and enhancing natural resources (including mitigating climate change) and reducing poverty.

Several years ago, agricultural development stakeholders including farmers, ministries, NGOs, research centres, community-based and farmers' organizations began testing and/or implementing a number of indigenous and innovative sustainable land management (SLM) techniques aimed at mitigating land degradation and rehabilitating degraded soils. A diversity of techniques including non-exhaustive zaï (planting pits), semi-circular hoops (half-moons), stone ridges, farmer managed natural regeneration of trees, utilization of animal and green manure have been promoted, which have generated some benefits to farmers. Nevertheless, most of these technologies showed some shortcomings regarding their sustainability either because they are labour intensive (e.g. maintenance of zaï pits) or because of the scarcity of the resources needed for their establishment (e.g. stone used in the construction of stone bunds).

Conservation Agriculture (CA) is increasingly envisaged by WCA agricultural development stakeholders as a more relevant, comprehensive and effective approach to the restoration of natural resources that integrates environmental, technical, economic and social dimensions, hence enabling farmers to face efficiently current challenges and thus ensuring the sustainability of their activity (Djamen *et al.*, 2005; Tittonell *et al.*, 2012). Conservation Agriculture is also considered as an approach to managing agroecosystems for improved and sustained productivity, increased profits and food security, while preserving and enhancing the resource base and the environment (Hobbs *et al.*, 2007; Kassam *et al.*, 2009a). It is defined as a generic concept that refers to a family of cropping systems in which three fundamental principles are implemented simultaneously at farm level based on locally formulated adaptive practices (Dumanski *et al.*, 2006; FAO, 2008): no or minimum mechanical soil disturbance, maintenance of permanent soil cover, and diversification of crop rotations and associations including with legumes.

Conservation Agriculture principles are considered as appropriate to various types of farms and environments (Friedrich *et al.*, 2012), but there is scepticism from some researchers about its applicability in smallholder farming in Africa (Giller *et al.*, 2009; Serpantié, 2009; Anderson and Giller, 2012). Criticisms are based on the poor access of farmers to farm inputs (e.g. herbicides often widely used in direct seeding), the lack of empirical evidence and consistency on the effects of CA on improvement of soil fertility and yield, the reduction of labour requirements and the difficulty to achieve adequate soil cover because of livestock pressure on crop residues. Moreover, Naudin (2012) highlights that African smallholders are facing the biggest problems with permanent soil cover and crop diversification when they want to use CA. The awareness and adoption of CA in the African continent is rather considered to be on the increase (Derpsch and Friedrich, 2010; Friedrich *et al.*, 2012), but the contribution of Africa in the total area under CA in the world is still very low (1%; about 1,012,840 ha).

At the continental level, WCA is the sub-region with the least contribution. Ghana, with a total estimated area of 30,000 ha under CA, is the only WCA country in the list of African countries where the adoption of CA is recorded (Friedrich *et al.*, 2012).

This situation is paradoxical as WCA is considered as part of the areas where the potential benefits of CA are believed to be the highest (Lal, 2007). Hence the present negligible adoption of CA raises the question whether CA is really relevant and accessible for smallholder farming in WCA. The weak contribution of WCA and more generally of Africa to global figures on CA might be seen as the consequence of major difficulties CA adoption is facing, leading farmers to prefer other soil fertility management techniques. However, it is also important to take into account factors like the newness of CA in WCA, which implies insufficient reliable evidence, inadequate awareness and mastery by stakeholders concerned or the lack of appropriate support at both technical and policy levels for an effective up-scaling of CA. Initial research operations and recommendations on the usefulness of CA from some researchers for farming in WCA were made two decades ago (Boli, 1996; Lal, 1989), but it is only recently that meaningful efforts are being made to explore further the potentials of CA and engage progressively in its effective and appropriate dissemination, taking into account the diversified biophysical and socio-economic conditions (Kassam *et al.*, 2009b; ACT *et al.*, 2012; Lahmar *et al.*, 2012; Tittonell *et al.*, 2012). Moreover, while there are several examples of past and ongoing CA development or research projects in WCA, the updated status of CA, synthesis and assessment of existing knowledge that could really allow the appraisal of the interest of CA for WCA farmers is lacking.

A first inventory of CA in WCA (Djamen *et al.*, 2005) was made in 2005 within the framework of the Third World Congress on Conservation Agriculture held in Nairobi (Kenya) and hosted by African Conservation Tillage Network (ACT). This inventory showed that CA was still new in WCA. Since 2005, new CA initiatives have emerged while some of the existing ones have been renewed or consolidated (Tittonell *et al.*, 2012). These initiatives have been going on in several countries with little collaboration among them. The current situation of CA in WCA is not well known. This gap is likely to hamper the elaboration and implementation of future CA research agenda and dissemination in WCA, but also the decision-making process of stakeholders who might be willing to mobilize CA to meet challenges of the agricultural sector in the sub-region. Moreover, an updated status of CA in WCA could serve as a basis to enhance synergies and complementarities between current and future initiatives both at national and sub-regional levels, to understand better determinants of the current situation and to envisage prospects accordingly. This chapter aims to contribute to fill the CA knowledge gap through the search of a comprehensive answer to the following questions: is CA a promising alternative to ensure the sustainability of farms in WCA? What are the evidence, knowledge gaps and prospects for CA in WCA? The methodology used is based mainly on literature review, lessons and preliminary results of several past and ongoing CA operations in which the authors have been involved in WCA. The chapter is organized into four main sections: history and overview of CA in WCA; available knowledge on CA in WCA; adoption of CA in WCA; conclusion and prospects. The review mostly focuses on the situation where CA actually has been implemented.

13.2 History and Overview of Conservation Agriculture in West and Central Africa

13.2.1 History and current status

In many areas of WCA, there are farmers implementing farming practices more or less based on indigenous knowledge, which are already in line or convergent with CA principles. However, these existing CA-favourable practices are not always done with the objective of SLM. They are rather opportunistic practices or caused by socio-economic constraints. For instance, in the semi-arid area of Burkina Faso, it was noted that direct seeding, which is a common practice, is not only dictated by the erratic rainfall but also by the lack of tillage equipment (Essecofy, 2011). Crop association is also widespread and reflects both the increasing pressure on land and also a strategy to diversify production systems and to manage climatic risks. In northern Cameroon, for example, the rapid expansion of direct seeding, favoured by the better access to herbicides (Dugué and Guyotte, 1996), is related to the willingness of producers to alleviate constraints due to lack of tillage equipment and timely seeding of their crops.

These practices that can now be used as entry points to a gradual shift towards CA systems were heavily discouraged by extension services that were targeting green revolution models, hence recommending practices such as soil tillage and monoculture. The difficulty often mentioned to change the mindset to move towards CA is largely due to the fact that during past decades farmers and technicians associated ploughing and monocropping to modern farming. Presently, they see the eventual return to direct seeding, crop association or rotation as a regression. Nevertheless, the first scientific research operation on no-tillage in WCA was reported in the early 1960s when work was conducted in Ghana on direct seeding (Mrabet, 2002). From the 1970s, several research activities on direct seeding were conducted in Nigeria at the International Institute of Tropical Agriculture (IITA), which also led the experimentation for the introduction of mucuna as a cover crop in farming systems in the sub-humid area of Benin from the mid-1990s (Lal, 1986). Mucuna was grown in rotation or in combination with cereals with the objective of improving soil fertility, controlling weeds and reducing soil erosion. Similar research operations were launched in Togo and Côte d'Ivoire where unsuccessful attempts were made to introduce *Pueraria phaseoloides* intercropping for the improvement of short fallows (Autfray, 1997). In Benin, Houndékon *et al.* (1998) noted that farmers who adopted the production of mucuna were mainly those whose land was very degraded and infested with weeds (*Imperata cylindrica*) and those who managed to sell mucuna seeds.

In Ghana, Boahen *et al.* (2007) reported that the development of CA is closely related to the increasing land pressure, which in the 1980s forced a number of farmers to abandon the traditional system of shifting cultivation that was previously used to restore soil fertility. Slash-and-burn and then continuous cropping on the same piece of land with reduced fallow periods (duration of fallows was cut from 5 years to just 3 years) made it necessary to search for technologies that would increase yields. The government urged and supported national research institutes to develop SLM techniques including CA principles, e.g. minimum tillage, mulching and use of cover crops. Research findings on minimum tillage and direct-planting techniques were promoted with herbicides from the early 1990s by the Ghana Grain Development Project in collaboration with Monsanto, Sasakawa Global 2000 and the Ministry of Food and Agriculture (MOFA). Generally, the process of adaptation and dissemination of CA in Ghana was done through a series of projects implemented with the participation of MOFA, but with the determining support of development or cooperation organizations including German Technical Cooperation (GTZ), Danida, Sasakawa Global 2000 and Monsanto. Conservation agriculture practices introduced and promoted in the study areas included slash-and-mulch without burning, use of cover crops, and minimum tillage with herbicides and direct planting.

Moreover, during the first years of the 2000s several research projects on the development of CA systems were launched in the cotton production areas of several French-speaking WCA countries including Burkina Faso, Cameroon, Mali and Chad (Djamen *et al.*, 2005; Abba *et al.*, 2006; Sissoko, 2009). These operations were funded mainly by the French Development Agency and were aimed at developing and widely disseminating innovative and effective SLM technologies to recover soil fertility and stop the downward trend of cotton production. During the same period, research activities were conducted in Guinea on the use of direct seeding with herbicides (Kourouma and Bozza, 2005). Cropping systems including direct seeding plus cover crops were also developed in Benin from the early 1980s to the mid-2000s. In Burkina Faso, FAO, through a pilot project entitled 'integrated agricultural production systems (PRODS) as a priority area for interdisciplinary actions (PAIA) approach PRODS/PAIA', have developed since 2000 an innovative CA system characterized by improved biomass production and a better integration of livestock in the system (Kassam *et al.*, 2009b). But this initiative has been very localized, hence has a limited outreach.

Cameroon is probably the only WCA francophone where substantial efforts have been undertaken continuously over time to refine CA systems and engage with the dissemination phase. Screening of cover crops was made, different options for biomass production were explored, on-station experiments started in 2000 and were followed and complemented by on-farm tests until 2007. Effective CA systems adapted to the agroecological diversity of the cotton zone of north Cameroon were designed and validated (M'Biandoun *et al.*, 2010; Naudin *et al.*, 2010). The 'Projet Conservation des Sols – Eau Sol Arbre (PCS ESA)' started in 2007 for the out-scaling of CA. HORUS Entreprise (2009) reported that the main CA systems disseminated in 2008 in northern Cameroon included cereal plus cowpea grain (33% of the area), cereals plus crotalaria (29%) and cereals + brachiaria (26%), cereals + mucuna (9%) and direct seeding of cotton under mulch (3%).

At present, accurate and updated data on the status of CA in WCA are lacking. But it is empirically known that the north Cameroon and Ghana are areas where CA is the most developed in WCA. In 2009 the area under CA in Cameroon was estimated to be 2541 ha (HORUS Entreprise, 2009). However, Seguy (2008) reported by Serpantié (2009) noted that less than 10% of the declared area was really under CA *stricto sensu*, the other 90% was under simplified cultivation techniques that can be considered as transitional steps towards CA. Nevertheless, the objective, rather ambitious, of the project was to reach 14,125 ha in 2009 and 25,000 ha in 2010 (HORUS Entreprise, 2009). The level of achievement of this objective is not yet known. In Ghana, areas under direct seeding + herbicides were estimated at about 300,000 ha in 2009 (Serpantié, 2009). In 2012, the area under CA was evaluated at 30,000 ha (Friedrich *et al.*, 2012).

Stakeholders actually involved in the development of CA in WCA can be grouped into two categories according to the geographical zones where they are operating: (i) local stakeholders including farmers and their organizations, cotton companies, private sector for input supply, governments through development projects, local universities and research centres; and (ii) international organizations including: international NGOs (African Conservation Tillage Network (ACT), Catholic Relief Services (CRS)), research centres (International Center of Collaboration on Agricultural Research for Development (CIRAD); the World Agroforestry Center (ICRAF)) and international development agencies (FAO, IFAD).

There is a growing interest of agricultural development stakeholders in CA. While CA was usually a part of SLM projects, during the last few years there has been an increasing number of projects and programmes primarily focused on CA (Table 13.1). Among these projects, there is notably the Smallholder Conservation Agriculture Project in WCA (SCAP), which in a coordinated manner managed to strengthen collaboration between SLM stakeholders and most importantly made a major breakthrough in awareness creation and the exploration of the potential of CA in WCA in a wide rainfall gradient ranging from 350 mm year^{-1} in the Maradi area (Niger) to about 3000 mm year^{-1} near Dubreka (Guinea). Based on the outcomes of SCAP, several IFAD[1] investment programmes that worked with SCAP in Burkina Faso (PDRD, PICOFA), Guinea (PADER/BGN) and Niger (PPILDA) have included a CA component in forthcoming activities/phases (ACT *et al.*, 2012).

To date, CA is not yet part of the main focus of the agricultural development policies

Table 13.1. Major recent or ongoing Conservation Agriculture projects in West and Central Africa.

Projects	Countries/areas of implementation	Donors	Duration (years)
Agroecology-Based Aggradation-Conservation agriculture **(ABACO)**: targeting innovations to combat soil degradation and food insecurity in semi-arid Africa	Burkina Faso, Kenya, Madagascar, Mozambique Tanzania, Zimbabwe,	European Commission	4 (2011–2014)
Smallholders Conservation Agriculture Promotion in Western and Central Africa **(SCAP)**	Burkina Faso, Guinea, Niger	IFAD, AFD	3.5 (2008–2012)
Projet de Conservation des Sols Eau Sol Abre II (second phase) **PCS – ESA II**	Cameroon (Northern Region)	AFD	5 (2007–2012)
Conservation Agriculture in Africa: Analysing and Foreseeing its Impact, Comprehending its Adoption **CA2Africa**	Africa (selected case study all over Africa)	European Commission	3 (2010–2012)

of WCA countries. Nevertheless, there is an increasing number of stakeholders including farmer organizations (FOs), NGOs, private sector and to a lesser extent the States through the development projects involved in the promotion of the CA. If this trend is strengthened and continues over time, it will likely help to overcome shortcomings of projects activities that are limited in a specific period and sites; it will also lead to a real dynamic, which could be monitored and evaluated at the medium- or long-term basis better to identify and assess the impact of CA on livelihoods and resilience of WCA farmers.

13.2.2 Typology of Conservation Agriculture-based cropping systems in West and Central Africa

West and Central Africa presents a great diversity of socio-economic and biophysical conditions. Though CA principles are often supposed to be applicable in all contexts (Anderson and Giller, 2012), it is obviously not reasonable to consider only one CA-based cropping system could be suitable for every area. It is therefore necessary that CA be tailored to local conditions by identifying CA options that can perform better taking into consideration the characteristics of the area where they are implemented.

We distinguish four types of CA-based cropping systems (CA-CS) for WCA smallholder farming using the following criteria that are considered to be somehow influencing the modalities of application of the three CA principles: rainfall, socio-economic conditions (land pressure, food security, cattle rearing, access to market etc.) and existing farming practices prior to the introduction of CA. The typology also takes into account CA systems under development or already promoted. The four types of CA-based cropping systems include (Table 13.2):

1. CA-CS 1: CA featuring agroforestry, direct seeding under mulch from tree/shrub pruning, cereal in association with leguminous food crop.
2. CA-CS 2: CA featuring crop residue retention; direct seeding under mulch of crop straws, cereal in association preferably with leguminous food crop.
3. CA-CS 3: CA featuring cover crops, direct seeding under biomass of cover crops, cereal grown in association or rotation with fodder crops.
4. CA-CS 4: CA featuring cover crops, direct seeding under mulch of cover crops, cereal grown in rotation with fodder crops or improved fallow.

Table 13.2. Potential Conservation Agriculture-based cropping systems for West and Central African smallholders.

	CA-CS 1	CA-CS 2	CA-CS 3	CA-CS 4
Soil tillage	Direct seeding/ ripping	Direct seeding/ ripping	Direct seeding + herbicide	Direct seeding + herbicide
Material for organic soil cover	Biomass of shrubs (*Piliostigma reticulatum, Guiera senegalensis, Hyphaene thebaïca* etc.) + cereal straw	Mulch of cereal eventually complemented with biomass of shrubs or grass	Biomass of cover crops + straws of cereal	Biomass of cover crops + grasses
Main crop	Millet/sorghum	Sorghum/millet	Maize, sorghum, cotton	Rice/maize
Cover crops/ associated crops	Cowpea/groundnuts	Cowpea/ groundnuts	Fodder crops (brachiaria, mucuna, dolichos etc.)/leguminous food crops	Fodder crops (brachiaria, pigeon pea, *Stylosanthes* sp., mucuna, dolichos etc.)
Crop association/ rotation	Association	Association	Association/ rotation	Rotation
Average accessible soil cover rate (%)	30–50	50–70	80–100	100

CA-CS 1 and CA-CS 2 are designed preferably for semi-arid zones, where the low rainfall is a limiting factor for biomass production and diversification of cover crops. Population density is very high and food insecurity is frequent, hence the cover crops selected by farmers are leguminous food crops and mainly cowpea and groundnuts. Crop diversification, the third principle of CA, is based on crop association rather than on crop rotation because of the land scarcity. Crop varieties to be used in association with the cereal should have a short cycle, and if possible be resistant to drought and pests, and less demanding regarding water and nutrients. Direct seeding, generally without application of herbicide, is already well practised taking advantage of the shallow nature of soils, but also to meet the challenges of lack of equipment and the poor and unpredictable rainfall. Some farmers do practise minimum soil disturbance through shallow ripping.

CA-CS 1 is designed for semi-arid areas with very low rainfall (<500 mm year^{-1}) and high population density (>70 inhabitants km^{-2}). The production and conservation of biomass on the plot is a major issue because of the low rainfall and the high pressure of livestock. However, the minimum of 30% soil cover recommended by FAO to fulfil the principle of soil cover can be reached by mobilizing millet straw and mostly biomass from prunings from native shrubs such as *Piliostigma reticulatum* and *Guiera senegalensis*. This CA-CS is already part of the traditional cropping systems in many villages of the WCA Sudano-Sahelian zone (ACT *et al.*, 2012) where farmers managed trees/shrubs as coppiced stumps at a density of a least several hundred per hectare (Garrity *et al.*, 2010). In fact, CA-CS 1 relies on the improvement of the existing farming practices through the co-building of appropriate technical (agronomic) and socio-economic (management) innovations that will enable a higher production of shrub biomass *in situ* without constraining the realization of crop management operations nor increasing the competition for water and nutrients between the crop and shrubs. This CA-CS can be considered to some extent as a concrete example of the emerging concept of 'evergreen agriculture' which is defined as the integration of particular tree species into annual food-crop systems (Garrity *et al.*, 2010).

CA-CS 2 is designed for the semi-arid areas, but with slightly better rainfall (600–800 mm year^{-1}) than in zones for CA-CS 1. Thanks to the relatively good rainfall, the main food crop grown here is sorghum or even maize in some cases. Some cover crops including *Mucuna* sp. and *Brachiaria* sp. among others can be grown, but farmers tend to prefer a food leguminous crop because of the rampant risk of food insecurity and also the land scarcity. Cereal straws are the main materials used for soil cover; these residues can eventually be complemented with shrub biomass or grasses collected on other plots or in the bush. It is possible to achieve a soil coverage of 50 to 70%. As in CA-CS 1, crop diversification is achieved mainly through crop associations because of high pressure on land.

CA-CS 3 and CA-CS 4 are quite similar as they are designed for areas with medium to high rainfall. Crop production is diversified including a cash crop such as cotton, rice and maize unlike in the semi-arid area. The relatively low population density makes farmers more eager to practise not only crop rotation but also the cultivation of fodder crops, which could be an entry point for the development of CA.

CA-CS 3 is tailored for areas with an average rainfall of about 800–1200 mm year^{-1}; this rainfall is enough for the production of a wide variety of cover crops that can be also used for human or animal feeding. A rate of 100% soil coverage is possible as the rainfall enables production of biomass. The pressure on land is not very high because of a medium population density (20–70 inhabitants km^{-2}). The third principle of CA can be achieved either through crop associations or crop rotations. It was noticed that farmers prefer to practise crop association with leguminous food crops (cowpea, groundnuts) and use crop rotations with fodder crops like *Mucuna* sp. or *Brachiaria* sp., which they consider difficult to manage when cultivated in association with cereal on the same field. In these sub-humid areas, it is necessary to grow cover crops to produce complementary biomass for soil cover as biomass of the main crops (maize, cotton) does not always last throughout the dry season or is easily decomposed by termites as compared to sorghum or millet straws. Soil tillage is practised by a majority of farmers to control weeds that grow rapidly ahead of the sowing of the crop. Hence, a transformation of existing farming practices into CA will require at least at the beginning of the process the use of herbicides for weed control when direct seeding is practised.

CA-CS 3 is the most common type of CA found in several areas of WCA where CA development or research operations have been conducted or are still ongoing (Boahen *et al.*, 2007; Naudin *et al.*, 2010; ACT *et al.*, 2012; Dugué *et al.*, 2012b; Lamantia, 2012). A wide variety of cover crops can be cultivated according to the agroclimatic conditions, but also according to farmers' preference. Cover-crop species usually found include *Brachiaria* sp., *Mucuna* sp., *Stylosanthes* sp., *Crotalaria* sp., *Dolichos lablab* and *Cajanus cajan* amongst others.

CA-CS 4 has almost the same characteristics as CA-CS 3, except that it is more appropriate for areas with high rainfall (>1400 mm year^{-1}) and a low population density (<20 inhabitants km^{-2}). The high rainfall is favourable for the cultivation of a wide variety of cover crops. In some cases, the rainfall pattern might be bimodal, allowing two cropping seasons per year. Furthermore, it is possible to produce a large volume of crop biomass to ensure 100% soil cover. The low population density enables the implementation of the practice of fallow, which can be improved by the introduction of cover crops including shrubs that can bolster nutrient supply through nitrogen fixation and nutrient cycling. Herbicide is used to control weeds.

Further research operations are still needed to provide comprehensive and reliable knowledge on the evidence and performance of all the four types of CA-CS. Boundaries of areas of the different CA systems are not rigid. Depending on the socio-economic and agronomic conditions, more than one type of CA system can be practised in the same area (Fig. 13.1).

Sub-humid areas with better rainfall and low population density offer more favourable conditions for the production and

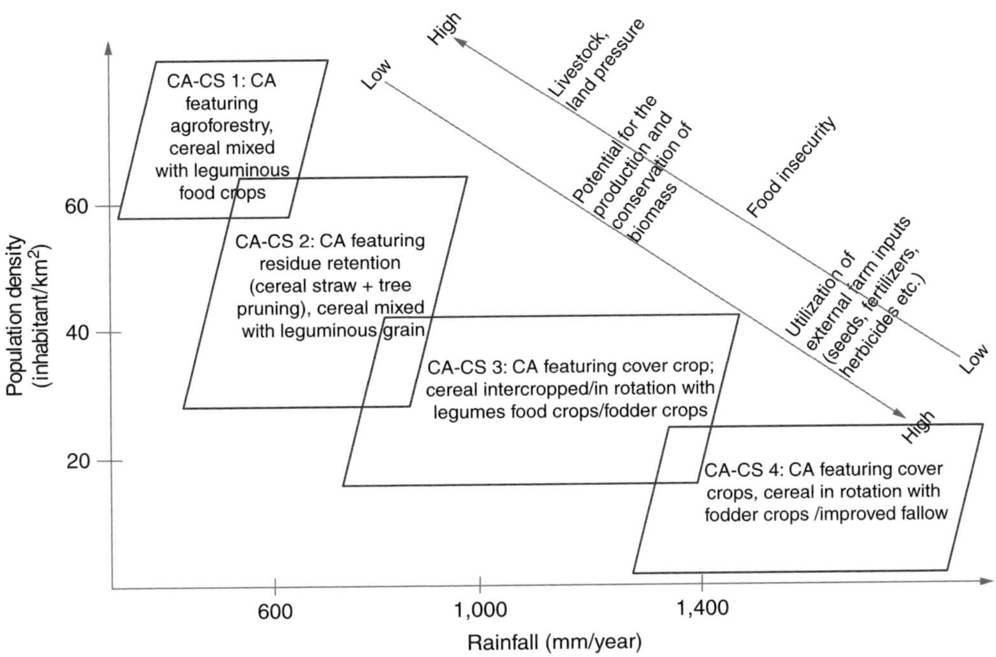

Fig. 13.1. Conceptual distribution of different types of Conservation Agriculture systems in West and Central Africa according to rainfall and population.

conservation of biomass. However, the implementation of CA in these areas (CA-CS 3 and CA-CS 4) seems more costly because of the high dependence on external farm inputs including mainly herbicides for weeding, seed of cover crops and fertilizers. It appears from the characteristics of different types of CA-CS that when markets for farm inputs and products are well developed, areas with high rainfall and low population density are the most favourable for the full implementation of CA principles. However, despite some agroecological and socio-economic challenges, it is observed that there is room for CA in semi-arid zones with even some innovations that can contribute to enriching the implementation of the concept of CA.

13.3 Available Knowledge of Conservation Agriculture in West and Central Africa

Existing scientific evidence on CA-related topics in WCA are still rare. Most research conducted so far has dealt mainly with biophysical aspects. Data on socio-economic effects of CA in the region are very few and recent. Most of the available data were generated from on-station research.

13.3.1 Biophysical aspects

Organic matter and carbon sequestration

Soil organic matter (SOM) has tremendous benefits on soil chemical, biological and physical properties, which in turn affect the productive capacity of soils (Mrabet, 2002). The improvement of SOM is one of the major benefits expected from the implementation of CA under the tropical climate of WCA where mineralization of organic matter is very rapid. It recognized that it is very difficult to restore the initial rate of carbon once the soil is degraded; hence, it is better whenever possible to prevent the deterioration or to undertake necessary efforts to reduce the depletion rate of soil fertility.

Tillage pattern plays an important role in carbon sequestration. Existing evidence on the effects of CA on SOM and carbon sequestration are contradictory. The results of several studies show a slower decrease rate and better stability of C and SOM under CA as compared to conventional farming.

Various examples support this assertion: in upland rice production systems in the sub-humid area of Benin, Saito *et al.* (2010) observed that after 3 years of experimentation, there was no significant effect of tillage management on soil organic carbon (SOC). In contrast, Boli *et al.* (1996) in their trial in the northern Cameroon savannah zone noted a 50% decrease of carbon within 3 years in conventional farming. The decrease stabilized at C= 0.25%. The depletion rate was lower under CA, and the carbon stabilized at 0.35% after 3 years of cultivation. In western Nigeria, Lal (1976) found higher SOC in the surface soil horizon (0–10 cm) of no-tillage (NT: 1.30%) than of conventional tillage (ConvT; 0.86%) after 5 years of experimentation. SOM declined by 0.03% per month under ConvT for the first 12 months of conversion from fallow to continuous maize. Research results show that organic matter depletion could be slowed and reversed through reduction in tillage operations or their elimination. Soil structure could improve with conservation tillage allowing for better water infiltration and retention. Citing Lal (1986), Mrabet (2002) indicated that ConvT with mouldboard plough and harrows reduced SOC by 72% while a 7% increase in SOC was recorded under NT during a period of 6 years. Moreover, in western Nigeria, Lal (1976) found that after 5 years of experimentation, SOC in the surface soil horizon (0–10 cm) was higher under NT (1.30%) as compared to ConvT (0.86%). SOM declined by 0.03% per month under ConvT for the first 12 months of conversion from fallow to continuous maize. The decline rate was only 0.004% month^{-1} under NT. No significant difference was noted in SOM content between tillage systems in the 10–20 cm soil depth, but it was observed that that NT stratified SOC (stratification ratio of 1.8, calculated as SOC of 0–10 cm divided by SOC of 10–20 cm depth), while the soil under ConvT had a uniform distribution of SOC (ratio of 1.1). Moreover, the benefits of CA on soil quality cannot only be ascribed to NT, they might also come from the implementation of other CA principles such as crop rotation. For example, Tarawali and Ikwuegbu (1993) reported by Tarawali *et al.* (1998) observed in Nigeria that the improvement of short fallow with *Stylonsanthes* significantly decreased the soil's bulk density and increased its porosity (capacity to retain moisture), its CEC, and its organic C and N contents (Table 13.3).

Water infiltration, runoff and erosion

Existing research results highlight the positive effects of the CA in the improvement of water infiltration and control of runoff and erosion (Table 13.4). The presence of soil cover and NT improve the stock of available water through better infiltration and reduction of evaporation. In the cotton production zone of northern Cameroon, Soutou (2004) observed significant differences in soil moisture: 0.15% and 0.20% in conventional agriculture (ConvA) and in CA, respectively. Moreover, the wetting front was 10 cm deeper under CA as compared to conventional farming. The increase in the depth of the wetting front improves the stock of groundwater, leading to better yield particularly in areas where water is a limiting factor.

The absence of a plough pan in soils under CA increases the volume of soil explored by the roots, and thus water available to the plant. Improved water supply at the beginning of the crop cycle (depth of

Table 13.3. Soil chemical and physical properties under *Stylosanthes* and natural fallow (Tarawali and Ikwuegbu, 1993).

Property	*Stylosanthes* (3 years)	Natural fallow (>4 years)
N content (g kg^{-1})	1.14	0.87
CEC (cmol kg^{-1})	3.24	2.22
Organic C (g kg^{-1})	4.31	2.70
Bulk density (g cm^{-3})	1.51	1.66
Total porosity (%)	43.1	37.4
Macroporosity (%)	42.1	36.4
Microorganisms ($n \times 10^7$ g^{-1})	34	12

Table 13.4. Comparative results on water infiltration, runoff and erosion in conventional farming and Conservation Agriculture systems.

	Conventional farming	Conservation Agriculture
Soil moisture (%)[a]	0.15	0.20
Depth of the wetting front (cm)[a]	26.7	36.6
Actual evapotranspiration (mm)[a]	327	277
Water infiltration (%)[b]	65	94
Drainage (mm)[a]	49	87
Runoff (%)[b]	16–40	0.3–9
Erosion (t ha^{-1} $year^{-1}$)[b]	5–30	0.2–3
Loss of solid suspensions (t ha^{-1} $year^{-1}$)[b]	3–9	0.2

[a]Soutou (2004); [b]Boli (1996)

wetting front) during vegetative growth and flowering stages was noted under CA systems. The positive effect of this increase in the supply of water is likely to positively affect water consumption by the crops and final yields. Similarly, Boli (1996) observed that water infiltration in CA was 94% for the amount of rainfall as against 65% in conventional farming. Conservation Agriculture significantly reduced runoff. Under CA, runoff had a very low variability according to rainfall, unlike in ConvA. These results are consistent with those of Mrabet (2002), who noted that there is generally a direct relationship between ConvT and erosion as inappropriate use of tillage implements certainly enhances erosion. Sissoko (2009) observed that soil cover significantly reduces water runoff, notably at the beginning of the rainy season. Conservation Agriculture systems were found very effective not only for the control of infiltration, but also in the conservation of the productive capacity of the soil. In fact, the amount of solid suspension (organic matter, clay and silt) was significantly less in CA than in ConvT, 0.2 against up to 9 t ha^{-1} $year^{-1}$ (Boli, 1996). Soil loss observed in CA consisted primarily by erosion of the edge of the plot along the sedimentation channel. It was noted that the capacity of CA systems to reduce runoff and erosion depends on the level of degradation of the soil. The maximum erosion was 2 and 4 t ha^{-1} $year^{-1}$ for CA system on non-degraded land and degraded land, respectively.

However, CA might show some shortcomings in certain climatic and soil conditions. Because of its high infiltrability, CA increases drainage and causes asphyxia of roots and unavailability of nitrogen during abundant or frequent rains. Boli (1996) suggested that an adaptation of the application of fertilizer will be necessary; he recommended a supplement of 20 units of N to be applied after the wettest period of the year for the sandy soils of the northern Cameroon savannah.

Biodiversity, pest and disease dynamics

Significant effects of CA on the improvement of biodiversity were recorded from empirical observations and scientific research operations in WCA. According to Boahen *et al.* (2007), CA farmers in Ghana noted that using cover crops without the traditional burning has increased the population of pests such as leaf borers, millipedes, caterpillars and grasshoppers. The cover-crop canopy created a good microclimate for them.

Moreover, it was found in northern Cameroon that while restoring cotton production, soil management systems based on a NT with mulch approach intercropped with cereals also enhanced the diversity of microfauna and their biological activity (Brévault *et al.*, 2007). No tillage with grass mulch (*Brachiaria ruziziensis*) (NTG) and no-tillage with legume mulch (*Crotalaria retusa* L. or *Mucuna pruriens*) (NTL) had an impact on the abundance, diversity and functional role of soil invertebrates compared to ConvT and NT without mulch. Examination of the soil macrofauna pattern revealed that the abundance and diversity of soil arthropods were significantly higher in NTG and NTL than in ConvT plots (+103 and +79%, respectively), while that of NT plots was in-between the NT groups and ConvT (+37%). Formicidae (53.6%), Termitidae (24.7%) and Lumbricidae (9.4%) were the most abundant detritivores while Julidae (46.1%), Coleoptera larvae (22.1%) and Pyrrhocoridae or Reduviidae (11.8%) were the dominant herbivores. The major constituents of the predatory group were Araneae (33.8%), Carabidae (24.6%), Staphylinidae (15.7%) and Scolopendridae (10.3%).

In the northern Cameroon savannah, Boli (1996) reported the positive effect of mulching on the density and biological activity of earthworms whose role is decisive on soil macroporosity, which is a key factor of water infiltration. Plots under CA had eight times more castings than tilled plots. Moreover, in an Ultisol in the Guinean zone of Burkina Faso, Bado *et al.* (2011) found that crop rotation, the third principle of CA, influenced nematode infestation, but the effects on soil and sorghum root infestation were different according to the rotation. The cowpea–sorghum rotation increased soil and sorghum-root infestation by nematodes, while groundnut–sorghum decreased the nematode population. The soil of the cowpea–sorghum rotation contained 1.5 to 2 times more nematodes than the soil from monocropping of sorghum. In contrast, the soil of the groundnut–sorghum rotation contained from 17 to 19 times fewer nematodes than that of the monocropping of sorghum. However, nematode infestation did not affect yield of the succeeding sorghum. It was concluded that the parasitic effect of nematodes was limited by the predominance of positive N-effects.

If the positive impact of CA on soil biodiversity is recognized, this biodiversity also appears as a key factor for the effectiveness of CA in some cases such as in crusted soil. In northern Burkina Faso, Mando (1997) observed that termite activity in mulch was significantly contributing to improve water conservation of crusted soils. Mulch without termites did not have a statistically significant effect on the water status of structurally crusted soil.

Although CA-based cropping systems tend to improve soil biodiversity, their net effect can be difficult to predict. Furthermore, they can kill or serve as refuge for pests and beneficial organisms, thus requiring farmers to modify their crop management systems. Results of experiments (Brévault *et al.*, 2009) conducted in the cotton production zone of northern Cameroon show that the presence of mulch (of *Brachiaria ruziziensis* and *Calopogonium mucunoides*) negatively affected cotton seedling stand by 13–14% compared to non-mulched plots; and the proportion of damaged seedlings was higher in mulched than in non-mulched plots supporting the hypothesis that mulch favoured soil pest damage. It was found that the use of insecticidal seed dressing increased the seedling stand and the number of dead millipedes collected, and fungicide had little or no effect on seedling stand and vigour. Moreover, there is a potential effect of mulch on the natural inhibition of fungi. Early sowing and adequate systemic seed protection are suggested to reduce risk of aphid infestation (Deguine *et al.*, 1994).

13.3.2 Yield

Depending on the duration and agroclimatic conditions of the areas, CA may have a positive or negative impact on yield. In northern Cameroon, M'Biandoun *et al.* (2010) noted that cotton yields were higher under CA as compared to under ConvT or direct seeding without mulch (Table 13.5). Moreover, they observed that over the years the yields fluctuated under direct seeding (DS) and ConvT, while they were more stable and tended to increase under CA.

The magnitude of positive effect of CA on yield is more visible and significant when water is a limiting factor for crop production as in the semi-arid zones or during cropping seasons with rainfall deficit. In the

Sudano-Sahelian zone of northern Cameroon, cotton yield under CA was 12% and 24% higher compared to tillage and NT, respectively (Naudin *et al.*, 2010). Differences between treatments were more significant in areas with a rainfall deficit. Differences between treatments were not significant in sites with average rainfall (900–1200 mm). In the semi-arid zone of Burkina Faso, Bougoum (2012) noted that during the 2011 cropping season characterized by a 30% rainfall deficit, sorghum yield under CA system significantly increased proportionally to the density of soil cover, moving from 414 to 874 kg ha^{-1} under direct seeding (DS) without mulch and direct seeding with a mulch density of 4 t ha^{-1}, respectively (Fig. 13.2).

Table 13.5. Yields of cotton (averaged over 5 years) under different management systems in northern Cameroon (Extracted from M'Biandoun *et al.*, 2010).

Management system	Yield (kg ha^{-1})
Conventional tillage	1143
Direct seeding	880
Conservation agriculture	1472

However, in some instances a yield penalty has been recorded at the beginning of the transformation process of conventional farming to CA. The duration of this period might vary from 3 to 5 years (Boli, 1996; Giller *et al.*, 2009). The yield penalty is probably one of the reasons why some farmers prefer to practise CA on soil with good fertility when possible (cf. section 13.4 on adoption). Delayed positive impact of CA on crop performance is sometimes attributed to the time necessary to rebuild or improve SOM in degraded soils.

However, existing evidence in WCA does not clearly confirm nor negate the popular statement on initial yield penalty. There are experiences showing yield penalties, but the duration of these experiences is not long enough really to allow the identification of the time required and the magnitude of yield increase penalty period. In Mali, Sissoko (2009) noted that the yield of cotton under CA was slightly lower. Nevertheless, it was expected that in the long term, the increase of SOM under CA will lead to increased nutrient availability for crops and consequently better yields.

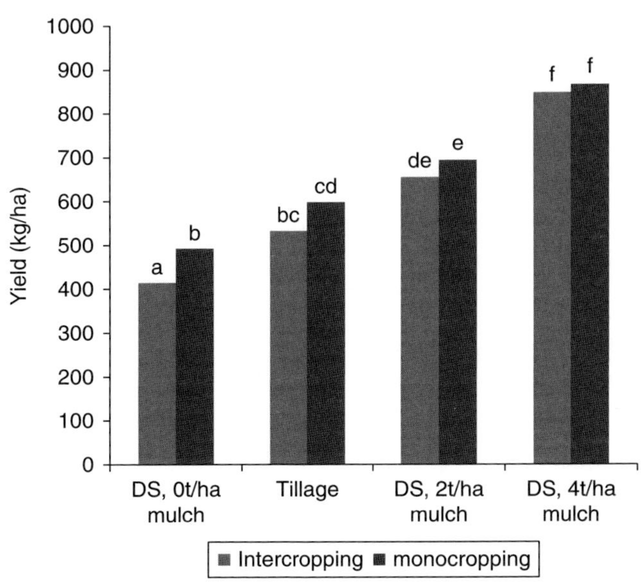

Fig. 13.2. Sorghum yield (kg ha^{-1}) in intercropping and monocropping systems as affected by the amount of mulch used as ground cover. Means between treatments followed by different letters differ significantly at $p<0.05$ (Bougoum, 2012).

In fact, it was found that yield penalty under CA is positively correlated with the level of soil fertility. In northern Cameroon, cotton yield under CA decreased from 1.9 to 1.2 t ha^{-1} on fertile and unfertile soils, respectively. The difference was larger in maize production with a gap of about 1.5 t ha^{-1} (Boli, 1996). The latter noted that maize yield in northern Cameroon was 20 to 40% higher in ConvT as compared to CA during cropping seasons with high rainfall. He explained the difference in yield was due to the leaching of nutrients under CA systems, waterlogging due to abundant and frequent rainfall or due to the compaction of the degraded soil, which is not favourable for the development of roots of the crops. Lal (1977) recommended as a precondition to the installation of CA in degraded and compacted soil, a deep ploughing ranging from 26–35 cm when the degradation of the soil was caused by tillage using animal-drawn equipment, and from 30–35 cm if the compaction occurred due to the use of heavy machinery. For crop rotation or crop association, selected suitable crop species are important as they may help to reduce yield penalty in CA systems. From research conducted in sub-humid area of Burkina Faso, Bado *et al.* (2011) noted that monocropping of sorghum produced the lowest yield, while legume (*Vigna unguiculata, Arachis hypogea*)–sorghum rotations increased sorghum yields by 50% to 300%. Groundnut–sorghum and cowpea–sorghum rotations increased soil mineral N by 36% and 52%, respectively.

Issues related to the delay in the immediate positive impact of CA on yields show that CA practices should not be considered as a set of stand-alone technologies. Rather its implementation goes together with good farming practices (e.g. production of green manure) and sometimes even CA-antagonistic operations such as an occasional tillage/sub-soiling of the compacted soils. Furthermore, it should be acknowledged that yield penalties are not systematic in all conditions, and also that CA is not only for farmers with infertile soil but also for producers with fertile land who can use the technology to maintain and improve soil quality and ultimately crop yields and production factor productivities.

13.3.3 Socio-economic effects of Conservation Agriculture in West and Central Africa

Labour

Labour requirement is a part of the major challenges faced by smallholder farmers in WCA. Available research results on the effects of CA on labour requirement are quite divergent or even contradictory depending on the context. In Ghana, Boahen *et al.* (2007) reported a 30% labour reduction with CA as compared to the traditional slash-and-burn. The reduction was attributed to less labour required for land preparation and weeding operations. Similarly, a statistically significant 66.7% decrease in labour requirements during weeding management operations under CA systems was noted in Mali (Autfray and Sissoko, 2011). The gain obtained was mainly by savings in tillage operations, which are more time-consuming than herbicide application. However, it was noted that the reduction of labour requirements tended to decrease during subsequent years, resulting in no statistically significant difference between CA and conventional farming. This was attributed to a lower soil cover after the first year, thus highlighting the challenge of production and conservation of enough biomass on the farm. Naudin (2012) noted in north Cameroon that weed control was improved with CA, resulting in the reduction of number of weedings required and the delay in the date of first weeding.

Unlike in Mali and Ghana, the implementation of CA in the semi-arid zone of Burkina Faso resulted in a 71% increase in labour requirements (Table 13.6). The increase was due to the fact that farmers were not able to store enough biomass on the plot, thus had to spend additional time (about 10 man days ha^{-1}) to collect and spread further material for soil cover. Further, crop diversification, which is the third principle of CA, was based on crop

Table 13.6. Labour requirement (man days ha^{-1}) in conventional farming and Conservation Agriculture-based cropping system in northern Burkina Faso (Bougoum, 2012).

Operations	Conventional farming	Conservation Agriculture
Tillage	4.1	
Collection and spreading of additional material for soil cover		9.7
Sowing of sorghum	5.1	7.8
Thinning	4.9	6.17
Seeding of cowpea		4.3
First weeding	8.7	9.7
Second weeding	5.9	6.6
Application of NPK	2.5	3.4
Application of urea	1.8	2.7
Harvesting of sorghum	7.6	9.3
Harvesting of cover crop		10.2
Total	40.7a	69.8b

Values sharing different letters differ significantly at $p<0.05$, according to Duncan's multiple range test.

association rather than rotation. Crop association increased labour requirement per hectare, particularly for seeding and harvesting operations of the cover crops. Moreover, a higher labour use was recorded under CA because of difficulties farmers faced to sow crops in a soil covered with mulch. This raised the need to introduce appropriate direct seeding equipment such as a jab planter or animal-drawn direct seeder.

Research results available so far in WCA are convergent regarding the positive effects of CA on labour productivity (Boahen *et al.*, 2007; M'Biandoum *et al.*, 2010; Bougoum, 2012). Results obtained in northern Cameroon highlight that labour productivity under CA systems is on average 20% higher compared to conventional farming (Table 13.7). The magnitude of increase in labour productivity depends on several factors, including mainly the type of crop grown, yields and the strategy for crop diversification. In the semi-arid zone of Burkina Faso, it was

Table 13.7. Labour productivity (FCFA/man day) in conventional farming and Conservation Agriculture in northern Cameroon (M'Biandoun *et al.*, 2010).

Crop	Conventional farming	Conservation Agriculture	Percentage of change (%)
Cotton	2210	2920	32.1
Maize	3778	4460	18.0
Sorghum	2032	2630	29.4

noted that the increase in labour productivity was about +258.5% and only +54.8% under CA if crop diversification was achieved through association or rotation, respectively (Bougoum, 2012). The difference was due to the fact that with crop association, two crops including a cereal and a leguminous grain crop were grown simultaneously while in crop rotation the plot was bearing only one of the two crops and only once.

Productivity and margins

Conservation Agriculture tends to have a positive impact on total productivity and economic margins. In the northern area of Burkina Faso (rainfall <700 mm $year^{-1}$), preliminary results of a trial to evaluate the specific and combined effects of CA principles on technical and economic results of sorghum production in association with cowpea highlighted the positive effects of CA on gross margin, gross product and labour productivity (Bougoum, 2012). The highest effect was noted on gross margin resulting from yield increase (+41.6%) and particularly of the economic value of cover crop (cowpea) grown in association with cereal (Table 13.8). The study also showed that effects of CA on productivity and margins were variable based on the number and the combination of CA principles implemented by the farmer. Organic soil cover was the CA principle having the highest impact on sorghum yield, labour productivity and return in investment (Table 13.8). However, soil cover also increased

Table 13.8. Rate (%) of technical and economic performance of specific and combined effects of Conservation Agriculture components/principles compared to conventional agriculture practices (Bougoum, 2012).

CA component applied	Sorghum yield	Labour	Gross product	Labour productivity	Gross margin
Organic soil cover	76.0	18.6	76.0	183.3	236.0
Direct seeding	−17.7	9.6	−17.7	−45.3	−40.1
Crop association	−10.9	46.2	110.1	105.5	116.9
Direct seeding + crop association	−30.7	41.5	73.3	53.5	56.8
Direct seeding + organic soil cover	44.8	30.0	44.8	54.9	101.4
Direct seeding, soil cover and crop association	41.6	71.5	173.2	158.4	219.9

labour requirements because farmers had to spend additional time (about 11 days ha^{-1}) to collect straw and other biomass to cover the soil, furthermore they experienced difficulties to perform manual sowing under mulch.

Crop association tends to increase labour requirement per hectare, particularly for seeding and harvesting. Crop association also causes a decrease of about 11% of sorghum yield, but this decrease was not statistically significant. The reduction of sorghum yield caused by intercropping is lower when the density of mulch is higher. On the other hand, crop association increases land and labour productivity.

Direct seeding applied solely without the two other CA components had a negative effect on both technical and economic results except for labour, where a slight reduction of about 10% was noted. Poor results of direct seeding were attributed to the fact that farmers did not use herbicides to control weed, thus having considerable difficulty in their control. Further, the effect of soil surface capping is severe in the study area, leading to poor emergence and growth of plants on plots where direct seeding was practised. However, it was noted that the results obtained under direct seeding can be significantly improved if organic soil cover is practised. These observations tend to confirm the importance of simultaneous implementation of the three CA principles for a significant effect on productivity and margins.

13.3.4 Policy support

The awareness and interest in CA is growing, since agricultural stakeholders are progressively observing CA as an alternative to face the issue of climate change productivity and competitiveness. An increasing number of farmers' organizations and local and international NGOs are now engaged in the promotion of CA though mostly at the district or village levels. These efforts to transform existing farming practices are not supported with appropriate policy framework. In fact, it is acknowledged that the process for the sustainable management of natural resources requires both technical innovation and institutional innovation as highlighted by Balarabé *et al.* (2012). This is not yet the case in WCA where policy support for the promotion of CA both at the national and sub-regional levels is still weak despite increasing knowledge generated on the potential of the technology and growing interest. Most CA experiences in WCA are donor-funded projects, and hence they are very limited in time and in number of farmers reached. A better ownership of these experiences by institutional stakeholders will foster their implementation and impact, and most importantly for spreading the coverage by such projects.

National and sub-regional frameworks for the promotion of SLM and adaptation and mitigation of climate change exist, but unlike in Eastern or Southern Africa, CA is not yet clearly included. In Ghana, Boahen *et al.* (2007) observed that although some experience and knowledge on Conservation Agriculture was available, there was no conscious effort to promote it for large-scale adoption. This situation is quite similar in several countries where the impact of such projects on CA were questioned as they are sceptical and prefer to rely on the old model of green revolution based on the utilization of external inputs (mineral fertilizer, improved seed, herbicides, tractor, etc.) and tillage.

In WCA, diverse stakeholders are involved, contributing with their own specific capacity. However, a framework that may improve this collaboration and create innovative synergies and complementarities does not yet exist. Ghana might be an exception, but there is some inertia even there. Boahen *et al.* (2007) noted that a National Conservation Agriculture Team was established. It was composed of representatives from the ministry of agriculture, research institutions, universities, international organizations and other companies. Its role was to coordinate the CA programme in Ghana by facilitating collaboration and building synergy among CA practitioners. Individual projects provided funds for their representatives to attend meetings. However, the team was dormant for several years; and in 2007 the CA project was no longer active in Ghana, except for demonstrations sponsored by Monsanto for the purpose of selling Round-Up.

The full participation of all concerned stakeholders in the design, implementation, follow-up and assessment of CA activities is presently seen as a key requirement not only for the success of the project, but also for its sustainability. The emerging concept of the co-innovation platform is considered as an appropriate tool to meet this requirement. It is presently being used in Burkina Faso in the framework of the ABACO project (Tittonell *et al.*, 2012). The assessment of the achievements of that project will generate useful information on the effectiveness of innovation platforms in the promotion of CA.

13.3.5 Dissemination approach

Taking into account the complexity of the CA technology, most of the approaches used for its dissemination in WCA are participatory to facilitate progressive learning and mastery of the technology by farmers but also to allow technicians and research to conduct efficiently their activities on the adaptation of CA systems. Three main approaches can be distinguished.

1. The farmer field school (FFS): this approach was used mainly by the SCAP project in Burkina Faso, Guinea and Niger. It consists of bringing together a group of farmers who improve their skills on CA by conducting three or four CA treatments on a communal plot with the support of facilitators who in most cases are field technician of the ministry of agriculture. Through the FFS, the selection, assessment and comparison of CA treatments are decided with the participation of farmers. Moreover, the latter are urged to implement the lessons learned from activities conducted in the communal plot. FFS approach is not used only for collective learning, but also for the co-building of locally suited CA-based cropping systems (ACT *et al.*, 2012).

2. The farmer trainer approach: farmer trainers (FT) are lead farmers who have been initially trained on CA and are receiving support from the project/NGOs to backstop other farmers interested in introducing CA in their own farms. Sometimes, farmer trainers are involved in the implementation of a CA demonstration or on farm research operations on their farms with the support of research centres or extension services. This is a strategy for improving the mastery of the technology by trainer and to update their knowledge and skills so that the farmer can better answer queries raised by trainees. In the framework of the SCAP project, it was noted that some farmer trainers later on became facilitators of FFS groups. Others, mostly in Niger, decided to come together to create a community-based organization providing services (mainly training) on payment to other farmers.

3. Public extension: this is found mostly in cases where a project hires field staff of the ministry of agriculture. For instance, the scaling-up of CA in northern Cameroon started in 2007. It is conducted by extension staff of SODECOTON, the cotton company that is hosting the project. Agronomists of the extension unit of the project are also involved in the dissemination activities. However, except in the cotton production area of Cameroon, CA is not yet fully included in the SLM alternatives disseminated by public extension services in WCA.

Dissemination of CA in WCA has been mainly by targeting individual farmers. However, in some areas, a community-based approach is also being used. The project PCS ESA II in northern Cameroon is testing an integrated approach of natural resource management with the objective to disseminate CA while addressing through the participation of all village stakeholders the crucial issues of sharing of crop residues between livestock and crop production (Dugué *et al.*, 2012b). The vision is also to take into consideration all the stakes relative to local socio-economic and agro-ecological aspects, and to target territories more favourable to the dissemination of CA.

Further to activities of extension staff, promotional events and participatory learning sessions are organized to create awareness of stakeholders. For instance, in the study area of the SCAP and ABACO projects in Burkina Faso, assessment and learning sessions are organized after harvesting to carry out a comprehensive evaluation of activities implemented and to harness data and information necessary for the consolidation of CA techniques tested and to discuss with farmers strategies for the next stage of introduction of CA in their farming practices. Farmers' indicators and their appreciation emphasized during the assessment and learning sessions were used for monitoring and consolidation of tested CA systems (ACT *et al.*, 2012).

Inter- and intra-village exchange visits are organized. The objective of inter-village exchange visits is to increase awareness about CA and also to enable those already testing CA in different villages to share their experiences and assessments. During the visit, host farmers present to visitors CA demonstration and action research activities they implemented, preliminary lessons and difficulties they experienced. Intra-village visits are organized to enable CA-farmers to present and share their experience with their colleague farmers living in the same village, but not yet participating in CA activities in the village. These visits give the opportunity to farmers to interact and to discuss some collective issues regarding the implementation of CA at the village level, such as crop residue management. Boahen *et al.* (2007) reported that farmer exchange visits were a major factor of adoption of CA in Ghana. After seeing benefits from other farmers' fields, they decided to experiment on small plots, 10 m × 10 m, and later extended to an average land area of a quarter of hectare.

The lack of sufficient and adequate human resource is one of the key challenges the dissemination of CA in facing in WCA where most of public extension staff is ageing and not used to participatory approaches required for properly introducing and scaling of a holistic innovation such as CA.

13.4 Adoption of Conservation Agriculture in West and Central Africa

Though CA research and development work has been conducted or is ongoing in several WCA countries, precise, easily accessible and updated official data on the adoption of CA in this sub-region are rare. There are no effective mechanisms for the monitoring on areas and number of CA farmers at national and sub-regional levels. Furthermore, the results of earlier studies on the long-term adoption in WCA are not available. Existing results are mostly those of adoption measured at the end of projects. Hence these results need to be consolidated with impact studies.

13.4.1 Rate and intensity of adoption of Conservation Agriculture in West and Central Africa

It was found that though farmers are interested in CA-cropping systems involving the practices that correspond to the three CA principles, the general tendency is a partial adoption of CA, meaning the three principles are not always adopted together. In Ghana, Boahen *et al.* (2007) reported that the interest of farmers in CA systems was growing with the lifetime of the CA project, resulting in an increase in the number of farmers practising minimum soil disturbance, and demonstration plots increased from 170 in 1996 to 440 in 2000. Further, it was noted that 76% of the farmers participating in CA projects were practising at least one of the three CA components being promoted. The main reasons for the adoption of CA were the reduced cost of production and increased yields.

In Burkina Faso, a study was conducted on the adoption of CA in the sub-humid (900 mm annual rainfall with cereals, cotton and livestock) zone of the Eastern Region; this was one of the intervention sites of the SCAP project in WCA. The rate and spread of adoption of CA were determined by calculating: (i) the proportion of farmers who adopt at least of the three principles; and (ii) the percentage of area on which they implemented the adopted CA principle(s). The results showed that the adoption rate varies with the principles of CA (Ganou, 2012). The 63.5% of farmers who adopted CA were practising it in less than 20% of their hectarage. Crop diversification, generally achieved through crop associations and in a few cases by crop rotations, has the higher adoption rate (Table 13.9). In contrast, direct seeding is the CA principle least adopted. There is a difference between the rate of adoption and intensity of adoption. This difference characterizes the early phase of adoption and dissemination (Knowler and Bradshaw, 2007).

The intensity of adoption of a CA principle depends more on the farmer's ability to manage constraints arising from the application of this principle than the benefits he can obtain. Thus, crop diversification has the highest intensity as it does not pose major problems to farmers, except an increase in labour requirement, which is generally well paid. Direct seeding and soil cover have lower adoption intensities because their implementation generates more challenges to farmers. Direct seeding requires additional costs for the procurement of sprayers and herbicides. Difficulties to produce and keep enough biomass on the plot limit the area where soil cover is achieved. It was observed that as in other sub-regions of Africa, WCA smallholder farmers adopt CA progressively starting by one or two CA principles that are easy to implement according to their situation and that are likely to generate immediate effects. This farmers' behaviour of disassembling technology packages, and initially adopting the most relevant or doable component, was reported in Zimbabwe (Mazvimavi and Twomlow, 2009). Thus WCA CA farmers engaged in partial adoption could keep utilizing the adopted principles and later on adopt the additional component if they have a better understanding of their benefits and management skills and capacity to overcome eventual challenges for the implementation of others.

Table 13.9. Rate and intensity of adoption of Conservation Agriculture practices in eastern Burkina Faso (Ganou, 2012).

	Direct seeding	Soil cover	Crop diversification	Conservation Agriculture
Adoption rate (%)	50.8	80.7	98.9	63.5
Adoption intensity (%)	19.6	16.3	34.5	15.4

13.4.2 Factors determining the adoption of Conservation Agriculture in West and Central Africa

After a comprehensive synthesis of factors that likely affect the farmer's decision to adopt CA, Knowler and Bradshaw (2007) concluded that there are no universal variables, hence the factors considered are site specific. In the literature, there are several studies highlighting the potential factors that can hamper the adoption of CA in sub-Saharan Africa, and probably in WCA. Lal (2007) cited some of the key factors and they include amongst others: (i) biophysical constraints translated by yield penalty registered in some climates and soil types; (ii) poverty, leading to poor accessibility of farm inputs to the smallholder; (iii) high pressure and competitive use of biomass; and (iv) lack of appropriate equipment for direct seeding operations.

In Ghana where access to land is an issue for most farmers, Boahen *et al.* (2007) observed that for a farmer to adopt a cover crop and use it, the farmer needs at least 2 years of user rights of the land to benefit from the investment made to improve soil fertility. Furthermore, farmers owning their own land clearly preferred CA practices, while for farmers using hired land, an important determinant to adopt CA was how long the farmer has access to the land.

Moreover, in sub-humid area of Burkina Faso, it was found from an adoption survey using a logit regression model that the decision of farmers to adopt CA depends on five main factors: the hectarage, the number of cattle, the percentage of degraded land, the existence of support for the implementation of soil conservation techniques and the participation in CA-farmers field school activities (Ganou, 2012).

1. Hectarage (–). This result can be understood by considering the adoption costs of CA (effort to keep straw on the plots and/or to collect additional residues for soil cover, procurement of herbicides, etc.). The larger the area cultivated, the more farmers will consider the cost of adoption of CA.

2. The number of cattle (+). The probability for a producer to adopt CA is even more important for the farmer who has a large number of cattle. In rural areas the number of cattle is a sign of wealth. Hence a producer with a large number of cattle can cope with the initial costs associated with the adoption of CA. Moreover, CA may enable farmers to improve fodder supply for livestock particularly through the introduction of fodder crops either as intercrops or in crop rotation. This observation was also made in northern Cameroon where the growing of *Brachiaria*, a fodder crop, is emerging as a drive for the adoption of CA (Lamantia, 2012).

3. The percentage of degraded land (–). Farmers with a large area of degraded lands were less likely to adopt CA. A high proportion of degraded land requires significant investment by farmers to shift to CA. Hence, without support of projects by other agricultural development organizations, farmers have difficulties to meet additional costs needed in the implementation of CA. Further, obtaining an observable positive impact of CA on soil fertility in the medium or longer term as opposed to shorter term is a discouraging factor. Farmers are more interested in immediate gain or solution of the problem. Therefore, they prefer to use fertilizers (mineral or organic) or to cultivate new lands whenever it would be possible for them to access such resources.

4. The existence of support for the implementation of soil conservation techniques (+). Thus, the adoption of CA is an increasing function of assistance for soil conservation. Undoubtedly, increased assistance to farmers for implementing soil conservation practices increases their likelihood to adopt CA.

5. Frequent attendance at FFS activities (+). The probability of adoption of CA for a farmer is higher if he participates in FFS activities. Attendance at field school activities is more frequent. Attendance at field school activities reflects the farmer's desire to test new techniques and receive information necessary to address the depletion of his soil; this attendance is also necessary to have a good mastery of CA and its benefits, but also to develop adequate strategies to address challenges emerging during its implementation.

It appears that farmers adopt CA mostly for the immediate effects from which they can benefit and not for the expected mid- or long-term improvement of soil properties. Availability and accessibility to agricultural services are also a key component of the adoption. Hence there is a concern about long-term adoption as generally access to agricultural services is not always adequate after the project terminates as it is during its lifetime, when the maximum services are supplied to participating farmers.

13.4.3 Impacts of the adoption of Conservation Agriculture on-farm

The results of the studies show that the effects of CA on the performance and functioning of farms are rather positive. Conservation Agriculture contributes to the improvement of food security and farm income. In the Democratic Republic of Congo, the implementation of a No-Till Agriculture (NTA) project had positive impact in reducing food insecurity amongst targeted households. The proportion of targeted households that could access only one meal per day dropped from 14% to 3% during the project lifetime (2009–2012); during the same period the proportion of farmers taking three meals a day increased from 20% to 43% (Mkomwa *et al.*, 2012).

In Burkina Faso, the effects of the adoption of CA were evaluated *ex ante* from an extrapolation of results of CA tests from plot level to farm scale using Olympe and Cikeda models (Ouédraogo, 2012; Zerbo, 2012). The CA system considered was characterized by direct seeding of sorghum in association with cowpea, with soil cover with an average of 4 t straw ha^{-1}. Results of simulations showed that the adoption of CA has positive effects on the cereal balance (+61%), gross margin (+345%) and the feed balance operations (+23%), but increased the labour requirement (+71%). The highest effects were noted in farms where the existing farming practices were dissimilar or even antagonist to CA principles. The increased demand for labour is mainly due to the fact that crop diversification in the CA system is considered based on crop association and not on crop rotation. Crop association leads to increased labour requirements for planting, weeding and harvesting operations. In addition, the labour requirement for sowing of crops by hoe increases by 54% when the soil is covered (with about 4 t mulch ha^{-1}). Farmers plan to use the cash surplus generated by CA to develop a small business for livestock, and strengthen availability of agricultural equipment. Farm performance (yield, gross margin, fodder) evolves with time and area under CA at the farm-level. However, farmers initially do not have enough resources to apply CA on more than 60% of their farm area. The increased requirements for labour and the difficulty to keep enough biomass on the plots are the main barriers for increasing area under CA. To meet the shortage of labour, farmers plan to increase the mobilization of existing family labour; however, this strategy could penalize activities conducted by women and youth. The proposed solution for the conservation of crop residues is more collective than individual: increasing public awareness, adaptation of guidelines for access to resources and transformation of the organization of the territory in order to find the spatial allocation of cropping that can be more favourable to CA.

Research on the effects of CA tend to confirm the usefulness of this technology as a means to ensure the sustainability of farms and highlight the need for farmers to combine technical innovations and collective organizational innovations to foster its adoption.

13.5 General Conclusion and Prospects

The general objective of this chapter was to find out from the existing scientific evidence in WCA how far CA is pertinent and accessible to farmers of this sub-region as a means to achieve the sustainable intensification of their farming systems. The methodology was based mainly on the analysis of the available scientific publications dealing with the issue of CA in WCA. This choice was dictated by the need to base the analysis as much as possible on reliable evidence. Grey literature, often rich but potentially subjective, generated from activities of past or ongoing CA projects in the sub-region has not been much considered. The analysis showed that the issue related to what has become known as CA in WCA has been treated for over 40 years by researchers who have explored several aspects. Available results tend to confirm the expected potential of CA systems, but the existing policy framework is not yet very conducive for the promotion of CA. Moreover, there are still some grey areas that need to be clarified to have comprehensive knowledge on the potential of CA and modalities for its implementation in WCA.

13.5.1 Encouraging results, but inadequate policy framework

Available evidence tends to confirm the expected positive effects of CA in combating soil degradation (improvement of soil properties, biodiversity, control of runoff and erosion) and its potential in the betterment of farmers livelihoods (increased yield, productivity and margins) in the longer term. These results are in concordance with data on the benefits generated by CA in various areas in the world (Hobbs, 2007; Lal, 2008; Kassam *et al.*, 2009a) and confirm the option to consider CA as a possible alternative for improving the sustainability of farming in WCA. However, reported failure of CA systems or emerging issues in some cases show that CA should be promoted with some site-specific recommendations. On degraded and compacted soils for instance, sub-soiling may be necessary before starting the establishment of CA to facilitate water infiltration thus avoiding the risk of waterlogging during years with abundant rainfall.

Conservation Agriculture is less developed in WCA as compared to other sub-regions of Africa (Friedrich *et al.*, 2012). This situation reflects less the difficulties encountered by producers in the implementation of the CA than the relative newness of the technology and most importantly the little effort made so far for its promotion in the sub-region. These efforts have tended to increase in recent years but they are not yet up to the complexity level needed for CA systems to harness the medium- and long-term benefits. Decision makers do not seem well aware of CA and tend to rely on the conventional green revolution model based on high use of external farm inputs (fertilizers, improved seeds, herbicides) and tillage. The low availability and poor access to agricultural services (inputs, equipment, rural credit, extension, counselling, marketing etc.) are part of general constraints facing the African agricultural sector (Mutsaers and Kleene, 2012), but these challenges significantly affect the promotion of CA. In fact, a farmer who practises CA needs to be closely backstopped by a skilled technician in order to conduct successfully the learning process necessary to have a better mastery of CA. He also needs to have easy access to equipment for direct seeding or to good-quality herbicide to control weeds. These services are generally provided to farmers by the project during its lifetime. However, when the project closes the sustainability and dissemination of achievements are no more assured either by the public services or by other rural development stakeholders.

Nevertheless, it is worth noting that while WCA states were having increasing difficulty to provide agricultural services to new actors, including mainly farmers' organizations, NGOs and the private sector (input sellers) are emerging. These new organizations are now playing an important role in the production and supply of agricultural services. Increasing their awareness and developing their capacities on CA are important elements of the strategy for dissemination of CA. Moreover, in WCA cotton-production areas, agricultural services are relatively well established. There are relatively well structured and functional supply chains and marketing channels of agricultural inputs and products though these channels are relying on the cotton value chain (Devèze, 2006). An increased awareness and interest of the cotton companies on CA would lead these cotton production areas to become potential growth poles for CA dissemination.

13.5.2 Needs for further and operational knowledge

There are still knowledge gaps on several aspects of CA, although early research on NT and mulch crops dates back more than 40 years (Lal, 1976). Available knowledge remains fragmented and incomplete. Giller *et al.* (2011) developed a research agenda to evaluate the potential of CA for smallholder farming in Africa. Further to their findings and based on the current situation of CA in WCA, we have identified three areas of research: (i) comprehensive characteristics of different types of CA systems in WCA; (ii) adoption process of CA and dynamics of cropping systems; and (iii) organizational innovations required for the promotion of CA. Further knowledge on these issues will enable a comprehensive assessment of the role of CA in WCA and provide adequate tools to leverage CA for the improvement of the sustainability of farming in WCA.

13.5.3 What are the performance and operational modalities of different Conservation Agriculture systems?

The typology of CA systems presented at the start of this chapter provides a conceptual framework to the diversity of situations when building or promoting CA in WCA. However, knowledge on the performance and operational modalities of different types of CA systems identified are yet to be completed. Important research investments are particularly required on CA systems identified for semi-arid areas (CA-CS 1 and CA-CS 2), and the effects of various intensity of application of the different principles of CA components.

13.5.4 What are the consequences of partial adoption of Conservation Agriculture on the dynamics of cropping systems?

The few available research results showed that the adoption of CA in WCA is done partially with the variable introduction of one or two of the three CA components depending on the farmer. The rate and intensity of adoption of CA and its components are variable depending on the context. This trend towards partial adoption raises the question of the divisibility of the CA and the conditions necessary for a successful adoption process. Furthermore, the features and comparative performances of transitional cropping systems emerging from the partial adoption of CA are still not well known. A long-term study is necessary to generate knowledge on the fate of emerging cropping systems. Theoretically, three scenarios are possible: (i) stabilization and therefore confirmation of a new cropping system; (ii) regression (dis-adoption) and return to traditional cropping system; or (iii) move towards complete CA systems. Precise knowledge about the frequency, conditions and modalities for the realization of these scenarios would enable the assessment of the influence of the adoption of CA in the evolution of farming systems in study areas of CA projects. Furthermore, such knowledge would help in the elaboration of appropriate strategy for the promotion of CA and its smooth integration in a comprehensive sustainable land management policy including other land management techniques as an alternative, complementary or extension to CA. Finally, it is necessary to develop and establish a monitoring system that will allow assessing the medium- and long-term dynamics of the CA in the project sites and in countries of the sub-region.

13.5.5 What are the potential costs and benefits of organizational reforms required for the scaling of Conservation Agriculture?

Researches on CA in WCA were mainly conducted at the plot scale. Data on the effects of implementation of CA at farm scale are still rare. These data are rather confirming the potential positive effects of CA on the betterment of the livelihoods of farmers but they also highlight existing and potential issues that may arise during adoption and dissemination of CA. These difficulties are mainly organizational and can be appraised at farm, village and national levels. Research and appropriate innovations must be developed at these various levels to overcome the existing or arising challenges.

1. At farm level, CA changes the overall functioning of the farm. The farmer needs adequate information and support to appropriately plan crop rotations, find new strategies for feeding livestock, work organization, management of weeds and pests in general.
2. At regional scale, the spatial organization of different activities, rules of access and community management of resources including land and crop residues as well as the interrelationships between their different uses should be reviewed in the context of CA. Mixed crop–livestock production systems dominate in WCA. The transition towards CA will require changes in livestock management practices without reducing or increasing forage resources. Hence, it is important that together with other actors, the research develops: (i) complementary innovations that solve the issue of feed; and (ii) tools for dialogue and negotiation to address the question of access and management of resources with all different categories of stakeholders.
3. At the national level, there is the question of agricultural policy. Current agricultural policies focused on the conventional green revolution models seem relatively easy to implement and have the potential to respond more quickly to the urgent needs of food security. However, it is increasingly recognized that this model is not sustainable with regards to the present environmental, agronomic and socio-economic challenges. It is important to provide decision makers with appropriate macro-economic data on costs, comparative benefits and implementation modalities of a policy more favourable for the adoption and scaling of CA as a mean to insure the sustainability of farming in WCA.

Acknowledgements

This work is an outcome of activities carried out in the framework of two IFAD-AFD and EU-funded projects, respectively: (i) the Smallholders Conservation Agriculture Promotion (SCAP) in Western and Central Africa; and (ii) the ABACO (Agroecology-based Aggradation–Conservation Agriculture: Tailoring innovations to combat food insecurity in semiarid Africa). The authors are grateful to the funders and would like also to thank Drs Kanwar Sahrawat, Ram A. Jat and Amir Kassam for their valuable comments and suggestions in the finalization of this article.

Note

[1] IFAD, International Fund for Agricultural Development; PADER/BGN, Programme d'Appui au Développement Rural en Basse Guinée Nord; PDRD, Programme de Développement Rural Durable; PICOFA, Programme d'Investissement Communautaire en Fertilité Agricole; PPILDA, Programme Promotion des Initiatives Locales de Développement à Aguie.

References

Abba, A.A., Hofs, J.-L. and Mergeai, G. (2006) Relever les défis environnementaux pour les filières cotonnières d'Afrique de l'Ouest et du Centre. *Biotechnology, Agronomy, Society and Environment* 10, 351–359.

ACT, CIRAD and ICRAF (2012) Final report of the project Smallholder Conservation Agriculture Promotion in Western and Central Africa (SCAP). ACT, Nairobi, Kenya.

Akponikpè, P.B.I., Minet, J., Gérard, B., Defourny, P. and Bielders, C.L. (2011) Spatial fields' dispersion as a farmer strategy to reduce agro-climatic risk at the household level in pearl millet-based systems in the Sahel: A modeling perspective. *Agricultural and Forest Meteorology* 151, 215–227.

Anderson, J.A. and Giller, K. (2012) On heretic's and God's blanket Salemen: contested claim for Conservation agriculture and the politics for its promotion in African Smallholder farming. In: Sumberg, J. and Thompson, J. (eds) *Contested Agronomy: Agricultural research in a changing world.* Earthscan, London.

Autfray, P. (1997) Fixation de l'agriculture à base de vivrier en zone forestière de Côte d'Ivoire. Création de systèmes de culture avec des légumineuses de couverture sur les dispositifs d'Oumé de 1994 à 1996. Cirad/Idessa, Bouaké, Côte d'Ivoire.

Autfray, P. and Sissoko, F. (2011) Conservation agriculture advances with permanent rainfed cotton based cultivation with animal traction in Southern Mali. Communication at the 5th World Congress of Conservation Agriculture incorporating 3rd Farming Systems Design Conference, September 2011 Brisbane, Australia. Available at: http://www.wcca2011.org (accessed 3 December 2012).

Bado, V., Sawadogo, A., Thio, B., Bationo, A., Traoré, K. and Cescas, M. (2011) Nematode infestation and N-effect of legumes on soil and crop yields in legume-sorghum rotations. *Agricultural Sciences* 2, 49–55.

Balarabé, O., Dugué, P. and Lifran, R. (2012) Capital sol et innovation institutionnelle. *Economies et Sociétés, Série Systèmes agroalimentaires* 34, 1927–1944.

Boahen, P., Dartey, B.A., Dogbe, G.D., Boadi, E.A., Triomphe, B., Daamgard-Larsen, S. and Ashburner, J. (2007) *Conservation Agriculture as Practised in Ghana.* Nairobi. African Conservation Tillage Network (ACT), Centre de Coopération Internationale de Recherche Agronomique pour le Développement (CIRAD), Food and Agriculture Organization of the United Nations (FAO), Rome, Italy.

Boli, B.Z. (1996) Fonctionnement des sols sableux et optimisation des pratiques culturales en zone soudanienne humide du Nord-Cameroun. Thèse de doctorat en Sciences de la terre, Option Sciences du Sol et Production Végétales. Université de Bourgogne en Sciences de la Terre, Dijon, France.

Boli, B.Z., Roose, E., Bep-Aziem, B., Sanon, K. and Waechter, F. (1996) Effets des techniques culturales sur le ruissellement, l'érosion et la production de coton et maïs sur un sol ferrugineux tropical sableux. Recherche de systèmes de culture intensifs et durables en région soudanienne du Cameroun (Mbissiri, 1991-92). *Cahiers ORSTOM Pédol* 28, 309–326.

Bougoum, H. (2012) Analyse des effets spécifiques et combinés des principes de l'agriculture de conservation sur la conduite et les performances technico-économiques des parcelles de sorgho (Sorghum bicolor (L). Moench). Mémoire d'ingénieur Agronome, Institut du Développement Rural/Université Polytechnique de Bobo Dioulasso (UPB), Bobo-Dioulasso, Burkina Faso.

Brévault, T., Bikay, S., Maldès, J.M. and Naudin, K. (2007) Impact of a no-till with mulch soil management strategy on soil macrofauna communities in a cotton cropping systems. *Soil & Tillage Research* 97, 140–149.

Brévault, T., Guibert, H. and Naudin, K. (2009) Preliminary studies of pest constraints to cotton seedlings in a direct seeding mulch-based system in Cameroon. *Experimental Agriculture* 45, 25–33.

Collomb, P. (1999) Une voie étroite pour la sécurité alimentaire d'ici à 2050. Economica, FAO, Paris.

Deguine, J.P., Goze, E. and Leclant, F. (1994) Incidence of early outbreaks of the aphid *Aphis gossypii* Glover in cotton growing in Cameroon. *International Journal of Pest Management* 40, 132–140.

Derpsch, R. and Friedrich, T. (2010) Global Overview of Conservation Agriculture Adoption. Proceedings, Lead Papers, 4th World Congress on Conservation Agriculture, 4–7 February 2009, New Delhi, India, pp. 429–438. Available at: http://www.fao.org/ag/ca/6c.html (accessed 19 November 2012).

Devèze, J.-C. (2006) Le coton, moteur du développement et facteur de stabilité au Cameroun du Nord? *Afrique Contemporaine* 217, 107–120.

Djamen, N.P., Maraux, F., Ashburner, J., Triomphe, B. and Kienzle, J. (2005) L'agriculture de conservation en Afrique francophone de l'ouest et du centre: état des lieux, enjeux et défis. In: *Regards sur l'agriculture de conservation en Afrique de l'ouest et du centre et ses perspectives.* CIRAD, AFD, FFEM, FIDA, FAO, Rome, pp. 63–75.

Dugué, P. and Guyotte, K. (1996) Semis direct et désherbage chimique en zone cotonnière du Cameroun. *Agriculture et développement* 11, 3–15.

Dugué, P., Autfray, P., Blanchard, M., Djamen, N.P., Dongmo, A.L., Girard, P., Olina, J.-P., Ouédraogo, S., Sissoko, F. and Vall, E. (2012a) L'agroécologie pour l'agriculture familiale dans les pays du Sud: impasse ou voie d'avenir? Le cas des zones de savane cotonnière de l'Afrique de l'Ouest et du Centre. Communication présentée au Colloque René Dumont, Paris, 15 et 16 Novembre 2012. GRET, AUF, Fondation René Dumont. Available at: http://www.gret.org/2012/10/colloque-rene-dumont-revisite-et-les-politiques-agricoles-africaines (accessed 10 December 2012).

Dugué, P., Balabaré, O., Olina, J.P. and Kossoumna-Liba'a, N. (2012b) Etude de cas n°8: Agriculture de conservation, production fourragère et sécurité alimentaire. Le cas de l'introduction de *Brachiaria ruziensis* dans les systèmes de production de la zone des savanes du Cameroun. Available at: http://www.diplomatie.gouv.fr/fr/enjeux-internationaux/securite-alimentaire-mondiale-et/agriculture-et-changement/article/systemes-de-production-durables-en (accessed 18 January 2013).

Dumanski, J., Peiretti, R., Benetis, J., McGarry, D. and Pieri, C. (2006) The paradigm of conservation tillage. *Proceedings of the World Association of Soil and Water Conservation* 1, 58–64.

Essecofy, G.F. (2011) Potentiel de développement de l'agriculture de conservation des petites exploitations agricoles familiales: étude de cas à Gori et Kompienbiga (Burkina Faso). Thèse de Master of Science du CIHEAM. CIHEAM/IAM Montpellier, Montpellier, France.

FAO (2008) Conservation Agriculture. Available at: http://www.fao.org/ag/ca/index.htm (accessed 11 November 2012).

FAO (2011) *Save and Grow, a policymaker's guide to sustainable intensification of smallholder crop production*. Food and Agriculture Organization of the United Nations, Rome, Italy.

FAO, WFP and IFAD (2012) *The State of Food Insecurity in the World 2012. Economic growth is necessary but not sufficient to accelerate reduction of hunger and malnutrition*. Food and Agriculture Organization of the United Nations, Rome, Italy.

Friedrich, T., Derpsch, R. and Kassam, A. (2012) Overview of the Global Spread of Conservation Agriculture. *Field Actions Science Reports*. [Online], Special Issue 6. Available at: http://factsreports.revues.org/1941 (accessed 21 December 2012).

Ganou, S.O. (2012) Analyse des performances technico-économiques et de l'adoption des systèmes d'agriculture de conservation dans la région de l'Est du Burkina Faso. Mémoire de Master II en Innovation et Développement en milieu rural. Université de Ouagadougou, Burkina Faso.

Garrity, D.P., Akinnifesi, F.K., Ajayi, O.C., Weldesemayat, S.G., Mowo, J.G., Kalinganire, A., Larwanou, M. and Bayala, J. (2010) Evergreen agriculture: a robust approach to sustainable food security in Africa. *Food Security* 2, 197–214.

Giller, K.E., Witter, E., Corbeels, M. and Tittonell, P. (2009) Conservation agriculture and smallholder farming in Africa: The heretics' view. *Field Crops Research* 114, 23–34.

Giller, K.E., Corbeels, M., Nyamangara, J., Triomphe, B., Affholder, F., Scopel E. and Tittonell, P. (2011) A research agenda to explore the role of conservation agriculture in African smallholder farming systems. *Field Crops Research* 124, 468–472.

Hobbs, P.R. (2007) Conservation agriculture: What is it and why is it important for future sustainable food production? *The Journal of Agricultural Science* 145(2), 127–137.

Hobbs, P.R., Sayre, K. and Gupta, R. (2007) The role of conservation agriculture in sustainable agriculture. *Philosophical Transactions of the Royal Society, Series B* 363, 543–555.

HORUS Entreprise (2009) Mission d'appui à Sodécoton en vue d'améliorer son système d'appui aux producteurs. Partie 2 – aspects relatifs aux SCV. Paris, France

Houndékon, V., Manyong, V.A., Gogan, C.A. and Versteeg, M.N. (1998) Déterminants del'adoption de *Mucuna* dans le département de Mono au Bénin. In: Buckles, D., Eteka, A., Osiname, O., Galiba, M. and Galiano, G. (eds) *Plantes de couverture en Afrique de l'Ouest - Une contribution à l'agriculture durable*. Crdi, Ottawa, Canada, pp. 45–54.

IPCC (2007) Fourth Assessment Report of the Intergovernmental Panel on Climate Change. IPCC.

Kassam, A., Friedrich, T., Shaxson, F. and Pretty, J. (2009a) The spread of Conservation Agriculture: justification, sustainability and uptake. *International Journal of Agricultural Sustainability* 7, 292–320.

Kassam, A., Kueneman, E., Kebe, B., Ouedraogo, S. and Youdeowei, A. (2009b) Enhancing Crop-Livestock systems in Conservation Agriculture for Sustainable Production Intensification. A farmer Discovery Process Going to Scale in Burkina Faso. *Integrated Crop Management* 7. FAO, Rome 42 pp.

Knowler, D. and Bradshaw, B. (2007) Farmers' adoption of Conservation Agriculture: A review and synthesis of recent research. *Food Policy* 32, 25–48.

Kourouma, M. and Bozza, J. (2005) Le semis direct au secours de l'exploitation de Solo Koulibaly. In: *Regards sur l'agriculture de conservation en Afrique de l'ouest et du centre et ses perspectives*. CIRAD, AFD, FFEM, FIDA, FAO, Rome, pp. 39–49.

Lahmar, R., Bationo, B.A., Lamso, N.D., Guéro, Y. and Tittonell, A. (2012) Tailoring conservation agriculture technologies to West Africa semi-arid zones: Building on traditional local practices for soil restoration. *Field Crops Research* 132, 158–167.

Lal, R. (1976) No-tillage effects on soil properties under different crops in western Nigeria. *Proceedings of Soil Science Society of America* 40, 762–768.

Lal, R. (1977) Importance of tillage systems in soil and water management in the tropics. In: Lal, R. (ed.) *Soil Tillage and Crop Production*. IITA, Ibadan, Nigeria, pp. 25–32.

Lal, R. (1986) No-tillage and surface-tillage systems to alleviate soil-related constraints in the tropics. In: Sprague, M.A. and Triplett, G.B. (eds) *No-tillage and Surface-tillage Agriculture: the Tillage Revolution*. Wiley, New York, pp. 261–317.

Lal, R. (1989) Conservation tillage for sustainable agriculture: tropics versus temperate environments. *Adv. Agron.* 42, 85–197.

Lal, R. (2007) Constraints to adopting no-till farming in developing countries. *Soil and Tillage Research* 94, 1–3.

Lal, R. (2008) Soils and sustainable agriculture. A review. *Agronomy of Sustainable Development* 28(1), 57–64.

Lamantia, A. (2012) Analyse comparative des processus d'adoption et des impacts du Semis direct sur Couverture Végétale permanente (SCV) sur les exploitations agricoles familiales dans 3 régions tropicales: Madagascar, Cameroun et Laos. Mémoire de Fin d'Études. Agrocampus Ouest, Rennes, France.

M'Biandoun, M., Dongmo, A.L. and Balarabe, O. (2010) Systèmes de culture sur couverture végétale en Afrique centrale: conditions techniques et socioéconomique pour son développement. In:Seiny-Boukar, L. and Boumard, P. (eds) *Actes du colloque 'Savanes africaines en développement: innover pour durer'*, 20–23 April 2009, Garoua, Cameroun. Prasac, N'Djaména, Tchad; Cirad, Montpellier, France, CDROM. 10 pp.

Mando, A. (1997) The impact of termites and mulch on the water balance of crusted Sahelian soil. *Soil Technology* 11(2), 121–138.

Mazvimavi, K. and Twomlow, S. (2009) Socioeconomic and institutional factors influencing adoption of conservation farming by vulnerable households in Zimbabwe. *Agricultural Systems* 101, 20–29.

Mkomwa, S., Beguin, E., Mulimbi, W., Birindwa, D. and Sadiki, J. (2012) No-Tillage Agriculture (NTA) in Kailo, Kasongo and Kabambare Territories of Maniema Province (DR Congo), Final Evaluation Report. Catholic Relief Services (CRS), Kinshasa, Democratic Republic of Congo.

Mrabet, R. (2002) Stratification of soil aggregation and organic matter under conservation tillage systems in Africa. *Soil and Tillage Research* 66(2), 119–128.

Mutsaers, H.J.W. and Kleene, P.W.M. (2012) *What is the Matter with African Agriculture? Veteran's view between past and future*. KIT Publishers, Amsterdam, the Netherlands.

Naudin, K. (2012) You can't eat mulch and have it too. Cropping system design and trade-offs around biomass use for Conservation Agriculture in Cameroon and Madagascar. PhD Thesis, Wageningen University, Wageningen, the Netherlands.

Naudin, K., Gozé, E., Balarabe, O., Giller, K.E. and Scopel, E. (2010) Impact of no tillage and mulching practices on cotton production in North Cameroon: A multi-locational on-farm assessment. *Soil & Tillage Research* 108, 68–76.

Ouédraogo, Y. (2012) Analyse 'ex-ante' des effets de l'agriculture de conservation sur le fonctionnement et les performances technico-économiques des exploitations agricoles à l'aide de la modélisation: cas de Koumbia. Mémoire d'ingénieur, UPB, Bobo-Dioulasso, Burkina Faso.

Pieri, C. (1989) *Fertilité des terres de savanes. Bilan de trente ans de recherché et de développement agricole au sud du Sahara*. Ministère de la Coopération et du Développement, CIRAD, Montpellier, France.

Saito, K., Azoma, K. and Oikeh, S.O. (2010) Combined effects of *Stylosanthes guianensis* fallow and tillage management on upland rice yield, weeds and soils in southern Benin. *Soil and Tillage Research* 2, 57–63.

Séguy, L. (2008) *Rapport de mission au Cameroun*. Cirad, Montpellier, France.

Serpantié, G. (2009) L'agriculture de conservation à la croisée des chemins en Afrique et à Madagascar, VertigO - la revue électronique en sciences de l'environnement 9, 3 [Online]. Available at: http://vertigo.revues.org/9290 (accessed 15 February 2010).

Sissoko, F. (2009) Analyse des flux d'eau dans les systèmes de culture sous couverture végétale en zone soudano sahélienne: cas du coton semé après une culture de sorgho/brachiaria au sud du mali. Thèse de Doctorat de SupAgro Montpellier, Discipline: Sciences du sol – Agronomie. SupAgroMontpellier, Montpellier, France.

Soutou, G. (2004) Modifications du bilan hydrique par les systèmes de culture sur couverture végétale: Cas du cotonnier et du sorgho dans l'Extrême Nord du Cameroun. Mémoire de fin d'études d'Agronomie approfondie. École Nationale Supérieure agronomique de Montpellier (ENSAM), Montpellier, France.

Tarawali, G. and Ikwuegbu, O.A. (1993) The potential of forage legumes in soil management for sustainable livestock and crop production in the subhumid zone of Nigeria. In: Cook, H.F. and Lee, H.C. (eds) *Proceedings, 3rd International Conference on Sustainable Agriculture*, Wye College, University of London. Wye College Press, London, pp. 435–443.

Tarawali, G., Dembélè, E., N'Guessan, B. and Youri, A. (1998) Smallholders' use of Stylosanthes for sustainable food production in subhumid West Africa. In: Buckles, D., Eteka, A., Osiname, O., Galiba, M. and Galiano, G. (eds) *Plantes de couverture en Afrique de l'Ouest - Une contribution à l'agriculture durable*. Crdi, Ottawa, Canada, pp. 107–170.

Tittonell, P. and Giller, K.E. (2012) When yield gaps are poverty traps: The paradigm of ecological intensification in African smallholder agriculture. *Field Crops Research*, 1–15.

Tittonell, P., Scopel, E., Andrieu, N., Posthumus, H., Mapfumo, P., Corbeels, M., van Halsemaf, G.E., Lahmar, R., Lugandu, S., Rakotoarisoa, J., Mtambanengwe, F., Pound, B., Chikowo, R., Naudin, K., Triomphe, B. and Mkomwa, S. (2012) Agroecology-based aggradation-conservation agriculture (ABACO): Targeting innovations to combat soil degradation and food insecurity in semi-arid Africa. *Field Crops Research* 132, 168–174.

World Bank (2012) *Turn Down the Heat. Why a 4°C warmer world should be avoided*. World Bank, Washington, DC.

Zerbo, I. (2012) Analyse des effets potentiels de l'agriculture de conservation sur les performances technico-économiques des exploitations agricoles de Sindri (Province du Bam, Burkina Faso). Mémoire d'ingénieur, UPB, Bobo-Dioulasso, ACT, Ouagadougou, Burkina Faso.

14 Conservation Agriculture in Southern Africa

Justice Nyamangara,[1] Regis Chikowo,[2] Leonard Rusinamhodzi[3] and Kizito Mazvimavi[4]

[1]International Crops Research Institute for the Semi-Arid Tropics, Matopos Research Station, Bulawayo, Zimbabwe; [2]University of Zimbabwe, Harare, Zimbabwe; [3]Centro Internacional de Agricultura Tropical, Harare, Zimbabwe; [4]International Crops Research Institute for the Semi-Arid Tropics, Patancheru, Andhra Pradesh, India

14.1 Introduction

Conservation Agriculture (CA) based on minimum mechanical soil disturbance has been widely promoted in the smallholder farming sector in southern Africa over the last decade, as a means to increase productivity, improve food security, primarily through donor/non-governmental organizations. Although the number of farmers exposed to CA has significantly increased on the back of promotional programmes, the area allocated to CA remains small, as farmers generally adapt and adopt new technologies slowly due to increased labour demand associated with the manual form of the technology. The majority of the smallholder farmers have adopted minimum tillage but integration of crop rotation and achieving the minimum 30% organic mulching components have remained the least adopted. The low adoption of mulching is due to low crop productivity with crop residues barely reaching 1 t ha^{-1} and competing uses for the residues as livestock feed, firewood and building material. The lack of incorporation of legumes into the cropping systems is attributed to limited or non-existent markets for the legumes. Research in CA in southern Africa is limited and most promotion programmes are based on experiences from outside sub-Saharan Africa. There is need to conduct research that enables extension agents to appropriately target CA both technically and socio-economically. To increase the adoption of CA there is a need to increase crop productivity through increasing smallholder access to soil fertility inputs, introduction of mechanized implements for tillage and introduction of herbicides for weed control. Conservation Agriculture has done well in increasing productivity at many locations. Although most governments in Southern Africa support CA still, there is a need to streamline policies to avoid potential conflicts with conventional tillage (ConvT) and livestock production.

14.2 History of Conservation Agriculture in Southern Africa

Productivity of rainfed cropping systems in Southern Africa is far below potential, mainly due to inherent poor fertility of the soils, paltry external resource use and erratic rainfall, often associated with prolonged mid-season dry spells. Resource-conserving

 Conservation Agriculture: Global Prospects and Challenges (eds R.A. Jat, K.L. Sahrawat and A.H. Kassam)

management systems that positively alter the soil crop environment are perceived to mitigate the negative impacts of limited access to production resources and climate change. In Southern Africa, rainfall is projected to decline by an estimated 30% while temperature will increase by an estimated 3°C by 2050 (IPCC, 2007). The utility of CA in the sub-region should therefore be evaluated against its ability to reduce climatic risk in the face of substantial crop yield losses linked to soil moisture deficits.

The initial version of CA in Southern Africa was first implemented at a commercial farm in the sub-humid area of Zimbabwe in the 1980s. The objective then was to reduce soil erosion and stabilize crop yields by minimizing soil tillage and retaining crop residues and increase profits. The current version of CA is broader and based on three principles: (i) minimum soil disturbance; (ii) permanent soil cover provided by a mulch of organic residues or cover crops; and (iii) crop rotations and associations, preferably with legumes (http://www.fao.org/ag/ca). Conservation Agriculture was introduced to smallholder farmers in Zambia by the Conservation Farming Unit (CFU) in 1996 (Haggblade and Tembo, 2003; Umar *et al.*, 2011) and in Zimbabwe by non-governmental organizations (NGOs) in the 2003/4 cropping season (Marongwe *et al.*, 2011). In Malawi, CA was initially introduced by Sassakawa Global 2000 (SG 2000) in 1998, but large scale promotion started in 2004 when the International Maize and Wheat Improvement Center CIMMYT introduced it in Balaka, Dowa and Mzimba districts. In South Africa, CA is mainly practised by large-scale commercial farmers, and in Mozambique it is still in its infancy with promotion taking place in central and northern parts of the country.

14.3 Current Status

The majority of smallholder farmers practising CA in Southern Africa have mostly adopted minimum tillage but integration of crop rotation with legumes and achieving the minimum 30% organic mulching component is still low. Strong mixed crop–livestock interactions in the smallholder areas imply that residues are not available for mulching in preference to livestock feed, e.g. in dry regions of Zambia, Zimbabwe and South Africa's Limpopo Province. However, in Malawi and Mozambique crop production dominates livestock production and the potential for mulching at recommended crop residue intensities is higher. In Zimbabwe and Zambia, about 90% of the over 130,000 farmers practising CA are using planting basins, and at different residue use or crop rotation intensities. However, panel surveys conducted in Zimbabwe over 6 years have shown that the benefits of CA both in terms of yield and soil quality are largest when all three principles are used (Nyamangara *et al.*, 2012a).

In Zimbabwe, numerous organizations such as international agriculture research institutions (e.g. CIMMYT, ICRISAT, ICRAF) and national and international NGOs (e.g. Catholic Relief Services, CARE International, Concern, Oxfam and World Vision amongst others) and governmental extension services are promoting CA (Marongwe *et al.*, 2011). Conservation Agriculture is being promoted mainly through specific projects intended to promote its uptake among smallholder farmers. The planting basin tillage option is targeted to areas of limited draught power; preparation of planting basins (Fig. 14.1) as well as application of organic sources of nutrients occurs in the dry season, giving farmers an opportunity to plant early (Twomlow and Hove, 2006). In 2008/09 data from a yield survey conducted by ICRISAT across agroecologically contrasting districts in Zimbabwe showed that farmers implementing CA achieved 25–65% higher yields in their CA plots compared to conventional tillage (ConvT) plots (Table 14.1). By the 2009/10 cropping season about 88,262 households were practising one or more of the CA principles in parts of their fields in Zimbabwe (Marongwe *et al.*, 2011).

Fig. 14.1. Smallholder farmer fields under CA in Southern Africa: (a) crop residue mulch spread on the soil surface in a field in Lilongwe District, Malawi; (b) hand-hoe-made planting basins in Nkayi District, Zimbabwe; and (c) demonstration of sowing using jab planter in Masvingo District, southern Zimbabwe.

The promotion of CA in Zambia involves several stakeholders from the private sector, government and the donor community (Umar *et al.*, 2011). Incentives are at the core of its promotional efforts through NGOs such as Care International. Planting basins and ripping are the most promoted technologies and to a lesser extent the dibble stick. The Conservation Farming Unit (CFU, Zambia) has been promoting dry-season land preparation using minimum tillage systems, crop residue retention, crop rotations and cover crops, precise input application, agroforestry and incorporation of perennial crops (CFU, 2006). Zambia is considered a country with the largest CA area in Southern Africa but most farmers only use part of the CA system on part of their land (Baudron *et al.*, 2007).

Table 14.1. Yield trends in Conservation Agriculture and conventionally tilled fields across 13 districts located in contrasting agroecological natural regions in Zimbabwe. Data collected from same farmers as part of a panel survey conducted by ICRISAT.

		2008/09		2009/10		2010/11	
Natural region	District	CA	ConvT	CA	ConvT	CA	ConvT
NR II	Bindura	1490	1208	1686	1302	1203	1150
	Murewa	2132	1412	2372	1738	2437	1861
	Seke	1635	962	884	856	1627	1238
	Average	1752	1194	1647	1299	1756	1416
NR III	Chirumhanzu	1428	914	751	663	1120	1120
	Masvingo	2439	1355	1386	1199	2031	1043
	Mt Darwin	1190	877	926	1115	1172	769
	Average	1685	1048	1021	992	1441	977
NR IV	Gokwe South	1433	713	1972	1059	1559	1559
	Insiza	1646	1105	551	370	1007	384
	Nkayi	1579	792	2581	1540	1497	897
	Nyanga	1308	874	741	677	1694	1231
	Average	1492	871	1461	912	1439	1018
NR V	Binga	1384	868	1034	649	1267	1329
	Chivi	1658	874	228	168	1063	512
	Hwange	1563	713	1241	1737	958	1145
	Average	1535	818	834	851	1096	995
Combined average		1607	974	1258	1006	1433	1095

CA, Conservation Agriculture field; ConvT, conventionally tilled field; NR, natural region.

In Malawi, CIMMYT in partnership with Total Land Care (TLC) and the government extension department are the major agents promoting CA. Conservation Agriculture was introduced in Malawi in 1998 by Sassakawa Global 2000 (SG 2000) supported by the Malawian Government through a targeted input programme (TIP) funded by various donor organizations (Ito *et al.*, 2007). The major driver behind this initiative was a set of management practices such as improved recommendations on plant populations, herbicides for weed control (supported by Monsanto) and adequate fertilization, which was closely associated with an emphasis on input support. The approach was not sustainable because much of the SG 2000 promotion was conducted in a linear top-down approach ignoring the need to build supportive bottom-up networks that would facilitate improved and sustained access to inputs. In 2004, CA was reintroduced in some target communities around Balaka (south), Dowa (central) and Mzimba (north) through collaborative efforts between the International Maize and Wheat Improvement Centre CIMMYT and the Research and Extension Departments of the Malawi government. This work was later expanded to other districts in collaboration with TLC, a NGO registered in Malawi, Tanzania, Mozambique, Zambia and Switzerland.

In Mozambique, since 2007, the International Centre for Tropical Agriculture (CIAT), CIMMYT, and development agencies from both government and non-governmental organizations have been involved in the promotion of CA in central and northern Mozambique. Demonstrating that agricultural development interventions and research can improve livelihoods of smallholder farmers has been the main objective of these interventions (Nkala *et al.*, 2011). The most promoted options were the jab-planters and planting basins (Fig. 14.1) due to limited availability of draught power. A national CA taskforce has recently been set up to drive the uptake of CA in Mozambique. The CA taskforce is comprised of representatives from farmer organizations, donor agencies, local and international NGOs, UN agencies and key ministries such as Ministry of Agriculture.

14.4 Prospects

The initial move for reduced tillage on large-scale commercial farms the world-over was driven by the need to minimize crop production costs (fuel, wear and tear of machinery).

Inherent poor soil fertility and unreliable rainfall have entrenched poor productivity in smallholder farms across Southern Africa. This has been aggravated by application of suboptimal fertilizer rates (Palm *et al.*, 2004) by farmers resulting in substantial nutrient mining. Low fertilizer application rates are largely driven by poverty and unavailability of the fertilizers. Overlaying these socio-economic and inherent biophysical challenges is the human-induced land degradation due to prolonged continuous cultivation. The mouldboard plough, a basic mechanization tool that is used by smallholder farmers in many parts of southern Africa, has been linked to destruction of soil structure, accelerated depletion of soil organic matter (SOM) and sheet erosion. Recently, there have been compounding impacts of the HIV/AIDS pandemic on the capacity of households to produce food. Rural-to-urban migration has also negatively impacted on labour availability in the smallholder farming areas. In the face of these many challenges, several initiatives have been put forward to ensure sustained crop production in the smallholder sector. Among such initiatives has been the promotion of CA and integrated soil fertility management (ISFM).

Conservation Agriculture has gained much publicity in Southern Africa, and has been actively promoted by the NGO community, farmer organizations, research institutions, FAO and regional institutions such as AU-NEPAD[1] and COMESA, as an ecologically sound vehicle towards food security. With CA, positive mutually reinforcing feedbacks are expected such as increased water productivity in the face of increasing risks associated with climate change. Conservation Agriculture creates more conducive conditions for farmers to invest and reverse years of soil physico-chemical degradation. Cropping systems in Southern Africa are particularly at risk from unreliable rainfall, due to more frequent El Niño phenomenon development in the eastern tropical Pacific. It is projected that rainfall in most of Southern Africa will decline by about 30% during the next 50 years (IPCC, 2007).

Experience in the region with smallholder farmers already indicates that strict adherence to the three key principles in implementing CA is usually constrained by several factors, including the evolution of current farming systems over the past 50 years, the past farmer training that rewarded excellent tillage (e.g. criteria for master farmer certification in Zimbabwe) and the extent of crop–livestock integration. Therefore, adapted and niche-based CA alternatives are needed to increase water productivity, mitigate climatic risks, restore and maintain soil fertility, and provide agro-ecological functions. Simply put, CA provides farmers with the opportunity to regenerate rather than exploit the environments in which they derive their livelihoods.

14.5 Research Results Reported

14.5.1 Effect of Conservation Agriculture on yield

Most studies on CA have focused on yield effects of the practice, especially on the staple maize crop (Table 14.2). Most of the studies were implemented as part of relief programmes and therefore focused on resource-constrained households. Consequently in some cases fertilizer effects were reported as CA effects because fertilizer was only applied to CA plots and farmers did not have fertilizer inputs for the ConvT plots. Also due to a strong interaction between cropping and communally grazed livestock, the spreading of crop residues on soil surface as mulch was not done, or the

Table 14.2. Effects of Conservation Agriculture (CA) on the yield of maize, sorghum, cowpea and cotton compared with conventional tillage (ConvT) in southern Africa.

Crop	ConvT yield (t ha^{-1})	CA yield (t ha^{-1})	Yield advantage (%)	CA options Duration (seasons)	Cropping system	Companion/rotational crop	Reference
Maize	3.2	6.7	111	2	Monocrop	–	Rockstrom *et al.* (2009)
Maize	3.2	6.8	117	2	Monocrop	–	Rockstrom *et al.* (2009)
Maize	0.8	5.9	638	3	Rotation	Pigeonpea	Rusinamhodzi *et al.* (2012b)
Maize	0.8	5.8	625	3	Intercrop	Pigeonpea	Rusinamhodzi *et al.* (2012b)
Maize	0.8	2.8	250	3	Intercrop	Pigeonpea	Rusinamhodzi *et al.* (2012b)
Maize	3.4	4.9	41	3	Monocrop	–	Ngwira *et al.* (2012)
Maize	3.4	4.2	21	3	Intercrop	Lablab	Ngwira *et al.* (2012)
Maize	3.4	4.2	22	3	Intercrop	Mucuna	Ngwira *et al.* (2012)
Maize	3.4	4.3	25	3	Intercrop	Pigeonpea	Ngwira *et al.* (2012)
Cotton	0.5	0.4	–13	3	Monocrop	–	Baudron *et al.* (2012)
Sorghum	0.4–1	0.4–1.1	10–11	2	Intercrop	Cowpea, pigeonpea, jackbean, sunnhemp, velvet bean	Baudron *et al.* (2012)
Maize	4.4	5.4	23	7	Monocrop	–	Thierfelder *et al.* (2012)
Maize	4.4	7.3	66	4	Rotation	Sunnhemp	Thierfelder *et al.* (2012)
Maize	3.1	4.5	45	6	Monocrop	–	Thierfelder *et al.* (2012)
Maize	3.1	4.2	35	5	Rotation	Cotton	Thierfelder *et al.* (2012)
Maize	3.1	6.1	97	5	Rotation	Cotton–Sunnhemp	Thierfelder *et al.* (2012)
Maize	5.0	4.7	6	5	Monocrop	–	Thierfelder *et al.* (2012)
Maize	5.0	6.4	28	4	Rotation	Sunflower	Thierfelder *et al.* (2012)
Maize	5.0	6.4	28	4	Rotation	Sunflower–Beans	Thierfelder *et al.* (2012)
Maize	6.3	6.4	2	4	Monocrop	–	Thierfelder *et al.* (2012)
Maize	6.3	8.1	29	3	Rotation	Cowpea	Thierfelder *et al.* (2012)
Cowpea	0.4	0.2	–44	5	Rotation	Maize–Sorghum	Mashingaidze *et al.* (2012)
Cowpea	0.4	0.3	–34	5	Rotation	Maize–Sorghum	Mashingaidze *et al.* (2012)
Sorghum	4.2	2.6	–37	6	Rotation	Maize–Cowpea	Mashingaidze *et al.* (2012)
Sorghum	4.2	3.7	–12	6	Rotation	Maize–Cowpea	Mashingaidze *et al.* (2012)
Maize	0.92	0.86	–6	2	Monocrop	–	Masvaya *et al.* (unpublished)
Maize	0.92	0.81	–12	2	Rotation	Cowpea	Masvaya *et al.* (unpublished)

residues were removed and only applied at the start of the cropping season. This implied that the mulching effect of water conservation was only achieved during the cropping season.

Early field testing of CA focused on the basin tillage package, which has potential to be adopted by resource-constrained farmers who do not have access to draught power. The central component of the basin tillage package is the planting basin, which is prepared using hand hoes (Fig. 14.1). Field testing conducted by ICRISAT in 2004–2006 in farmers' fields across 11 districts under semi-arid conditions in Zimbabwe showed that additional grain yield benefits were gained under CA despite the higher labour demand (Twomlow and Hove, 2006). However, meta-analyses have shown that yield benefits of CA take several years to show (5–9 years) compared to soil erosion and runoff control effects, which are immediate (Rusinamhodzi *et al.*, 2011; Nyamangara *et al.*, 2012c). Research results from experiments conducted in Southern Africa also confirmed the trend (Table 14.2). Thierfelder and Wall (2012) reported maize yield benefits on CA plots, compared with ConvT, after several seasons under sub-humid conditions in central Zimbabwe. In Malawi, Ngwira *et al.* (2012) reported no significant maize grain yield results in CA plots in the first 4 years and significant yield benefits were only recorded in the fifth (29–48%) and sixth (43–51%) years in Zidyana smallholder area, central Malawi.

Limited studies have been conducted to assess the effect of targeting CA according to socio-ecological conditions. Based on worldwide published data on CA research, Rusinamhodzi *et al.* (2011) reported that: (i) 92% of that data showed that mulch cover in high rainfall areas leads to lower yields due to waterlogging caused by reduced evaporation; (ii) 85% of the data showed that soil texture is important and improved yields are likely in well-drained soils; (iii) 73% of the data showed that CA practices require high inputs especially N for improved yield; and (iv) 63% of the data show increased yields are obtained with crop rotation. A meta-analysis of CA trials conducted by ICRISAT (2004–2011) showed similar results (Nyamangara *et al.*, 2012c). In a related study, Nyamangara *et al.* (2012a) reported that basin tillage + mulching without fertilizer depressed yield by 48% whereas with fertilizer addition it increased yield by 20–33%; basin tillage + rotation without fertilizer depressed yield by 28% whereas with fertilizer addition it increased yield by 7–9%; and all the three principles without fertilizer depressed yield by 36% whereas with fertilizer application yield was increased by 58–69%. However, both Rusinamhodzi *et al.* (2011) and Nyamangara *et al.* (2012c) reported strong correlation between yield from CA treatments and environmental mean implying that CA does not have the potential to adequately address the challenges associated with poor rainfall distribution and therefore should be promoted in combination with other water conservation/drought mitigation strategies.

14.5.2 Effect of Conservation Agriculture on soil properties

In Southern Africa, initial experimentation with reduced tillage systems on research stations and smallholder farms was based on ripping systems using ox-drawn ploughs mounted with ripper tines, and ridging using animal-drawn high wing ridgers. Legume–cereal sequences, a key component of what currently defines CA, were seldom integrated in such experimentation. Results of years of such studies showed that CA systems did not necessarily improve crop yields compared to conventional ploughing, but reduced soil loss and increased soil aggregate stability, infiltration rate and moisture retention were observed (Vogel, 1993; Munyati, 1997; Nyagumbo, 2008).

More recently, field-level (Thierfelder *et al.*, 2012; Table 14.3) studies have shown that CA consistently improves soil physical properties compared with ConvT, but effects on chemical properties are less convincing (Nyamangara *et al.*, 2012a). A study by CIMMYT at Henderson Research Station, central Zimbabwe, reported 38–65% increase in infiltration on CA plots on a clay soil compared with ConvT (Thierfelder and

Table 14.3. Effect of Conservation Agriculture on selected soil physical properties in fields of smallholder farmers in Zimbabwe compared with conventional tillage.

Parameter	Soil clay content (%)	Conservation Agriculture	Conventional tillage
Soil organic C (g kg^{-1})	Low	73.5	41.2
	High	80.7	47.0
Bulk density (g cm^{-3})	Low	1.40	1.44
	High	1.45	1.49
Aggregate stability (Ima)[a]	Low	102.2	97.7
	High	112.3	95.1
Pore volume	Low	15.9	7.7
	High	9.0	7.6

Low clay content, 13–18%; high clay content, 18–45%. [a]Ima is the stable macro-aggregation index

Wall, 2012). Using a rainfall simulator, the authors reported 45–90% and 47–151% more time to ponding on CA plots at Henderson Research Station and Hereford smallholder area, respectively, compared with ConvT. Thierfelder and Wall (2012) also reported 123–168% and 11–24% greater aggregate stability on CA plots at Hereford (clay soil) and Chikato (sandy soil) smallholder areas, respectively, in central Zimbabwe. The study demonstrated that clay soils were more responsive to CA than sandy soils in terms of soil physical properties.

In a study by ICRISAT covering nine districts of contrasting agroecological conditions (<450 – 900 mm rainfall $year^{-1}$) demonstrated that CA significantly reduced soil bulk density, and increased water-holding capacity, pore volumes, organic carbon and aggregate stability in both low clay (13–18%) and high clay (18–45%) soils, compared with ConvT (Table 14.3). In Nkhotakhota District of Malawi, Ngwira *et al.* (2012) reported no significant effect of CA compared with ConvT on soil organic carbon (SOC) over six cropping seasons on sandy soils under both low and high rainfall conditions.

14.5.3 Modelling Conservation Agriculture effects and meta-analysis studies

Modelling exercises undertaken using the Agricultural Production Systems Simulator (APSIM) model with main forcing variables as (i) soil texture (clay or sandy soils), (ii) presence or absence of residues and (iii) early or late planting, confirmed the benefits of CA components in drier seasons, especially on soils with >30% clay (Chikowo, 2011). However, additional water infiltration in seasons with >700 mm rainfall was not associated with additional crop yield gains. Currently, APSIM routines for simulating waterlogged conditions tend to be inadequate. The challenge to effectively manage risk using CA in a practical way revolves around accessing accurate weather forecasts, at a scale that can aid decision making, for at least the agricultural extension systems and the NGO community.

14.6 Conservation Agriculture and Labour Needs in Southern Africa

Surveys conducted in Zimbabwe have attributed the limited expansion in area under CA to larger labour demand associated with preparing planting basins and weed pressure compared to ConvT. A study conducted across five districts in Zimbabwe showed that labour demand for CA based on hand-hoe prepared planting basins was more than double (84.7 labour days ha^{-1}) compared with ConvT (38.6 labour day ha^{-1}) (Nyamangara *et al.*, 2012b). Basins are destroyed by grazing livestock during the dry season and so farmers have to construct new basins except where the fields are securely fenced. However, the introduction of animal-drawn rippers to open up planting lines reduced the labour demand by 44% to 47.6 labour days ha^{-1}, which was however still larger compared with ConvT. In a sub-humid area in north-eastern Zimbabwe, CA required 2.5 to 4.5 times more labour in land preparation compared to ConvT depending on soil texture (Rusinamhodzi *et al.*, 2012a). However, in central Malawi, Ngwira *et al.* (2012) reported the opposite, labour requirement for CA (47 labour days per ha^{-1}) was 38.3% smaller compared with ConvT (65 labour days ha^{-1}). In Malawi ConvT refers to ridges and furrows constructed by manually operated hand-hoes, whereas in Zimbabwe a mouldboard plough is used for tillage using mostly cattle for traction.

The studies in Zimbabwe consistently showed larger yields under CA compared to ConvT (Table 14.1). Consequently gross margin and returns to labour were larger under CA (US$149.95 and US$1.77 day^{-1}, respectively) compared to ConvT (US$48.02 and US$1.24 day^{-1}, respectively) (Nyamangara *et al.*, 2012b). In Malawi, Ngwira *et al.* (2012) reported 105% larger gross margin under CA (maize/mucuna intercrop) (US$704 ha^{-1}) compared to a maize monocrop under ConvT (US$344 ha^{-1}). In central Mozambique, Rusinamhodzi *et al.* (2012b) reported a marginal rate of return of at least 343% when maize and pigeonpea were intercropped under no-till. However, labour productivity studies in CA systems in Southern Africa are still limited and do not cover all possible variations of CA practices.

14.7 Conservation Agriculture in Southern Africa: Potential Pitfalls

14.7.1 Conservation Agriculture in Zambia

In Zambia, the Zambia National Farmers Union (ZNFU) established the Conservation Farming Unit (CFU) in 1995 to develop and promote the adoption of Conservation Farming (CF that later transformed into CA) practices by Zambia's small-scale farming community. In 1999 the government of Zambia endorsed the promotion of CF as part of national extension policy. Since then, there has been proliferation of CA, promoted by NGOs funded by different agencies. The FAO has been instrumental in extending CA to thousands of farmers through working closely with the Ministry of Agriculture and Cooperatives (MACO) extension structures in 14 districts. The Golden Valley Agricultural Research Trust (GART), in conjunction with the University of Zambia, has provided backstopping with vital research components, albeit at a small scale. Most of the financial resources used so far on CA in the region have largely bypassed the research community, a situation that has created major knowledge gaps. Some key questions recently raised by Andersson and Giller (2012) could have been answered had the research component been prioritized from the onset.

Conservation Agriculture in Zambia has had more favourable conditions for its uptake than elsewhere, and is being adopted mainly in regions I and II. Region I is characterized by flat and steep topography, soils are predominantly Haplic Luvisols and the area receives annual rainfall of less than 700 mm $year^{-1}$. Region IIa covers the central plateau and receives rainfall of 800–1000 mm, soils are mainly Haplic Lixisols. Region IIb is the western plateau with rainfall of 800–1000 mm; soils are the infertile coarse sands, Ferrallic Arenosols (FAO, 1973). In much of eastern Zambia where cattle have been decimated by diseases, the traditional farming practices have had a lot in common with CA with critical adjustments required on crop rotations and residue management to align with CA (farmers must stop burning residues). There was explicit recognition of cattle ownership in the development of CA guidelines. Cattle owners could use the Magoye ripper while an elaborate basin technology using the narrow-bladed Chaka hoe was designed for non-owners to suit all crop types. Tillage was restricted to the precise area where the crop was to be sown (5–15% of the surface area) with tilling depth only sufficient to break through plough or hoe pans. Land preparation commenced soon after harvest and was ideally completed in advance of the rainy season. A key requirement to reap the benefits of CA is the establishment of a precise and permanent grid of planting basins or planting furrows, within which successive crops are planted each year and within which mineral and/or organic nutrient resources (and lime at times) are accurately applied (http://www.conservationagriculture.org). However, the precise location of basins on sandy soils is not guaranteed as they are prone to complete destruction during manual weeding operations, raising questions on potentially missing opportunities for capturing residual effects of previous year's applied nutrients.

In both categories (basins or rippers) the CFU through GART has developed a diversified production model of CA that integrates *Faidherbia albida* trees in the fields in order to reduce reliance on external fertilizer inputs. *Faidherbia albida* is an indigenous leguminous tree that has unique reverse phenology characteristics, shedding leaves during the rainy season. The crops directly benefit from the decomposing high-nitrogen leaves, with the trees minimally extracting soil water due to limited leaf transpiration. Data from GART indicate that soils sampled under mature *F. albida* in farmers' fields had more than double SOC and available phosphorus under the canopy than away from it. To date, over 120,000 smallholder farmers are reported to have already benefited from the use of CA in Zambia. In such a country where oil prices are comparatively high, with associated knock-down effect on prices of many goods and services (transport, fertilizers and equipment), CA principles will most likely appeal to mechanized large-scale farmers who are eager to make a profit and ensure sustained production.

14.7.2 Conservation Agriculture in Zimbabwe

In Zimbabwe, CA out-scaling started as a relief package, targeting vulnerable households. The technology was promoted extensively through the protracted relief programme (PRP, EU and others) by more than 25 NGOs, reaching out to almost 130,000 farmers over a 4-year period, mainly across four agroecological regions. Linking inputs to CA for vulnerable households provided a large pull factor, a tag that to date confounds objective evaluation of the CA technology. Training services of the NGO field staff has been provided by various NGOs (including River of Life and ICRISAT) for the semi-arid regions of the country, while the Foundations for Farming (formerly known as River of Life) has provided CA training to organizations implementing CA with communities in the relatively higher rainfall zones of the country. ICRISAT advocates for complementing CA with micro-dosing, an approach that involves use at most of one-third of the traditionally recommended fertilizer rates in an efficient manner. But this introduces an important dilemma in managing low rates of mineral N fertilizer in combination with mostly cereal crop residues. Micro-doses of N fertilizer in the presence of residues are likely to result in a prolonged period of immobilization as mineral N will be limited in relation to C availability. Therefore, it becomes a logical requirement to exclude residues in the basin technology if micro-doses of mineral-N fertilizer are used, a twist that introduces an unwelcome trade-off that is in conflict with core CA principles.

In Zimbabwe, FAO maintains a database on NGO interventions, assesses areas that need to be explored and coordinates key initiatives for efficient resource use by different players. Indications are that over 90% of smallholder farmers practising CA use basins. The Zimbabwe CA national taskforce has produced extensive guidelines on CA, with the Zimbabwean basins configuration of 15 cm × 15 cm × 15 cm different from the 15 width × 20 cm depth × 30 cm length basin design being promoted in Zambia. While mentioned in the manual, the crop rotation component of CA was vaguely addressed by the implementing NGOs. This is probably not surprising as an analysis by Mapfumo and Giller (2001) showed that although farmers indicated their normal crop sequence was legume–cereal–cereal, the actual area sown to legumes was less than 5% of the cropped area. This can be partially explained by a combination of lack of viable legume grain markets and the need by farmers to satisfy their staple food before they can diversify to legume crops and other crops.

14.7.3 Conservation Agriculture in Malawi

Smallholder farmers in Malawi largely till the land using the hand-hoe. The system is based on planting on ridges, which are completely destroyed during the next cropping season and shifted to the previous year furrow position. Crop residues are buried on the ridge position as it is formed. This practice results in 100% soil disturbance, in sharp contrast to the prerequisites of CA. Also the system results in the development of hoe pans. Currently CA is mainly being promoted by CIMMYT and TLC and studies by Ngwira *et al.* (2012) have shown that labour demand under CA is smaller than under ConvT due to the laborious nature of preparing ridges in the latter method. A baseline survey report published by the Ministry of Agriculture showed that 58% of the CA promoters in Malawi had a clear understanding of what CA is (Ministry of Agriculture, Irrigation and Water Development, 2012). The same report indicated that the main methods used to promote CA were extension workers (23.4%), lead farmers (18.5%), demonstration sites (18.5%), farmer field schools (16.6%) and field days (15%).

14.8 Government Policies

Implementation of CA in Southern African countries is at different stages, driven by significant policy shifts in some countries and indifference in others. The overarching factor seems to be the level of activity of the NGO community, which has played an advocacy role as well as strategic lobbying for external funding for the technology.

The government policies around CA are still fragmented. The conditional technical performances of CA often reported have not done much to influence agricultural policy in Southern Africa. In Zimbabwe, the government has adopted CA as one of the sustainable technologies that can increase productivity and production and a CA up-scaling framework that targets at least 500,000 farmers practising CA on at least 250,000 ha by the year 2015 with an average yield of 1.5 t ha^{-1} on CA fields has been published (AMID, 2012). Across the border in Zambia, the Ministry of Agriculture in 1998 formally embraced conservation farming as an official policy of the Zambian government (GART, 2002). Organizations such as CLUSA have required all farmers in their programmes in central and southern provinces to plant in CF basins as a condition for receiving input credit and marketing support. In Mozambique and Malawi, CA has not been formally included in the agricultural policies of these countries but sustained lobbying is underway.

14.9 Concluding Remarks

There is an urgent need to harmonize CA packages targeting different typologies of farms and farming systems' 'niches', at least at country level. The inadequately trained field workers pose a major risk in CA dissemination, as they train farmers based on inadequately grasped CA concepts. A well-formulated strategy of training for transformation to take both farmers and extension personnel through a transitional phase and evaluating alternative methods for the different CA components is necessary. Adopting a flexible system where CA initiatives are coupled with alternative resource management strategies to allow farmers and development partners to explore and capture diversity of technical innovations generated through participatory and empirical research will likely stimulate spontaneous uptake of CA. There is need for concerted investment in integrated soil fertility research and management in order to revitalize degraded soils in order to optimize the benefits of CA. Development of a targeted CA curriculum that adequately addresses the theory of the subject together with the practical side, as currently constituted, must be integrated into such a strategy. There are also compelling grounds that point to strategic policy shifts to facilitate the medium- to long-term benefits of CA to be realized as CA is not a quick-fix strategy. It is, therefore, recommended that CA initiatives should be at least 5 years to allow farmers to gain confidence during the often prolonged transitional phase from conventional ploughing. A major factor that handicaps reviews on CA in Southern Africa is the paucity in empirical data, with most 'grey' data originating from NGOs being of poor quality and therefore scientifically unpalatable. Future CA research initiatives should, therefore, aim to generate more useful empirical data through carefully designed adaptive experiments with communities, so as to guide development in the absence of undue input inducements as currently depicted in the environments from which the bulk of the current CA statistics are drawn.

Note

[1] African Union's New Partnership for Africa's Development.

References

AMID (2012) Approaches to the implementation of conservation agriculture among promoters in Malawi – baseline study. Ministry of Agriculture, Irrigation and Water Development (AMID), Malawi, 45 pp.

Andersson, J.A. and Giller, K.E. (2012) On heretics and God's blanket salesmen: contested claims for conservation agriculture and the politics of its promotion in African smallholder farming. In: Sumberg, J. and Thompson, J. (eds) *Contested Agronomy: Agricultural Research in a Changing World.* Routledge, London.

Baudron, F., Mwanza, H.M., Triomphe, B. and Bwalya, M. (2007) *Conservation agriculture in Zambia: A case study of Southern Province*. African Conservation Tillage Network, Centre de Coopération Internationale de Recherche Agronomique pour le Dévelopment, Food and Agricultural Organization of the United Nations, Nairobi.

Baudron, F., Andersson, J.A., Corbeels, M. and Giller, K.E. (2012) Failing to yield? Ploughs, Conservation Agriculture and the problem of agricultural intensification: an example from the Zambezi Valley, Zimbabwe. *The Journal of Development Studies* 48, 393–412.

Chikowo, R. (2011) *Climatic risk analysis on conservation agriculture in varied biophysical and socio-economic settings in southern Africa*. Food and Agriculture Organisation (FAO), Johannesburg.

CFU (2006) *Reversing food insecurity and environmental degradation in Zambia through conservation agriculture*. Conservation Farming Unit, Lusaka.

FAO (1973) Luangwa Valley Conservation and Development Project. Report on project results, conclusions and recommendations. *FO:DP/ZAM/68/510 Terminal Report*. Food and Agriculture Organization of the United Nations, Rome, Italy.

GART (2002) Accelerated transfer of conservation farming technologies: a concept note. Golden Valley Agricultural Research Trust (GART), Lusaka.

Haggblade, S. and Tembo, G. (2003) Conservation farming in Zambia: EPTD Discussion Paper No. 108. IFPRI, Washington, DC.

IPCC (2007) *Climate Change. The Physical Science Basis*. Contribution of Working Group I to the Fourth Assessment Report of the Intergovernmental Panel on Climate Change. Cambridge University Press, Cambridge, UK.

Ito, M., Matsumoto, T. and Quinones, M.A. (2007) Conservation tillage practices in sub-Saharan Africa: The experience of Sassakawa Global 2000. *Crop Protection* 26, 417–423.

Mapfumo, P. and Giller, K.E. (2001) *Soil Fertility Management Strategies and Practices by Smallholder Farmers in Semi-arid Areas of Zimbabwe*. ICRASAT/FAO, Patancheru, India.

Marongwe, L.S., Kwazira, K., Jenrich, M., Thierfelder, C., Kassam, A. and Friedrich, T. (2011) An African success: the case of conservation agriculture in Zimbabwe. *International Journal of Agricultural Sustainability* 9, 153–161.

Mashingaidze, N., Madakadze, C., Twomlow, S., Nyamangara, J. and Hove, L. (2012) Crop yield and weed growth under conservation agriculture in semi-arid Zimbabwe. *Soil Tillage Research* 124, 102–110.

Ministry of Agriculture, Irrigation and Water Development (2012) Approaches to the implementation of conservation agriculture among promoters in Malawi – baseline study. Ministry of Agriculture, Irrigation and Water Development, Malawi, 45 pp.

Munyati, M. (1997) Conservation tillage for sustainable crop production systems: results and experiences from on-station and on-farm research. *The Zimbabwe Science News* 31, 27–33.

Ngwira, A.R., Thierfelder, C. and Lambert, D.M. (2012) Conservation agriculture systems for Malawian smallholder farmers: long-term effects on crop productivity, profitability and soil quality. *Renewable Agriculture and Food Systems*, 1–12.

Nkala, P., Mango, N. and Zikhali, P. (2011) Conservation Agriculture and livelihoods of smallholder farmers in Central Mozambique. *Journal of Sustainable Agriculture* 35, 757–779.

Nyagumbo, I. (2008) A review of experiences and developments towards conservation agriculture and related systems in Zimbabwe. In: Goddard, T., Zoebisch, M.A., Gan, Y.T., Ellis, W., Watson, A. and Sombatpanit, S. (eds) *No-till Farming Systems*. Special publication No. 3. World Association of Soil and Water Conservation, Bangkok, pp. 345–372.

Nyamangara, J., Masvaya, E.N., Tirivavi, R. and Nyengerai, K. (2012a) Effects of hand-hoe conservation agriculture on soil fertility and maize yield in selected smallholder farming areas in Zimbabwe. *Soil and Tillage Research* 126, 19–25.

Nyamangara, J., Mazvimavi, K., Kunzekweguta, M. and Pedzisa, T. (2012b) *Labour requirement for conservation agriculture: a case of smallholder farmers in Zimbabwe*. ICRISAT, Bulawayo, Zimbabwe, 12 pp.

Nyamangara, J., Nyengerai, K., Masvaya, E.N., Tirivavi, R., Mashingaidze, N., Mupangwa, W., Dimes, J., Hove, L. and Twomlow, S. (2012c) *Effect of conservation agriculture on maize yield in the semi-arid areas of Zimbabwe*. ICRISAT-Bulawayo, 42 pp.

Palm, C.A., Machado, P.O.A., Mahmood, T., Melillo, J., Murrel, S.T., Nyamangara, J., Scholes, M., Sisworo, E., Olesen, J.E., Pender, J., Stewart, J. and Galloway, J.N. (2004) Societal responses for addressing nitrogen fertilizer needs: Balancing food production and environmental concerns. In: Mosier, A.R., Syers, J.K. and Freney, J.R. (eds) *Agriculture and the Nitrogen Cycle: Assessing the impacts of fertilizer use on food production and the environment*. SCOPE/Island Press, Washington, DC, pp. 71–89.

Rockstrom, J., Kaumbutho, P., Mwalley, J., Nzabi, A.W., Temesgen, M., Mawenya, J., Barron, J., Mutua, J. and Damgaard-Larsen, S. (2009) Conservation farming strategies in East and Southern Africa: yields and rainwater productivity from on-farm action research. *Soil and Tillage Research* 103, 23–32.

Rusinamhodzi, L., Corbeels, M., van Wijk, M.T., Rufino, M.C., Nyamangara, J. and Giller, K.E. (2011) A meta-analysis of long-term effects of conservation agriculture practices on maize grain yield under rain-fed conditions: lessons from southern Africa. *Agronomy for Sustainable Development* 31, 657–673.

Rusinamhodzi, L., Corbeels, M., Nyamangara, J. and Giller, K.E. (2012a) Labour burden not crop productivity increased under no-till planting basins on smallholder farms in Murehwa district, Zimbabwe. *ISFM2012 Abstracts*. CIAT-TSBF Nairobi, pp. 22–26.

Rusinamhodzi, L., Corbeels, M., Nyamangara, J. and Giller, K.E. (2012b) Maize-grain legume intercropping is an attractive option for ecological intensification that reduces climatic risk for smallholder farmers in central Mozambique. *Field Crops Research* 136, 12–22.

Thierfelder, C. and Wall, P.C. (2012) Effects of conservation agriculture on soil quality and productivity in contrasting agro-ecological environments of Zimbabwe. *Soil Use and Management* 28, 209–220.

Thierfelder, C., Cheesman, S. and Rusinamhodzi, L. (2012) Benefits and challenges of crop rotations in maize-based conservation agriculture (CA) cropping systems of southern Africa. *International Journal of Agricultural Sustainability* 1–17.

Twomlow, S. and Hove, L. (2006) Is conservation agriculture an option for vulnerable households. Briefing note no. 4, ICRISAT-Bulawayo, Zimbabwe, 4 pp.

Umar, B.B., Aune, J.B., Johnsen, F.H. and Lungu, O.I. (2011) Options for improving smallholder conservation agriculture in Zambia. *Journal of Agriculture Sciences* 3(3), 50–62.

Vogel, H. (1993) Tillage effects on maize yield, rooting depth and soil water content on sandy soils in Zimbabwe. *Field Crops Research* 33, 367–384.

15 Conservation Agriculture in Argentina

Juliana Albertengo,[1] César Belloso,[1] María Beatriz Giraudo,[1] Roberto Peiretti,[1] Hugo Permingeat [1,2] and Luis Wall[3]

[1]*Asociación Argentina de Productores de Siembra Directa (Aapresid), Santa Fe, Argentina;* [2]*Universidad Nacional de Rosario, Santa Fe, Argentina;* [3]*Universidad Nacional de Quilmes, Buenos Aires and CONICET, Buenos Aires, Argentina*

15.1 Introduction: Questing for More Productive and Less Damaging Farming Systems

Since agriculture began ten thousand years ago, humanity has been able to produce sufficient agricultural products to sustain population growth and development. During this period, traditional farming methods (with some variations) were almost all based on a notion that associated soil tillage with the cultivation of crops. This way of practising agriculture – known as the 'tillage based paradigm' – generated high costs in terms of general agroecosystem deterioration since it caused soil erosion and degradation, organic matter and nutrients depletion and desertification, among others.

If we wish to continue satisfying the increasing demand for agricultural products in the future, while reducing environmental costs, we need to find alternative ways to develop agriculture that are more productive and balanced. In this context, the no-till (NT) system, including Good Agricultural Practices (GAPs) related to soil management, crop rotation, balanced crop nutrition, integrated pest management, responsible use of phytosanitary products and integrated cattle management, was introduced in Argentina as an entirely new way to conceptualize and carry out agricultural production.

The NT system or Conservation Agriculture (CA) leaves behind the ancestral idea that to cultivate a crop there is a need to till the soil. In fact, under CA (and NT system), soil disturbance is virtually eliminated. Only a tiny slot is made through the soil mulch cover during the planting or seeding operation so that the seed (and, eventually, starter fertilizers) can be placed in close proximity within the soil in order to allow germination and the development of the new crop. All crop residues are left on the soil surface and only the grains, oilseeds, etc. are harvested and taken away from the field.

Together with this new concept of NT, there has also been an important conceptual evolution towards the idea that by applying fertilizers we aim to 'feed the soil rather than fertilizing the crop'. This strategy aims at enhancing crop performance based on the improvement of the soil and the agroecosystem itself in the short, medium and longer term. To achieve this goal, it will be necessary to achieve at least a proper soil nutritional balance. Sufficient amounts of nutrients should be given to the system to replace the amounts taken away by the harvested grains and other parts of the plants.

Also, since soil carbon content represents the most important element for an adequate soil nutrition and agroecosystem functioning, at least its level should be maintained and, where possible, increased.

The pampas in Argentina (30–40°S, 58–68°W) is a vast plain of some 50 Mha, with a flat or slightly rolling relief, and grasses as natural vegetation. The evolution of Argentine agriculture with increased yields and low fertilization rates in central Argentina revealed the practice of nutrient mining in most of the pampean soils. These soils are the most important from the point of view of agriculture, developed in quaternary sediments covering the central plains in the country. This pampean loess is rich in minerals, high in soil organic matter (SOM) content and with physical characteristics for the formation of well-structured surface horizons, deep, dark, suitable for crop root development and water storage for basically rainfed agriculture.

The conventional production model is not sustainable from an environmental viewpoint. Production objectives must include not only improving the yield with lower external demand, but also maintain nutrient levels to obtain the best yields possible. By increasing the cultivated area with grasses and also rational amount of fertilizers will benefit the soil and production. CA, involving a proper crop rotation and plant nutrition strategy, represents a powerful tool to maintain good soil carbon levels and even increase them in many cases.

Another important benefit of CA is that, when fully and correctly adopted, it is possible to reach a 'beyond-sustainability stage'. At this stage, important elements such as soils, soil biotic load and biodiversity, water availability and quality improve considerably. This in turn permits higher and less variable yields as well as larger profits, which usually increase the farmer's wealth benefiting their families, communities, regions and countries. On a broader scale, it also helps humanity in general to meet the growing demand for agricultural products (Peiretti, 2003). It is worth noting that soil biology under CA management is not deeply understood and this knowledge will certainly help to improve agricultural management in terms of sustainable production. In this regard, the interaction between scientists and farmers to exchange opinions and experiences and to develop new knowledge is essential (Wall, 2011).

For Argentina, the challenge is to adequately fit the CA system and complementary GAPs into an agricultural growth plan for the near future, taking into account its shape, characteristics and functionality. Indeed, the success of such an agricultural growth model will not only depend on the local factors and circumstances but also on international conditions arising from globalization, since nowadays global and local forces intensely interact generating long-term changes.

15.2 AAPRESID – The Argentine No-Till Farmers Association

AAPRESID is an NGO established in 1989 to create an interactive network of innovative farmers seeking practical ways to develop, adapt and adopt a new farming paradigm based on the NT system. The premise 'The Challenge is to Innovate' has remained the general guideline for AAPRESID's actions over the years.

At the beginning, pioneer farmers were concerned about water and wind erosion and how to stop it. In addition, they wanted to capture and store more water as it is a key factor in rainfed agriculture. The adoption of CA as a complete cropping system based on permanent NT and its massive diffusion was based on a network of information, and on exchange of experiences and challenges, where connectivity and generosity were, are and will be responsible for this highly expansive phenomenon in Argentina. What is more, the innovation and the development of agricultural machinery as a consequence of new needs helped to create a machinery industry centre for national and international markets.

In the pursuit of this objective, AAPRESID organizes many different activities, among them: the Annual National Congress (the latest congress attracted 5000 attendees, 30% of them by online participation); the development of trials; the promotion of field days and seminars for technological exchange between national and international farmers; and the promotion of joint actions with universities, agricultural research centres, NGOs and private companies to test new machinery and all kinds of technologies.

15.3 Conservation Agriculture System in Argentina: Past, Present and Future

15.3.1 Early experience in Conservation Agriculture in Argentina from the mid-1970s

For almost a century, but particularly between 1940 and 1980, very important political, economic and technological limitations inhibited a progressive growth of the total Argentine agricultural output. Production did not grow, and if it did, it was only by a fraction of its true potential. In addition, agricultural production was carried out in a non-sustainable way and hence agricultural development during these years was not free of cost in terms of soil and agro-ecosystem degradation.

Among the local referents who took the first steps in soil conservation and conservation practices to stop the problem of soil erosion is Dr Jorge S. Molina, a professor at the University of Buenos Aires. Technicians from the Agricultural Experimental Stations of INTA – Marcos Juárez and Pergamino – carried out some trials showing the efficient management of water resources and erosion control conducted on farmer fields under NT. The main advantage was that NT was related to better economy, reduced water loss through evaporation and runoff and bringing forward the sowing date.

The benefits were not only agricultural rather the advantages were wide-ranging: in addition to increased feed production, livestock could graze for longer periods, as the soil conditions after rain were firmer than under tillage-based systems.

CA systems started to become widely adopted 15 years later, due to the problem of soil erosion, increased operating costs and the emergence of herbicides at lower prices that allowed for more effective weed control compared to mechanical weed-control technology, which made the new system eventually economically viable. Innovation in the design of new agricultural machinery was also a challenge in those years while today there are several brands of equipment in the country.

The NT system occurred first in the farm then timidly moved to the scientific organizations. At the beginning the local development was not the result of research sponsored by government agencies or universities but the result of some producers' need, many of whom were also technicians, who committed themselves to the experience despite the fact they did not have a strong academic background that supported them. Sometimes it involved a huge risk.

No-till adoption involved science and innovation, infinite generosity and openness to share experience and knowledge. Its momentum and subsequent dissemination were possible because of AAPRESID pioneers, their empowerment and their determination to face difficulties, to produce information and to share it. It was a trial and error process, marked by successes and failures – a collective intelligence example.

The system was gradually adopted, but with the onset of genetically modified crops the expansion was boosted. However, this is not a requirement for NT.

The NT system also allowed for the expansion of agricultural frontiers in the country, where ecology and production turned to be compatible.

Argentina has never had specific policies related to CA, except for a brief period (1999–2003) in Santa Fe province. It is a key essential and an opportunity for Argentine agribusinesses to implement government policies promoting CA for sustainable agricultural activities (economic, environmental and social) based on innovation (technological, organizational and institutional), assuming the commitment to interact with public and private organizations in order to achieve a comprehensive development of the nation and to be able to satisfy the demands of humanity.

The Conservation Agriculture or NT system is adopted in response to the need to prevent soil loss by water and wind erosion, improve water infiltration into the soil and its storage.

In Argentina, the adoption of the NT system has continued to grow steadily during the first 8 years (up to 2008) of this millennium, although turbulent economic times during 2001–2002 have caused some fluctuations. The alternate paradigm has clearly rendered its fruits: the total production of cereals and oilseeds peaked close to 100 Mt during the 2007/2008 cropping season in Argentina.

Since its creation, AAPRESID has actively promoted the establishment of regional groups among its members. These groups, around 30 at the present time, are scattered across the different production areas of the country and generate many benefits. The dynamism and proactive functioning of these regional groups played a key role in promoting a progressive adoption of the NT system by Argentine farmers in the past 20 years (Fig. 15.1), mainly

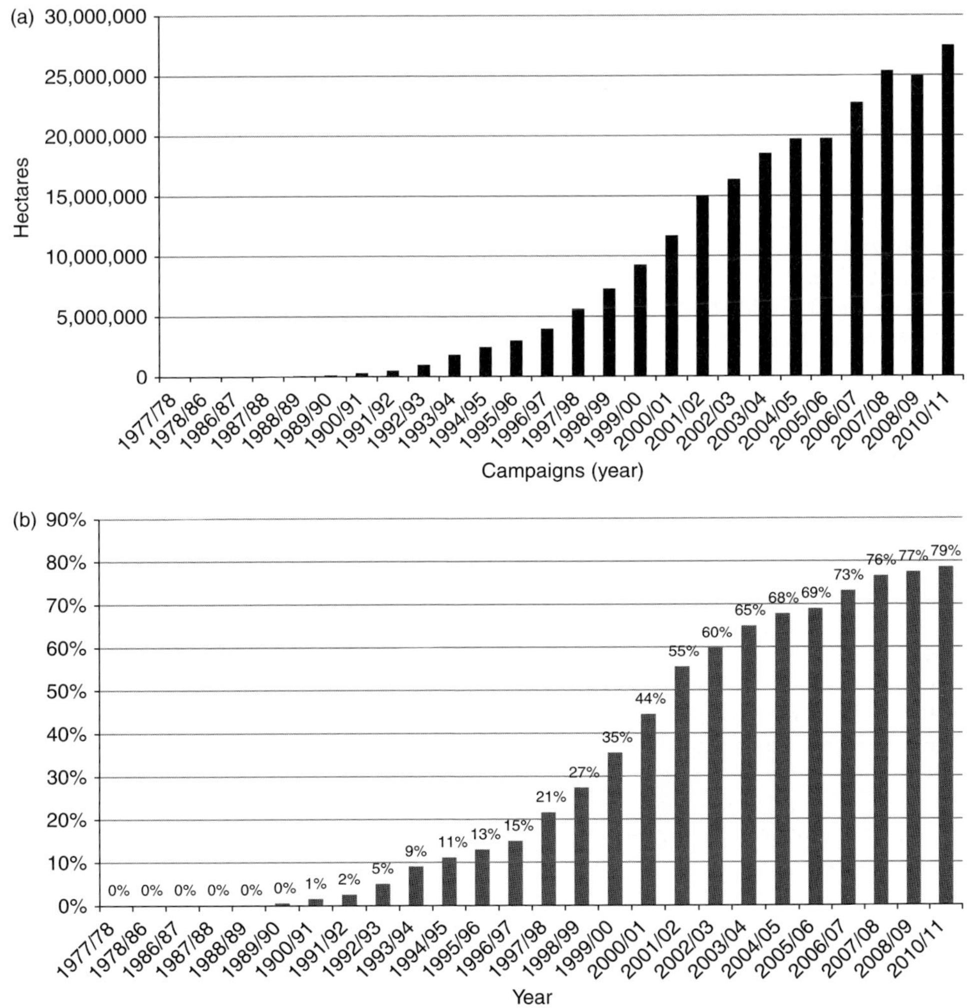

Fig. 15.1. Evolution of no-till area in Argentina in hectares (a) and percentage (b) during the 1977–2011 period (AAPRESID, 2012).

growing simultaneously with the adoption of genetically modified crops (Fig. 15.2), although in different proportions the adoption process has included all major crops in Argentina. Figure 15.3 shows adoption of CA systems in different provinces of Argentina.

At the international level, the actions of AAPRESID have been aimed at fostering greater interaction with other countries, continents and cultures. In this context, AAPRESID was, for example, one of the founder members of the American Confederation of No Till Farmers Associations working for a Sustainable Agriculture (CAAPAS). This institution was established in 1992 and currently comprises seven American countries (Canada, Mexico, Paraguay, Uruguay, Chile, Brazil and Argentina) among its permanent and invited members. At present, the CAAPAS country members represent around 60% of the world's NT agricultural area (Friedrich *et al.*, 2012).

Another strategic area in CA is nutritional balance of the soils and crops. At present, Argentine agriculture uses on average around 100 kg fertilizer ha^{-1} $year^{-1}$ (Fig. 15.4). This rate of fertilization is still insufficient to secure adequate nutritional soil levels. Complete information about the use of fertilizers in Argentina as well as some other related data can be found in Fertilizer 2012 (http://www.fertilizar.org.ar). The adoption of bio-fertilizers in the equation of NT paradigm is still an important challenge for the future; and it is necessary to understand the correct nutrient cycles to avoid negative environmental impacts by chemical fertilization.

15.3.2 Evolution of Conservation Agriculture in Argentina

No-till was the first stage of the adoption of a series of GAPs, but time proved the need for a methodology to transform these GAPs in a systems approach such as CA.

Further developments were the Agricultura Certificada™ (AC; Certified Agriculture) and the Chacras (Ch) systems, to accompany the evolution of a sustainable agricultural management.

Agricultura Certificada is an environmental and agronomic quality management system of production processes under NT and comprises the management of an interrelated network of processes within an organization aiming to satisfy all the involved interests. It is a quality system that considers the characteristics and aspects of a product, process or service, related to the capacity to

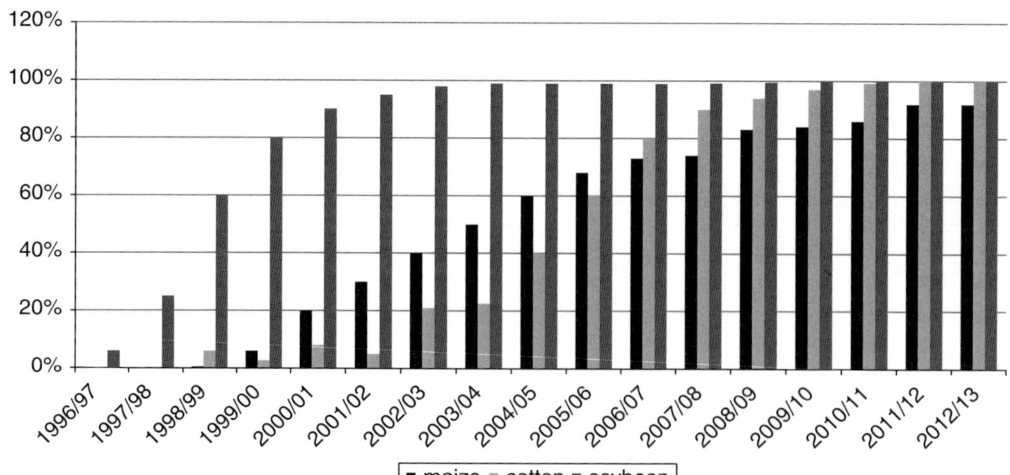

Fig. 15.2. Evolution of cultivated area under GM crops in Argentina (Argenbio, 2013).

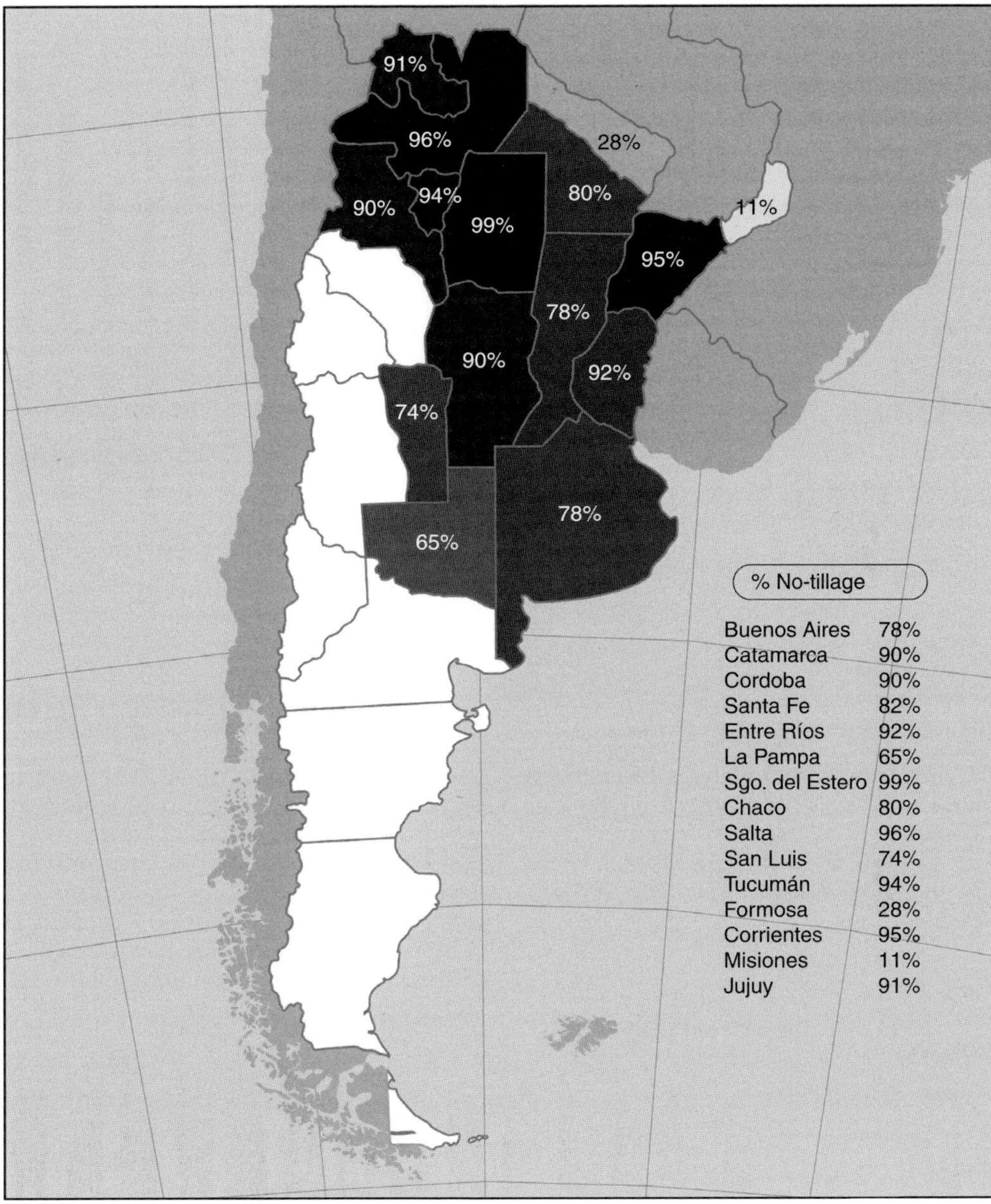

Fig. 15.3. Percentage of cultivated area under no-tillage systems in different provinces of Argentina.

satisfy the clients' needs. The components consist of GAPs protocols and manual and edaphic and efficiency management indicators, and is one way to show to the society that agricultural production is taking care of the environment, energies, economics and ethics.

The Ch system encourages the development of sustainable technologies adapted to each particular environment, based on an open network between producers, technicians and researchers, both public and private.

AAPRESID CLASSROOM is an open-share project focused on trying to communicate

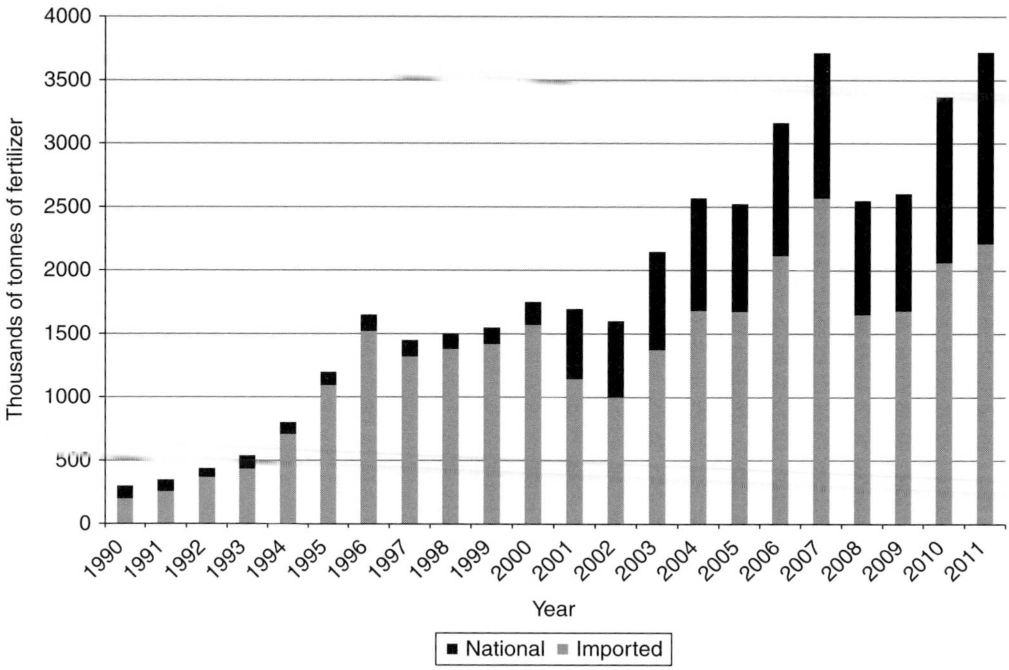

Fig. 15.4. Evolution of use of fertilizers in Argentina (Fertilizar Asociación Civil, 2012).

the agricultural production system to the whole society, working with local communities and, mainly, with primary and secondary schools to impart knowledge about how to produce food and energy in a sustainable system.

15.3.3 The next ten years

Looking at future scenarios for Argentinean agricultural production, three key questions arise. How will production levels evolve during the coming years? Which are the possible obstacles to its growth? And what would be a realistic production target for the mentioned period? As previously mentioned, the production of grains and oilseeds in Argentina has grown significantly in recent times by use of CA system. Production levels have more than tripled from around 30 Mt in 1990/1991 to around 97 Mt during the 2007/2008 cropping season (Fig. 15.5). Time and again, earlier projections have been surpassed by reality.

As for the future, it is reasonable to assume that production could grow between 20% and 40% above current levels. These projections in fact can be realized but depends, at the same time, both on global and domestic circumstances. Government policies are crucial to accompany sustainable agricultural production, under CA including GAPs, to maintain the soils to meet global food demand.

Among several other activities of AAPRESID oriented towards contributing to higher yields on a sustainable basis, the most important one is the AC project, which was launched 5 years ago.

15.4 Research Results Reported in Argentina

15.4.1 Effect on soil quality (physical, chemical and biological)

Soil organic matter (SOM) is a key component that affects soil physical, chemical and biological properties, and it is also a requirement to obtain high crop productivity. Many national and international studies refer to the effects of soil tillage, crop rotation and crop residue management relative to the consequent changes in SOM, and soil physical, chemical and biological properties.

Traditional agricultural systems – under continuous tillage – have led to reduced levels of SOM in the entire pampas region. Some of the soil carbon was lost due to erosive processes while some was lost due to carbon dioxide emissions to the atmosphere. Literature shows that NT systems – since they do not disturb the soil and maintain the natural order of its solid components – generate an accumulation of SOM in the topsoil (3–10 cm depending on weather conditions, soil type, crops practised and time; Fig. 15.6) (Alvarez and Barraco, 2005).

Some studies conducted by INTA EEA Marcos Juárez show that NT achieves better

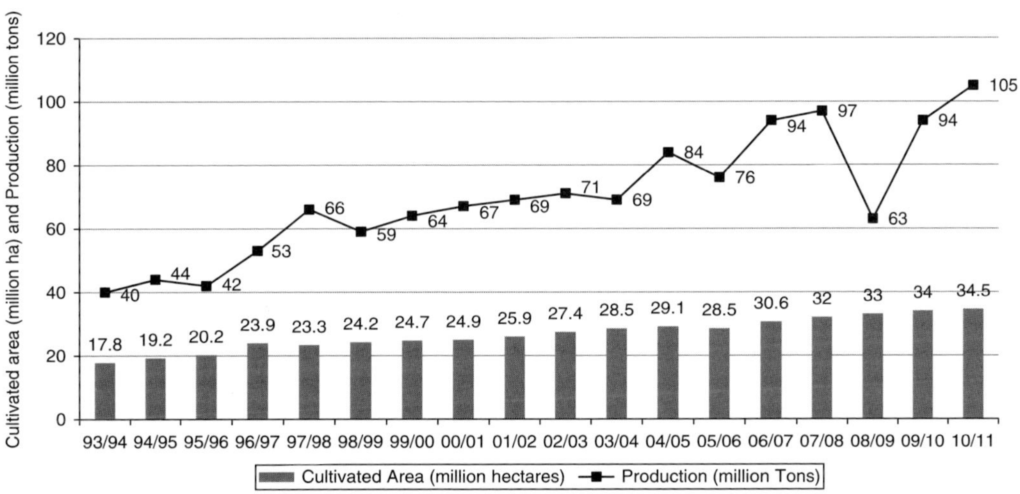

Fig. 15.5. Evolution of cropped area and production in Argentina from 1970 to 2008 (AAPRESID, 2008).

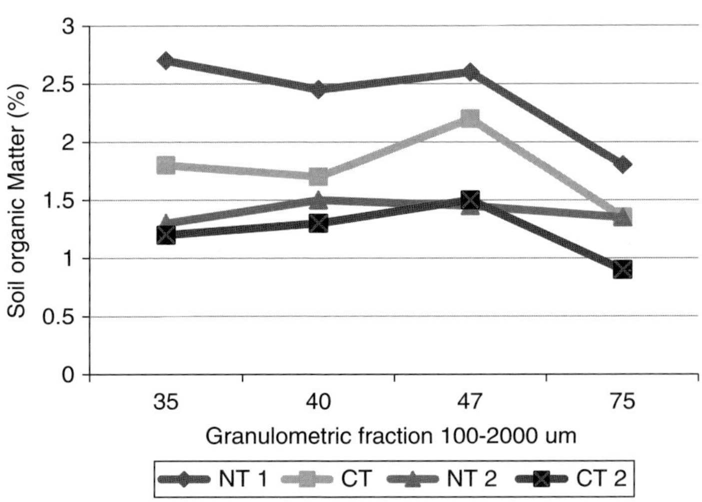

Fig. 15.6. Soil organic matter content in a granulometric fraction of 100–200 μm of different textured soils under no-till and conventional tillage. Locations: Macachin (M), Dorila 1(D1), Dorila 3 (D3), Anguil (A) in La Pampa Province.

total soil organic carbon (SOC) and total soil nitrogen (TSN) contents, especially on the soil surface and in the topsoil; however, after 14 years it does not reach the values of virgin soils (VS). Minimum tillage combined with NT (CombTS) causes SOC losses in comparison with the values obtained under NT (Tables 15.1, 15.2) (Gudelj and Musiero, 2000; Morón *et al.*, 2005).

No-till promotes the development of soil microorganisms because it causes less soil disturbance (Fig. 15.7). The presence of crop residue over the soil surface helps to stabilize soil temperature variations, conserves soil moisture and increase nutrient availability. The results obtained also indicate that both total fungi and actinomycetes populations are good biological indicators that reflect the beneficial effect of soil management based on CA (Pérez Brandán *et al.*, 2010).

The soil biology is definitely one of the main frontiers of human knowledge since the discovery of the 'plate count anomaly' (Staley and Konopka, 1985). The huge quantitative difference between the culturable microorganisms and the microorganisms that can be detected by direct observation methods point out that our knowledge on

Table 15.1. Measured values of the indicators evaluated in different fields at 0–2.95 cm depth; Marcos Juárez, Córdoba Province.

Indicator	SD	LComb	LC	CV
COT (g C kg^{-1} soil)	19.3 c	14.9 b	13.3 a	22.5 d
NT (g N kg^{-1} soil)	1.8 c	1.4 c	1.3 d	2.2 d
C-POM 212 (g C kg^{-1} soil)	4.5 b	2.8 a	2.6 a	6.1 c
C-POM 53 (g C kg^{-1} soil)	3.5 b	2.2 a	2.1 a	7.9 c
C-MAOM (g C kg^{-1} soil)	11.3 b	9.9 a	8.6 a	8.4 a
N-POM 212 (g N kg^{-1} soil)	0.18 b	0.11 a	0.08 a	0.25 c
N-POM 53 (g N kg^{-1} soil)	0.25 b	0.15 a	0.13 a	0.31 c
N-MAOM (g N kg^{-1} soil)	1.40 c	1.18 b	1.06 a	1.68 d
PMN (Mg N-NH$_4$ kg^{-1} soil)	39 b	35 b	18 a	80 c

SD, no tillage; LComb, combined tillage systems; LC, conventional tillage; CV, virgin soils; COT, total soil organic carbon; NT, total nitrogen; C-POM, particulate soil organic matter carbon; N-POM, particulate soil organic matter nitrogen; C-MAOM, mineral associated organic matter carbon; N-MAOM, mineral associated organic matter carbon nitrogen; PMN, potentially mineralizable nitrogen.
In an indicator, the values followed by the same letter do not significantly differ from P>0.10 MDS.

Table 15.2. Measured values of the indicators evaluated in different fields at 7.5–15 cm depth; Marcos Juárez, Córdoba Province.

Indicator	SD	LComb	LC	CV
COT (g C kg^{-1} soil)	11.6 a	12.4 b	12.0 a	13.9 b
NT (g N kg^{-1} soil)	1.1 a	1.2 b	1.1 a	1.5 c
C-POM 212 (g C kg^{-1} soil)	0.2 a	0.7 a	0.4 a	0.6 a
C-POM 53 (g C kg^{-1} soil)	0.5 a	0.7 a	0.8 a	1.6 b
C-MAOM (g C kg^{-1} soil)	10.8 a	11.4 a	10.8 a	14.2 b
N-POM 212 (g N kg^{-1} soil)	0.01 a	0.04 b	0.04 b	0.03 b
N-POM 53 (g N kg^{-1} soil)	0.06 a	0.08 ab	0.08 a	0.10 b
N-MAOM (g N kg^{-1} soil)	1.02 a	1.08 a	0.98 a	1.60 b
PMN (Mg N-NH$_4$ kg^{-1} soil)	8 a	16 b	9 a	25 c

SD, no tillage; LComb, combined tillage systems; LC, conventional tillage; CV, virgin soils; COT, total soil organic carbon; NT, total nitrogen; C-POM, particulate soil organic matter carbon; N-POM, particulate soil organic matter nitrogen; C-MAOM, mineral associated organic matter carbon; N-MAOM, mineral associated organic matter carbon nitrogen; PMN, potentially mineralizable nitrogen.
Note: In an indicator, the values followed by the same letter do not significantly differ from P>0.10 MDS.

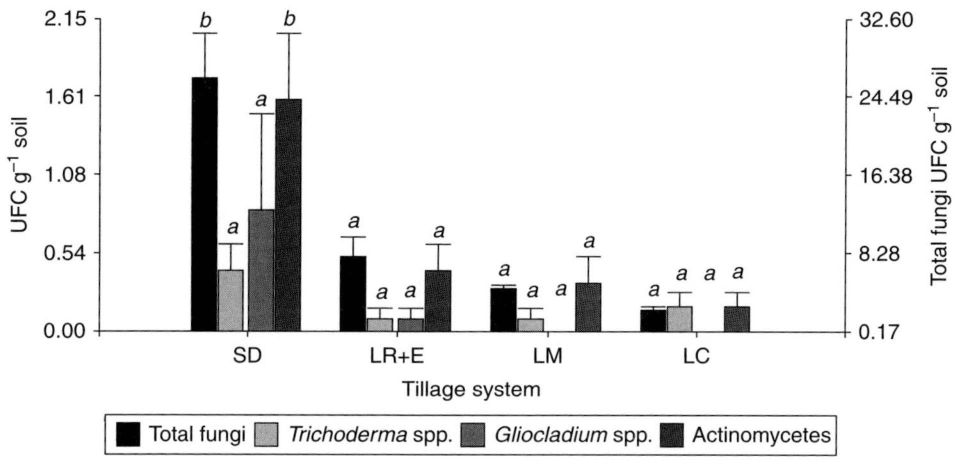

Fig. 15.7. Actinomycetes, *Trichoderma* spp., *Gliocladium* spp. and total fungi populations under different tillage systems (Ferraris, 2009; Ferraris and Mousegne, 2009). The bars having the same letter do not differ significantly based on LSD test ($P<0.05$). UFC, colony-forming units; SD, no-tillage; LR+E, reduced tillage using harrow; LM, minimum tillage using disc; LC, conventional tillage.

soil microbiology based on pure culturable microorganisms, in a Pasteur point of view and following Kosch postulates for microbial activities, is definitely limited and only covers 1% of the existing microbial life in the soil. New research approaches using biochemical tools and systemic point of view are helping us to start to describe and understand the soil microbiology.

In this context, at the initiative of AAPRESID, the Soil Biology and Sustainable Agricultural Production – BIOSPAS (acronym of its Spanish name; http://www.biospas.org/en) – was created in 2007, and it started work in the field in 2009. BIOSPAS was organized under the auspices of the Argentine Ministry of Science, Technology and Productive Innovation through its Strategic Area Projects. The BIOSPAS project combines the involvement of 12 research laboratories located at different universities or public sector research institutes, AAPRESID and two private companies, La Lucía S.A. and Rizobacter Argentina S.A. (Wall, 2011).

The aim of BIOSPAS is to find potential biological indexes that could discriminate between soils under GAPs and soils under non-sustainable agricultural practices (that is, soils managed without following GAP protocols, which basically means monoculture agriculture without nutrients reposition, done with a short term point of view).

After 4 years of work, it is possible to find many different biochemical and biological characteristics that discriminate between GAP and non-GAP managed soils. As an example, using molecular tools it is possible to find bacterial groups that appear to be markers of GAP and non-GAP soil managements independently of the soil texture and geographical environment (Figuerola *et al.*, 2012) or to show changes in a particular bacterial group such as the *Pseudomonas* group which characterize GAP versus non-GAP soils (Agaras *et al.*, 2012).

The already described situation of NT agriculture in Argentina and South America offers a unique opportunity of study for scientists, whose interaction with farmers is definitely a major determinant to succeed in the development of new and useful knowledge in agriculture soil biology.

15.4.2 Carbon sequestration

As is well-known, the higher the yield, the greater the soil nutrients removal, and carbon (C), 'the mega-nutrient', forms 58% of the edaphic SOM. Soil C cannot be bought or applied to extensive crops; it must be produced in the fields (Cordone *et al.*, 2004). Carbon inputs or system gains are due to the transformation of crop residues into organic matter, through the humification process.

The productive capacity of soils under continuous agriculture in the central pampas region can be maintained if the soil is managed within NT systems with soil cover so as to reduce C losses, if it is properly fertilized so as to increase yields while fixing more C from the air, and if a crop rotation with enough proportion of cereal crops (Cordone *et al.*, 2004) is practised so as to add a larger amount of crop residues (high C:N ratio) (Table 15.3)

15.4.3 Production levels and yields under different managements

The pampas region of Argentina includes most of the annual cropping area of the country with almost 30 Mha of cropped land. Cropping is relatively recent; it started only 100 to 120 years ago for the oldest fields. Low fertilizer use and continuous nutrient removal, with increasing crop yields in the last years, have resulted in deficiencies of N, P and S in most of the region.

Soybean monoculture is very common nowadays in Argentina; about 70% of the pampas lands have a soybean crop, 20% double crop and 10% maize. The average production during the last 5 years was less than 5 t ha^{-1} $year^{-1}$. An AAPRESID reference farmer sowed 50% double crop and 50% maize. The average production during the last 5 years was more than 10 t ha^{-1} $year^{-1}$, corresponding to more than 200% increase in yield (same water, land and radiation) (Table 15.4).

When dealing with agricultural practices, we should not consider the yield achieved by one crop; we should analyse the whole crop rotation. In this regard, CA enables compliance with production programmes while maintaining production levels. 'As farmers, we should not consider the individual production of one year, but production over the years' (Ghio, 2011).

In the south-east region of Córdoba Province, more specifically in Monte Buey City, Grupo Romagnoli together with IPNI and AAPRESID have been comparing two crop rotation plans: standard (wheat/soybean–maize–first planting date soybean) and intensive (double annual crops including wheat and barley in winter, combined with sorghum, maize, sunflower and second date planting soybean).

After 13 years, the standard crop rotation with no fertilization (after 13 crops practised) had an average annual yield of 5491 kg grain ha^{-1}, whereas the fertilized one with NPS (nitrogen, phosphorus and sulphur) yielded 8321 kg grain ha^{-1}. On the other hand, intensive crop rotation (24 crops practised in 13 years) with no fertilization achieved an average annual yield of 5604 kg grain ha^{-1}, whereas that fertilized with NPS yielded 9588 kg grain ha^{-1} (Fig. 15.8). This means that intensive crop rotation yielded almost twice as much compared to the non-fertilized standard crop rotation, and 15% more than fertilized standard crop rotation. It is also worth mentioning

Table 15.3. Carbon content in crop residues.

Crop	Yield (kg ha^{-1})	Harvest index (%)	Residue (kg ha^{-1})	C contributed (kg ha^{-1})	Annual C contributed (kg ha^{-1})
Wheat	3,500	37	5,959	2,682	4,517
Second soybean	2,500	38	4,079	1,836	–
Maize	10,000	45	12,222	5,500	5,500
First soybean	3,500	38	5,711	2,570	2,570

Carbon concentration in the residues of the three crops is supposed to be 45%.

Table 15.4. Annual average yield levels (of three crops) under no-till during 1997/98 to 2010/11 at H. Ghio Farm, Marcos Juarez County (2-year crop rotation: maize–wheat/soybean DC).

Yearly production average (3 crops)														
Campaign	1997/98	1998/99	1999/00	2000/01	2001/02	2002/03	2003/04	2004/05	2005/06	2006/07	2007/08	2008/09	2009/10	2010/11
Yield t ha^{-1} year^{-1}	8.86	8.61	8.82	9.35	9.34	8.91	10.04	11.30	10.37	10.85	10.20	9.10	8.90	11.00
	Production average (last 5 years): 10.02 t ha^{-1} year^{-1}													

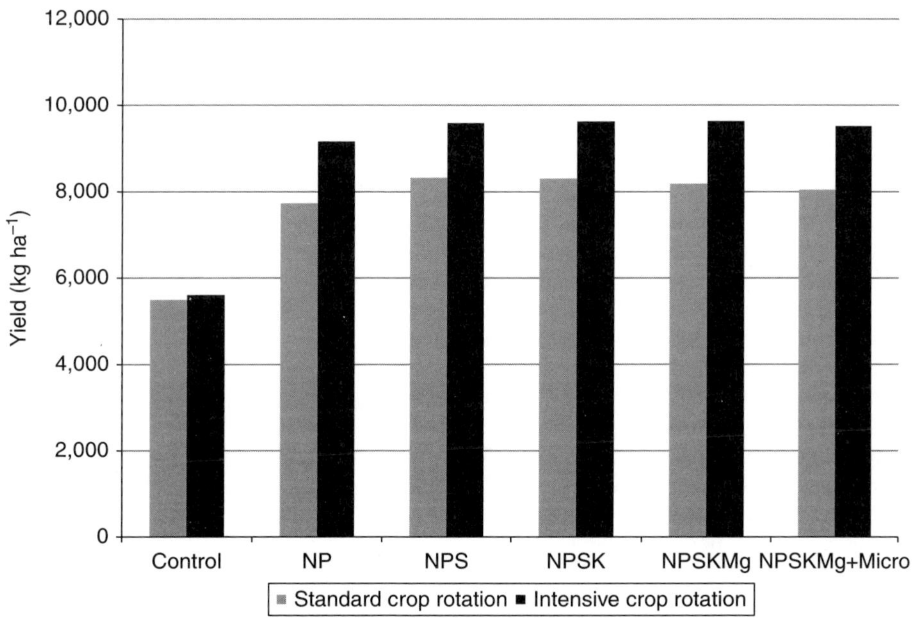

Fig. 15.8. Average crop yields under standard crop rotation versus intensive crop rotation under different nutritional management options (13 years) (Ghio, 2009; Ghio *et al.*, 2010). N, nitrogen; P, phosphorus; K, potash; S, sulfur; Mg, magnesium; Micro, micronutrients.

the positive impact on the soil due to a greater amount of crop residues over its surface.

A long-term trial developed in Marcos Juárez, Córdoba Province, compared continuous NT with combined tillage (from 1986/87 to 1989/90 NT was practised only with second planting date soybean; from 1989/90 onwards minimal tillage was done before wheat sowing and NT with the other crops). The results showed that maize was the only crop that expressed greater differences in comparison with combined tillage – 619 kg ha^{-1} more under continuous NT (Table 15.5).

15.4.4 Water runoff, infiltration, soil water content; soil conservation

The most significant impact of CA on crop production is the beneficial modification of soil water balance. The most affected components of this balance are water losses due to evaporation and runoff, and the water gain due to infiltration.

It is commonly observed that in CA fields there is decreased water evaporation due to lower soil temperature and the cushioning effect of crop residue over the surface that increases the resistance to the

Table 15.5. Average yield achieved in 13 agricultural years under two tillage systems (1986/87–1998/99), 'Don Osvaldo' Farm (Camilo Aldao, Córdoba Province). Typical Argiudol soil, Hansen series (Class I).

Crops	Continuous no-tillage (kg ha^{-1})	Combined tillage (kg ha^{-1})
5 years maize	7578	6959
6 years wheat	3536	3461
6 years soybean as second crop	2838	2814
2 years soybean as first crop	3014	2956
Total average (13 years)	6320	6027

diffusion of water vapour. More infiltrated water in CA fields is a consequence of the following: (i) crop residue reduces the impact of raindrops, avoiding soil particle disaggregation and, therefore, preventing soil sealing; (ii) crop residues help rainwater stay longer on the surface; and (iii) the presence of continuous and stable biopores increases soil saturated conductance. At the same time, water runoff decreases. In brief, the most remarkable immediate consequences of water balance are increased water infiltration and reduced water losses within the system, resulting in a net gain of available water for plants (Figs 15.9–15.11).

15.4.5 Climate change mitigation and adaptation

Basically, there are two ways of mitigating climate change: reducing emission sources and increasing the amount of greenhouse gases (GHG) stored in ground systems, for instance, through C sequestration in the soil.

Conservation Agriculture, by reducing fossil fuel consumption, together with lowering carbon dioxide emissions (due to the absence of tillage) and carbon sequestration (due to the organic matter increase) helps to mitigate the greenhouse effect.

Conservation Agriculture reduces 40% of fossil fuel consumption due to the absence of tillage. With 1 l of fuel, it is possible to produce 50 kg of grain under conventional tillage (ConvT), whereas, under CA, it is possible to yield 123 kg (Lorenzatti, 2006).

Transforming all the pampas cropland into CA would increase organic carbon by 74 Mt, which is approximately twice the annual C emissions from the consumption of fossil fuels in Argentina (Alvarez and Steinbach, 2006a, b).

15.4.6 Off-site environmental benefits

More than 20 years ago, Argentinean farmers considered CA as the opportunity to turn agriculture into a more productive and sustainable system. Nowadays, the same principles of that fusion between production and environment constitute the foundations of a global opportunity: to produce more, with less water, less soil erosion and less contamination.

No-till system environmental benefits include:

- Reduced wind erosion by 96%;
- Increased soil biological activity and biodiversity;
- Minimized soil degradation and improved soil fertility;
- Reduced water evaporation by 70% (Sinclair *et al.*, 2007);
- Less water use to produce the same amount of food;
- Less fossil fuels for machinery.

Other NT system or CA benefits include:

- Increased yield stability;
- New production areas;
- Lower production costs;
- Less working hours, more time to plan and become trained.

For efficient use of water under CA there are two key GAPs, namely: crop rotation adjusted to crop diversity and intensity based on the environment, and the associated nutritional management. Those practices enable production

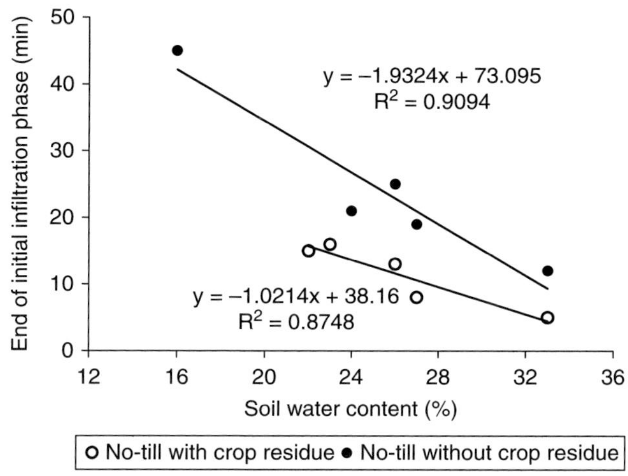

Fig. 15.9. Initial infiltration phase and initial soil water content (0–5 cm) (De La Vega *et al.*, 2004).

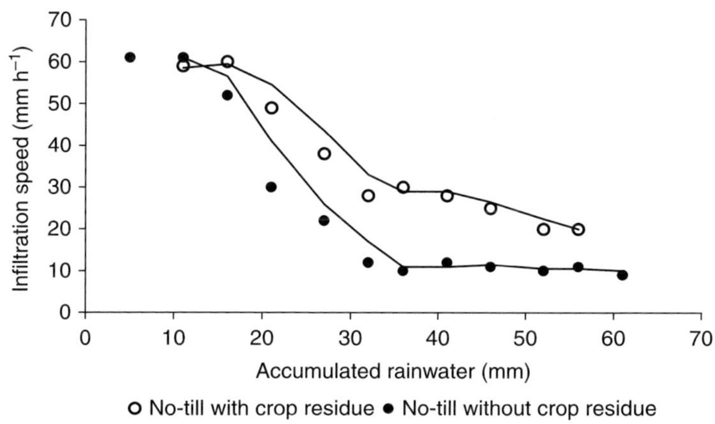

Fig. 15.10. Linear regressions describing the relationship between infiltration rate and cumulative rainfall for each treatment. The plotted values are the average of five replicates.

to be maximized according to the environmental potential, which is expected to result in better returns for the producers.

15.4.7 Biodiversity

Under CA, in the topsoil there is great biological activity and diversity responsible for most SOM mineralization, formation and recycling, and nutrient availability. Crop rotation with different amounts of crop residue provides the substrate that will feed the soil microorganisms, establishing a population balance similar to that of natural environments but with a predominance of other species adapted to agroecosystems.

15.4.8 Economic returns

Crop rotation with predominance of cereals has many advantages compared to monocropping (Fig. 15.12). Crop rotation also provides economic benefits (Table 15.6) since it improves production levels, produces more net income per hectare and different incomes from the different crops per campaign.

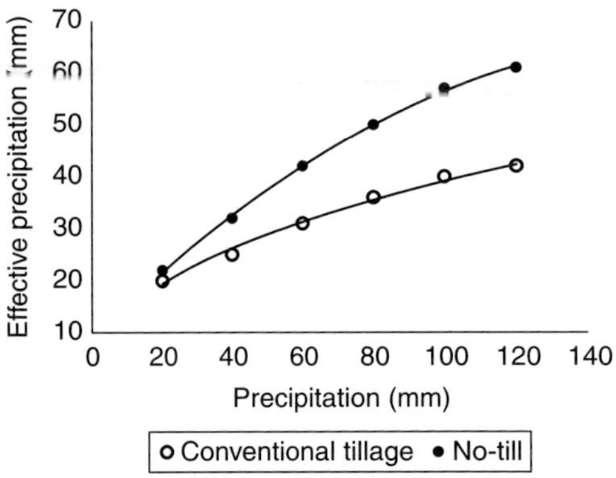

Fig. 15.11. Relation between precipitation and effective precipitation under no-till and conventional tillage in an entic Haplustoll soil in Manfredi, Córdoba Province (Micucci, 2003).

15.5 Constraints Encountered in Scaling-up Conservation Agriculture in Argentina

15.5.1 Residue management and supply

Residue distribution is a key aspect to take into account to avoid obstructions when planting the following crop. To reach that goal after harvest, the combine harvester needs to have a facility to grind pieces and also a good spreading device with variable width of residue cover on the soil. To maintain soil cover over time in regions with high temperatures, it is important to slow residue decomposition. Residue size is the key. Long, big pieces are required and the priority is a good and even distribution. Sometimes it is also advisable to replace or modify the combine devices to achieve this goal. It is very important to check the operation of every combine piece to avoid accumulation of residue in the lines across the field, which will have later effects on the following crop.

15.5.2 Soil compaction

Even if NT adoption helped to reduce and limit erosion processes, physical degradation due to machinery transit across wet fields and a soybean monocrop became a major risk in NT systems. Moreover, increased agricultural production brought the need for higher working capacity equipment and as a consequence an increase in weight. In addition, fertility and crop protection treatments increased in number, bringing more traffic to the field, which contributes to the problem (Gerster and Bacigaluppo, 2012).

Gerster and Bacigaluppo (2004, 2008) while working on tracks generated by traditional dual axle hoppers on typical Argiudol soils, observed that places where machinery traffic had been more intensive presented an increase in bulk density, reduced basic infiltration rates, lower soil profile scan by roots and poorer biological activity. Consequently, soybean and maize yielded up to 28 and 15% less, respectively (Fig. 15.13). In addition, they assessed the presence of nodules on soybean and observed a significant reduction both in weight and amount on primary and secondary roots.

The alternatives to reduce or limit negative traffic effects consist of using lightweight machinery and/or changing the wheels to reduce their specific pressure on the soil, or applying satellite-guided controlled traffic systems with permanent tramlines.

On the other hand, NT soils with a proper cereal crop rotation are able to recover structure. This process depends on the features of each field (soil cover, organic matter, etc.) and can last for several years. Cereal crop inclusion in the crop rotation also helps for breaking compact soil layers with roots and biological activity since they generate pores and channels that boost infiltration speed and other crop root development (Gerster and Bacigaluppo, 2012).

15.5.3 Weed infestation

The accumulation of crop residues over the soil surface alters the environmental factors that function as signals for weed germination and establishment (Tuesca, 2009). The absence of soil disturbance also brings about

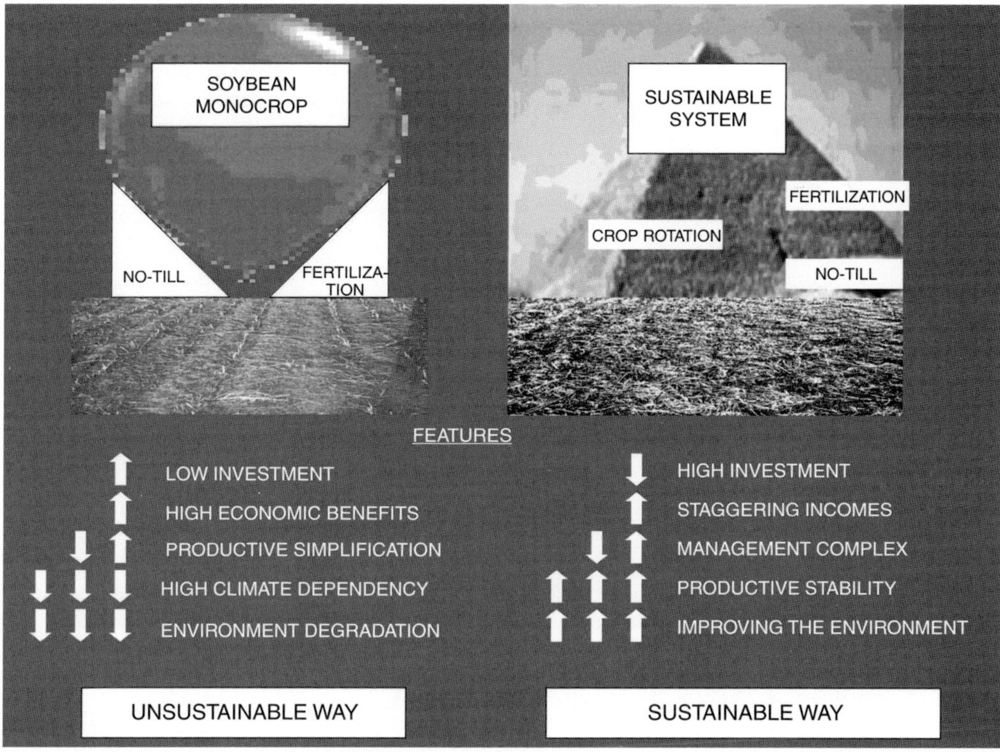

Fig. 15.12. Sustainable crop production model based on Conservation Agriculture versus soybean as monocrop (Ghio, 2006).

Table 15.6. Economic benefits obtained from rotating maize–wheat/soybean compared with first planting date soybean in the south-east region of Córdoba Province (Ghio, 2011).

Crop	Wheat	Second soybean	Maize	First soybean
[a]Yield t ha^{-1}	4.30	3.40	1.23	5.00
Market Price (US$ t^{-1})	1.85	3.00	1.70	3.00
Planting, harvesting and insurance costs (US$ ha^{-1})	349.54	307.98	632.64	339.83
Production costs (US$ t^{-1})	0.80	0.90	0.51	0.68
Indirect cost (US$ ha^{-1}) own field	50.00	50.00	100.00	100.00
Net income (US$ ha^{-1})	281.18	567.16	1015.12	922.67
Indifference yield (t ha^{-1})	2.22	1.13	4.43	1.25
Net income (US$ ha^{-1}) wheat–soybean	848.34			
Net income (US$ ha^{-1}) from crop rotation		931.73		

[a]Average of ten campaigns

changes in the vertical distribution of weed seeds in the soil profile, prolonging their longevity and affecting their management.

Some trials in the south-west region of Buenos Aires Province (Gigón *et al.*, 2013) show that weed population under ConvT is greater compared to that under NT (Table 15.7). It was observed that some species become strongly adapted to established NT systems whereas other species decrease in number.

(a)
(b)

Fig. 15.13. (a) Soybean root affected by soil compaction and (b) soybean root not affected by soil compaction in Argiudol soil.

15.5.4 Insect-pest and disease problems

Successive NT years have generated beneficial environmental conditions for some diseases and pest proliferation. In Argentina, NT has brought about an increase in the amount and diversity of the soil fauna, mainly composed of invertebrate

Table 15.7. Weed diversity parameters assessed in 78 commercial wheat fields in the south-west of Buenos Aires Province under conventional tillage and no-till.

	ConvT	NT	Significant difference
Richness (number of species per field)	8.31	6.71	$p<0.05$
Shannon diversity	1.92	1.67	ns
Richness (total number of species)	63	50	

molluscs (snails and slugs), annelids (worms) and insects. Some of these organisms behave like pests, while others regulate pests or help in their decomposition, mineralization and humification, macro- and micro-nutrient movement and structure and aggregation of the soil.

The major issue faced is the damage caused by slugs to soybean and sunflower. The development of this pest depends mainly on large amounts of residue and the humidity of the soil.

For slug management, the fields need to be monitored and searched for the eggs. Slugs have food preferences, thus toxic baits (methaldehide) are very useful for control while no crops or weeds are on the field. Toxic baits applied with a special device over the planting line reduced the costs of application in half (Fernández Palma, Necochea, 2006, pers. comm.).

Amongst the difficulties that NT farmers face, diseases are most important. No-till and minimum tillage practices severely affect necrotrophic pathogens, which are residue dependent, and it helps them propagate, which may generate epidemics. When crop rotations are followed we are acting on the survival phase, diminishing inoculum quantity and, as a consequence, the intensity of the disease over the crop. Therefore, crop rotation with non-host crops is very important to avoid linking NT with epidemics as a generator of them (Carmona and Melo Reis, 2012).

Necrotrophic pathogens (fungi and bacteria) cause rotting and spotting in different plant organs, resulting in severe reductions on yields and frequent lowering in grain quality.

End of growth cycle diseases such as soybean Asiatic rust and frog-eye spot turned out to be the more important diseases in the past 10 years, causing major losses. They were favoured by genotype susceptibility, NT and monocropping (Carmona *et al.*, 2011).

Early disease detection and integrated disease management (including genetic, chemical and cultural control) are the key to reduce the contact between the pathogenic agent and the susceptible host to diminish infection rate and further epidemic progress.

15.6 Efforts/Policies Required for Scaling-Up Conservation Agriculture

15.6.1 Government support and policy towards Conservation Agriculture; incentives in the form of subsidy on implements

In 2000, the natural resources sub-secretary of MAGIC launched a project for sustainable development called 'Soil 2000' (Subsecretaría de Recursos Naturales, 1996). In Santa Fe province (Argentina), this project aimed to reverse the degradation processes and subsequent losses of productivity caused by tillage and monocropping and to spread a management system in harmony with nature to protect and improve soil quality and boost yields in a sustainable way with a NT system.

'Soil 2000' allowed farmers to participate in a network of sustainable production demonstration units for their region and receive tax stimulus (Act No. 10552) upon exchange of a sustainable production commitment.

15.6.2 Identifying suitable cover crops/increasing crop residue supply

As for crop rotation, cover crops are a technological tool that has gained great relevance in many cutting-edge technological systems in Argentina's agriculture. Cover crops are practised in a specific time and space window, generally with no other crop, not to produce grains but to create biomass. Cover crops favour C input and soil cover, and improve soil edaphic conditions, seeking higher yields for the subsequent crops in the rotation and better water use efficiency.

It was long believed that in order to store water in the soil it was important to have prolonged clean fallow periods (with no weeds). However, this reasoning did not consider the amount of water lost through direct evaporation. According to Gil (2002), in the northern area of the country, water efficiencies in fields under NT but poorly rotated, and consequently with poor soil cover, do not exceed 20%: about 8 out of 10 mm of rainwater are lost.

In Argentina, including legumes (for instance, *Vicia villosa*) as cover crops for biological fixation of nitrogen is becoming a common practice.

Some experiments done by Grupo Romagnoli produced 4000–6000 kg ha^{-1} dry matter with a N concentration of 5–6.9%. Figure 15.14 shows the impact of using vetch before maize in the rotation (Lorenzatti and Romagnoli, 2010, 2012).

15.6.3 Developing and supplying suitable machinery

In the past 20 years Argentina had a continuous advancement in planters, tractors, combines and sprayers with characteristics according to the country's needs.

Changes in machinery development in Argentina include:

- Higher number of rows;
- Lower inter-row distance;
- Auto-trailer transport system;
- Replacement of Brazilian planters/seeders by Argentinean ones;
- American NT planters adoption and introduction of the mono-disc sower and double fertilization (in row and between rows);
- Introduction of electronic devices on planters: seeding monitors, yield maps;
- Variable rate seed and fertilizer application;

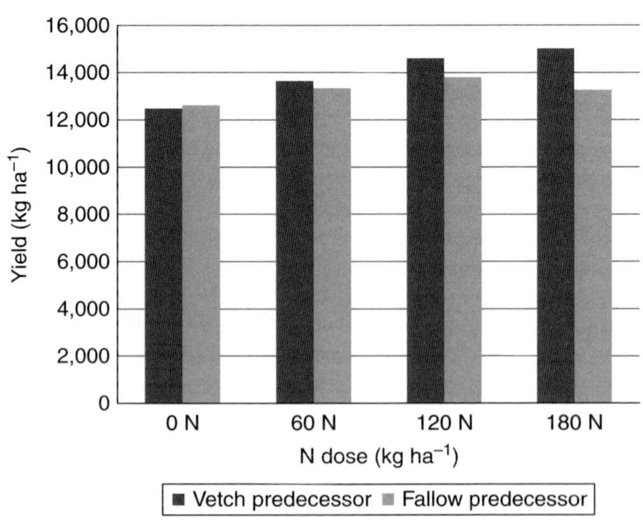

Fig. 15.14. Maize yield with chemical fallow and culture of vetch as predecessor, 2009/10 campaign (Monte Buey, Córdoba Province).

- Automatic pilot, satellite markers;
- Air seeders.

15.6.4 Developing effective and integrated insect-pest and disease management techniques

Incorrect, unmeasured and irrational use of phytosanitary products generates negative effects that can harm human health and the environment and generate resistance in insects. However, the integrated pest management (IPM) approach aims to reduce these problems through different techniques and steps (Igarzábal, 2004). It considers economic, social and environmental factors. It also claims for a deeper knowledge about pest biology, life cycles, food habits and reproduction time, not only about the pest, but also of predators and parasites that perform natural control, and abiotic factors for its reduction such as temperature changes, rainfall, droughts, etc. The main strategies of IPM are the use of different combinations according to the situation, but including genetic, cultural, biologic, ethologic, physical, legal and chemical control (Cobbe, 1998).

IPM is especially important in countries where agricultural production is a major industry, such as Argentina (AAPRESID, 2009).

15.6.5 Technology dissemination through trainings/field days/media

In the same spirit and pioneer's will, AAPRESID continues to promote very open knowledge exchange.

Field days

'A farmer in action' is the brand name of AAPRESID's field days. They are organized by regional groups in different parts of the country. These meetings allow farmers to open the doors of their farms and share experiences, realities and problems about the NT system.

AAPRESID Annual Congress

This is the annual and most important event. It is based on technological, organizational and institutional innovations. Management practices and strategies to boost productivity in a sustainable way, economic, political and social perspectives are taken into account, since they are in every farmer's mind.

Well-known specialists, national and international speakers share information with farmers. Every year, AAPRESID gathers more than 3000 farmers, scientists, engineers, professors, students, agri-input companies and members of civil society to find a space where innovation is the key.

Internships

Senior students of the agronomy courses have the opportunity to apply their theoretical knowledge and receive hands-on training in NT systems, introducing future professionals into the working environment related to sustainable agricultural industry.

Publications

This is how AAPRESID periodically communicates scientific knowledge, the most up-to-date technological information, experiences, GAPs and all the institutional life of AAPRESID.

Website

Being a network aimed at sharing experiences and knowledge, the website (http://www.aapresid.org.ar) is a fundamental space for AAPRESID. Members can have access to information related to the institution, various educational packages and many agronomical and business-type documents.

15.7 Conclusions

Argentina has a great potential to significantly increase its agricultural production during the next 10 years and take full advantage of the opportunity provided by the ever-increasing demand for food to sell its agricultural surpluses. A proper understanding of this opportunity by the political and urban sectors is an absolutely necessary condition to create an adequate socio-economic environment in which the rural sector can maximize its output. If we recall that only around 8% of the agricultural lands worldwide are cultivated utilizing CA, we become aware that there is still a long way to go in the process of transforming the world's farming system. Some of the achievements are: 96% less soil erosion, 66% less fuel use, maintenance or improvement of the organic matter in soil, higher water use efficiency, increase in soil fertility, lower production costs, higher production stability and higher potential yield.

Let us imagine for a moment, what a huge step forward it would be for humanity if significant parts of the world would progressively transform their agricultural system through the adoption of NT, thus achieving far higher outputs. Imagine also what it would mean for Argentina to lead this worldwide agricultural revolution aimed at sustaining our civilization.

Indeed Argentina has all the qualities – the soil, the technology, the human resources – to become one of the leaders in this process. Argentina bears a huge responsibility, to produce food. The country can produce food for more than 400 million people and the population in the country is only 40 million, so...

The challenges are enormous, but so will be the rewards.

References

AAPRESID (2012) Institutional magazine, 116.

AAPRESID (2012, 2008) http://www.aapresid.org.ar. Information upon request from Aapresid's Digital Library.

AAPRESID (2009) *Agricultura Certificada*. Manual de Buenas Prácticas Agrícolas.

Agaras, B., Wall, L.G. and Valverde, C. (2012) Specific enumeration and analysis of the community structure of culturable Pseudomonads in agricultural soils under no-till management in Argentina. *Applied Soil Ecology* 61, 305–319.

Alvarez, C. and Barraco, M. (2005) Efecto de los sistemas de labranzas en las propiedades edáficas y rendimiento de los cultivos. Indicadores de calidad física de suelos. Boletín Técnico No 4. 2005. INTA EEA Gral. Villegas.

Alvarez, R. and Steinbach, H. (2006a) Valor agronómico de la materia orgánica. Capitulo 2 Materia Organica. In: Pascale, A.J. (ed.) *Valor Agronómico y Dinámica en Suelos Pampeamos*. ISBN 950-29-0911-9, pp. 13–27.

Alvarez, R. and Steinbach, H. (2006b) Factores climáticos y edáficos reguladores del nivel de material orgánica. Capítulo 3 Materia Orgánica. In: Pascale, A.J. (ed.) *Valor Agronómico y Dinámica en Suelos Pampeamos*. ISBN 950-29-0911-9, pp. 31–54.

Argenbio (2013) Los cultivos transgénicos y su adopción. Available at: http://www.argenbio.org (accessed June 2013).

Carmona, M. and Melo Reis, E. (2012) Enfermedades en cultivos bajo siembra directa en Argentina y Brasil: pasado, presente y prospectivas de manejo. *XX Congreso Aapresid. Aapresid 2012*. Rosario, Santa Fe, Argentina.

Carmona, M., Gally, M., Sautua, F., Abello, A. and Lopez, P. (2011) *Uso de mezclas de azoxistrobina y triazoles para controlar enfermedades de fin de ciclo de la soja Cátedra de Fitopatología*. FAUBA. Buenos Aires, Argentina.

Cobbe, V. (1998) Capacitación participativa en el manejo integrado de plagas (MIP). Una propuesta para América Latina. Available at: http://www.fao.org/Regional/LAmerica/prior/segalim/prodalim/protfito/cobbe.pdf.

Cordone, G., Martínez, F., Andriulo, A. and Ghio, H. (2004) *El balance de carbono del suelo*. EEA INTA, Pergamino.

De la Vega, G., Castiglioni, M.G., Massobrio, M.J., Chagas, C.I., Santanatoglia, O.J. and Irurtia, C. (2004) Infiltración en un argiudol vertico bajo siembra directa en condiciones variables de cobertura humedad inicial. Asociación Argentina de la Ciencia del Suelo.

Ferraris, G. (2009) Microorganismos con efecto promotor de crecimiento (PGPM) en cultivos extensivos. Impacto sobre los rendimientos, la eficiencia de uso de los nutrientes y otros caracteres de interés agronómico. Resúmenes. In: *II Jornadas Bonaerenses de Microbiología de Suelos. 'Herramientas Microbiológicas para una Agricultura Sustentable'* UNICEN, 2009. Azul, Buenos Aires, pp. 8–9.

Ferraris, G. and Mousegne, F. (2009) Inoculación de soja en el norte, centro y oeste de buenos aires. Resultados de experiencias y prácticas de manejo para mejorar su eficiencia. Revista de Soja en siembra directa. Aapresid, 2009. Rosario, Santa Fe Argentina.

Fertilizar Asociación Civil (2011) Investigación. Available at: http://www.fertilizar.org.ar/2011/estadisticas.php (accessed May 2012).

Figuerola, E.L.M., Guerrero, L.D., Rosa, S.M., Simonetti, L., Duval, M.E., Galantini, J.A., Bedano, J.C., Wall, L.G. and Erijman, L. (2012) Bacterial indicator of agricultural management for soil under no-till crop production. *PLoS ONE* 7(11), e51075.

Friedrich, T., Derpsch, R. and Kassam, A. (2012) Overview of the global spread of conservation agriculture. Field Actions Science Reports [Online], Special Issue 6. Available at: http://factsreports.revues.org/1941 (accessed 6 November 2012).

Gerster, G. and Bacigaluppo, S. (2004) Consecuencias de la densificación por tránsito en Argiudoles del sur de Santa Fe. *XIX Congreso Argentino de la Ciencia del Suelo*, Paraná, Entre Ríos, Argentina.

Gerster, G. and Bacigaluppo, S. (2008) Distribución de la compactación en el perfil del suelo utilizando diferentes neumáticos. Consecuencias sobre el enraizamiento del cultivo de soja. *Para mejorar la producción No 39*. EEA INTA Oliveros, pp. 68–72.

Gerster, G. and Bacigaluppo, S. (2012) *Manejo de suelo*. La Soja, 2010.

Ghio, H. (2006) Herramienta clave para articular la rotación. Panel de Trigo. *XIV Congreso Aapresid*. Rosario Santa Fe, Argentina.

Ghio, H. (2009) Manejo integral del agua en sistemas intensivos de producción. *La era del Ecoprogreso XVII Congreso Aapresid*. Rosario Santa Fe, Argentina.

Ghio, H. (2011) Parcelas de fertilización a largo plazo en dos ambientes. Aapresid, EEA INTA Marcos Juárez e IPNI.

Ghio, H., Gudelj, V., Boll, M. and García, F. (2010) Long-term on-farm demonstrations in the Central pampas of Argentina: A case study. Corral de Bustos-Camilo Aldao, Córdoba, Argentina. *Better Crops* 94, 1.

Gigón, R., Vigna, M.R. and López, R.L. (2013) Efectos del sistema de siembra sobre la comunidad de malezas en cultivos de trigo del sudoeste de la provincia de Buenos Aires. EEA INTA Bordenave. *Revista Cultivos Invernales en siembra directa*. Aapresid, Rosario Santa Fe, Argentina.

Gil, R.C. (2002) *La siembra directa y la conservación del suelo*. Instituto de Suelos, Centro de Investigaciones de Recursos Naturales, INTA, Argentina.

Gudelj, O. and Masiero, B. (2000) *Efecto de diferentes practicas de labranza sobre la estabilidad estructural y la densidad aparente de un suelo argiudol típico de la pampa húmeda argentina*. INTA EEA, Marcos Juárez.

Igarzábal, D. (2004) Nuevos servicios agropecuarios: el monitoreo de plagas. *XII Congreso de Aapresid 'La hora del Empowerment'*. Aapresid, Rosario Santa Fe, Argentina.

Lorenzatti, S. (2006) Factibilidad de implementación de un certificado de agricultura sustentable como herramienta de diferenciación del proceso productivo de siembra directa. Tesis Maestría en Agronegocios, University of Buenos Aires.

Lorenzatti, S. and Romagnoli, J. (2010) La evolución silenciosa de la siembra directa y el desafío de las rotaciones. *Revista Especial en Siembra Directa, Rotaciones*. Aapresid, Rosario, Santa Fe, Argentina.

Lorenzatti, S. and Romagnoli, J. (2012) Agua: Maximización de la eficiencia de uso del agua con rotación y manejo nutricional. *Revista Especial en Siembra Directa*. Aapresid, Rosario Santa Fe, Argentina.

Micucci, F.G. (2003) *Impacto de las prácticas de manejo sobre la eficiencia de uso del agua en los cultivos extensivos de la Región Pampeana Argentina*. Cátedra de Fertilidad de Suelos, Facultad de Agronomía, Universidad de Buenos Aires.

Morón, A., Gudelj, V., Sawchik, J., Galarza, C., Marelli, H. and Arce, J. (2005) *Indicadores de la calidad de suelo en lotes de producción agrícola con labranzas contrastantes en Córdoba, Argentina*. INTA EEA, Marcos Juárez.

Peiretti, R.A. (2003) The CAAPAS actions and the development of the MOSHPA model. II *World Congress on Conservation Agriculture*. Foz de Iguazú, Brazil, 11–15 August 2003.

Pérez Brandán, C., Huidobro, J., Conforto, C., Arzeno, J.L., March, G., Meriles, J. and Vargas Gil, S. (2010) *Impacto de los sistemas de labranza sobre indicadores biológicos de calidad de suelo*. INTA EEA, Salta.

Sinclair, T.R., Salado-Navarro, L.R., Salas, G. and Purcell, L.C. (2007) Soybean yields and soil water status in Argentina: Simulation analysis Research Article. *Agricultural Systems* 94(2), 471–477.

Staley, J.T. and Konopka, A. (1985) Measurement of in-situ activities of nonphotosynthetic microorganisms in aquatic and terrestrial habitats. *Annual Review in Microbiology* 39, 321–346.

Subsecretaría de Recursos Naturales (1996) 'Suelo 2000'. *Programa de desarrollo sustentable. Subsecretaría de Recursos Naturales del MAGIC.* Gobierno de la Provincia de Santa Fe.

Tuesca, D. (2009) Cambios en las comunidades de malezas asociados con el sistema de labranza y el uso intensivo de glifosato. *Cátedra de Malezas, Facultad de Ciencias Agrarias. UNR. XV Congreso de Aapresid.* Rosario Santa Fe, Argentina.

Wall, L.G. (2011) The BIOSPAS consortium: soil biology and agricultural production. In: de Bruijn, F.J. (ed.) *Handbook of Molecular Microbial Ecology I: Metagenomics and Complementary Approaches.* John Wiley & Sons, New York.

16 Summing Up

Amir H. Kassam,[1,2] Theodor Friedrich[1] and Ram A. Jat[3,4]
[1]*Plant Production and Protection Division, Food and Agriculture Organization of the United Nations, Rome, Italy;* [2]*School of Agriculture, Policy and Development, University of Reading, UK;* [3]*International Crops Research Institute for the Semi-Arid Tropics, Patancheru, India;* [4]*Directorate of Groundnut Research, Junagadh, India*

Reducing soil disturbance by tillage began in the USA in the 1930s as a response to the 'dust bowls' in the Great Plains. Research on 'conservation' or reduced tillage, with early versions of a chisel plough, was initiated in the Great Plains in the 1930s to alleviate wind erosion of soil that was being pulverized by tillage and left exposed to wind and rain. Stubble mulch farming was also developed, and can be seen as a forerunner of no-tillage (NT) farming. This collection of practices led to what became known as conservation tillage. The modern successor of NT farming – now generally known as Conservation Agriculture (CA) – goes much further. It involves the simultaneous application of three practical principles based on locally formulated practices: minimizing soil disturbance (NT seeding); maintaining a continuous soil cover of organic mulch and plants (crop residues, stubbles and cover crops including legumes); and cultivation of diverse plant species that, in different farming systems, can include annual or perennial crops, trees, shrubs and pastures in associations, sequences or rotations, all contributing to enhancing system resilience. Conservation Agriculture, in conjunction with good crop, nutrient, weed and water management, is at the heart of FAO's new sustainable agricultural intensification strategy. Several organizations with global reach, such as CIMMYT, ICARDA, ICRISAT, CIRAD, ACIAR, AFD besides NARS institutions, universities, NGOs and farmer associations, are working to promote CA in different parts of the world.

Worldwide, CA is now practised on an estimated 125 Mha of arable cropland, mainly in North and South America, particularly the USA, Canada, Brazil, Argentina and Paraguay, and in Australia and New Zealand, but also increasingly in China, Kazakhstan, Ukraine and Russia. During the past decade, it has begun to spread in Asia more generally (including on the Indo-Gangetic Plains), in Europe (including in the UK) and in Africa. Conservation Agriculture has now spread over 1 Mha in Africa, including in South Africa, Mozambique, Zambia, Zimbabwe, Malawi, Madagascar, Kenya, Sudan, Ghana, Tunisia and Morocco, and some two-thirds of the area is under smallholder production. Much of the latter adoption has occurred in recent years as a result of more policy and extension attention and development resources being directed towards the promotion of CA through participatory dissemination and up-scaling approaches. Over the past decade the area of CA has increased at an average rate of 7 Mha year^{-1}, but in recent years the annual rate of spread has increased to some 10 Mha.

The above pattern of adoption and spread of CA is reflected across most of the chapters of this book but each chapter tells a country-specific or region-specific story of why, how and when it all began, what is the current status of adoption and how it is spreading, and what the future prospects are. In the USA, the initial impetus to reduce soil disturbance and adopt NT farming arose in response to the 'dust bowls' of the 1930s. In the case of countries such as Brazil, Argentina and Paraguay, where NT farming started in the 1970s and 1980s, the main initial driver was soil degradation due to water erosion from rainfall of exposed and loose topsoil from intensive tillage, in addition to low profitability of farming. In Canada and Australia, the initial driver towards CA was wind and water erosion. Subsequently, other factors such as the possibility of greater productivity and profit through greater adaptability to drier or wetter conditions, as well as reduction in production inputs of seeds, fertilizer, pesticide, energy and time also became important drivers for transformation from tillage farming to CA. More recently, CA has also begun to spread in a number of countries in Africa, Asia and Europe, the main drivers being the loss of or stagnating productivity due to soil degradation from erosion, loss of organic matter and soil structure, soil compaction, as well as rising costs of production. Conservation Agriculture is also being recognized as contributing to longer-term sustainability and resilience of crops and cropping systems, and of food and agriculture systems, against increased climatic variability and climate change. Although in some countries CA is still limited to the research sector, it is increasingly seen as an appropriate practical concept to promote in the future to achieve sustainable production intensification and to rehabilitate degraded agricultural lands and ecosystem services. While CA has its share of critics, differences in perspectives and appropriateness of CA are not over the efficacy of locally formulated CA practices but rather more with process of deciding where and how to promote the adoption and spread of CA.

What is now becoming increasingly clear is that because tillage-based agriculture at any level of technological development disrupts soil-mediated ecosystem functions and reduces soil productive capacity, it is not ecologically or economically capable of sustaining current production levels or production intensification. Further, tillage agriculture is not capable of fully harnessing necessary ecosystem services such as clean water, carbon sequestration, water and nutrient cycling, climate regulation and erosion control. Being a net emitter of greenhouse gases, tillage agriculture is also unable to mitigate climate change. In contrast, CA not only offers an approach to intensify production in an ecologically sustainable way, it is far less costly, economically and socially, than tillage agriculture. CA utilizes the whole ecosystem and the natural biodiversity including soil microorganisms and soil meso-fauna to build soil health and productive capacity and protect crops from weeds, insects and pathogens. Given CA's ability to improve rainfall infiltration and soil moisture storage as well as an increase in soil and root volume, there are improved interactions between plant roots and soil nutrients, and between plant roots and soil microorganisms such that there is greater resilience to biotic and abiotic stresses in CA systems compared with tillage systems.

Conservation Agriculture also allows greater precision with farm operations and higher efficiencies of input use in smallholder farms. This is particularly important in pro-poor development projects where purchased production inputs are not only scarce but must be made affordable. Higher input factor productivities with low levels of inputs in CA systems can provide a greater return to investment and a more robust basis for sustainable production intensification. On large farms with CA, it becomes possible to overlay controlled traffic farming and GPS-based precision farming to operate with best efficiencies of energy and input use. For several years now a carbon offset credit scheme has been operating in Alberta, Canada, based on CA to which controlled traffic farming and GPS-based precision farming are being added. Similarly, in Brazil, a programme called 'cultivating good water' has been operating in the Paraná

3 basin based on CA on large and small farms in order to improve the quality and quantity of clean water feeding into the Itaipu Dam whose working life has been extended considerably as a result. Elsewhere, in China, the spread of CA on small farms has helped in reducing the dust in the atmosphere in Beijing. In Spain, CA-based olive orchards have reduced soil erosion and flood risks in some 30% of the olive groves. In Western Australia, due to the adoption of CA in the semi-arid winter rainfall areas, there has been a significant reduction in land degradation and rehabilitation of degraded land from previous misuse with tillage agriculture. Such large scale ecosystem services of carbon sequestration, watershed services, cleaner air and reduced flood risks are not possible with tillage agriculture. Harnessing such services can be promoted through schemes in which farmers can receive payments for improved environmental and biodiversity management in agricultural landscapes.

When farmers decide to switch to CA from tillage farming, the expected mix of economic and environmental benefits manifests itself over time. The benefit mix varies in make-up and time scale depending on several factors including: agroclimatic conditions and variability within and between seasons; initial status of soil health and drainage under tillage systems; farm size and source of farm power; cropping system sophistication; yield levels under tillage systems; farmer expertise and experience of CA systems; access to production inputs, equipment and machinery; and competition for crop residues as livestock feed, and farm- and community-level arrangements for its enhancement and management. Given the infinite number of possible permutations in farm ecological and socio-economic conditions and social arrangements for changing from tillage-based systems to CA, a pattern of economic and environmental benefits can be recognized, which is increasingly supported not only by farmer performance but also by on-farm and on-station research across all continents and agroecologies.

In general, CA benefits can include: increased factor productivities and yields (depending on prevailing yield levels and extent of soil degradation); up to 70% decrease in fuel energy or manual labour; up to 50% less fertilizer use; 20% or more reduction in pesticide and herbicide use; some 30–50% less water requirement; and reduced cost outlay on farm machinery. Further, with CA it is possible to enhance climate change adaptability of cropping systems, farms and landscapes because of improved soil–plant moisture relations while at the same time mitigating climate change through greater carbon sequestration and lower emissions of greenhouse gases of CO_2, N_2O and CH_4. Due to much greater rainfall infiltration and reduced runoff and soil erosion, CA can also decrease flood risks, raise water resource quality and quantities as well as reduce infrastructure maintenance costs.

Conservation Agriculture does not provide a solution to all farming problems, although it does offer an alternative approach ecologically to underpin crop production systems so that they are sustainable and resource enhancing and conserving, offering on-farm productivity benefits and landscape-level ecosystem services. FAO refers to this as the 'Save and Grow' approach to sustainable production intensification with an ecosystem approach. However, like with any farming system, adoption of CA has its constraints that must be overcome for large-scale dissemination. The establishment of CA methods can be difficult in the initial years in some semi-arid areas and on heavy clay soils, compact soils and poorly drained land. Control of pests and diseases can be a concern in some instances where crop residues are left on the soil, and pesticides/herbicides may be required, at least in the initial years. Leaving crop residues on fields as mulch would eliminate an important source of animal fodder in areas where livestock play an important role in farm economies. There can be other location-specific socio-economic issues that must be addressed such as perceived risk of loss in productivity in initial years or possible displacement of paid farm labour. On larger farms, the lack of appropriate equipment for seeding and fertilizer placement through surface mulches can be problematic.

Adoption and spread of CA internationally offers lessons that show that the above-mentioned challenges can be and are being overcome by farmers, rich and poor, small and large, through locally formulated solutions involving a range of public and private sector stakeholders working together with farmers along different pathways of adoption and transformation. The negative effects of difficult biophysical conditions can be reduced as improved, physical and biological soil conditions are established through CA practices, and diversified crop rotations and associations can keep crop pest/disease risks low. Integrated weed management is easier where hand cultivation is practised; and the use of an initial herbicide application followed by crop rotations and maintenance of a continuous soil cover by plants and mulch can eventually reduce weed competition. Crops whose yield is located below ground, such as white potato, sweet potato, cassava, groundnut and sugarbeet, can also be planted into untilled soil, and harvested with minimal soil disturbance using appropriate harvesting equipment or changes in cropping practices. Rice too is produced without puddling the soil. In CA systems with livestock husbandry, total biomass production is increased over time so that it is possible to manage on-farm residue allocation between livestock feed and soil protection dynamically in order to satisfy both goals. Where communal grazing of crop residues is a constraint in maintaining soil cover, a community-based solution can be found so that some crop residue is retained. The constraint of lack of suitable mechanical equipment diminishes over time as a sufficient market develops for the local manufacturer.

In the coming decades, every effort by all concerned must be made to transform tillage agriculture to CA. There are several ways to support immediate and widespread up-scaling of CA:

- In all new agriculture development projects, include CA as the basis for sustainable production intensification and engage all the relevant stakeholders to ensure success.
- Revise universities' agriculture curricula to include teaching the next generation of farmers and agricultural development practitioners about CA as an alternative and sustainable way of farming.
- Fund more innovative practical research to tackle soil, agronomic and livestock husbandry challenges through our universities and research centres.
- Advocate for initial government support in terms of subsidies to make appropriate farm equipments more readily accessible and to reduce any risks of possible productivity losses during the initial years of switching to CA.
- Encourage governments to update their agricultural policies and bring institutional reforms that support the up-scaling of CA, especially in Asia, Africa and Europe – where it is perhaps most urgently needed.
- Develop large-scale programmes that would offer payments to CA farmers for harnessing ecosystem services such as carbon sequestration, watershed services for increasing the quality and quantity of water resources, control of soil erosion and reduction in flood risks, and enhancing pollination services.

Fuller advantage of the benefits offered by CA can be taken if all stakeholders become involved in facilitating the transformation process as is happening in countries such as Brazil, Argentina, Paraguay, the USA, Canada and Australia. This is also beginning to occur in countries in Europe (e.g. Finland, Spain), Africa (e.g. Zambia, Zimbabwe) and Asia (e.g. Kazakhstan, China). However, a more structural response to the opportunities presented by CA calls for a realignment of agricultural institutions, including research, extension and education, as well as agriculture development policies to enable CA to become the preferred agriculture paradigm choice around which to strengthen national and international food and agriculture systems. As a result of the process of World Congresses on Conservation Agriculture, there is now a global multi-stakeholder CA Community of Practice (CA-CoP) that is facilitating the uptake and spread of CA internationally. During the past decade, the effort to promote CA has become increasingly better organized, and donor agencies, governments, national research and extension systems, private sector firms, NGOs and farmers themselves are engaged in finding ways and means to introduce and spread CA.

The future requires that farming and agricultural landscapes everywhere must be multi-functional, ecologically sustainable and integrate into the greater ecosystems alongside non-agricultural land uses. This means that any agricultural production enhancement must go hand in hand with the enhancement and delivery of desired ecosystem services, and production systems must be efficient with high production factor productivities as well as resilient in on-farm performance and in their socio-economic development at the civil society level. Food and agriculture systems internationally need effectively to address local, national and international challenges, which include: food, water and energy insecurity, climate change, pervasive rural poverty and degradation of natural resources. As this volume of national and regional assessments clearly shows that the principles of CA and their locally formulated adapted practices, with their potential capacity to slow and even reverse productivity losses and environmental damages, appear to offer an entirely appropriate solution to all types of farms in all agroecologies. While to some readers this statement may sound overly optimistic, to all the authors who have contributed their practical expertise to this book, CA-based farming systems appear to be the best available option for meeting future food security needs sustainably, while alleviating poverty and building livelihoods, and rehabilitating and enhancing ecosystem functions and services.

Index